ROUTLEDGE HANDBOOK OF DISINFORMATION AND NATIONAL SECURITY

This interdisciplinary Handbook provides an in-depth analysis of the complex security phenomenon of disinformation and offers a toolkit to counter such tactics.

Disinformation used to propagate false, inexact or out of context information is today a frequently used tool of political manipulation and information warfare, both online and offline. This Handbook evidences a historical thread of continuing practices and modus operandi in overt state propaganda and covert information operations. Further, it attempts to unveil current methods used by propaganda actors, the inherent vulnerabilities they exploit in the fabric of democratic societies and, last but not least, to highlight current practices in countering disinformation and building resilient audiences.

The Handbook is divided into six thematic sections. The first part provides a set of theoretical approaches to hostile influencing, disinformation and covert information operations. The second part looks at disinformation and propaganda in historical perspective offering case study analysis of disinformation, and the third focuses on providing understanding of the contemporary challenges posed by disinformation and hostile influencing. The fourth part examines information and communication practices used for countering disinformation and building resilience. The fifth part analyses specific regional experiences in countering and deterring disinformation, as well as international policy responses from transnational institutions and security practitioners. Finally, the sixth part offers a practical toolkit for practitioners to counter disinformation and hostile influencing.

This handbook will be of much interest to students of national security, propaganda studies, media and communications studies, intelligence studies and International Relations in general.

Rubén Arcos is a senior lecturer in communication sciences at University Rey Juan Carlos (URJC) in Madrid, Spain, and a researcher of the Cyberimaginario research group. He is a cofounder and codirector of IntelHub—International Online Intelligence Hub.

Irena Chiru is a professor of intelligence studies at 'Mihai Viteazul' National Intelligence Academy, Romania. She is a member of the editorial advisory board of the International Journal of Intelligence and Counterintelligence.

Cristina Ivan is a researcher at the National Institute for Intelligence Studies, "Mihai Viteazul" National Intelligence Academy in Romania. She holds a PhD in cultural studies.

ROUTLEDGE HANDBOOK OF DISINFORMATION AND NATIONAL SECURITY

Edited by Rubén Arcos, Irena Chiru and Cristina Ivan

Routledge
Taylor & Francis Group

LONDON AND NEW YORK

Cover image: Getty Images © Arkadiusz Warguła

First published 2024
by Routledge
4 Park Square, Milton Park, Abingdon, Oxon OX14 4RN

and by Routledge
605 Third Avenue, New York, NY 10158

Routledge is an imprint of the Taylor & Francis Group, an informa business

British Library Cataloguing-in-Publication Data
A catalogue record for this book is available from the British Library

Library of Congress Cataloging-in-Publication Data
A catalog record has been requested for this book

ISBN: 978-1-032-04050-9 (hbk)
ISBN: 978-1-032-04054-7 (pbk)
ISBN: 978-1-003-19036-3 (ebk)

DOI: 10.4324/9781003190363

Typeset in Bembo
by Taylor & Francis Books

CONTENTS

ILLUSTRATIONS

Figures

Tables

ABOUT THE CONTRIBUTORS

Jamil Ammar Jamil Ammar research focuses on the intersection of free speech, religious freedom, digital intermediaries, countering violent extremism and disinformation leading to violence. Currently, he serves as adjunct professor at Osgoode Hall Law School. He is also senior research affiliate at the Canadian Network for Research on Terrorism, Security and Society.

https://orcid.org/0000-0003-2185-602X
jamil.ammar@trios.com

Rubén Arcos Rubén Arcos, PhD, is a senior lecturer in communication sciences at University Rey Juan Carlos and member of the research group Cyber-imaginary. He serves as Vice Chair of the Intelligence Studies Section at the International Studies association (ISA). His research is focused on intelligence services and intelligence analysis, foreign disinformation, and hybrid threats. He has been appointed national member for the NATO Science & Technology Organization's research task group SAS-189 "Anticipatory Intelligence for Superior Decision Making".

https://orcid.org/0000-0002-9665-5874
ruben.arcos@urjc.es

Cristina M. Arribas Cristina María Arribas is a Ph. candidate and researcher in the Department of Audiovisual Communication and Advertising at Universidad Rey Juan Carlos. She holds a bachelor's degree in Contemporary History, a bachelor's Degree in East Asian Studies, and an MA in Intelligence Analysis. Currently, she participates as a researcher in projects EU-HYBNET and DOMINOES, both funded by EU programmes. Earlier she has worked as senior intelligence analysts in multinational companies in the cybersecurity sector. Her last article is "Responses to digital disinformation as part of hybrid threats: a systematic review on the effects of disinformation and the effectiveness of fact-checking/debunking".

https://orcid.org/0000-0002-6079-5832
Email: cristina.arribas@urjc.es

Alina Bârgăoanu Alina Bârgăoanu, PhD, is member of the Advisory Board of European Digital Media Observatory (EDMO) and of EDMO's task force on the war in Ukraine; President of the Board, European Institute of Romania (since 2015); member of the High-Level Expert Group on Fake News and Online Disinformation, EC (2018); founder and editor-in-chief of the fact-checking portal Antifake.ro; media commentator and columnist on topics related to disinformation, resilience and the role of technology in shaping the information ecosystem.

https://orcid.org/0000-0003-3912-8442
abargaoanu@comunicare.ro

Hamilton Bean Hamilton Bean is Associate Professor of Communication at the University of Colorado Denver, where he conducts research at the intersection of communication, organization, and security. He serves as a Non-Resident Fellow for the U.S. Joint Special Operations University (JSOU) and is the author of No More Secrets: Open Source Information and the Reshaping of U.S. Intelligence.

http://orcid.org/0000-0002-0987-4076
HAMILTON.BEAN@ucdenver.edu

Corneliu Bjola Corneliu Bjola is Associate Professor in Diplomatic Studies at the University of Oxford and Head of the Oxford Digital Diplomacy Research Group. He also serves as a Faculty Fellow at the Center on Public Diplomacy at the University of Southern California and as a Professorial Lecturer at the Diplomatic Academy of Vienna. He has published extensively on issues related to the impact of digital technology on the conduct of diplomacy with a recent focus on public diplomacy, international negotiations, and methods for countering digital propaganda.

https://orcid.org/0000-0003-3609-4240
corneliu.bjola@qeh.ox.ac.uk

Ruxandra Buluc Ruxandra Buluc is a senior researcher in the National Institute for Intelligence Studies, "Mihai Viteazul" National Intelligence Academy, Romania. Her research focuses on argumentation, discursivity, narrativity, and their use in building security culture, strategic communication as well as in disinformation and propaganda campaigns. She works in European-funded projects which are aimed at building security culture and resilience to disinformation and radicalization.

https://orcid.org/0000-0003-0587-7779
buluc.ruxandra@animv.eu

Joseph Cannataci Prof. Cannataci is head of the Department of Information Policy; Governance at the Faculty of Media; Knowledge Sciences of the University of Malta. He also co-founded and continues as Co-director of STeP, the Security, Technology; e-Privacy Research Group at the University of Groningen in the Netherlands, where he is Full Professor, holding the Chair of European Information Policy & Technology Law.

https://orcid.org/0000-0001-6309-0734
jcannataci@sec.research.um.edu.mt

Teresa Capelos Tereza Capelos is Associate Professor in Political Psychology and Director of the Institute for Conflict Cooperation and Security at the University of Birmingham. Her research examines the role of uncertainty and affective ambivalence in establishing trust during national and international crises and tensions, and the psychological structure of reactionary orientations and ressentimentful affect in the context of contemporary grievance politics.

https://orcid.org/0000-0002-9371-4509
t.capelos@bham.ac.uk

Irena Chiru Irena Chiru is a professor of intelligence studies at "Mihai Viteazul" National Intelligence Academy, Romania. Currently she serves as the chair of the International Association for Intelligence Education – European Chapter, a position from which she is advocating for building bridges between scholars and practitioners in intelligence. Her main research interests are intelligence analysis, critical intelligence studies, intelligence cultures, and strategic communication.

0000-0002-4202-4209
chiru.irena@animv.eu

Abby DiOrio Abby DiOrio is a law student at The Catholic University of America, Columbus School of Law in Washington, D.C. She received a B.A. in Political Science with a focus in American Politics from the University of Virginia. She served as Lead Producer for Pherson's Intelligence Analysis Professional Certification workshops and was a primary developer for Pherson's Critical Thinking Fundamentals online training program.

0000-0002-9794-6438
abby.diorio@gmail.com

Dan Dungaciu Dan Dungaciu is the general manager of the Institute of Political Sciences and International Relations of the Romanian Academy. He is Professor at the Faculty of Sociology, University of Bucharest, where he coordinates the Security Studies and Information Analysis Master Programme.

0000-0002-0761-4983
dan.dungaciu2012@gmail.com

Lucian Dumitrescu Lucian Dumitrescu works as a researcher at the Institute of Political Sciences and International Relations of the Romanian Academy, and associate lecturer at the Faculty of Sociology, University of Bucharest. He holds a Ph. D. in Sociology and another Ph. D. in Philosophy. Main areas of interest: comparative grand strategy, strategic narratives, political underdevelopment and institutional weakness, historical sociology.

0000-0001-8165-285X
dulust@gmail.com

Flavia Durach Flavia Durach, PhD, is Associate Professor at the National University of Political Studies and Public Administration, Bucharest. Her research interests include the study of media effects. She was involved in international and national projects dedicated to countering fake news and online disinformation, among which the fact-checking portal Antifake.ro (as editor), and several media literacy initiatives founded by NATO Public Diplomacy Programmes and the European Commission.

https://orcid.org/0000-0002-4405-8605
flavia.durach@comunicare.ro

Arhan S. Ertan is an assistant professor of economics & finance at the department of International Trade at Boğaziçi University. He received his PhD in Economics from Brown University. His academic interests span various topics related to economic development (with a long-term and institutional perspective), behavioral economics (focusing on the roles of personal characteristics and institutions of cooperation), international economics (trade and finance) and migration (domestic and international).

arhan.ertan@boun.edu.tr

Jan Goldman Jan Goldman is Professor of Intelligence and Security Studies at The Citadel, Military College of South Carolina. He spent over 30 years as a practitioner and educator in the U.S. intelligence community, focusing his research on ethics and intelligence operations, secrecy, intelligence analysis, psychological operations, and intelligence education. He is the Editor-in-Chief of the International Journal of Intelligence and CounterIntelligence; and the Founding Editor of the Security Professionals Intelligence Education Series (SPIES) at Rowman and Littlefield Publishers.

jgold.ijic@gmail.com

Pablo Hernández He is Head of Academic Research at Maldita.es. He has collaborated with research journals such as adComunica or Dilemata on the problem of countering disinformation. He has spent the longest stage of his career at La Sexta where he has been the executive producer in several news shows between 2007 and 2020. He was also an executive producer at CNN+ and previously worked as a reporter in other TV stations, newspapers and news agencies.

0000-0003-1529-8578
phernandez@maldita.es

Sanshiro Hosaka Sanshiro Hosaka is a research fellow at the International Centre for Defence and Security (Tallinn) and a PhD student at the University of Tartu, Johan Skytte Institute of Political Studies. His current research interests include strategic narratives of non-democracies targeting academia, decolonization of Russian studies, political technology, Soviet/Russian covert actions, intelligence history, and Baltic security and defence. He received research awards from the Japanese Association for Russian and East European Studies (2017) and the Japanese Association for Ukrainian Studies (2022).

0000-0002-3258-0455
yokokumano2018@gmail.com

Cristina Ivan Cristina Ivan is researcher at the National Institute for Intelligence Studies, "Mihai Viteazul" National Intelligence Academy. Her work focuses on security and intelligence studies, as well as cultural studies. She has specialized in the cultural study of violence, radicalization and terrorism, propaganda and disinformation, critical intelligence studies and taken an active part in European projects targeting an enhanced understanding and early detection of radicalization, propaganda, and disinformation.

0000-0003-2624-5009
ivan.cristina@animv.eu

Veli Pekka Kivimaki Veli-Pekka Kivimäki is a Senior Analyst with the Finnish Security and Intelligence Service (Supo). He is also an adjunct lecturer in Geospatial Intelligence (GEOINT) at the Johns Hopkins University, and a visiting lecturer on open source intelligence (OSINT) at the University of Jyväskylä. Previously, he served in strategic analysis with the Finnish Defence Forces. Mr. Kivimäki's core research areas are the evolution of OSINT, and how emerging technologies impact national security. Before his work in government, he was also one of the original members of the award-winning Bellingcat Investigation Team.

https://orcid.org/0000-0003-3186-9262
vkivima1@jhu.edu

Grete Krisciunaite Grete Krisciunaite holds a First Class Honours for the Joint Degree in MA Economics and International Relations at University of Aberdeen, and a First Class Honours Degree in MSc Political Psychology of International Relations at University of Birmingham. She researches cognitive and affective features of political decision-making and diplomacy in the international arena.

0000-0002-2945-2102
grete.krisciunaite@gmail.com

Aleksandra Kuczy⊠ska-Zonik Aleksandra Kuczy⊠ska-Zonik is Assistant Professor at the John Paul II Catholic University of Lublin and the Head of Baltic Department of the Institute of Central Europe, Poland. She is a doctor in political science, doctor in humanities in the field of archaeology. Research interests: politics and security in East-Central Europe, the Baltic states, Russian minority, Soviet heritage.

0000-0002-5672-9613
kuczynska.a@gmail.com

Deanna Labriny Deanna Labriny received her B.A. in Political Science, Foreign Affairs with a concentration in the MENA region from the University of Virginia. She has completed courses in the Masters in Applied Intelligence Program at Georgetown University. As an analyst and researcher for Pherson Associates, Deanna conducted research and analysis on historical and current Russian disinformation campaigns targeting the US and democratic transatlantic states; edited and updated the 3 rd Edition of Critical Thinking for Strategic Intelligence and the 3 rd edition of Structured Analytic Techniques for

Intelligence Analysis. Deanna has produced threat analyses on environmental threats to US military and law enforcement operations.

0000-0002-9173-1794
dlabriny.dc@gmail.com

Josephine Lukito Dr. Josephine Lukito is an Assistant Processor in the School of Journalism and Media at the University of Texas at Austin. She is also a Senior Faculty Research Associate in the Center for Media Engagement.

https://orcid.org/0000-0002-0771-1070
jlukito@utexas.edu

Ilan Manor Ilan Manor (PhD Oxford University) is a digital diplomacy scholar at Ben Gurion University of the Negev. His 2019 book, The Digitalization of Public Diplomacy, was published by Palgrave Macmillan as was his 2021 co-edited volume, Public Diplomacy and the Politics of Uncertainty. Manor has contributed to many academic journals including International Affairs, The International Journal of Communication, International Studies Review, Policy & Internet, Global Policy and Media, War & Conflict.

https://orcid.org/0000-0003-2039-3721
manori@bgu.ac.il

Roger Mason Dr. Roger Mason is vice president of LECMgt LLC based in Camarillo, California USA. Dr. Mason has designed over fifty wargames and tabletop exercises for military, public safety, and emergency management applications. He is a published author on the topic of wargames. He has published two commercial wargames. Dr. Mason is a wargame instructor as a faculty member of the Military Operations Research Society. He has served as an adjunct faculty member for the postgraduate program in intelligence analysis at Rey Juan Carlos University in Madrid. His research interests include hybrid warfare, time and decision making, and the use of artificial life techniques in wargames.

Email: roger@lecmgt.com

Florina Cristiana (Cris) Matei Florina Cristiana (Cris) Matei is a Lecturer at the Naval Post-graduate School in Monterey, California, where she teaches M.A. courses for the Center for Homeland Security and Department of National Security Affairs, which cover such topics as security and political regimes, civil-military relations and democratic transition/consolidation, transnational security threats, and intelligence. She is the co-editor (with Bruneau) of The Routledge Handbook of Civil-Military Relations, published in 2012; (with Halladay) of The Conduct of Intelligence in Democracies: Processes, Practices, Cultures published in 2019; (with Halladay and Bruneau) of The Routledge Handbook of Civil-Military Relations—Second Edition, published in 2021; and (with Halladay, and Estevez) of the Handbook of Latin American and Caribbean Intelligence Cultures, published in 2022. Since January 2020, Cris has been associate editor of the International Journal of Intelligence and Counterintelligence. Cris also serves as Co-Section Chair for the Intelligence Studies Section (ISS) of the International Studies Association (ISA).

cmatei@nps.edu

Björn Palmertz Björn Palmertz is the Coordinator of the Center for Cybersecurity at RISE and Former Counter Influence Coordinator, Swedish Psychological Defence Agency. During the last 16 years, Bjorn has worked to analyze and counter hybrid threats through various roles in Swedish society. As chief analyst at the Armed Forces Psychological Operations Unit, as a teacher and researcher at the Defence University, and as counter influence coordinator at the Civil Contingencies Agency and Psychological Defence Agency. Prior to that he served in media production and marketing roles for, among others, Turner Broadcasting in London, the film studio Dreamworks SKG in Los Angeles and advertising agency Wildell in Stockholm. Today, Bjorn is tasked to develop the Center for Cybersecurity at the Research Institutes of Sweden. He holds the rank of Major in the reserve of the Swedish Amphibious Corps, has an MSc in Political Science from the Swedish Defence University and a BA in Communication Studies from UCLA.

palmertz@hotmail.com

James Pamment James Pamment (PhD Stockholm University, 2011) is Director of the Lund University Psychological Defence Research Institute. Pamment's main research interest is in the role of strategic communication in countering hostile foreign interference, such as information influence operations and hybrid threats. Academic publications include the books British Public Diplomacy; Soft Power: Diplomatic Influence; Digital Disruption and Countering Online Propaganda and Violent Extremism (with Corneliu Bjola). As a consultant, Pamment has worked with EU and UN institutions, several governments, the EU-NATO Hybrid Threats Centre of Excellence, and the NATO Strategic Communications Centre of Excellence among others. Previous affiliations include the Carnegie Endowment for International Peace, the Swedish Defence University, University of Texas at Austin, University of Southern California, and Oxford University.

https://orcid.org/0000-0001-5128-1007
james.pamment@isk.lu.se

Randolph H. Pherson Randolph Pherson is CEO, Globalytica, LLC. He has published several books, including Structured Analytic Techniques for Intelligence Analysis and Critical Thinking for Strategic Intelligence. Mr. Pherson was a career Central Intelligence Agency (CIA) intelligence analyst and manager, last serving as National Intelligence Officer (NIO) for Latin America. He is a recipient of the Distinguished Intelligence Medal for his service as NIO and the CIA's Distinguished Career Intelligence Medal. He received his AB from Dartmouth College and MA in International Relations from Yale University.

rpherson@pherson.org.

Aitana Radu Dr. Aitana Radu is the Security Research Coordinator within the Department of Information Policy & Governance. Her research focuses on different aspects of security science, from violent radicalisation to intelligence oversight. Since 2013, Dr Radu has carried out extensive EU-funded research focused on radicalization (ARMOUR and JP-COOPS projects), law enforcement practices (CITYCoP project), the implementation of the European Investigation Order (SAT-LAW and PRE-RIGHTS projects), developing security science (ESSENTIAL project) and intelligence analysis in the context of border security (MIRROR and CRITERIA projects). Dr. Radu obtained her M.A. (Comparative Political Science) from the University of

Bucharest with a thesis on democratic transitions in the Middle East, her M.A. in the Management of Intelligence Activities for National Security from the National Intelligence Academy with a thesis on the security risks posed by the radical Islamic discourse, and her PhD in Intelligence and National Security from the National Intelligence Academy with a thesis on the transformation of intelligence organizations. More information and publications at https://www.um.edu.mt/maks/ipg/staff/aitanaradu

aradu@sec.research.um.edu.mt

Vira Ratsiborynska Vira Ratsiborynska is an Adjunct Professor on NATO and transatlantic approaches to security and Global Politics at the Vrije Universiteit Brussel (VUB), Brussels School of Governance, Belgium. She is also a Senior Researcher and Associate Research Director at ISIFFA International, a Research analyst/ contractor working in support of the EU institutions and NATO, and a Member of the Jean Monnet Centre of Excellence on European Security and Disinformation in Multicultural Societies. Her research interests include international security, hybrid threats, the relations between the EU–NATO and the Eastern partners, Russia and Central Asia, e.g. in energy, trade, geopolitics, border security, conflict management, and peacekeeping. Vira Ratsiborynska holds a Ph.D. from the University of Strasbourg, France. She publishes extensively in English and French on the topics of the European Union and the Eastern flank's security, the relations with the Eastern Partnership countries, NATO and Russia. She obtained a Master's degree in EU Studies from the College of Europe (Belgium) and a Master's in Political and Social Studies from Sciences Po (France). Her professional experience includes working in support of such organizations as the European institutions and NATO entities.

0000-0002-6998-121X
viraratsiborynska@gmail.com

Julian Richards Julian Richards is a Professor of Politics at the University of Buckingham in the UK. He is Director of BUCSIS, the Centre for Security and Intelligence Studies at the University of Buckingham; and Associate Dean of the university's School of Humanities and Social Sciences. As well as teaching university programmes at various levels and supervising PhD research on all aspects of global security and intelligence issues, he has an extensive list of academic publications, and is a frequent commentator in national and international media on security and intelligence process and policy. Prior to academia, he enjoyed a long career in state intelligence for the UK government.

https://orcid.org/0000-0003-0613-4264
julian.richards@buckingham.ac.uk

Frédérique SEGOND Frédérique SEGOND, HDR, is Director of the Defense Security Mission at Inria and associated researcher with the ERTIM team at the Institut National des Langues et cultures Orientales (Inalco). Previously, she was Director of R&D and Intelligence products at Bertin IT and Director of Partnerships and Innovation at Grenoble Alpes University. In 2011, she joined VISEO to create and manage the company's Research Center. Frédérique worked for 18 years at the Xerox European Research Center in Grenoble where she was Principal Scientist and managed the Natural Language Processing research group. She has also spent 3 year as researcher at the IBM scientific center in France, and one year at the IBM Watson research center in Yorktown. She was a lecturer at Telecom Paris Sud. Frédérique is co-author of six books, more than 80 scientific articles and 5 patents. She was president of the Association pour le Traitement Automatique des Langues (ATALA), French academic organization focused on Natural Language processing. Frédérique serves in different scientific committees. She is member of the CLARIN (European research infrastructure for language as social and cultural data) scientific advisory board and was member of the Board of Directors of ELRA (European Language Resources Association).

0000-0001-9420-9654
frederique.segond@inria.fr

Valentin Stoian-Iordache Valentin Stoian is a researcher with the National Intelligence Studies Institute. He is specialised in intelligence and security theory. He holds an M.A. and PhD in political science from the Central European University, Budapest and a B.A. in political science from the University of Bucharest. He has published work on topics such as hybrid warfare, the securitization of corruption, critical security studies and the ethics of intelligence.

0000-0002-1393-8933
stoian.valentin@animv.eu

Bryan C. Taylor Bryan C. Taylor is Professor of Communication at the University of Colorado Boulder, and Director of its Peace, Conflict, and Security Program. His research interests include the communicative status of nuclear weapons, and the role of mimesis in articulations of media and security. His related research has been published in journals including Annals of the International Communication Association, Communication Theory, Critical Studies of Media Communication, and elsewhere. He is co-editor of the volume Nuclear Legacies: Communication, Controversy, and the U.S. Nuclear Weapons Complex.

Adrian Tudorache Adrian Tudorache is a Romanian diplomat, currently assigned at the Embassy of Romania to the United States of America, where he covers strategic and pol-mil affairs. He holds the diplomatic rank of Minister Counselor. Between 2013 and 2016, he served in the Ministry of Foreign Affairs of Romania, Division for Security Policy where his main tasks included: the advancing the Strategic Partnership between Romania and the United States in the security and defense dimension, the development of Romania's foreign and security policy towards the 2014 NATO Wales Summit and 2016 NATO Warsaw Summit, as well as in the context of NATO's adaptation to the post-2014 security environment, working on formats of regional security consultations, such as the Bucharest Format. His other assignments included public diplomacy and outreach in the Romanian Embassy in Washington (2008-2013), public diplomacy in Bucharest (2007-2008), as well as a social and labor attaché in the Embassy of Romania to Germany (2005-2006). He holds a PhD in sociology awarded by the University of Bucharest in 2014. He graduated from the Global Master of Arts Program at Fletcher Schools of Law and Diplomacy (2020), and the Master of Peace and Security Studies at the University of Hamburg (2004).

adrian_ot@yahoo.com

H. Akin Unver H. Akin Unver is an associate professor of International Relations at Ozyegin University, specializing in conflict research, computational methods and digital crisis communication. He is a fellow of Carnegie Endowment's Digital Democracy Network and serves as a member of TikTok's MENA-T Security Advisory Council. Previously he served as a Research Associate at the Center for Technology and Global Affairs, Oxford University and a Senior Research Fellow at GUARD (Global Urban Analytics for Resilient Defence) at the Alan Turing Institute. He is the author of 'Defining Turkey's Kurdish Question: Discourse and Politics Since 1990' (Routledge Series in Middle Eastern Politics). He is the Istanbul organizer of the Summer Institute in Computational Social Science (SICSS) and the founder of the Istanbul Twitter Developers' Community. Previously he has taught international security and political communication at Oxford, Princeton and University of Michigan.

akin.unver@ozyegin.edu.tr
0000-0002-6932-8325

Agnes Venema Agnes Venema is a researcher pursuing a dual doctorate in intelligence and national security at the "Mihai Viteazul" National Intelligence Academy in Bucharest, Romania and the University of Malta, specializing in Image Intelligence and the responsible use of emerging and disruptive technologies (EDTs) in security. Ms. Venema is an expert on synthetic media and deepfakes and has used her legal background to inform legal and political consultations in New Zealand, England and Wales, and at the European level on this topic. As a consultant, Ms. Venema frequently conducts capacity building training in the field of cybersecurity and synthetic media for clients such as the Diplomatic Academy Vienna and the Academy of European Law, and produces research and foresight policy products for Geneva Centre for Security Sector Governance and various governments. She also worked on the Security Sector Reform at the United Nations Mission in Timor-Leste (UNMIT) and was part of an Organization for Security and Cooperation in Europe (OSCE) election monitoring mission. Ms. Venema holds an LL.M. degree in Human Rights and Criminal Justice from the Irish Centre for Human Rights at NUI Galway and a Bachelor degree from University College Utrecht, where she opted for a semester abroad at the Masaryk University in Brno, Czech Republic.

0000-0002-5951-1986
agnesvenema@hotmail.com

James J. Wirtz Dr. James J. Wirtz has served as a professor, department chairman and dean at the Naval Postgraduate School (NPS), Monterey, California. He played a pivotal role in creating the Center for Homeland Defense and Security at NPS, which educates state, local, federal and tribal officials to better respond to natural and deliberate threats. Professor Wirtz's written work includes scores of books and articles on national security, intelligence affairs and foreign matters. His textbook published by Oxford University Press, Strategy in the Contemporary World, is in its 7 th edition and has introduced thousands of students to the study of strategy and international security. During his career, he has served as a consultant to U.S. agencies and allied governments and has led mobile education teams in locations as varied as Tegucigalpa, Honduras and Nur-Sultan, Kazakhstan. A native of New Jersey, Dr. Wirtz graduated from Columbia University and the University of Delaware and was a recipient of an Olin Fellowship at Harvard University. In 2016, the International Studies Association honored him as a Distinguished Scholar.

0000-0003-2184-0239
jwirtz@nps.edu

ACKNOWLEDGMENTS

We extend our deep gratitude to the many individuals who have contributed differently to this volume. Thanks so much to our contributing authors, whose work made the Handbook of Disinformation and National Security possible. The different academic and professional backgrounds of this line-up of international experts provided interdisciplinarity and diversity to the book.

We wish to express special thanks to our senior editor for Military, Strategic and Security Studies at Routledge, Andrew Humphreys, whose decisive stance to move the project forward and priceless professional advice at the different stages of the process made the Handbook possible. Many thanks to Devon Harvey, our editorial assistant, for her patience and support during the publication process. Thanks a lot also to our copyeditor Judit Varga, and to Daniela Amodeo for her support with the first proofs of the Handbook. We also wish to thank the anonymous reviewers of the Handbook proposal for their reviews and suggestions.

Last but not least, thanks so much to François Fisher, Director of the Permanent Secretariat of the Intelligence College in Europe, for his valuable comments and for writing the final concluding remarks to the Handbook.

The views expressed in this Handbook are those of the respective authors of contributing chapters and do not represent the stance, views, or policies of the author's parent organizations or national governments.

INTRODUCTION

Rubén Arcos, Irena Chiru and Cristina Ivan

The battle against disinformation and hostile influencing is global. Just a quick look at the major actors whose divergent voices are heard in this field from one side or another of the democratic spectrum makes the international dimension of the topic undisputable. The relevance of the topic of disinformation proves high not only for academics interested in the how and the why but also for citizens, experts, practitioners, and policy makers. Hence, the many talks, studies, debates on a topic that has become extensively transdisciplinary, claimed, and appropriated by a wide variety of disciplines. The danger is that in the medley of voices and perspectives, the red thread of disinformation as a practice favoured by intelligence and defence in waging hybrid conflicts, might be lost. This is the starting point of our own journey into the nexus forged between disinformation and national security, as well as its many facets. This handbook advances an interdisciplinary perspective that focuses on the need for knowledge of security and intelligence practitioners, as well as media and communication first line experts and any other public stakeholders and civil society representatives directly involved in creating societal resilience to these threats.

The main objective of the *Routledge Handbook of Disinformation and National Security* is hence to bring to focus this complex issue and recalibrate discussions back to their point of origin, which is intelligence and security studies. From this perspective, disinformation is not only a complex phenomenon, but also an instrument of influence at the hands of state and non-state actors to achieve political aims, and that, hence, requires the use of national security and intelligence machineries for understanding, anticipating, disrupting and neutralising it, particularly when foreign information manipulations are conducted clandestinely by covert state sponsors or by using the tools of stealth and secrecy. A competitive edge against authoritarian states engaged in hostile influencing through information will not be achieved only through in-depth assessments on the strategic environment including capabilities and intentions of state and non-state actors for destabilisation and interference. The gaining of competitive advantages also requires anticipatory approaches by identifying future trends and interdependencies, assessing risks and vulnerabilities to disinformation in democratic societies and through preventive strategic communications. Analyses and evaluations on disinformation and hostile influencing activities in the information domain can be conducted from two different perspectives: post-mortem or post-event, once malicious activities have been conducted by adversaries and situational analyses and attribution of manipulative and influencing activities are required for informing decisions the responses, and

DOI: 10.4324/9781003190363-1

also from a more anticipatory perspective that may enable authorities and relevant stakeholders to mitigate and impending threat and the potential effects of information manipulations in the perceptions and decisions of political actors and civilians.

The editors and authors of this handbook struggled to try to understand collectively how disinformation was and is used as an instrument of political warfare and influence by adversarial states and enemy actors. This was no simple task as it required a wide detour through historical cases, cultural filters applied to advance disinformation, traditional and new media representations, political warfare, use of new technologies to interfere with electoral processes and many more. The highly interdisciplinary nature of the studies dedicated to disinformation in the past two decades has also made this journey both difficult and rewarding. For in the end, putting together perspectives and findings brought forth by academic experts in intelligence and security studies, international relations, media studies, cultural studies, political psychology, IT, digital communication, public diplomacy has proven to be the right way to try and reconnect the dots back to the roots of the problem.

While security and intelligence studies have remained at the forefront of this initiative, what we hope to have achieved is a coherent and transdisciplinary approach that covers blank spots and articulates points of interference between the disciplines involved in studying the means, tools, role and impact of disinformation on national security. As a result, *the Routledge Handbook of Disinformation and National Security* surveys, assesses and debates the very recent developments in the field of disinformation impacting national security illustrating both the novelty that the use of technologies brought forth, but also the legacy provided by the instrumentalization of information (be it true or false) across centuries. What in the past was a powerful tool of political mischief is now a manual case for understanding how content manipulation works through distorted truths, half-reported facts and utter lies.

One major conclusion is that critical thinking, fact-checking and argument checking, media and digital competencies, quick debunking and building resilience are some of the most invoked means in fighting disinformation. To do that, the academic scholars and the public institutions via public policies have in recent years become very active in mapping the status quo and structuring solutions. However, as the contributions included in the *Handbook* make proof of, an attempt to sail the often uncharted waters of disinformation remains at times a discouraging initiative requiring an increasing level of skills and competences. The variety of definitions, the undue flexibility experts and practitioners alike take in using apparently interchangeable concepts, and the multiplicity of campaigns and appeals to retraining our brains and becoming anti-disinformation sensors complicate the picture.

The interdisciplinary perspective and the multiple methodological lenses applied to the studies incorporated in this handbook make it a useful tool for a rather wide audience of graduate students, academic researchers and professionals in the fields of security and intelligence studies, communication and international relations across the world. Our aim is to assist graduates and professionals in security, intelligence, media, communication, and education to understand, frame, assess, place in an accurate context and accordingly divert the wicked means and tools of disinformation. The *Handbook* integrates and articulates a comprehensive theoretical grounding (including history, communication and media studies) and it does so also by providing a close critique of existing literature. Finally, the handbook is centred on putting forward a practical oriented perspective combining theoretical knowledge with historical analysis and offering multiple methodological lenses that can be applied for accurately understanding and countering current propaganda, disinformation, covert information operations and information warfare. Its interdisciplinary character makes it useful in academic programs at both undergraduate and graduate level.

In order to make it a useful tool for experts, we have divided the Handbook in six main parts:

Part 1 is dedicated to updating experts on current theoretical perspectives, concepts and definitions applied to propaganda, disinformation, covert information operations and hybrid threats. In it, authors have taken both a chronological perspective on how concepts evolved and how they were reframed by major studies in the field, but also how these concepts enter into dialogue with each other and get to cross-pollinate meanings. **In the first chapter**, Cristina Ivan, Irena Chiru and Rubén Arcos look at hybrid security threats in the information domain and attempt a clear delineation of the informational spectre from the other facets of hybrid conflicts - be they related to infrastructure, cyber, space, economy, military/defence, culture, social/societal, public administration, legal, diplomacy, political and information. Furthermore, authors seek to strike a necessary balance between the Eastern and Western doctrines, highlighting points of confluence and most of all approaches towards new technologies and their use in waging hybrid warfare while at the same time building resilience to adversarial acts. **The second chapter** focuses on deterrence by denial and resilience building. In their own original and well-paced style, James Pamment and Bjorn Palmertz give a funny and clever explanation of choices that raise collective impact and form an approach to deterrence that is centred on raising the costs to the adversary. Last but not least, **chapter 3** gives an overview of military deception and perception management. In it, James Wirtz illustrates the application of military deception and perception management across the levels of war. The chapter also reviews how leading theorists assessed the importance and role of deception and perception management in battle and concludes by reflecting on deception in the Information Age.

Part 2 helps the reader locate concepts in a broader historical perspective, a much-needed endeavour for any reader that wants to understand disinformation, covert information operations and related terms in an accurate way. In **chapter 4**, drawing on the KGB training manuals, Sanshiro Hosaka looks at the fundamental role active measures played during the Cold War from the point of view of "external intelligence" and "intelligence from the territory". This essential study for anyone who wants to understand the origins of contemporary disinformation operations, is followed in **chapter 5** by a complementary look on Russian historical disinformation practices. Here, Randy Pherson, Deanna Labrini and Abby DiOrio focus on the period post-Cold War, when such campaigns are accelerated by the rapid growth of social media platforms. A detailed case study illustrates how Russia used the Five Ds to deflect blame for the shoot down of MH17 aircraft in Ukraine in 2014 and subsequent interferences in electoral processes in the US and Europe. In **chapter 6**, Jan Goldman narrows the lenses and provides a deep analysis of how information gets to be weaponized and how influencing operations targeting intelligence services operate. Finally, in **chapter 7**, Bjorn Palmertz and James Pamment give an overview of influence operations. This chapter examines the historical origins of Russian influence operations against the West, how these have shaped Russian perspectives on their use, and the methodological components they employ to attain their aims in the information environment. It also discusses a number of focus areas relevant for democracies when working to improve societal resilience and counter influence capability. And with this, we get closer to nowadays practices and challenges.

Part 3 then takes a broad survey on different contemporary challenges, starting with terrorist propaganda, then moving to computer mediated disinformation and malicious influencing in cyberspace, only to conclude with cognitive warfare and effects of conspiracy theories on the post-truth environments affecting contemporary democracies. In **chapter 8**, Jamil Ammar looks at the striking ways in which the new generation of Jihadi fighters nowadays were able to develop a new propaganda theory and test a number of measures that proved to be highly effective in spreading disinformation leading to violence. **Chapter 9** is first in a series of chapters dedicated to digital disinformation and their effects. Here, Josephine Lukito looks at the famous phenomenon of electoral interference and computer-mediated disinformation, insisting on that foreign actors seek to interfere in elections by injecting digital disinformation into social media ecosystems. In **chapter 10**, Adrian Tudorache adds to the discussion by focusing on propaganda as a tool of political warfare and what perception management can contribute to its understanding and mitigation. These considerations on contemporary challenges are complemented in the subsequent chapters 11 and 12 by a more granular perspective, in which disinformation is placed under the microscope via detailed case studies. In **chapter 11**, Corneliu Bjola and Ilan Manor offer a finely crafted analysis on how disinformation through digital and public diplomacy is used to besmirch the image of another country. The chapter contextualises the importance of historical narratives and takes a close look at Russia's use of digital diplomacy to refashion the contemporary image of the Baltic states by disseminating revisionist historical narratives. Subsequently, **chapter 12** advances a different case study, this time on foreign disinformation activities aimed at the United States that involve using social media to mimic domestic political discourse that exacerbates societal division and polarisation. Here, Hamilton Bean and Bryan Taylor use mimetic and rhetorical theory to offer an innovative critique of the personal vigilance approach, as well as to demonstrate how social media facilitates communicative processes of imitation and adaptation. The next two chapters focus on a tech savvy analysis of the tools and means employed by digital disinformation. In **chapter 13**, Agnes Venema discusses audio-visual forgeries and fake documentary evidence and their implications for national security, while H. Akin Unver and Arhan S. Ertan employ in **chapter 14** an innovative, data and cost oriented model of analysis of why countries engage in disinformation campaigns even though they know that they are likely to be debunked later on. The authors also highlight that deterrence in information warfare is attainable if the "defender" can signal its debunking and "naming-shaming" capacity prior to the disinformation campaign and if it can mobilise the support of the international audience against the attacker. The last two chapters dedicated to contemporary challenges focus on a more individual and psychological angle of the discussion, returning discussions on the effects of disinformation campaigns on individual cognitive processes and the dynamics of social polarisation. In **chapter 15**, Ruxandra Buluc takes a two-step approach to examine the influence and effects that conspiracy theories and unfounded rumours have on contemporary democracies and. Further on, Alina Bârgăoanu and Flavia Durach in **chapter 16** argue that disinformation has turned the human mind into an international battlefield best captured by the term "cognitive warfare". By conceptualising cognitive warfare as a new, technology-driven phenomenon, authors reflect on the implications this phenomenon has for research and policy making.

   Part 4 is dedicated to solutions investigated from multiple angles and perspectives. The main questions addressed in this section of the *Handbook* are how disinformation can be better identified, mitigated, countered, prevented and last but not least, how we can build resilience. **Chapter 17** tackles the journalistic approaches to information sources; fact-checking, verification and detached reporting. The author, Pablo Hernández-Escayola, insists on ethical journalism and the need to address the credibility of sources as an essential issue in the new chaotic and fragmented communication ecosystem. In **chapter 18**, Aitana Radu and Joseph Cannataci address informational leaks from a national security perspective, proposing a three-pronged approach to using legal reform to achieve business process re-engineering in the world of information leaks relevant to national security. The authors' aim is to provide a universally applicable framework for assessing such information leaks by including newly articulated legal principles which would need to be entrenched within national law and be applied by a specially designated and/or established court considering evidence which may inter alia also be presented or confirmed by an independent oversight authority. The legal perspective proposed by Radu and Cannataci is then complemented in **chapter 19** by Irena Chiru and Ruxandra Buluc who offer an ethics oriented understanding of military public relations, strategic communication and persuasion. This chapter aims to address the perpetual ethical dilemma: how can military public relations and strategic communication be effective and persuasive while at the same time being ethical? Tereza Capelos and Grete Krisciunaite shift the debate towards what they label as emotional diplomacy in times of uncertainty and disinformation and, as a result, **chapter 20** brings together theoretical insights from the fields of international relations, psychology, sociology, neuroscience and organisational studies to identify 11 types of uncertainty and their role in disinformation impact building. In **chapter 21**, Veli-Pekka Kivimäki addresses the value of open-source information for intelligence purposes with a particular focus on the challenge of disinformation. This chapter examines the evolution of open-source information available for intelligence, and how the new forms and methods of disinformation are challenging the intelligence production process. Last but not least, Cristina Ivan in **chapter 22** interrogates the newly emerged domain of protective factors against disinformation and aims to provide an interdisciplinary perspective on disinformation, its weaponization against democratic societies and ways to counteract it. The survey on current approaches is complemented by a cultural studies analysis of potential protective factors which could be integrated into an innovative approach. Finally, the case study proposes insight into protective factors to disinformation, as reflected by the Ukrainian/ global grassroots cultural response to Russian propaganda and disinformation.

Part 5 addresses regional specificities in countering disinformation. EU and NATO perspectives are complemented here by regional experiences from the Baltic countries, China and Latin America. Valentin Stoian, in **chapter 23**, looks at countering disinformation in the Eastern Neighbourhood with special focus in the discussion being placed on EU policies, practices and organisations. Conversely, Vira Ratsiborynska discusses in **chapter 24** the NATO's perspective in developing effective Strategic Communications. The next three chapters take a more granular perspective highlighting disinformation tactics and modus operandi at regional level. In **chapter 25**, Aleksandra Kuczyńska-Zonik explores methods to deal with Russian disinformation in Lithuania, Latvia and Estonia. The author offers an in depth overview of the Baltic states to explain the particularities of each state through the lens of security studies and political culture. Dan Dungaciu and Lucian Dumitrescu concentrate their analysis in **chapter 26** on the disinformation threat by tackling lessons from the Chinese experience, while Florina Cristiana Matei, in **chapter 27** sets out to capture lessons learnt from Latin America where the end of the Cold War coupled with the advent of internet and social media have brought about a surge of disinformation.

The **last section** of the *Handbook* sketches a toolkit for practitioners directly involved in preventing, early detecting, signalling and mitigating disinformation. **Chapter 28** discusses the use of discourse analysis as an analytical tool in propaganda detection and understanding. Julian Richards applies here the core principles of propaganda theory and discourse analysis method for a selected extremist texts. This generates a set of recommendations for policy makers as to the utility of such an approach in the contemporary counter-extremism environment. In close complementarity, **chapter 29** focuses on anticipatory approaches to disinformation, warning and supporting technologies. Here Rubén Arcos and Cristina María Arribas discuss on the use of issue management tools, indications and warning and technological support from developments in AI and machine learning, concluding that countering disinformation and influence activities in the infosphere requires not only reactive strategies when situations escalate and malicious contents have greater reach and potential to influence beliefs, attitudes and behaviours of target audiences, but also anticipatory approaches able to early identify emerging issues that can be the object of disinformation. In a close follow up, **chapter 30** delves deep into the technological aspects of disinformation. In it, Frédérique Segond takes a close look at technologies that can also be used to support the detection of information manipulation on the Internet, focusing on online media and social networks. **Chapter 31** concludes the series of chapters included in the Toolbox by adding the educational component of experiential learning. Roger Mason offers a close reflection on wargaming disinformation campaigns and provides a clear recipe on how a tabletop exercise on disinformation can be constructed from scratch.

We started plans for this Handbook in the Spring of 2020, in the midst of the COVID-19 pandemic and associated infodemic. At the same time, we were working on the research project CRESCENT—*Mind the gap in media CoveRagE and Strategic communication in CasE of security Threats – the development of critical thinking and responsible reaction*—under the auspices of the EU's Erasmus+ programme. During the writing and editing process, we were able to sadly observe how the Russian Federation made use of disinformation machineries and propaganda practices preceding and justifying its military aggression against Ukraine. For good or bad, these events and developments have brought a whole set of ongoing case studies in which disinformation and all sorts of information manipulations have been mobilised and available in full sight for the attentive academic observer. We have also witnessed how Western intelligence has taken a more active role in public communication by sharing information with the media on activities and intentions of the Kremlin before the military attacks against Ukraine, and also by publicly disseminating intelligence updates through social media platforms like Twitter in what seems an implicit acknowledgment that in our digital era of communicative overabundance and disinformation, addressing effectively security threats also requires influencing the public understanding of the threat. This represents an important turn in the (secretive) communicative practices of intelligence services, and suggests a new paradigmatic assumption: the public and relevant non-governmental stakeholders need to know to be safeguarded against information manipulations and hostile unacknowledged attempts to shape foreign public opinions.

Looking in retrospect at our efforts as editors, there were many other angles that we would have liked to include in this *Handbook* which does not succeed to exhaust all strands of debates on disinformation and the nexus created with national security. However, as time and resources are always limited for practitioners and experts, what we have tried to do is limit the *Handbook* to the utmost essential, so that, once readers have covered these bases, they can safely move on to their niche interests without losing any part of the essential map to guide their understanding. With this hope in mind, we hope readers will enjoy this essential handbook as much as we have enjoyed putting it together!

PART I

Theoretical Perspectives: Concepts and Definitions

1

HYBRID SECURITY THREATS AND THE INFORMATION DOMAIN: CONCEPTS AND DEFINITIONS

Cristina Ivan, Irena Chiru and Rubén Arcos

Hybrid warfare

The concept of hybrid warfare is generally acknowledged as being introduced in the academic debates on warfare by Frank Hoffman. Starting from the simple observation that we cannot limit ourselves to describe modern war tactics in terms of "big and conventional" versus "small and irregular"[1], Hoffman's definition is the first depiction of what today is labelled as hybrid warfare. He predicted that "future contingencies will more likely present unique combinational or hybrid threats that are specifically designed to target (…) vulnerabilities".[2] In this first conceptualization of hybrid warfare, there are two features that emerge as underlying characteristics that are somehow implicit to each other—the fusion of different war forms and the increasingly blurred distinctions between what was by then distinguished as conventional regular warfare and the irregular, asymmetric, non-conventional methods used by enemy agents, such as insurgent groups or terrorists. Later on, Hoffman also pinpointed to convergence and combination of modes of warfare, conventional capabilities, irregular tactics and formation, terrorist acts and organized crime, this being features of hybrid conflicts that might challenge the modern understanding of warfighting.[3]

Also, the term "hybrid" is mentioned seven times in the 2022 *Strategic Concept* adopted by NATO at the Madrid Summit, that includes specific references to hybrid tactics and operations used by the Russian Federation and the People's Republic of China's. The document clearly states that "hybrid operations against Allies could reach the level of armed attack and could lead the North Atlantic Council to invoke Article 5 of the North Atlantic Treaty."[4] While recognizing that non-military tools such as disinformation, subversion, sabotage and other hybrid methods have been used in the past with a destabilization purpose, it is also highlighted that information technologies and globalization facilitate the scale, speed and intensity of hybrid methods.[5]

The hybrid warfare concept did not come out of the blue, but rather in a logical sequence of multiple labels, each trying to catch a glimpse of the mutation produced at the beginning of the 21st century. Traditionally, when referring to the genealogy of modern conflict, we

DOI: 10.4324/9781003190363-3

refer to the introduction of irregular warfare at the beginning of the 20th century (which combined the classic confrontation between two armies with guerrilla tactics), asymmetric warfare (which added popular revolts, military insurrection, political assassinations and propaganda etc.)[6] and last but not least fourth generation warfare, new warfare, low intensity conflict, advanced technology warfare; all labels that attempted to detect and define the mutating warfare patterns that diverged from classic war.[7] Russian military scholars coined a similar term—"the new generation warfare" and later on "new type warfare"—in which, it was said, the role of military capabilities has been diminishing, and in which "differences between strategic, operational and tactical levels will be obliterated, as will be the difference between offensive and defensive activities".[8]

Going back to the definition of hybrid warfare, it must be mentioned that early definitions, precluding Hoffman, were mostly generated by the attempt to describe military and non-military tactics used in different conflict situations, such as the Second Chechen War. It was Nemeth who, five years before Hoffman, defined hybrid warfare carried out by the Chechens as "a type of war that involves the entire society and which combines tactics of conventional war with tactics used in irregular warfare and with information operations that make innovative use of technology".[9]

Fifteen years onwards, we can better appreciate Nemeth's and Hoffman's call for awareness and their prediction that this kind of warfare played on all strategic cultural weaknesses not just of the US, but of all democratic societies. A shift of paradigm was at the time in its early stages and required a shift in the cognitive landscape of those tasked to prevent and deter it. As hybrid warfare has interfered and impacted electoral processes, state strategic decisions and societal cohesion, cybersecurity of state and civilian infrastructure in the past two decades, we can now better grasp its evolution and fingerprint on our communities. In 2020, a report of the *Joint Research Centre* of the European Commission and the *European Centre of Excellence for Countering Hybrid Threats* illustrated the progress made in defining and addressing hybrid warfare in all its components. The rather scarce list of tactics evidenced by Hoffman in 2020 were detailed by the Hybrid CoE Report as tactics evidenced in the following 13 domains/ fields of intervention: infrastructure, cyber, space, economy, military/defence, culture, social/ societal, public administration, legal, diplomacy, political and information.[10] The mentioned report and conceptual framework insists on the fact that in its nature, hybrid warfare is not new, but rather a combination of old methods which are "repackaged and empowered by changing security dynamics, new tools, concepts and technologies targeting vulnerabilities in several domains in an unprecedented manner".[11] In a nutshell, what the authors reiterate is that the shift of paradigm announced by Hoffman is brought about not by a changing nature of the conflict, but rather by the opportunistic and clever combination of technological innovations and virtual environments that complement an old reality and that allow for a far larger impact across society. At the same time, features evidenced in the report also made reference to keeping actions "under the radar" (especially in the early phases of the conflict) and blurring boundaries. Once again, 15 years into the new reality, we may notice that hybrid warfare is still being defined in terms of its doings and impact, rather than in terms of conceptual configuration. Hence, we can argue,

> while remaining highly *underconceptualised,* the term hybrid threats demonstrates an excess of contextual features—any new instrument, social behaviour or practice of consumption related to the use and abuse of new technologies is then described as part of the new range of competitive and confrontational tactics characterizing hybrid warfare.[12]

The prevailing feature we need to highlight is that of an active incorporation and opportunistic use of any new technology that may create a competitive advantage and stimulate compromising of democratic processes, bringing about distrust, confusion and social polarization.

Another aspect worth mentioning is the distinction between hybrid warfare and hybrid threats. Some authors argue that using the term warfare is unnecessarily weaponizing and/ or militarizing our understanding of the state of affairs and that reference should rather be made to hybrid threats. Mikael Wigell, for instance, dedicated a full study to distinguishing between hybrid warfare and what he labelled as "hybrid interference", hence clarifying boundaries of acts located in the grey zone between war and peace.[13] While warfare may be a term that unnecessarily places interventions in a conflict mode, what we have seen throughout 2014, when the annexation of Crimea by the Russian Federation occurred, to the 2022 invasion of Ukraine and the subsequent war, is the emergence of a different strategy; one in which peace and war are placed in a continuum. Wartime hybrid tactics are not inherently different from the "regular" ones used during peace time, but only intensified and expanded to encompass a larger sphere of impact. Hence, whether a distinction might be useful for policy purposes, the features of hybrid threats do not distinguish from those of hybrid warfare which simply takes intensity and volume to a higher level.

However, we should note Wigell's contribution to highlighting the "informational" nature of hybrid interference: "hybrid interference sees liberal democratic values as vulnerabilities that can be exploited to drive wedges through democratic societies and undermine governability. As such, hybrid interference can be used as a priming phase as part of planning to escalate conflict activities".[14] Wiggell's conceptualization of the interference helps maintain a distinction between different levels of intensity and nuances the priming phases of a conflict that may either precede the range of overt and covert military, paramilitary and civilian measures outlined for instance by the NATO doctrine of hybrid warfare. The fact that hybrid interferences may exists in themselves, without necessarily leading to conventional war is indeed a fact, yet one must also acknowledge that the very emergence of such interferences alert to the emergence of a cultural, social and political divide and that adversarial strategies are already in place, seeking to weaken both social cohesion and state capability of response. Therefore, what distinguishes hybrid warfare from hybrid interference, in our understanding, is the level of intensity, of preoccupation for plausible deniability and non-attribution, and not the very nature of the threat or action pattern.

Out of the 13 fields of action mentioned above, that form the landscape of hybrid threats according the Hybrid CoE Report (infrastructure, cyber, space, economy, military/defence, culture, social/societal, public administration, legal, diplomacy, political and information), the most heavily researched and investigated by security, intelligence, international relations, philologists and sociologists alike, not to name journalists and political scientists, is the information domain which gave birth to a significant amount of literature, topics including propaganda, fake-news, disinformation, covert information operations etc. It is this particular field that the mentioned report describes as "the hallmark of hybrid threats,"[15] and which we shall detail below for the purpose of the present handbook, precisely because of its potential for weaponization against liberal democracies.

Hybrid security threats are built by propaganda and disinformation as well as covert information operations aimed to manipulate perceptions over the reality of the adversarial country's citizens. Such threats are effective as long as they remain undiscovered, denied and refer to plausible interpretations. One must remark that even though they are not apparently the most stringent threats to address, their potential for harm is significant both in the short and medium terms. Playing on identity politics and social causes, tackling cultural, societal, political values and norms, hybrid security threats in the information domain have the potential to create social unrest, polarize society, insularise communities, sow distrust in own government and state legitimacy and hence attack the very fabric of democratic governance and liberal society.

> The information domain is strongly linked to the culture and society domain (...) because disinformation campaigns and other tools in this domain seek to affect the homogeneity of the target state's culture and society. It is also influenced by the intelligence domain (...), as information obtained via cyber or traditional espionage can be leaked to influence public opinion, perceptions and discourse.[16]

To do justice to the concept of hybrid security threat in the information domain we must also mention its counterpart in the Russian literature, namely the concept of active measures (*aktivinyye meropriatia*) which also heavily rely on propaganda, disinformation, forgeries, while also comprising non-information methods such as political repression, assassination etc.[17] Just as with hybrid interference, the objective of active measures is to advance strategic influence by impacting the way events unfold and the way they are perceived, as well as their consequences.

Another concept closely related to active measures, and which we can equate to hybrid warfare, is the information war, which in Russian literature is defined as "a combination of military and non-military means to influence the informational-psychological space of a targeted audience to achieve certain political goals".[18]

Hence, when addressing concepts related to hybrid threats, especially in the information domain, one must be aware that there is a significant overlapping of terminology between different schools of thought and that while the West focuses on defining hybrid threats as aggression coming from the outside, perpetrated mostly by adversarial entities, in the East, Russian and Chinese doctrines focus on hybrid actions, especially in the information domain, that equally target in and out groups and audiences. Historical studies of the Cold War era, for instance, highlight the use ambivalent use of propaganda and disinformation by the Soviet regime both internally and in the near abroad represented by central and south-eastern Europe countries part of the communist bloc: "examining influence operations across varied geographic areas over the past century reveals that internal domestic considerations of regime survival, more than external foreign policy interests, were the primary drivers of resorting to active measures".[19] Furthermore, recent studies on the Chinese propaganda system have also highlighted the concept of power and its use for both foreign and domestic propaganda.[20]

There are however a number of historical evidence that also pinpoint to the use of covert information operations by the West, while some of the tools evidenced resemble quite strikingly the use of active measures in the East. Historian Rory Cormac, for instance, has looked into the British thinking about covert propaganda and illustrated his study with a tranche of information that the UK Information Research Department (IRD) files declassified in early 2019. As a result, Cormac comes to the conclusion that IRD, while created to counter Soviet propaganda in late 1940's, it used tools that included unattributable propaganda, forgery, bribery, covert political funding and orchestrating coups.[21]

Propaganda

Propaganda represents the intensive manipulation of information so as to influence perceptions and the ability of the target audience to make objective decisions. The overall aim of propaganda is to obtain strategic advantages, political and financial capital brand and image promotion etc. Nowadays, threats generated by propaganda and disinformation are predominantly circulated and made effective by exploiting the systemic vulnerabilities of the digital ecosystem. Propaganda was used historically in order to legitimate political regimes, advance certain ideological causes, and also as a way to mobilize the masses in case of armed conflict. For Lenin, propaganda had, at the beginning of the XX century, two main functions: to inform and mobilize own military troops and to undermine the morale and the trust of the opposite army forces.[22] Lenin associated propaganda with the need to adapt the message to the historical, cultural, and social environment and with "agitation measures".

The authors of the Russian current doctrine of new generation information warfare[23] consider information campaigns and hence propaganda as well, as one of the tools used in asymmetric warfare, together with active economic, political, technological, and ecological measures etc. In a more empirical approach, during the Cold War, L. John Martin defined propaganda as "something carried out by enemies which can be called propaganda only if that something is unsuccessful (...) and for a message to be considered propaganda, it is essential that its audience perceives but does not accept its intention to influence opinions".[24] Same author notes that, by contrast, in order to be successful, propaganda must be perceived as an appeal to rationality or as information that sheds lights a new perspective on a given problem. And, in order to produce the mentioned effect, propaganda often appeals to new and latest *information* produced via disinformation.

According to Cormarc[25], propaganda was used by British propagandists as:

- white (openly promoted by state affiliated channels of communication);
- grey (unattributable material which can be based on sanitized intelligence and distributed into foreign media outlets through trusted contacts);
- and black propaganda associated to forgeries, black radio, whispering campaigns, and that could be attached to either "disavowable operations, which could be denied but with potential embarrassment and black operations which should have no evidence".

On the reception side, in communication studies, a different distinction is being made between the three, the underlying criteria being not the attributable or unattributable character, but the level of transparency of the source. While in white propaganda the producer of the material is clearly indicated, in the gray propaganda the producer remains unclear, and the black propaganda it can be totally covert, and the public is deceived to believe in a fake author.[26]

Nevertheless, when discussing hybrid threats in the information domain, one needs to refer to historical examples, documented in the archives, which show that between the gray and black spectrum of propaganda there are carried a series of actions classified under the range of covert information operations. Discussing the British—Indonesian—Malaysian confrontation in the 1960's, which led to the destruction of the Indonesian communist Party, David Easter shows that enemy sides operated black radios, black newspapers, spreading rumours and writing slogans on walls, simply to try and influence perceptions of adversarial sides.[27] The specificity of unattributable, disavowable or black propaganda is that it targets key segments of the target population (young army conscripts, students, minorities etc.) that could be influenced through subversion and psychological warfare to act in the benefit of the propaganda agent.

Disinformation

Disinformation refers to an entire array of tactics and strategies used to propagate false, inexact or out of context information (therefore hijacked from their real meaning). Its intention is to provoke damages and/or profit. Continuous disinformation can severely affect democratic processes, national security, and social cohesion. In the long run, it undermines citizens' trust in legitimate authorities, the democratic system, and the benefits of the information society, thus diminishing citizens' permeability to information, knowledge, and progress.

Experts also separate conventional propaganda from disinformation by making reference to the source. If propaganda, or more specifically white propaganda, is accomplished via state institutions, state owned TV and radio channels, media outlets, government affiliated press and news portals, whose ties are theoretically easily traceable, disinformation operations take a great deal of effort in covering primary sources, while operations remain, at least partially, clandestine. Disinformation hence represents a form of propaganda whose main purpose remains illegal and clandestine, the message it puts forth being an intentionally manipulated account of a real situation or of a legitimate action carried out by an entity perceived as an opponent.[28]

Disinformation includes a series of practices such as: counterfeited documents, fake news, grey or black propaganda, foreign media source manipulation or control, operations carried out by agents of influence, radio stations, TV channels, sites, blogs, "clandestine" social media accounts, the use of non-governmental associations, political parties with kindred ideologies, protest manipulation or setting up, and, in extremis, blackmail and kidnapping.[29] Hence, disinformation as a concept overlaps over what has been historically defined in intelligence documents such as the UK Information Research Department mentioned before, as grey and even black propaganda, black radios or newspapers of the past being similar in concept and use to the fake social media of the present.

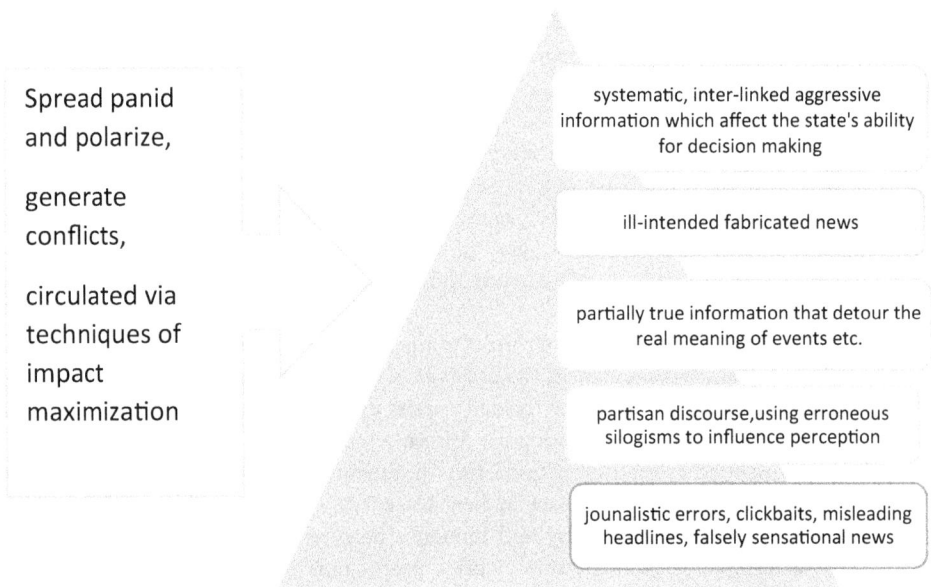

Spread panid and polarize,	systematic, inter-linked aggressive information which affect the state's ability for decision making
generate conflicts,	ill-intended fabricated news
circulated via techniques of impact maximization	partially true information that detour the real meaning of events etc.
	partisan discourse, using erroneous silogisms to influence perception
	jounalistic errors, clickbaits, misleading headlines, falsely sensational news

Figure 1.1 Understanding what disinformation does
Source: Author's creation

Perception management and intelligence deception

Perception management and intelligence deception are two other relevant concepts for the discussion. In their remarkable volume published during the Cold War on Soviet Strategic Deception, Dailey and Parker noted the distinction between perception management activities and intelligence deception, presenting a taxonomy of these two broad categories as part of deception. While perception management uses a complex set of hostile activities targeting mainly policy-makers, opinion leaders and the public, intelligence deception is directed at intelligence agencies and their officers and analysts by using controlled collection channels.[30] The Kremlin's communication practices and denials during the last years echoes what Daily and Parker pointed out for the Soviet case: that "Soviet perception management in peacetime is pervasive".[31]

Cyber information operations

Hybrid threats may yet to achieve their full capacity to influence society through information operations in the digital world. The online environment has given a boost to propaganda and disinformation generation and dissemination, with trolls, bots and honey pots virilising topics of interest and influencing perceptions through minute microtargeting. Influence of information operations during electoral processes in the past decade has been studied extensively and will also be detailed in the following chapters of this handbook. What is yet to reveal its

Table 1.1 Overview on perception management and deception.

	Perception management	*Intelligence deception*
To Whom	Policy-makers and decision-makers; opinion leaders, general public, analysts (to a certain extent and as long as content in open channels impact intelligence products).	Intelligence officers/analysts
Which Channels (note that some of these channels are also covert political techniques in active measures as described by Shultz and Godson in their work *Dezinformatsia*)	Public statements of government authorities and diplomats. Articles in journals and other authoritative sources. Information disseminated to foreign correspondents and foreign service personnel Overt propaganda, front organizations, conferences and events, foreign visitors. Press placements in controlled media Forgeries Agents of influence	Controlled human intelligence sources Technical intelligence collection apparatus of the targeted country
With what intended objective and **effects**	Influence policies and opinions in foreign countries	Confuse, misled, or distort the analyses and assessments of intelligence services and influence judgment in intelligence products, particularly (but not limited to) those dealing with military issues

Source: Adapted from Dailey and Parker, 1987

full force, however, is the impact of AI, big data and machine learning upon the way information is produced, disseminated, verified, manipulated and/ or forged. In the information field in particular, "a new term is needed to characterize the combination of information warfare with hacking and traditional use of agents of influence" as well as "the potent admixture of classic (Russian) human intelligence (HUMINT) tradecraft with cyber operations and information warfare."[32] It is in this nexus between HUMINT, cyber operations, information operations and the capacity for surveillance and control facilitated by AI tools that we shall see the true potency and danger of hybrid warfare in the years to come. And unless understood and regulated, this might prove the Achile's heel of our democratic systems. China's latest example of the new Internet governance system offers another example of how, in a dystopian perspective, in which social control, information operations and mass surveillance could be steered towards dismissing barriers to centralized control of data. As Rogier Creemers puts it, the Chinese internet governance and social credit scheme "will extend the logic of the dang'an system to every connected member of society, tempting towards compliance through the promise of middle-class aspirations, backed up with the threat of ostracism, joblessness or reduced access to public good".[33]

Infodemics and disinformation

During the COVID-19 disease outbreak, health authorities have struggled to manage the additional challenge posed by the overabundance of information, misleading and false content, and conspiracy theories, such as the 5G/COVID-19 conspiracy theory, related to the virus SARS-CoV-2. In July 2020, the World Health Organization (WHO) organized the 1st Infodemiology Conference. An infodemic is defined as "an overabundance of information—some accurate and some not—occurring during an epidemic. In a similar manner to an epidemic, it spreads between humans through digital and physical information systems. It makes it hard for people to find trustworthy sources and reliable guidance when they need it".[34] Hostile actors can seize the opportunity presented by an epidemic to advance their geopolitical interest by disseminating disinformation and narratives aiming to harm the credibility of institutions handling the situation or even the relationships between allies or partnering countries. For instance, the database of disinformation cases of the EUvsDisinfo project of the European External Action Service includes debunks of narratives spread in Russian outlets claiming that "Poland did not let Russian aircraft carrying aid to Italy pass through its airspace"[35] during the pandemic. In June 2020, the European Commission's Joint Communication on "Tackling COVID-19 disinformation" asserted that "in particular, Russia and China, have engaged in targeted influence operations and disinformation campaigns around COVID-19 in the EU, its neighborhood and globally."[36] As some experts have noted, the analysis of information on the tactics employed by hostile actors should result in the design of counterintelligence strategies in the post-COVID era.[37]

Conclusions

The several attempts of defining and explaining these inter-related concepts provide a complex and multi-layered view of old and new, Eastern and Western operations as systematic and concerted outreach towards diverse target audiences that need to be persuaded to act in favour of state strategic interests. Any endeavour aiming to understand and/ or to perform in the information domain should start by shedding light on the conceptual complexity and the dimensions of strategic influencing and action involved, which seem to range from diplomacy to the media, from intelligence to defence and from economic to social psychology. Only with this terminological framework in mind, the interdependences created between these multidimensional concepts can be apprehended, and the references to the methods and tools involved can be deciphered.

Notes

1 Hoffman 2007, 5. See also: Mattis and Hoffman 2005.
2 Hoffman 2007, 7
3 Hoffman 2007, 14
4 NATO 2022 Strategic Concept, 7.
5 Ibid.
6 Engels apud Ivan, Dobre and Popescu 2020, 22
7 Nemeth, 2002
8 Thomas, 2020, 14
9 Nemeth, 3
10 Giannopoulos et al., 2020
11 Giannopoulos et al., 6
12 Ivan et al., 2020.
13 Wigell, 2019
14 Wigell, 2
15 Giannopoulos et al., 32
16 Giannopoulos et al., 33
17 Gioe et al., 2020, 2
18 Fridman 2020, 2
19 Gioe et al., 2020, 3)
20 Edney, 2012
21 Cormac, 2019
22 Lenin, 1908
23 Valery Gerasimov, Sergeo Chekinov, and Sergei Bogdanov
24 Martin 2010, 49
25 Cormac, 1
26 Nabb Research Center Online Exhibits, n.d., American foreign relations n.d.
27 Easter 2010, 9
28 Martin 2010, 58
29 Idem.
30 Dailey and Parker, Soviet Strategic Deception, xv-xx
31 Ibid., xviii
32 Gioe, 2018
33 Creemers, 2016
34 WHO Infodemic Management. 2020. 1st WHO infodemiology Conference: how infodemics affect the world and how they can be managed, 30 June & 1, 7, 9, 14, 16 July 2020. https://www.who.int/newsroom/events/detail/2020/06/30/default-calendar/1st-who-infodemiology-conference
35 EUvsDisinfo. DISINFO: Controlled by the US, EU didn't react when Poland refused to aid Italy to cross its airspace. https://euvsdisinfo.eu/report/poland-refused-humanitarian-aid-to-italy-to-cross-their-national-airspace
36 European Commission 2020, Joint Communication, 3.
37 Gradon, COVID-19 and the information ecosystem: Lessons from Russian malign influence campaigns for the post-COVID-19 world, 211.

Bibliography

***American foreign relations. n.d. *"Propaganda—Types of propaganda"*, available at https://www.americanforeignrelations.com/O-W/Propaganda-Types-of-propaganda.html, last accessed October 15th 2022.

Cormac, Rory (2019). "Techniques of covert propaganda: the British approach in the mid-1960s", *Intelligence and National Security*, 34:7, 1064–1069, DOI: doi:10.1080/02684527.2019.1645434.

Creemers, Rogier. 2016. "Cyber China: Upgrading Propaganda, Public Opinion Work and Social Management for the Twenty-First Century", *Journal of Contemporary China* 26:103, 85–100, DOI: doi:10.1080/10670564.2016.1206281.

Dailey, Brian D. and Parker, Patrick. J. (eds.) (1987). *Soviet Strategic Deception*. Massachusets/Toronto: D.C Heath and Company/Lexington.

Desai, Neal, Andre Pineda, Majken Runquist, and Mark Fusunyan (2010). "Torture at times: water-boarding in the media", Joan Shorenstein Center on the Press, Politics, and Public Policy, Harvard Student Paper, available at http://nrs.harvard.edu/urn-3:HUL.InstRepos:4420886, last accessed October 15th 2022.

Easter, David (2010). "British Intelligence and Propaganda during the 'Confrontation', 1963–1966", *Intelligence and National Security*, 16:2, 83–102, DOI: doi:10.1080/714002893.

Edney, Kinglsey. 2012. "Soft Power and the Chinese Propaganda System", *Journal of Contemporary China*, 21:78, 899–914, DOI: doi:10.1080/10670564.2012.701031.

Edward, Herman, and Noam Chomsky (1988). "A Propaganda Model" in *Manufacturing Consent: The Political Economy of the Mass Media* by Herman Edward and Noam Chomsky, Pantheon.

***EUvsDisinfo. *DISINFO: Controlled by the US, EU didn't react when Poland refused to aid Italy to cross its airspace*, available at https://euvsdisinfo.eu/report/poland-refused-humanitarian-aid-to-italy-to-cross-their-national-airspace, last accessed October 10th 2022.

***European Commission (2020). *Tackling COVID-19 disinformation—Getting the facts right*. Brussels, 10.6.2020. JOIN 8 final., available at https://eur-lex.europa.eu/legal-content/EN/TXT/PDF/?uri=CELEX:52020JC0008&from=EN, last accessed October 10th 2022.

Fridman, Ofer (2020). "'Information War' as the Russian Conceptualisation of Strategic Communications", *The RUSI Journal*, 165:1, 44–53, DOI: doi:10.1080/03071847.2020.1740494.

Giannopoulos, G., H. Smith, and M. Theocharidou (2020). *The Landscape of Hybrid Threats: Conceptual Model Public Version*. European Commission, Ispra.

Gioe, David V. (2018). "Cyber operations and useful fools: the approach of Russian hybrid intelligence", *Intelligence and National Security*, 33:7, 954–973, DOI: doi:10.1080/02684527.2018.1479345.

Gioe, David V., Richard Lovering, and Tyler Pachesny (2020). "The Soviet Legacy of Russian Active Measures: New Vodka from Old Stills?", *International Journal of Intelligence and CounterIntelligence*, 33:3, 514–539, DOI: doi:10.1080/08850607.2020.1725364.

Gradon, Kacper T. (2022). "COVID-19 and the information ecosystem: Lessons from Russian malign influence campaigns for the post-covid-19 world" in Stacey E. Pollard and Lawrence A. Kuznar (eds.) *A World emerging from pandemic: implications for Intelligence and National Security*, Bethesda, MD: National Intelligence University, 207–227, available at https://ni-u.edu/wp/wp-content/uploads/2022/06/A-World-Emerging-From-Pandemic.pdf, last accessed October 10th 2022.

Hoffman, Frank G. (2007). *Conflict in the 21st century: the rise of hybrid wars*. Potomac Institute for Policy Studies, available at https://www.potomacinstitute.org/, last accessed October 10th 2022.

Ivan, Cristina (2017). "Conflict hibrid, propaganda si dezinformare—obiectiv strategic al Federatiei Ruse în vecinătatea apropiată și îndepărtată", *Romanian Intelligence Studies Review*, 17–18: 7–15.

Ivan, Cristina, Teodora Dobre, and Alexandra Popescu (2020). "Război hibrid azi, de la definiție la acțiune" in Cristina Ivan, Valentin Stoian-Iordache, Dana Sîrbu, Mihaela Teodor, Teodora Dobre, and Alexandra Popescu (eds.) *Amenințări hibride. Detecție timpurie, tipare acționale, reziliență*, 19–55, București, Editura ANIMV.

Lenin, V.V. 2018. "*V. I. Lenin, Lessons of the Moscow Uprising.*" https://www.marxists.org/archive/lenin/.

Martin, L. John. 2010. "Disinformation: an instrumentality in the propaganda arsenal." *Political Communication* 47–64.

Mattis, James N. and Hoffman, Frank (2005). "Future warfare: the rise of hybrid wars", U.S. Naval Institute Proceedings, November 2005, pp. 18–19.

Nabb Research Center Online Exhibits. n.d. "*The Colors of Propaganda.*" https://libapps.salisbury.edu/nabb-online/exhibits/show/propaganda/what-is-propaganda-/the-colors-of-propaganda.

NATO. 2022. *NATO 2022 Strategic Concept*. Adopted by Heads of State and Government at the NATO Summit in Madrid, 29 June 2022. https://www.nato.int/strategic-concept/.

NATO. 2018. *NATO Encyclopedia 2018*, December 2018. Brussels: NATO Public Diplomacy Division.

Nemeth, William J. 2002. *Future war and Chechnya: a case for hybrid warfare*. Calhoun: The NPS Institutional Archive.

Schultz, Richard H. and Godson. Roy. 1984. *Dezinformatsia: active measures in soviet strategy*. Washington: Pergamon-Brassey's International Defense Publishers.

Spicer, R.N. 2018. "Lies, Damn Lies, Alternative Facts, Fake News, Propaganda, Pinocchios, Pants, on Fire, Disinformation, Misinformation, Post-Truth, Data, and Statistics." In *Free Speech and False Speech*, by R. N. Spicer. doi:https://doi.org/10.1007/978-3-319-69820-5_1.

Thomas, Timothy. 2020. *The Chekinov-Bogdanov Commentaries of 2010–2017: What Did They Teach Us about Russia's New Way of War?* MITRE.

Wigell, Mikael. 2019. "Hybrid interference as a wedge strategy: a theory of external interference in liberal democracy." *International Affairs*. Oxford University Press.

WHO Infodemic Managment. 2020. 1st WHO infodemiology Conference: How Infodemics Affect The World And How They Can Be Managed, 30 June & 1, 7, 9, 14, 16 July 2020. https://www.who.int/news-room/events/detail/2020/06/30/default-calendar/1st-who-infodemiology-conference.

2

DETERRENCE BY DENIAL AND RESILIENCE BUILDING

James Pamment and Bjorn Palmertz

Introduction

How do you stop a bully from verbally attacking you? In the school playground, kids have a few effective options. First, they would likely test the age-old proverb: "sticks and stones may break my bones, but names will never hurt me." In doing so, they make use of a deterrence strategy aimed at affecting the adversary's calculus. "Names will never hurt me" is what is known in deterrence theory as *denial of benefit*[1]; the bully will not receive satisfaction from name-calling because it will not achieve the desired effect on the subject. The bully is thereby forced into a decision: either escalate to the use of force, or take the opportunity provided for both parties to walk away in a status quo (a so-called *off-ramp*[2]).

If the bully decides to escalate the conflict by threatening the use of force, one might hope for a group of friends to have your back. "You and whose army?" a brave child might be expected to retort. This is known in deterrence theory as *denial of capability*.[3] The bully's capability for violence will be nullified either by their own defensive skills or those of a group of friends and allies, who can for example physically restrain the adversary. Again, the bully is faced with a conundrum, leading either to further escalation or an off-ramp. Either gather an equivalent group of friends and initiate a full-scale playground brawl, or simply walk away.

Perhaps the bully has a group of equally unenlightened friends who are eager to join the fight. A third, last ditch option remains in case the adversary is not deterred: threaten to tell the teacher. As any veteran of the playground will tell you, this is a risky move. It might save you for the moment, but it will likely contribute to grievances that fuel further conflicts in the future. Plus, you will probably get called a snitch. Within deterrence theory, this would be considered *denial by punishment*.[4] Demonstrating to an adversary that their intended path of action will inevitably lead to punishment, such as in this case detention, provides a further junction at which the bully can either escalate or take the off-ramp.

The collective impact of these choices comprises an approach to deterrence that is centred on *raising the costs to the adversary*. For each aggressive action that the bully attempts to take, s/he is met with additional costs that force a re-evaluation of the rationale for aggression. Good deterrence marries these escalating costs with off-ramps that grant opportunities for both parties to de-escalate while saving face. When you do thought experiments[5] to map out likely cause

DOI: 10.4324/9781003190363-4

and effect in this manner, it is known as an *escalation ladder*. By denying benefits, capabilities, and with the support of appropriate punishments, the adversary is compelled to reconsider whether a conflict is worthwhile. It may be at this point that the bully feels that they have achieved their objective of testing your resolve, demonstrating your fear of conflict, or that their objectives of beating you up are not worth the cost. The fight is avoided, at least for today.

Deterrence as a tool of statecraft essentially works in the same way as deterrence in the playground. Societal institutions are responsible for maintaining routines that are supposed to be able to handle threats from all kinds of adversaries. A certain readiness, so to say, for the bullies that lurk in every corner of the geopolitical playground. When this becomes a natural, bureaucratic, and common-sense routine, it is called resilience. Returning to the playground metaphor, resilience might come from the bullied child seeking advice from a parent, carrying some spare lunch money, learning a defensive martial art such as aikido, getting better at sensing the early warning signals of when the bully is in a bad mood, or even carrying some band aids in their backpack. Resilience is ultimately about mitigating one's own vulnerabilities in preparedness for known risks and threats, enabling a continued capability and functionality despite being faced by challenges or external threats. It can also involve developing offensive capabilities, such as learning an offensive martial art, which can augment the deterrence effect and further raises the costs of use of force.

As we shall discuss in this chapter, deterrence and resilience are intimately connected. Deterrence is primarily about understanding threats and using levers that might force an adversary to reconsider their objectives and actions against you. Resilience is about understanding yourself and the measures you can take to become a less attractive target, or to bounce back more quickly from a setback. This combination of deterrence and resilience has developed into one of the key emerging areas of 21st century governance. Societal institutions are expected to perform a key role in deterring adversaries of all kinds, by building resilience and wielding a combination of denial and cost imposition strategies as a method of shaping collective security. However, the theories and practices governing such activities are still in their relative infancy. As this chapter will discuss, the contemporary application of deterrence theory upon so-called hybrid threats demonstrates both the vitality and endemic weaknesses of this under-researched area of security policy.

Modern deterrence theory

Modern deterrence may be defined at the broadest possible level as *the rational process by which a society identifies threats and develops strategies to limit the risks posed by them.* Often, the process centres on a specific adversary, which requires some specialist understanding of the adversary's intent, capabilities, and opportunities to cause harm. The adversary is typically a hostile foreign state actor, a nonstate actor such as a terrorist cell, or a group of some other kind that engages in dangerous activities. The goal of deterrence is behaviour change; as such it is a form of *strategic communication (in the broadest sense including symbolic actions) aimed at affecting the decision-making processes of an adversary.*[6]

Deterrence is usually divided into two areas: denial and costs. *Deterrence by denial* involves the reduction or removal of an adversary's capabilities and/or their intended effects. The three fundamental denial areas are:

- *Denial of benefit*: reducing or removing the rewards anticipated from adversarial behaviour.

- *Denial of capabilities*: restricting or nullifying the threatening capabilities that the adversary can bring to bear.
- *Denial by punishment*: levelling punitive measures upon the adversary.

These *denial* options contribute to an overall approach to *deterrence by imposition of costs*. Deterrence by imposition of costs is a mindset and form of strategizing based upon the assumption that the collective impact of denial measures on the adversary's cost/benefit analysis will lead them to conclude that their aggressive actions are not worth it. In essence, it asks the question, *can we make this type of attack more costly to carry out?* Those costs might for example be in terms of a need for increased resources to carry out the harm (e.g. more people, advanced tools, and work hours are required), unanticipated costs to reputations (e.g. an increased risk of attribution and exposure[7]), or highly damaging countermeasures (e.g. likelihood of counter-manoeuvres including offensive responses). While it is not always the case that the adversary acts rationally, at its core, deterrence by imposition of costs tries to make the harmful activity more costly (both metaphorically and actually) than the rewards the adversary anticipates from its problematic behaviour.

It is therefore essential to have some understanding of the adversary's decision-making processes, the resources they consider proportionate, and their own red lines or limitations. In other words, security policy experts need some understanding of the underlying historical and cultural context and frame by which the adversary views the world and interprets the strategies and actions of themselves and others. Some adversaries are more sensitive to certain types of costs than others; for example, it is often assumed that oligarchs are sensitive to economic sanctions and travel bans because of the expectation that wealth enables a luxurious international lifestyle. Others are acutely sensitive to attribution since they like to maintain a strongly positive reputation in public perceptions. Terrorists might be entirely unmoved by punitive or financial measures but respond to theological reasoning. Understanding what makes the adversary tick requires intelligence about their psychological make-up, motivations, information sources, and decision-making processes.

Deterrence by imposition of costs is usually seen as unfolding in several steps according to a strategic plan, into which are built escalatory and de-escalatory moves.[8] The escalation ladder involves "gaming out" anticipated adversary responses to denial and cost-raising measures. Experts working in security policy do their best to work through complex scenarios trying to figure out the consequences of different actions and reactions numerous steps ahead of the current situation. This gives the impression of deterrence as a game of chess in which strategies are mapped out by geniuses in ivory towers. However, in reality many escalation ladders are ad hoc and take the form of a series of tit-for-tat measures gradually escalating into a conflict, in which uncoordinated deterrence actions lead to one unanticipated crisis after another.

If deterrence seeks to understand a threat actor for the purpose of developing strategies to change their calculus, resilience tends to focus on one's own proactive and defensive capabilities for mitigating risk. Resilience refers to establishing *routines, processes, and practices that empower the whole-of-society to participate in collective security*. It is *actor-agnostic* in terms of threat; its focus is on the self. Resilience is key to strengthening how a society bounces back from adversity, and its resources support, inform, and strengthen all other deterrence actions. It is key to developing long-term societal processes that guide strategic, rather than haphazard, responses to threats. Resilience is, in other words, a complement to deterrence centred on developing societal institutions that can embody and enact a society's resolve. This offers credibility to the deterrence posture, which is especially important to show for capable adversaries that have continuous intelligence collection and a good analytical grasp of their targeted society.

Traditional deterrence theory

It is important to remember, however, that there is little consensus over these modern definitions of deterrence and resilience. As deterrence theory developed during the Cold War, it took the form of grand strategy underpinned by the looming threat of total annihilation: Mutually Assured Destruction (MAD), also known as the nuclear deterrent.[9] From this perspective, the various denial strategies fit within a hierarchical escalation ladder with nuclear war at the summit. This supported a cost/benefit calculus in which aggression could be managed between two rational parties with relative geopolitical parity (the West and the Soviet Union) on the assumption that either could invoke MAD in the final instance to keep the other in check. With the possible exception of the Cuban Missile Crisis, deterrence theory seemed to explain relative stability in the adversarial geopolitical relationship between East and West for around forty years.

In the past two decades, geopolitical analysts and governments have faced different kinds of problems than MAD. Asymmetric threats from state and substate actors such as terrorism, nuclear proliferation, and organised crime have dominated the geopolitical agenda. Hostile actors have hidden bombs inside shoes and flown airplanes into buildings; poisoned defectors with weaponised biohazards; they have used cyberattacks to hack, leak, and blackmail; they have used methods which blend tools as diverse as espionage, lawfare, election interference, and economic leverage to destabilise democracies; and they have made extensive use of informational threats such as propaganda, trolling, phishing, and disinformation. Often referred to as *hybrid threats*, these varied but coordinated activities target vulnerabilities under the threshold of what would lead to the use of a kinetic response, let alone invocation of MAD.[10]

Hence, modern deterrence is increasingly preoccupied with developing tools to manage coordinated lower-level threats and provocations on an escalation ladder that will likely never reach the dizzying heights of NATO's Article 5. or MAD. It is in this context that deterrence theory has been reworked in recent years from a means of managing strategic parity, to a theory centred on building resilience to the coordinated exploitation of societal vulnerabilities to the benefit of an adversary using asymmetric interventions and provocations (hybrid threats).[11] In theory, modern deterrence is a proactive means of compelling the adversary to change their behaviour without escalating to use of force. In practice, however, this paradigm shift has not gone smoothly due to flaws in the conceptualisation and practice of modern deterrence.[12]

Deterrence of hybrid threats

The shift in how deterrence theory is conceptualised and used may be anticipated to generate some practical and theoretical problems. We argue in this chapter that modern deterrence theory and practice is undermined by a misunderstanding of the role of information influence as it relates to hybrid threats. Information influence operations refers to *activities that aim to influence the perceptions, behaviour, and decisions of target groups to the benefit of hostile foreign powers.* The conceptual problem of how influence over perceptions, behaviour, and decision-making fits within theories of hybrid has caused some consternation. In our opinion, this confusion has led to steps being taken that should significantly improve deterrence and resilience within democratic societies in theory but face serious hurdles in practice. To explore this problem and its possible solutions, we first need to better understand how deterrence is applied to hybrid threats, at least at a general level.

As mentioned above, deterrence may be defined as *the rational process by which a society identifies threats and develops strategies to limit the risks posed by them.* An assumption within this definition is the existence of priorities. Some risks are more serious than others. In the case of a motivated adversary with significant resources and opportunities to cause harm, deterrence strategy is expected to prioritise the most serious threats and where possible channel adversarial behaviour toward less harmful outcomes. Limited resources, and an inability to entirely stop an adversary's threatening behaviour, leads to deterrence strategies that enforce a series of *red lines* or *trip wires* that would motivate escalation. These are known as *thresholds*.

Modern deterrence has built institutions that are specialised in monitoring high priority societal vulnerabilities and developing thresholds to inform about evolving threats. For example, an adversary might be expected to test computer networks at regular intervals, looking for avenues to infiltrate sensitive systems. A certain amount of adversarial testing can be tolerated; expected even. Most countries quietly monitor these efforts and establish thresholds that would be triggered in case of a sudden intensification of hostile activity. In the average western democracy, dozens of sensitive areas ranging from critical infrastructure to airspace violations to public debate are monitored, with thresholds in place to warn of sudden escalation. However, with limited resources, there is a natural tendency to focus on democracies' priorities at the most serious end of the adversarial spectrum: specifically, indicators that might suggest the imminent use of force or other immediate threats to public safety.

To offer an example, the Enhanced Forward Presence (EFP) is a multinational combat force deployed by NATO at the request of Estonia, Latvia, Lithuania, and Poland to deter the Russian Federation from border incursions and other provocations that involve kinetic force.[13] In 2014, parts of Ukraine were invaded by Russia under a deniable grey-zone operation, in which so-called little green men seemingly without national affiliation captured areas of Ukrainian territory. The EFP raises the threshold for such grey-zone attacks. In deciding whether to conduct illegal border incursions in an EFP country, Russia is forced to weigh up the denial of any benefits from using little green men against trained NATO battlegroups, and the eventual punitive measures that NATO would level at Russia should the activities be successfully attributed to the Kremlin. Consequently, the costs of grey-zone incursions into EFP-supported countries are raised significantly.

The decision to counteract the most serious strategic choices for adversaries, most obviously use of force, has the consequence of shifting the adversary's attention to lesser goals. For example, while the EFP is likely to successfully deter an invading force of little green men, it should expect adversarial narratives (such as the spurious claim that the EFP equates to NATO colonisation), disinformation (such as targeting perceptions of the conduct of NATO troops on the EFP country's soil)[14], and historical revisionism (such as efforts to redefine the relationships between the EFP country and the Soviet Union/ Russia). All of these are a lesser harm than border incursions and are therefore also more tolerable. One could therefore argue that the priority of deterring Russia from use of force has successfully shifted the adversary's calculus to the lesser harm of information influence. Modern deterrence works.

The problem, however, is that many "lesser" threats, if numerous and occurring over long periods of time, can grow into serious societal challenges. Sometimes their effects are more difficult to counter or manage than larger singular threats. Such effects are often delayed, making them harder to detect and manage through the use of indicators and thresholds.[15] Their cumulative impact over time may for example come to limit the freedom of action for governments, erode trust in institutions tasked with public safety, or undermine societal cohesion. Coordinated hybrid methods aimed at reducing public confidence in the EFP both in source and host countries could, over time, end up with the initiative scrapped due to negative perceptions of cost, a perceived lack of gratitude from the hosts, or improved perceptions of Russia's intentions; any of these examples are a far better outcome for an adversary than attempting to remove the battlegroups by diplomacy or force. *Coordinated sub-threshold activities* at the intersection of hybrid threats, espionage, and information influence aim to secure exactly these kinds of outcomes. Any of the activities taken individually might be mistaken for being irrational, harmless, and a minor inconvenience rather than part of a wider coordinated threat. It is for this reason that the hybrid approach is sometimes referred to as *death by a thousand cuts*.[16]

Information influence as a unique vulnerability

According to the basic principles of modern deterrence, democracies have funnelled resources into stopping war and other serious destabilising activities, while turning a blind eye to lesser threats such as social media manipulation, online trolling and harassment, and contemporary propaganda activities. This leads to two problems. First, information influence, by its nature, is susceptible to the death by a thousand cuts approach. Each word, each message, each narrative may appear relatively harmless, but is in fact cumulative in impact. The second problem is a miscalculation in the unique role of information influence in the conceptualisation of hybrid. Information is key to generating cognitive impacts that affect not just one domain (the *information environment*[17]) but that rather affects every domain the people and institutions impacted by the influence operations act within. It is a force multiplier, a skeleton key to opening all the doors and windows of the hybrid space.

Imagine the bullied child in the playground once again, who received the advice to ignore name-calling because it is less of a problem than physical violence. What costs might the repeated lies and psychological abuse have on how others view them (e.g. as an obvious target, victim or "loser"), their ability to concentrate in class (e.g. their performance and prospects), and their mental wellbeing (e.g. morale, self-confidence and psychological state)? How does it affect their relationships, their behaviour and actions, and their perceptions of the world? For countries, these problems are just as acute. Information influence may not be as serious as a kinetic attack, but its effects over time can be substantial. Indeed, while the shock of Russia's intervention into the 2016 US Presidential election opened the eyes of many to the immediate threat of information influence, the long-term and rather slow-burning impact of these activities across areas of society is still broadly unknown.

Information is highly fungible. While currency (or gold) is perhaps traditionally considered the most valuable fungible resource, it is worth underscoring that information can be used to achieve all kinds of outcomes including making money. It is peculiarly convertible. Channelling the behaviour of an adversary from use of force to use of information might seem logical, but democracies are vulnerable to information manipulation since the availability of accurate information is fundamental to democracy. Propaganda, conspiracies, and agitation can, with some simple online manipulation, be converted into demonstrations that descend into riots, as was for example seen in the January 6 attacks on the US Capitol building.[18] Other outcomes might include erosion of trust between government, social groups, media, and experts; a scenario which significantly undermines both political legitimacy and the possibility for constructive and rational political deliberation.

Information opens a strategic window into every hybrid domain because each domain has a communicative or informational element. This means that information is not simply a component or domain of hybrid threats but is more importantly *convertible between domains*. While hybrid refers to a variety of activities that are individually below the threshold for a response, their coordination makes them dangerous. Knowledge of a person's hobbies gained through digital surveillance can be used to generate system intrusions (cyberattacks). Knowledge of their financial situation opens the opportunity for manipulation using economic means (bribery or blackmail). Information accessed through hacking can be used to gain leverage over individuals with access to critical infrastructure or military secrets (espionage). Manipulating users to offer access to their private data can be exploited to conduct targeted election interference (social engineering). Often, the initial vulnerability is established through informational means. Information is like money in that it can be exchanged for almost anything the heart desires. Its coordinated use in hybrid domains sets it apart as a persistent and convertible resource rather than just another domain.

Our argument is therefore that information should be theorised as a uniquely fungible resource that can be converted across and between hybrid domains. This is important for making sense of why resilience against information influence must now be considered among the highest priorities for democratic societies battling hybrid threats. For while modern deterrence theory has established a logical framework for channelling adversarial behaviour into less harmful areas, the information environment has not been adequately prepared to cope with an influx of adversarial methods targeted upon it. During the last two decades we have increasingly seen combined operations in the hybrid threats space, using information's fungibility as a force multiplier. They employ a range of capabilities relating to for example intelligence, cyber, influence, economic and kinetic operations with the primary aim to achieve cognitive effects upon a target audience. The public sphere, upon which democracies depend, has become a central arena in which hybrid threats play out, leading to a host of chronic problems embedded across and between hybrid domains.

Unlike the "…names will never hurt me" proverb suggests, democratic societies are peculiarly vulnerable to information influence activities. Name-calling disproportionately affects democracies since trust and reputation are core elements of democratic debate and democratic community. News media's painful shift to digital services, changes in advertising and online consumption, and the business models of digital platforms have enabled the development of adversarial capabilities that merge cyberattacks, psychological operations, espionage, and "dark" public relations aimed at civilians in peacetime. Free and open public deliberation, upon which democracies depend, has been woefully prepared to cope with coordinated hybrid attacks. In effect, deterrence theory has channelled adversaries into these behaviour patterns with only limited efforts to build institutional capacity to deal with the problem.

Resilience to information influence as a core component of modern deterrence

Practitioners of deterrence and resilience are therefore faced with a seemingly insoluble problem. *How do we deter adversaries from the most serious forms of threatening behaviour without opening ourselves up to new strategic vulnerabilities within information influence?* A big part of the answer lies in developing societal resilience to information influence. In recent years, significant work has been carried out in this area in many countries; arguably the most interesting and successful interventions are from Northern Europe, where the threat of information influence has been treated most seriously. In part this is due to historical experiences enabling threat recognition and analysis, as well as strong public institutions and relatively strong social trust within society. European countries with close proximity to Russia have been quick to join NATO and the EU because their knowledge of Russian behaviour over the long durée compels them to remember that Ukraine's fate could befall any of them.

Information influence is a particularly difficult area to address in this manner since the barriers to entry for an adversary are so low. In deterrence terms, it is challenging to raise costs. First, lying is a protected freedom. In most liberal democracies, it is not illegal to lie, and hence nor is it illegal to spread many (but not all) types of falsehoods and misleading ideas, whether intentionally or unintentionally. Second, public deliberation does not only tolerate false and misleading ideas, but it is also built to debate them. Democracy depends on the free exchange of ideas, and checks and balances against lies are built into the system. Third, the arenas for public deliberation have changed radically in the digital era, with traditional, digital, and social media performing increasingly complex functions in (para-)social relationships.[19] We are more interconnected than ever before, but digital connections introduce meaningful changes to our (and our adversaries') ability to understand the intentions of online communities.[20]

One approach is to build deterrence and resilience around known vulnerabilities in the information environment that act as open windows to exploit other hybrid domains.[21] For example, a fake email that appears to be from e.g. Netflix uses information influence (in this case, credible images and text) to induce a target to click on a URL link. This URL link opens a number of possible exploits that could be utilised for a number of reasons in several hybrid domains, but all are clearly to the detriment of the target and the benefit of the hacker. If it is cheap and easy to send out thousands, even millions of phishing messages without any consequence or barriers, the problem can be defined as underestimating adversaries and lacking the ability to deter them. If the targeted Internet users cannot tell the difference between a false link, or their email providers' spam filters offer little protection, we have failed to prepare adequately for the threat, and hence resilience is insufficient. Both areas can and should be remedied, and the resources required are significantly less than the costs of inaction.

From a deterrence perspective, this resolves into the following questions: a) what can be done to deny adversaries opportunities to produce, distribute, and benefit from their harmful activities, and b) what can be done to raise the resilience of potential targets to these kinds of operations? The answers, in this case, are relatively straightforward to identify. On the deterrence side, governments and tech platforms can work together to raise the costs of phishing, by for example introducing verified email identities and blocking or warning when a user attempts to click on a suspicious link (denial of capability), and improving the ability to prosecute such crimes (denial by punishment). In terms of resilience, governments, universities, and tech platforms can collaborate over finding ways to strengthen known vulnerabilities, such as media literacy programmes to increase users' source criticism and cybersecurity enhancements to improve spam filtering. The point is that these efforts do not simply solve an information problem but resonate across all hybrid domains since the information gleaned from their exploits is a key vulnerability across domains.

Implicit in the above examples is recognition of the relative strengths of different societal actors. The public, private, and civil society sectors all have a role to play, and each brings something different to the table. Governments, particularly intelligence agencies and diplomats working with security policy, play a profoundly important role in the actor-specific calculations behind deterrence. But they are dependent on the support of private sector intelligence and cybersecurity firms, as well as the threat intelligence teams connected to critical infrastructure and tech platforms. Building situational awareness involves sharing information among stakeholders and allies both within the country and internationally in order to establish consensus around adversary intent, capabilities, and opportunities. Cutting-edge research and development, whether in the private sector, think tanks, or at universities, is essential for understanding threats and developing policy recommendations and operational solutions. And of course, political support is required for deterrence policies to take shape, particularly if they involve attribution or any offensive countermeasures.

On the resilience side, cross-government working is a priority for reducing the gaps and seams between institutions which hybrid methods tend to target.[22] Improving government communication is essential, not just in terms of messaging, narratives, and counter-disin-formation capabilities, but also in under-invested areas such as digital monitoring, risk assess-ment from a communication perspective, and open-source intelligence capabilities. Government has a responsibility to work with the private sector, universities, think tanks, NGOs, and journalists to improve the public's media literacy, to provide fact-checking where appropriate, and to inoculate in areas such as public health where disinformation can be countered proactively.[23] The goal should be to improve the integrity of the information environment as a means of supporting a more robust society. This often comes together in a doctrine, concept, or policy of civil (or "total") defence, which offers coherence to resilience activities and provides clear links between actor-agnostic activities focused on strengthening society as a whole in addition to actor-specific deterrence.

However, this is challenging because separation of powers in democracies means that these actors do not (and should not) naturally want to work together. Politicians want to stay in power. The private sector generates value for shareholders. Journalists hold power to account. Researchers pursue knowledge. NGOs advocate for specific issues. The public seeks a good life. No single actor can become an arbiter of truth. If the need for coordination here is clearly spelt out, the solution cannot be taken for granted. Creating a sense of common purpose without compromising agendas and identities requires precisely the kind of public debate and engagement that becomes difficult when trust is eroded. In other words, hybrid attacks damage our ability to come together to stop them. Once again, the need for skilled communication capable of mitigating the effects of information influence comes to the fore.

Liberal democracies face a significant challenge from hybrid threats because we tend to perceive informational and reputational attacks as a low priority. Efforts to adapt deterrence theory to contemporary asymmetrical threats show some promise, particularly when deter-rence is combined with a concept of resilience broader than simply bouncing back from adversity. Coherent efforts to create cross-government, whole-of-society resilience and deterrence capabilities are an important means of preparing societies for a geopolitical situa-tion in which coordinated sub-threshold activities become the norm. However, such activ-ities risk compromising democratic values, which means that their solution is neither obvious nor straightforward.

Notes

1 UK Ministry of Defence (February 2019) Joint Doctrine Note 1/19, Deterrence: the Defence Con-tribution, p. 39.
2 Roberts. P. & Hardie, A. (August 2015) The Validity of Deterrence in the Twenty-First Century, Royal United Services Institute, p. 9.

3 Pamment, J., & Agardh-Twetman, H. (2019). Can there be a deterrence strategy for influence operations? Journal of Information Warfare, 18(3), [123–135].

4 Pamment, J., & Agardh-Twetman, H. (2019). Can there be a deterrence strategy for influence operations? Journal of Information Warfare, 18(3), [123–135].

5 A thought experiment is a mental assessment of the implications of a hypothesis.

6 Keršanskas, V. (2020) Deterrence: Proposing a more strategic approach to countering hybrid threats, Hybrid Center of Excellence, pp. 9–11.

7 Pamment, J. & Smith, V. (2022) *Attributing Influence Operations: toward a community framework*. Riga: NATO Strategic Communication Centre of Excellence & EU-NATO Hybrid Centre of Excellence.

8 Brodie, B 1970, "Anatomy of deterrence", Theories of peace and security: A Reader in contemporary strategic thought, ed. J Garnett, Macmillan St Martin's Press, New York, NY, US, pp.87–105.

9 Schelling, T. C. (1966), The Diplomacy of Violence, New Haven: Yale University Press, pp. 1–34. Elpech, T. (2012) Concepts. In Nuclear Deterrence in the 21st Century: Lessons from the Cold War for a New Era of Strategic Piracy, RAND Corporation, pp. 23–60.

10 Weissmann, M., Nilsson, N., Palmertz, B. and Thunholm, P. (eds). Hybrid Warfare: Security and Asymmetric Conflict in International Relations. London: I.B. Tauris

11 Lindsay, J. R.; Gartzke, E. (2019) Introduction: Cross-Domain Deterrence, from Practice to Theory, Oxford University Press. Cullen, P & Reichborn-Kjennerud, E 2017, Multinational Capability Development Campaign (MCDC) countering hybrid warfare project: Understanding hybrid warfare, UK.

12 Neal, J. J. (2020) *Deterrence in a Hybrid Environment*, per Concordiam: Journal of European Security Defense Issues 10, No. 1: 16–23.

13 NATO SHAPE (accessed October 15th, 2022) Enhanced Forward Presence, https://shape.nato.int/efp/efp/present-structure

14 Emmott, R. (February 18) Expect more fake news from Russia, top NATO general says, Reuters. https://www.reuters.com/article/uk-germany-security-russia-nato-idUKKBN15X08U

15 Yadav, K. (November 30, 2020) Countering Influence Operations: A Review of Policy Proposals Since 2016. Carnegie Endowment for International Peace. https://carnegieendowment.org/2020/11/30/countering-influence-operations-review-of-policy-proposals-since-2016-pub-83333

16 Wigell, M., 2019. Hybrid interference as a wedge strategy: a theory of external interference in liberal democracy. *International Affairs, 95*(2), pp.255–275.

17 The information environment is *the aggregate of individuals, organizations, and systems that collect, process, disseminate, or act on information.* Source: US Department of Commerce, National Institute of Standards and Technology.

18 United States Senate (June 8, 2021) Staff Report: Examining the U.S. Capitol Attack—A Review of the Security, Planning, and Response Failures on January 6. https://www.hsgac.senate.gov/imo/media/doc/HSGAC&RulesFullReport_ExaminingU.S.CapitolAttack.pdf

19 Pamment, J., Nothhaft, H., Agardh-Twetman, H. & Fjällhed, A. (2018). Countering Information Influence Activities—The State of the Art, Swedish Civil Contingencies Agency & Lund University, pp. 18–21.

20 Palmertz, B. (2021). Influence operations and the modern information environment. In M. Weissmann, N. Nilsson, B. Palmertz & P. Thunholm (Authors), Hybrid Warfare: Security and Asymmetric Conflict in International Relations, pp. 113–131. London: I.B. Tauris.

21 U.S. Cybersecurity & Infrastructure Security Agency (Last revised: August 25, 2020) Avoiding Social Engineering and Phishing Attacks. https://www.cisa.gov/uscert/ncas/tips/ST04-014

22 Wigell, M., Mikkola, H. & Juntunen, T. (2021) Best Practices in the whole-of-society approach in countering hybrid threats, Directorate General for External Policies—European Parliament.

23 Pamment, J. (2021) Resist 2—Counter disinformation toolkit, UK Government Communication Service, pp. 8–9.

Bibliography

Brodie, B 1970, '*Anatomy of deterrence*', *Theories of peace and security: A Reader in contemporary strategic thought*, ed. J Garnett, Macmillan St Martin's Press, New York, NY, US, pp. 87–105.

Cullen, P. & Reichborn-Kjennerud, E. (2017) *Multinational Capability Development Campaign (MCDC) countering hybrid warfare project: Understanding hybrid warfare, UK*.

Keršanskas, V. (2020) *Deterrence: Proposing a more strategic approach to countering hybrid threats*, Hybrid Center of Excellence, pp. 9–11.

Lindsay, J. R.; Gartzke, E. (2019) *Introduction: Cross-Domain Deterrence, from Practice to Theory*, Oxford University Press.

Neal, J. J. (2020) Deterrence in a Hybrid Environment, per Concordiam *Journal of European Security Defense Issues* 10, No. 1: 16–23.

Palmertz, B. (2021). Influence operations and the modern information environment. In M. Weissmann, N. Nilsson, B. Palmertz & P. Thunholm (Authors), *Hybrid Warfare: Security and Asymmetric Conflict in International Relations*, pp. 113–131. London: I.B. Tauris.

Pamment, J. & Smith, V. (2022) *Attributing Influence Operations: toward a community framework*. Riga: NATO Strategic Communication Centre of Excellence & EU-NATO Hybrid Centre of Excellence.

Pamment, J. (2021) *Resist 2—Counter disinformation toolkit*, UK Government Communication Service, pp. 8–9.

Pamment, J., & Agardh-Twetman, H. (2019). Can there be a deterrence strategy for influence operations? *Journal of Information Warfare*, 18(3), [123–135].

Pamment, J., Nothhaft, H., Agardh-Twetman, H. & Fjällhed, A. (2018). *Countering Information Influence Activities—The State of the Art*, Swedish Civil Contingencies Agency & Lund University, pp. 18–21.

Roberts. P. & Hardie, A. (August 2015) *The Validity of Deterrence in the Twenty-First Century*, Royal United Services Institute, p. 9.

Schelling, T. C. (1966), *The Diplomacy of Violence*, New Haven: Yale University Press, pp. 1–34.

Elpech, T. (2012) Concepts. In *Nuclear Deterrence in the 21st Century: Lessons from the Cold War for a New Era of Strategic Piracy*, RAND Corporation, pp. 23–60.

U.S. Cybersecurity & Infrastructure Security Agency (Last revised: August 25, 2020) *Avoiding Social Engineering and Phishing Attacks*. https://www.cisa.gov/uscert/ncas/tips/ST04-014.

UK Ministry of Defence (February2019) *Joint Doctrine Note 1/19, Deterrence: the Defence Contribution*, p. 39.

United States Senate (June 8, 2021) *Staff Report: Examining the U.S. Capitol Attack—A Review of the Security, Planning, and Response Failures on January 6*. https://www.hsgac.senate.gov/imo/media/doc/HSGAC&RulesFullReport_ExaminingU.S.CapitolAttack.pdf.

Weissmann, M., Nilsson, N., Palmertz, B. and Thunholm, P. (eds). *Hybrid Warfare: Security and Asymmetric Conflict in International Relations*. London: I.B. Tauris

Wigell, M., 2019. Hybrid interference as a wedge strategy: a theory of external interference in liberal democracy. *International Affairs*, 95(2), pp.255–275.

Wigell, M., Mikkola, H. & Juntunen, T. (2021) *Best Practices in the whole-of-society approach in countering hybrid threats*, Directorate General for External Policies——European Parliament.

Yadav, K. (November 30, 2020) *Countering Influence Operations: A Review of Policy Proposals Since 2016*. Carnegie Endowment for International Peace. https://carnegieendowment.org/2020/11/30/countering-influence-operations-review-of-policy-proposals-since-2016-pub-83333.

3

MILITARY DECEPTION AND PERCEPTION MANAGEMENT

James J. Wirtz

Introduction

Although it rarely draws much attention, there is a "sporting assumption" inherent in just about every contest between individuals, teams, militaries or nations. The parties in these contests are not obligated to make their real objectives, strategies, or tactical or technical innovations known to their opponents, especially when it comes to the tricks that might be up their sleeve. Instead, the participants in these games—that is, contests in which the outcomes are determined by the interaction of opponents—usually possess less than accurate and complete information about their competitors. As one sails beyond the purview of the Corinthian etiquette that supposedly guides sporting events, however, competitors can do more than just keep their plans to themselves. They can use devious activities to mislead opponents about their intentions and to shape their opponent's situational awareness. Stratagem, denial and deception (D&D), operational security (OPSEC) and information operations (IO) can create significant advantages over opponents by translating a victory of the intellect into success on the geopolitical battlefield. By using deception to gain information dominance over enemy commanders, one can open pathways to victory in war.

D&D, OPSEC and IO are all concepts embedded in the terms "military deception" and "perception management". Over millennia, military strategists, philosophers, magicians, and con artists have had much to say about creating and using perception management in general, and deception in particular, to prompt opposing commanders and other pigeons to act against their interests. Instead of opening this entire Pandora's box of trickery, secrecy, illusion, and deceit, this chapter provides a brief survey of the dominant ideas and application of deception and perception management to the realm of military affairs. It begins by defining common terms and briefly explaining deception theory, which was best articulated by Barton Whaley, the savant of deception.[1] It also illustrates the application of military deception and perception management across the levels of conflict. It then offers an overview of how important military thinkers assessed the importance and role of deception and perception management in war. It concludes with a few thoughts about deception in the Information Age.

DOI: 10.4324/9781003190363-5

Terms and Theory

Admittedly, the definitions supplied here have varied over time and are continually revised by commands and military services around the globe. With that caveat in mind, it is possible to define several important terms, beginning with the linchpin concept of operational security. OPSEC is a process whereby militaries minimise the information available to opponents to prevent observers from assembling bits and pieces of disparate information to gain a detailed understanding of one's strategies, doctrines, capabilities and standard operating procedures. It begins by identifying the information to be protected. It then estimates the capabilities of outsiders to gather that information and identifies the vulnerabilities that outsiders might exploit to gain access to data. It acts on these assessments by taking actions to minimise the leakage of information. In a sense, OPSEC keeps the "sporting assumption" sporting by preventing outsiders from reading military organisations like an open book.[2] No one observation might be critical, but given time, numerous observations and skilled analysis, opposing intelligence agencies can gain a profound understanding of a targeted military using a series of casually acquired observations.

Denial and deception takes the OPSEC process one step further by actively exposing and manipulating information made available to opponents. Unlike OPSEC, denial selectively limits information available for observation to allow the opponent's perceptions to be influenced by the data purposely made available. Deception, by contrast, is the provision of deliberately misleading or false information to opponents to further shape their situational awareness and expectations about the future in ways that suit the deceiver's interest. D&D thus works in tandem—denial allows information supporting deception to leak out and works especially hard to guard information that might cause the target to look *too closely* at the deceptive material that is offered.[3] As Whaley managed to document at the end of a lifetime of work, there are an infinite number of stratagems that utilise devilishly cunning approaches to D&D to fit particular circumstances.[4] For instance, during the interwar period, the Nazis gave western observers access to souped up versions of fighter-prototypes, while hiding production aircraft, to create an image of a technologically advanced Luftwaffe. Another such stratagem—"hiding in plain sight"—was employed by the perpetrators of the 9/11 tragedy. Despite their limited clandestine tradecraft, the hijackers managed to blend into society by "disguising" themselves as normal students. Like most undergraduates, they occasionally went to class, spent a good deal of time in bars, and lacked a visible means of support. Deception was used to generate an appearance of normality, while denial was used to hide their nefarious intentions and preparations from casual observers. Their D&D effort would not have held up to close scrutiny—flight instructors were already telling authorities about their strange requests to learn how to take off and fly, but not land, airliners—but Al-Qaeda's stratagem was good enough to overcome the level of scrutiny faced by its operatives.[5]

Information Operations is a term of more recent origin that describes a sort of whole of government effort to shape the information environment to support ongoing military activities and to achieve political objectives. It involves everything from maintaining the security of computer and communication systems, to traditional OPSEC, to the control of the content and flow of information to shape the opponent's behaviour.[6] IO often focuses on creating narratives to explain military activity and national objectives in a way that would appeal to both domestic and international audiences. For instance, Vladimir Putin explained his invasion of Ukraine as intended to destroy a Nazi clique in Kiev, while the rest of the world championed a competing narrative—that the Russian invasion was an example of unprovoked and unjustified aggression.[7] The essence of IO is to place military operations in a broader political context, thereby increasing the probability that observers globally will come to accept the political objectives used to justify military use. IO is a deliberate effort to explain how military operations are intended to achieve legitimate, desirable and positive political goals.

Deception theory exploits the opponent's quest to uncover the information hidden by OPSEC. Deception begins with accurate intelligence about the opponent: successful deception planners base their stratagem on an estimate of the opponent's beliefs, plans and expectations. D&D then provides opponents with the information they are looking for, so to speak, to confirm their beliefs and satisfy their thirst for information. Of course, deception is even more convincing if the target discovers deceptive information in a "natural" manner. For example, highly protected information that is "accidentally" uncovered by the target would appear to be highly credible to the receiver. Misleading information that is confirmed by unrelated sources, or from sources that appear resistant to manipulation, would also appear legitimate. For deception to succeed, however, false or misleading information has to be presented to the target in subtle ways. It is important that deceptive information does not arouse the target's suspicion because *ceteris paribus*, it will not stand up to close examination as it is not an accurate reflection of reality. For something to be real, it has to be real in all of its manifestations—D&D works by getting opponents to accept this incomplete reality by leading them to focus on the realistic information provided to them, not the information that is missing. The need for subtlety, however, greatly complicates the task faced by the deceiver. If the deceiver is too subtle, then there is a risk that the target might not notice the deceptive information at all. D&D cannot alter reality, but it can change the targets' perception of reality by leading them to rely on the information stream controlled by the deceiver and by getting them to accept what is in fact a less than complete, and misleading, understanding of current and potential events.

The best D&D planners are probably born, not made, because they have a talent for understanding human foibles, organizational pathologies, and how the world works in general. These keen students of human nature and bureaucratic behaviour understand that people wear their emotions, innermost thoughts, and ambitions on their sleeves and that bureaucracy is predictable and easily manipulated. They understand what motivates people in social and bureaucratic settings. Successful deception planners also understand how the biases inherent in human cognition facilitate D&D. People tend to pay more attention to information that confirms their view of reality; they also tend to make judgments on limited information, especially when the information confirms their expectations.[8] D&D is the art of capitalising on cognitive, organisational and social biases to provide "inaccurate" or false information that meets people's expectations.

Deception planners also have a keen sense of humour. Whaley notes that there is an element of humour behind all military deception; it can be thought of as a practical joke played on the opponent to get them to act according to one's wishes. For instance, R.V. Jones, a keen student and practitioner of deception, once bet a colleague that he could get a person to place their phone in a bucket of water. After repeatedly calling an Oxford professor on the phone in a way that simulated a problem that seemed to be partially resolved by minor movements of the device, he managed to get the gullible man to submerge the phone. The "deception" worked because the victim had come to rely on a single source (the voice at the other end of the line) for information about his situation.[9]

Intelligence analysts and commanders know that they have to be on the constant lookout for D&D. When things look too good to be true, sceptical analysts realise that they have to suspect trickery by wily opponents. Despite the fact that those targeted by D&D generally recognize that they are targets, Whaley estimates that D&D generally succeeds about ninety percent of the time it is attempted.[10] Once the deceiver identifies the target's expectations, it is difficult for the target to escape the trap. John Ferris, who has written extensively about the history of D&D, notes that only four attributes can help a target avoid deception: "superior power and initiative; intelligence of outstanding quality or else so poor that it cannot pick up misleading signals, [and] an inability or unwillingness to act on any knowledge, true or false."[11] Two of these attributes resemble sheer incompetence, while superior command and intelligence is rarely available when most needed.

Deception across the Levels of War

Although there are experts who share command or service-wide responsibilities for devising OPSEC procedures and assessments and IO strategies, the burden of fulfilling these tasks resides with virtually every member of a military organization. The adage "loose lips sink ships" conveys the idea that OPSEC is only as strong as its weakest link and that even the most innocent compromise of operationally significant information by the most junior serviceperson can lead to serious consequences. It also is imperative that all personnel live up to the highest professional and ethical standards at all times because incidents, criminal activity and gratuitous violence can undermine hard won gains in the realm of information operations. The foolish U.S. soldiers who photographed Iraqi detainees at Abu Ghraib and then circulated this material on the World Wide Web did enormous damage to the standing of U.S. forces in general, and to U.S. efforts to quell ethnic and political unrest in Iraq in particular.[12] Senior leaders can also turn successful IO narratives into shambles: public airing of the U.S. Navy's failure to deal calmly with a COVID-19 outbreak on the USS Teddy Roosevelt in Spring 2020 damaged its credibility when it came to responding to the pandemic.[13]

When it comes to D&D, by contrast, devising various stratagem and executing various deceptive measures are usually handled by staff officers and specialists if for no other reason that the strictest attention must be placed on denial if deception is to succeed. Nevertheless, officers are generally aware of the principles of deception and perception management because they are portrayed as force multipliers, that is, they are activities that can help overcome superior numbers, minimise friendly casualties and boost the chances of battlefield success. Officers embrace the use of force multipliers as a defining element of military professionalism because they can help minimise destruction and casualties in combat, especially by helping to end hostilities as quickly as possible.

In terms of tactics, D&D is usually intended to achieve numerical superiority on some area of the battlefield to facilitate offensive operations by leading an opponent to misidentify the main avenue of attack. All sorts of ruses can be used to achieve this objective. Faked radio communications between non-existent units, simulated signals indicating force movements (sound of tank and troop movements, lights, and fires), deliberately leaked documents and rumours can be used in various combinations. In defensive settings, efforts can be made to simulate strength or to simulate the presence of defences when they barely exist, the latter stratagem is sometimes used to cover a withdrawal by inducing caution in an advancing force. D&D can also be used to enable special operations. During the Ardennes Offensive of December 1944, Waffen-SS units composed mostly of English speaking Germans dressed in American and British uniforms operated behind allied lines to sow confusion in what was already a chaotic situation.[14]

The impact of this sort of tactical deception is generally limited to the vicinity of the units involved, but occasionally small tactical successes can produce strategic effects. The classic example is the Trojan Horse, which was used by the Greeks to end a ten-year siege of Troy. The Greeks hid a small force of sappers inside a large wooden horse and left it outside the gates of Troy and then boarded their ships and sailed away. Although some were leery of this Greek gift, the Trojans brought the horse, which happened to be the symbol of Troy (notice the joke), inside the gates of the city as a war trophy. Under the cover of darkness, the Greek fleet returned as Troy's gates were opened by the soldiers who had climbed out of the horse unopposed by the Trojans who had done too much celebrating. Today the term "Trojan" is most often used in the realm of computer security to denote a cyber threat that manages to bypass security measures to damage software, but it is inspired by this story of a Greek gift.

At the operational level of war, D&D efforts can become far more extensive, although they still generally take on the goal of leading the opponent to misestimate where the main attack will fall. Operation Fortitude, the D&D operation that supported the 6 June 1944 Allied invasion of Normandy, for instance, was intended to lead the Germans to conclude that the Normandy invasion was actually a diversionary operation for the "main" amphibious landings that would occur against Norway (Operation Fortitude North) and the Pas-de-Calais (Operations Fortitude South). A host of D&D activities were undertaken in the effort to create the appearance of competing amphibious invasions: commando and bombing raids, fake wireless transmissions, leaks through diplomatic channels, the use of double agents, and simulation of military formations (inflatable tanks and other vehicles) to be observed from the air. Important Allied Generals were also assigned to actual command of simulated armies— Lieutenant General George S. Patton was given command of the forces assigned to Fortitude South because the Germans respected his ability to lead in battle.[15] Even more resources were invested in the diversionary operations launched by the Vietcong and their North Vietnamese allies in the run-up to the 1968 Tet offensive. Instead of simulating a threat, the communists actually created a real threat by launching a series of attacks between September 1967 and January 1968 across the Central Highlands of South Vietnam, culminating in the siege of the Marine outpost at Khe Sanh. The attacks were intended to fix American units in their usual operating areas along the relatively unpopulated borders of South Vietnam, exposing the urban areas that were mostly protected by the Army, Republic of South Vietnam to the effects of an all-out attack intended to spark a popular uprising. These urban attacks materialised but the hoped for uprising never occurred. These diversionary attacks were so powerful and persistent, however, that scholars today continue to debate exactly which was the diversion and which was the "main event."[16] This sort of major operational diversion also can occur during a war at sea. During the June 1942 Battle of Midway, the Japanese Navy invaded the Aleutian islands in an attempt to draw major elements of the U.S. Navy away from Midway, leaving the remaining U.S. force badly outnumbered. Because U.S. Naval code breakers had determined Japan's real intentions, U.S. commanders ignored the Aleutian ruse and concentrated their forces against Japan's primary attack and won the Battle of Midway.[17]

Deception and perception management at a strategic level of war can manifest in a variety of ways. Strategic deception might enable the very existence of clandestine actors that could not survive for long if detected by intelligence, military or law enforcement agencies. Criminal gangs, terror cells or even large terror organizations like Al-Qaeda rely on D&D for their very existence. By contrast, strategic deception and information management might be intended to alter perceptions of long term efforts to alter the strategic setting by lulling targets into acceptance of a gradually changing geopolitical situation. In this regard, the effort to test, destabilise and de-sensitise targets in the lead up to the Russian invasion of the Ukraine in 2022 might be a manifestation of strategic D&D. Some also note that the People's Republic of China is engaged in a similar effort at strategic deception and perception management by carrying out a long-term hybrid warfare strategy in the Western Pacific.[18] Strategic surprise attacks—the 7 December 1941 Japanese attack on Pearl Harbour or the 1973 Yom Kippur war—also benefited from D&D and perception management. The Japanese, for instance, maintained exemplary operational security to hide their plans, preparations and positioning of forces to attack Pearl Harbour, all the while engaging in negotiations with the United States to find a solution to a deteriorating geopolitical situation in Asia.

Is Deception Important in War?

Across the panoply of strategic thinkers, opinions vary about the utility and role of military deception and perception management in war. These different assessments should not be thought of as correct or incorrect; instead, they reflect the theoretical perspective, historical context or tactical, operational or strategic interests of the philosopher or strategist under consideration. While space precludes discussion of everyone's perspective on the topic, what follows is a brief summary of the views of the Chinese philosopher Sun Tzu, the Prussian philosopher Carl von Clausewitz, and the American strategist William H. McRaven—an ancient, modern, and contemporary strategist—on the importance of deception in war.

Sun Tzu

In his work *The Art of War,* Sun Tzu seamlessly integrates deception and perception management into his discussion of war's dialectic (that is, outcomes are determined by the *interaction* of the parties involved). Written during China's Warring States period (about 500 BCE), *The Art of War* provides a series of observations and vignettes that illustrate the nature of war while providing advice on how to win on the battlefield. Sun Tzu gets down to brass tacks:

> All warfare is based on deception. Therefore, when capable, feign incapacity; when active, inactivity. When near, make it appear that you are far away; when far away, that you are near. Offer the enemy a bait to lure him; feign disorder and strike him. Pretend inferiority and encourage his arrogance.[19]

Sun Tzu is suggesting that the battle for information dominance occurs before actual combat is joined and that winning that intellectual engagement can set the stage for success in war. By manipulating the opponent's situational awareness, a commander can concentrate forces against unprepared defences or lead opponents to fall into traps. In modern parlance, Sun Tzu sees D&D as a critical force multiplier when it comes to tactics and operations. D&D is depicted as the starting point for successful military operations.

Sun Tzu also depicts deception and perception management as a key strategic instrument in war. He suggests that the commander should strive to attack the opponent's strategy, which again highlights the battle for information dominance embedded in war. This is a critical insight—commanders should tamper with the information the opponent needs to execute their chosen strategy. He also seems to be suggesting that the commander should strive to "get inside the opponent's head," so to speak, by using various stratagem to cause frustration, confusion and doubt. He even states that the ultimate success in war is to defeat the opponent without even fighting. Here Sun Tzu seems to be suggesting that there is a role for D&D in the strategies of deterrence and coercion, which are both intended to alter the opponent's behaviour with either no, or limited, use of force. By contrast, contemporary strategists seem to give D&D short shrift when it comes to devising strategies of deterrence or coercion.

Carl von Clausewitz

Carl von Clausewitz was a Prussian officer who campaigned extensively in the wars against Napoleon's revolutionary France in the early 1800s. The author of *On War*, the most influential book written on the subject in the Western intellectual tradition, Clausewitz developed a theory of war that focuses on its dialectical nature and the interaction of citizens, the military and government in the conduct of hostilities. His theory also highlights war's fundamental political nature. Although it involved the use of force, war in Clausewitz's view, was ultimately about getting others to adopt behaviours, policies and politics in line with ones interests, to come around to "seeing things our way." Important military concepts—the fog of war, friction, and the inherent superiority of the defence over the offence in land war—are all found in *On War*.

Like most of us, Clausewitz was a product of his time; his views were shaped by his personal encounters with the scale, carnage and persistence of Napoleonic warfare. These encounters were generally characterised by Napoleon's effort not to gain a minor tactical advantage over the opponent, but to completely destroy the force arrayed against him to compel his opponents to accept his terms. The massive forces involved in these campaigns moved only at the speed of horse drawn transport and march, possessed no electronic communications and generally could not outpace the speed of rumour and human intelligence, which could travel between 4 to 11 miles per hour.[20]

Because Napoleon's objectives were well understood and news of troop movements usually arrived well before forces on the march, technological and transportation constraints greatly limited D&D efforts at the strategic and operational levels of war in the early 1800s. Unlike Sun Tzu, it is not really clear if Clausewitz ever encountered the use of stratagem to "get inside the opponent's head". It also is unlikely he witnessed successful use of stratagem at the strategic or operational levels of war. As a result, his commentary on the use of D&D was limited to the tactical level of war and even in this regard, he was not a big supporter of stratagem. Clausewitz acknowledged that D&D could help hide attacks, weaken defences and divert defenders away from the main avenue of attack. In practice, however, he believed that diversionary efforts and various types of stratagem usually failed to work. He also believed that efforts at diversion tended to take on a life of their own, consuming more and more resources that might be better used along the main axis of attack.[21] Efforts at D&D not only might fail to facilitate operations; they might actually reduce the prospects for its success. For Clausewitz, when it came to D&D, the game was not worth the candle.

Contemporary scholars also increasingly recognize that the very recent IO concept aligns with many of the ideas presented in *On War*. For example, in a recent commentary on the relationship between the British government and British people in matters of national defence, Hew Strachan and Ruth Harris note that a government must have two conversations when it comes to making strategy: "one with its civil service and armed forces and the other with the electorate that places it in office."[22] Engaging the electorate is important because the citizens in a democracy are the source of energy behind national defence; they are a key component of the Clausewitzian trinity of the people, the government and the armed forces that must participate in the making and execution of defence strategy. Given Clausewitz's description of war as something guided by political objectives, there is every reason to believe that he would have endorsed the development of information campaigns and narratives to interpret unfolding military events to bolster their political impact both at home and abroad.

William H.McRaven

As a student at the U.S. Naval Postgraduate School in the early 1990s, William H. McRaven crafted a MA Theses entitled, "The Theory of Special Operations". Years later, as the commander of U.S. Special Forces Command, McRaven used the ideas contained in this thesis as the basis of Operation Neptune Spear, the mission that eliminated Osama Bin Laden in Abbottabad, Pakistan on 2 May 2011.[23] In his thesis, McRaven described the fundamental challenge encountered during small unit commando operations—the commandos themselves lacked combat capability compared to the enemy units they would likely face. This insight led McRaven to suggest that special operations cannot succeed without creating and exploiting a situation of "relative superiority," whereby they temporarily enjoy a decisive military advantage. Creating this relative superiority is no small feat in the sense that special operation teams often operate deep behind enemy lines or target facilities or individuals heavily defended by the adversary.

Although several factors are critical in creating this relative superiority, McRaven stated that D&D and OPSEC stand out as crucial enablers of special operations. At a minimum, the target cannot be on alert when the commando team arrives on the scene. That would doom the operation to failure at the outset. Ideally, the target should be caught completely unaware of the unfolding operation, allowing the special operations team to undertake its mission *without any resistance at all*. McRaven emphasises the importance of OPSEC during the planning, rehearsal and movement of the team towards the target because any compromise of the operation could easily alert the defender, resulting in disaster. The entire operation itself, from planning to execution, has to be devised with an eye towards OPSEC: the fewest number of individuals should be involved in the operation to minimise opportunities for curious opponents to notice that "something is up." D&D also can play a part in the overall mission by acting as a second line of defence in the event of a breach of OPSEC. D&D can be used to plant subtle and misleading explanations for minor lapses in OPSEC, providing mundane narratives to account for what are more than mundane activities.

At a tactical and operational level, McRaven's theory is actually seeking to temporarily suspend the dialectic inherent war, thereby removing an active opponent from the battlespace. By removing opposition, special operations units can achieve objectives that would appear impossible to accomplish in the face of even limited opposition. For instance, who would have imagined that two five man teams, armed only with box cutters and mace, could destroy the World Trade Centre in an operation that lasted only a few hours? Because they can suspend the dialectic of war at the tactical and operational levels, OPSEC and D&D actually become a strategic enabler of the concept of special operations itself. In other words, commando raids, *ceteris paribus*, will not succeed unless they benefit from the element of surprise, and surprise is unlikely to be achieved without outstanding OPSEC and a bit of D&D thrown in for good measure.

Conclusion

Military deception and perception management are integral to military operations. Military professionals treat OPSEC and D&D as important tactical and operational force multipliers because they allow commanders to concentrate forces at points of enemy weakness, while keeping that concentration of force and imminent attack hidden from the opponent. Similarly, OPSEC and D&D can be used in defence to hide weakness or to simulate the presence of stiff defences in the event of an organised withdrawal of real units from some sector of the battlefield. Others see OPSEC and D&D as a way to temporarily suspend war's dialectic at the tactical, operational or even strategic level, allowing units to undertake a mission without any opposition, which can sometimes even produce strategic effects. For clandestine non-state actors, deception and perception management enable their very existence—by hiding in plain sight they can avoid detection by the authorities, which would result in prompt and unwanted attention. These types of clandestine outfits would not exist for long without OPSEC and D&D.

IO is a relatively new concept in the lexicon of military ideas. It encompasses a panoply of activities to create narratives to shape the political environment in support of military operations. While IO might be undertaken to support some military evolution, information operations actually help translate military operations into political effects by explaining the reasons for conflict and how military operations are intended to achieve widely desired political goals. IO bolsters support for the military effort among a domestic audience, while explaining to an international audience why war has been necessary to end some dangerous or unjust action or policy on the part of the opponent. IO points to the role of deception and perception management as a strategic instrument to gain political objectives by shaping the intellectual battlefield to suit one's political interests.

Today, the data deluge produced by the Information Revolution, however, is casting doubt not on the conventional wisdom that has been recounted in this chapter, but on how OPSEC, D&D and IO will work in the future. For example, the number of sensors that digitise, record and submit data to the World Wide Web is growing exponentially—everything from cell phones to door bells, to automobiles, to micro satellites is accurately and diligently recording and storing some small piece of reality. Will OPSEC become virtually impossible given this increasingly pervasive and persistent surveillance? For D&D to succeed, deception planners have to get targets to acquire "naturally" deceptive information, but is it becoming more difficult to get the target to notice deceptive information at all in the data deluge. IO too becomes more problematic given the cacophony of narratives. As grassroots organisations, traditional interest groups, political parties and competing governments work to "politicise" the most mundane events for political messaging, it becomes increasingly difficult to gain an audience's attention. Maybe D&D, OPSEC and IO will become easier—targets might be attracted to the "information order" that is deliberately created by deception planners and IO professionals. Maybe the future will just be a blur as officials, officers and average citizens awash in the data deluge struggle to develop useful situational awareness. It is just too early to tell how the Information Revolution will shape military deception and perception management in the future.

Notes

1 Whaley, *Practice to Deceive*; Whaley, *Stratagem*; and, Whaley and Busby, "Detecting Deception."
2 *Operations Security*.
3 Handel, "Intelligence and Deception"; Godson and Wirtz, "Strategic Denial and Deception"; and Shulsky, "Elements of Strategic Denial and Deception."
4 Whaley, *Practice to Deceive*.
5 Wirtz, "Hiding in Plain Sight," 58–60.
6 Joint Publication 3–13.

7 Stanar, "Understanding War."
8 Jervis, *Perception and Misperception in International Politics*.117–216.
9 Whaley, *Practice to Deceive*, 50–52.
10 Whaley, *Stratagem*.
11 Ferris, "FORTITUDE in Context."
12 Bartone, "Lessons of Abu Ghraib."
13 O'Hanlon, "Why Crozier was correct."
14 Butler, *The Black Angels*, 180–188.
15 Brown, *Bodyguard of lies;* Ferris, "FORTITUDE in Context."
16 Wirtz, "Deception and the Tet Offensive."
17 Prange, et. al., *Miracle at Midway*.
18 Babbage, *Stealing a March*.
19 Sun Tzu, *The Art of War*.66–67.
20 Lamarkewicz, "In an age before telephones and telegraphs, how fast did news travel?"
21 Handel, *Masters of War*, 130–133.
22 Strachan and Harris, *The Utility of Military Force and Public Understanding in Today's Britain*, ii.
23 McRaven, "The Theory of Special Operations; and Wirtz, The Abbottabad raid and the theory of special operations."

Bibliography

Babbage, Ross. *Stealing a March: Chinese Hybrid Warfare in the Indo-Pacific Issues and Options for Defense Planners*, Washington, D.C. Center for Strategic and Budgetary Assessments, 2019.

Bartone, Paul T. "Lessons of Abu Ghraib: Understanding and Preventing Prisoner Abuse in Military Operations." *Defense Horizons* Washington, D.C. Center for Technology and National Security Policy, National Defense University. November 2008. https://permanent.fdlp.gov/LPS105635/LPS105635/www.ndu.edu/CTNSP/docUploaded/DefenseHprozpm64.pdf.

Brown, Anthony Cave. *Bodyguard of lies*. New York: Harper and Row 1975.

Butler, Rupert. *The Black Angels*. New York: St. Martin's, 1979.

Ferris, John. "FORTITUDE in Context: The Evolution of British Military Deception in Two World Wars." In *Paradoxes of Strategic Intelligence: Essays in Honor of Michael Handel* edited by Richard K. Betts and Thomas G. Mahnken, 117–165. London: Frank Cass, 2003.

Godson, Roy and James J. Wirtz. "Strategic Denial and Deception." *International Journal of Intelligence and Counterintelligence* Vol. 19 Iss.4 (2000): 424–437.

Handel, Michael I. "Intelligence and Deception." In *Deception and Strategic Surprise!* edited by John Gooch and Amos Perlmutter, 122–154. London: Routledge, 1982.

Handel, Michael I. *Masters of War*, 2nd rev. ed. London: Frank Cass, 1982.

Headquarters Department of the Army. *Operations Security*, Army Regulation 530–531. Washington, D.C.: Department of the Army, 26 September 2014.

Information Operations Joint Publication 3–13. 20 November 2014. https://www.jcs.mil/Portals/36/Documents/puybs/p3_13.pdf.

Jervis, Robert. *Perception and Misperception in International Politics* new edition Princeton: Princeton University Press, 2017.

Lamarkewicz. "In an age before telephones or telegraphs, how fast did news travel?" *History Research Shenanigans*. March 7, 2020. https://historyboots.wordpress.com/2020/03/07/in-an-age-before-teleophones-or-telegraphs-how-fast-did-news-travel/.

McRaven, William H. "*The Theory of Special Operations*." M.A. Thesis, Department of National Security Affairs, Naval Postgraduate School, Monterey, CA June1993. http://archive.org/details.thetheoryof sp ec1094514838.

O'Hanlon, Michael E. "Why Crozier was correct." Order from Chaos [Blog] Washington, D.C.: The Brookings Institution, April 7, 2020. https://www.brookings.edu/blog/order-from-chaos/2020/04/07/why-crozier-was correct/`

Prange, Gordon W. and Donald M. Goldstein, and Katherine V. Dillon. *Miracle at Midway*. New York: Penguin Books, 1982.

Shulsky, Abram. "Elements of Strategic Denial and Deception," In *Strategic Denial and Deception: The Twenty-First Century Challenge*, edited by Roy Godson and James J. Wirtz, 15–32. London: Routledge, 2002.

Stanar, Dragan. "Understanding War: Beyond Competing Narratives," *The International Society for Military Ethics in Europe*. 26 March2022. https://www.euroisme.eu/index.php/en/views-on-war-in-ukraine/241-understanding-war-beyond-competing-narratives.

Strachan Hew and Ruth Harris, *The Utility of Military Force and Public Understanding in Today's Britain*. Santa Monica: Rand, 2020. P. ii.

Sun Tzu. *The Art of War*, translated by Samuel B. Griffith. New York: Oxford University Press.

Whaley, Barton. *Practice to Deceive: Learning Curves of Military Deception Planners*. Annapolis: Naval Institute Press, 2020.

Whaley, Barton. *Stratagem: Deception and Surprise in War*. London: Artech House, 2007.

Whaley, Barton and Jeffrey Busby. "Detecting Deception: Practice, Practitioners, and Theory," In *Strategic Denial and Deception: The Twenty-First Century Challenge*, edited by Roy Godson and James J. Wirtz, 181–222. London: Routledge, 2002.

Wirtz, James J. "Deception and the Tet Offensive," *Journal of Strategic Studies* Vol. 13, Iss. 2 (1990) 82–98.

Wirtz, James J. "Hiding in Plain Sight: Denial, Deception and the Non-State Actor," *SAIS Review*, Vol. 28, No. 1 (Winter-Spring 2008), 55–63.

Wirtz, James J. "The Abbottabad raid and the theory of special operations," *Journal of Strategic Studies* 2021. doi:10.1080/01402390.2021.1933953.

PART II

Historical Perspectives

4

COLD WAR ACTIVE MEASURES

Sanshiro Hosaka

Introduction

In recent years, "disinformation" has received significant public attention, the term often used in popular discourse interchangeably with misinformation, fake news or propaganda. However, disinformation—*dezinformatsiya* in Russian—is not necessarily false information. Although most Soviet disinformation contains forgeries or conspiracy theories, some of the most effective disinformation campaigns dealt with accurate information such as a pamphlet that recounted the number of acts of racial violence against African Americans and distributed in a dozen African countries under the guise of activists, or genuine email data of the US Democratic National Committee hacked and exposed in Wikileaks.[1] Thus, a more important question that needs to be asked here is: in what circumstances, a piece of information is relayed through what channel and to whom. These largely covert, complex processes that influence the perception of reality and decision-making of a particular target group or individual are called active measures—*aktivnye meropriyatsia*. It is this context that makes disinformation an indispensable component of active measures, conceptually distinct from other information products.

In the mid-1980s, Denis Kux, head of the U.S. working group on "active measures," described Soviet active measures as "a wide span of practices including disinformation operations [and] political influence efforts" aimed at "enhancing Soviet influence, usually by tarnishing the image of opponents." These measures often employed Soviet front groups and foreign communist parties as well as "agents of influence."[2]

Importantly, active measures involve "elements of deception and often employ clandestine means to mask Moscow's hand in the operation."[3] The practice of disguising the actual source makes a major difference from overt propaganda that makes little effort to camouflage its sources. While the Soviet government agencies published information on their behalf—propaganda—the Committee of State Security (KGB) operated under a false flag, e.g., falsely attributing information to other government agencies, public organizations or media outlets—active measures.[4] For Soviet intelligence officers, especially the so-called officers of the active reserve, using false credentials such as diplomat, journalist and scholar was an established practice. Active measures, aimed to prompt someone into a particular action (or non-action), usually have carefully selected targets such as decision-makers and elites, whereas propaganda tended to be directed at broader audiences.[5] In practice, however, active measures are often employed in conjunction with overt

DOI: 10.4324/9781003190363-7

propaganda, amplifying the latter's effect. During the Cold War (and beyond), the main targets of such operations were the United States, the main adversary, and its allies. After the border conflict that erupted in the late 1960s, China was also added to the enemy list.

Overall directions and general themes of the active measures—part of Soviet political warfare against the West—were defined at the highest level, the Party's Politburo.[6] Since a large part of active measures comprised covert operations, the primary responsibility for these operations was born by the KGB. Its Service "A" (active measures) created within the First Chief Directorate (foreign intelligence) supervised active measures operations in close coordination with the International Information Department of the Communist Party's Central Committee, which set the agenda for overt propaganda for major Soviet outlets, including *Novosti Press Agency, Pravda, New Times.* The International Department of the Central Committee managed relations with non-ruling communist parties and revolutionary movements as well as international fronts such as the World Peace Council, World Federation of Trade Unions and other committees and associations under the various banners. Specific plans for active measures operations were proposed to the KGB's Service "A" for approval by responsible divisions of the KGB central apparatus, first departments (foreign intelligence) of the territorial offices and foreign residencies.[7] Some operations were conducted in cooperation with satellite Socialist intelligence services (e.g., East German HVA, Czechoslovak StB, Cuban DGI).[8]

As Shultz and Godson point out, the major traits that make the Soviet active measures unique are: 1) Moscow develops active measures with long-term global goals, while Western covert activities lack political consistency in the long run; and 2) the range of instruments for active measures is virtually unrestricted, while that of the Western actions is constrained by cultural, political and moral factors.[9]

The US government separated "covert" actions from "overt" foreign policy activities and ascribed the former to the parcel of the Central Intelligence Agency, while leaving open the question of the coherent political leadership's support and government-wide coordination. For the Soviets, with active measures as an integral part of Soviet foreign policy, the demarcation between covert and overt activities was blurred.[10]

While the CIA was a government agency, the KGB, which only formally subordinated to the Soviet government, was the "sword and shield" of the Communist Party. According to the classified *History of the Soviet Security Organs*, the Soviet state security agencies at the behest of the KGB were "political bodies by the nature of their activities." Other Soviet agencies such as the Defense, Foreign and Internal Affairs Ministries, the Court and the Prosecutor's Office were also enrolled in the protection of state security, but "direct political protection of state security" was a prerogative of the KGB as "a body of special competence" tasked to fight the subversive activities of domestic and foreign forces hostile to the USSR, "using specific means, forms and methods."[11]

This mission of the Soviet intelligence to preserve the Communist Party regime led to two fundamental underpinnings that affected all its activities. The first pillar is its emphasis on counterintelligence. Some scholars call the USSR a "counterintelligence state," a state with a gigantic intelligence agency imbued with the pathological search for conspiracies and the elimination of enemies both within and outside the country.[12] While the Western understanding of counterintelligence is to expose the espionage activities of foreign intelligence services in domestic settings, Soviet counterintelligence starts from abroad.[13] Thus, its counterintelligence active measures pursued "aggravating contradictions" among the enemies.[14] The second underpinning is the reverse of the first: Soviet foreign (external) intelligence starts from the domestic settings—the so-called "intelligence from the territory of the USSR." This blurred boundary between foreign and domestic services shapes the peculiar understanding and scope of Soviet active measures.

Concepts and Components of Active Measures

Methods and Forms

There are different methods of active measures depending on the nature of the tasks and the availability of agent-operational resources. In addition to the disinformation mentioned above, the most common ones include:

- "Exposure" (*razoblachenie*) is used to reveal to the international community or the public of a target country the enemy's anti-Soviet conspiracies, aggressive intentions and subversive plans. Exposure can help form foreign public opinion beneficial to the Soviet Union and strengthen anti-American sentiments in various countries.
- "Compromising" (*komprometatsiya*) is used to cause moral and political damage to the enemy, and to undermine the reputation of the state institutions or individual politicians and other prominent figures of the Western countries and anti-Soviet emigrant centres by publishing or bringing to the attention of target individuals specially selected (often, concocted) materials.
- "Special positive influence" (*spetsial'noe pozitivnoe vozdeistvie*) involves the exercise of influence on governments, parties, prominent figures, and business elites, as a rule, within the limits of the laws of the target country, using the false credentials and agent-operational capabilities.

These methods take various forms: "conversations of influence" with prominent figures; promotion of targeted disinformation including documentary materials; publication of books, brochures, and leaflets under the guise of foreign authors; organization of radio and television programs and press conferences with prominent politicians and scientists; orchestration of rallies and demonstrations, appeals to government and inquiries to parliament in foreign countries; and assistance in the adoption of resolutions favourable to the USSR at international conferences. In practice, these methods are used in combination, thus "increasing the effectiveness of the implemented actions."[15]

Agents of Influence

In active measures, it is a matter of critical importance through *what channel* a piece of (dis) information is delivered to targets. In this sense, the role of "agents of influence" is vital. The United States Information Agency defines this category of agents as "foreigners who have been recruited by the KGB to influence the opinions of foreign publics and governments" (though the KGB definition of agents of influence includes Soviet citizens as well). These agents were valued by the Soviet intelligence because targets perceived them as "loyal patriots of their respective countries who are simply expressing their own personal opinions, not scripts written by the KGB."[16] An agent of influence can be a prominent journalist, academic or government official. According to a KGB foreign intelligence manual, agents of influence may also be "those who are not so prominent but still have an influence on government and civic figures (their close relatives, lovers, priests)."[17] Agents of influence were regarded to have an edge over other ways of transmitting (dis)information, such as press and radio, given

that "well-prepared materials through agents of foreign intelligence [were] swiftly communicated to targets (government, headquarters etc.)."[18] Moscow used these agents in carefully
orchestrated operations called "combination" (*kombinatsiya*).[19]

Unlike other covert assistants of the KGB officers, Chekists, this category of agents has its
peculiarities. First, in active measures, the KGB pursued a concrete outcome, and it was a
secondary concern whether collaborating foreigners are witting or unwitting.[20] As a KGB
handbook shows, the relationship between Soviet intelligence and foreigners may begin
as "confidential contacts," who were then cultivated into agents.[21] In certain circumstances, active measures operations were accomplished without establishing operative-
agent relations with foreigners, but only by supplying confidential contacts with "information sources" or staging a "leak" of materials on issues of their interest.[22] Notably,
Mitrokhin's foreign intelligence lexicon includes not only "agent of influence," but also a
separate entry titled "confidential contacts of influence," who are "[c]onfidential contacts
of intelligence officers (and intelligence agents) in government and political circles, who
are used clandestinely by Intelligence to carry out active measures designed to exert the
required influence on government agencies and the public and political life of a target
country."[23]

Second, such a relationship is a result of the agreement between a KGB officer and an
agent of influence. According to a KGB handbook, an agent is usually recruited not by a
formal written agreement on cooperation, but "by secret information, sometimes documentary, which he communicates to intelligence" and for which the agent sometimes
receives financial remuneration and signs the receipt.[24] However, agents of influence are an
exception to this practice because the primary purpose of this category of agents is not
intelligence collection.[25] Then, what else might define and cement their relationship? Shultz
and Godson argue that the rewards for agents of influence are more likely to be the KGB's
assistance tailored to each individual's specific needs for achieving their political or other
personal goals.[26] For agents recruited from among politicians, for instance, the Soviets did
everything they could to magnify their prestige in their home countries.[27]

The KGB took advantage of the need of certain foreigners to maintain contacts with
the USSR or with specific Soviet counterparts. A Soviet intelligence textbook, for
example, states that "diplomats and journalists accredited in the USSR will not be able
to make a career without contacts in the authoritative political and public circles" of the
Soviet Union. Similarly, the KGB knew the weaknesses of the Western scholars pro-
fessionally involved with Soviet or Russian affairs. The textbook points out that "Sovie-
tologists and some other specialists in capitalist and developing nations may lose their
professional authority and influence if they lose the opportunity to visit the Soviet Union
regularly and communicate with Soviet colleagues." Chekists recruited foreign graduate
students and trainees who needed "access to the materials of their interest and the assistance of members of the teaching staff" and sometimes "the support of major scientific
authorities recognized internationally." These aspects created favourable conditions for
the Soviet intelligence to establish contacts with foreigners for further indoctrination and
recruitment.[28]

In advance of the academic conferences held in the USSR, the KGB, together with
relevant agencies, prepared disinformation materials and identified "the concrete path of
communicating disinformation to the adversary." When the Western scholars exhibited
a professional interest in specific issues, Chekists exploited such an opportunity to
"successfully promote disinformation materials for the adversary" through Soviet
scientists.[29]

External Counterintelligence

According to Kux, the origin of active measures is traced back to Bolsheviki's political warfare against émigré organizations by spreading disinformation, luring their leaders to the USSR and infiltrating agents to sow dissent among them.[30] This classic variant of active measures is defined in more detail in the KGB counterintelligence dictionary, rather than in its intelligence dictionary:

> [Active measures are] actions of counterintelligence, allowing it to penetrate the enemy's plans, timely warn its unwanted steps, mislead the enemy, seize its initiative, and thwart its subversive actions.
>
> Active measures, unlike protective measures, for example, to ensure secrecy and to preserve state and military secrets, are offensive in nature, make it possible to reveal and suppress hostile activities at the earliest stage of their occurrence, force the enemy to expose itself, impose own will on it, force it to act in unfavourable conditions and in ways desired by the counterintelligence agencies.
>
> In the practice of counterintelligence activities, active measures include measures for building up the position of agents in the enemy's camp and its surroundings, conducting operational games with the enemy, disinformation, compromising and decomposing of the enemy force, bringing persons of particular operational value to the territory of the USSR to obtain intelligence information, etc.

Thus, active measures were not the exclusive domain of foreign intelligence officers. In 1960, for "a broad, coordinated attack on the main enemy," a First Chief Directorate (foreign intelligence) official called for officers of counterintelligence divisions to "consider the conduct of intelligence work as their own business" along with their counterintelligence work.[31] And the opposite was also true: according to the KGB foreign intelligence handbook, "the intelligence agencies of the socialist countries are entrusted not only with intelligence functions. They are also engaged in counterintelligence work in the capitalist countries."[32]

The tenet of "external counterintelligence" was the Soviet perception that intelligence and counterintelligence agencies of the capitalist states were conducting subversive activities not only in the Eastern bloc countries but also against their missions and citizens abroad. Indeed, in the mid-1950s, the 15th department of the First Chief Directorate in the Ministry of State Security, later renamed Directorate "K" in the KGB's First Chief Directorate, was responsible for "external counterintelligence" (*vneshnyaya kontrarazvedka*), which provided "counterintelligence support" for Soviet delegations and tourist groups.[33] According to John Barron, many officers of the Directorate "K" came from the Second Chief Directorate (domestic counterintelligence) and regional organs.[34]

However, the Soviet external counterintelligence went further to decompose and paralyze the enemy's intelligence agencies and emigre organizations abroad. While the interception of hostile activities in the USSR territory was the task of the domestic counterintelligence services, the KGB saw the vital role of Soviet foreign intelligence in this struggle because "all the threads of the enemy's subversive actions [were] drawn from the capitalist countries." In this sense, Soviet counterintelligence agencies could work "with much greater efficiency" if they timely received the materials revealing the enemy's plans, intentions and specific activities from foreign intelligence.[35] Oleg Kalugin, a former high-ranking KGB general and Directorate "K" chief, testified that external counterintelligence of the KGB residency in the United States targeted the CIA, the FBI, and the National Security Agency, while political intelligence dealt with the State Department.[36]

External counterintelligence infiltrated its agents into Western intelligence and counter-intelligence agencies. A successful penetration by the so-called "active offensive methods" allowed the Chekists to study their structure, personnel and methods, expose the enemy's plots and channels of agent penetration and intercept the communication channels.[37] KGB agents penetrating an enemy's agent network were particularly valuable because they could cultivate enemy intelligence officers, disinform enemy agencies and divert their attention and resources to secondary issues. Another target of external counterintelligence infiltration was the enemy's intelligence and counterintelligence schools.[38]

Chekists perceived that many emigre organizations owned their intelligence services and received support from the Western intelligence agencies, thus making them "very important objects of agent penetration" for external counterintelligence.[39] With the help of agents, the Soviet intelligence carried out "measures to disintegrate these organizations, paralyze their work, expose and compromise their leaders."[40] The KGB also put a high priority on the deep agent penetration into "schools where spies, saboteurs and terrorists from emigres are trained," viewing the recruitment of teaching staff and students of these schools as "one of the important tasks of counterintelligence work abroad." Thus, active infiltration allowed Chekists to conduct operational games to reveal and take advantage of enemy communication channels with enemy intelligence services.[41]

Intelligence from the Territory

The so-called "Intelligence from the territory" also provides the context to the peculiar understanding of the nexus between foreign intelligence and counterintelligence in the Soviet active measures. During the Cold War, despite the rising number of intelligence tasks, the KGB was not permitted to expand "legal" residencies overseas due to the quotas established by the countries that hosted Soviet diplomatic missions. Furthermore, the Western countries significantly strengthened counterintelligence measures against Soviet spies with sophisticated surveillance and technical equipment. These circumstances complicated the operational settings of the KGB foreign intelligence activities. Seeking new opportunities for intelligence work, in 1970, the KGB Collegium (the highest decision-making body) instructed the security organs to strengthen intelligence activities from the territory of the Soviet Union.[42]

Thus, intelligence from the territory sought to recruit foreign diplomats, military interns, businesspersons and journalists staying in the USSR for an extended length of time. At the same time, short-term visitors through scientific and technical, trade, economic and cultural exchanges were also a "quantitatively more significant" category. Many of these foreigners held prominent positions in their respective countries, had a direct link to the targets of Soviet intelligence penetration, possessed significant information or had "the ability to exert a certain influence on the formation of the domestic and foreign policy of their govern-ments."[43] From the viewpoint of the Soviet intelligence, channels for international scientific exchange were "of particular importance for solving intelligence tasks from the territory of the USSR," because the KGB observed that in the West, scholars were increasingly appointed to important government and diplomatic positions or invited to government agencies as consultants.[44]

As the KGB intelligence textbook states, one of the objectives of intelligence from the territory was, alongside collecting political intelligence, to conduct active measures that would: support the implementation of the Soviet foreign policies; disrupt the enemy's "aggressive plans and aspirations" including economic and ideological sabotage against the

USSR; aggravate the contradictions among the Western countries and weaken their political and military-economic systems; support the communist and national liberation movements; and decompose and compromise anti-Soviet émigré organizations.[45]

Together with agents of influence recruited from among Soviets and foreigners, KGB operatives acting under the guise of the relevant Soviet institutions were directly involved in implementing active measures from the territory of the USSR. KGB officers and agents conducted "conversations of influence" with targeted foreigners. Communicating a well-prepared piece of (dis)information, they carefully studied how their interlocutors reacted. Chekists travelled abroad as members of various Soviet delegations and participated in international meetings and conferences and cultural and scientific exchanges.[46]

Due to the increasing importance and complexity of the tasks dealt with by the KGB foreign intelligence, carrying out a one-time single measure was not adequate. Thus, Chekists practised "complex active measures," which were "a combination of actions that are different in form, methods and scope, but subordinate to a single goal" and implemented "simultaneously or sequentially over a certain period, complementing each other and contributing to an increase in the efficiency of the operation as a whole." Many operations from the territory of the USSR were an organic part of such complex measures aimed at solving the highly important "political, strategic, scientific, technical and national economic problems." When planning complex measures, therefore, specific problems and final goals had to be coordinated with the Directorate "RT" (abbreviation: *razvedka s territorii*) of the First Chief Directorate.[47]

However, the introduction of intelligence from the territory caused friction between local first departments (foreign intelligence) and second departments (counterintelligence). For example, in 1971, the foreign intelligence chief of the Kazakhstan KGB argued that his first department had to concentrate all its operational resources on a certain foreign area designated by Lubyanka, complaining that "the tasks for other countries set out in the [Center's] order should, in fact, be solved by counterintelligence units of the republic's KGB and the regional KGB directorates and departments." According to the foreign intelligence chief, counterintelligence units paid little attention to the intelligence work and "insufficiently outlined measures to revive intelligence work from the position of Kazakhstan."[48]

Active Measures and Foreign Policies

Covert active measures were carefully correlated with the political leadership's foreign policy pronouncements. There was a tendency; when the Soviet Union sought an atmosphere of rapprochement with the West, it scaled down disinformation operations.[49]

In the mid-1980s, the Western media welcomed Mikhail Gorbachev, General Secretary of the Communist Party of the Soviet Union, as a "young" and "pro-Western" Soviet leader, hailing his "New Thinking" foreign policies and perestroika—the overhauling of the stagnant state system. If active measures prior to Gorbachev had been aggressive by attacking and denigrating the West, did the New Thinking policies affect the content and volume of Soviet active measures?

Peace offensive

Gorbachev's New Thinking changed the themes of active measures. According to Vladimir Bukovsky, a former Soviet dissident who gained temporary access to the Communist Party's secret documents after the fall of the USSR, Gorbachev's image was meticulously calculated to appeal to the Western audience, especially moderate leftists and social democrats. In fact, in April 1985, KGB Chairman Viktor Chebrikov reported to Gorbachev that the KGB "carried out a complex of measures to aid, through Chekist means, the all-round implementation of the decisions" of the Party leadership, by instructing European KGB residencies to activate their contacts on détente, which had been halted by the West after the Soviet invasion of Afghanistan in 1979.[50]

Under Gorbachev, striking a balance between operational work and new requirements became an essential challenge for Soviet intelligence officers. Active measures had to be aligned with the Soviet leadership's "peace offensive."[51] According to the statement of Chebrikov at an internal party meeting of the KGB in February 1987, the broadened cooperation with western countries "significantly increase[d] the prestige of the USSR in the international arena and [brought] great political benefits." Under these new circumstances, the KGB officers had to "take a fresh look at the methods of their work." In particular, Chebrikov stressed:

> Active measures should be carried out in such a way that they will maximally contribute to the establishment of international cooperation in the name of preserving peace, further deployment of the anti-war movement abroad, exposure of the aggressive nature of the US foreign policy, revealing the dangerous consequences of the so-called "strategic defence initiative."[52]

In this context, the KGB foreign intelligence service was cautioned to "strictly calibrate" active measures to avoid "possible negative consequences" for the successful implementation of the Party's efforts to expand international cooperation and Soviet peace initiatives.[53] This directive was congruent with the sudden suspension of a Soviet disinformation campaign about the US production of HIV in mid-1987. Earlier, in 1985, the KGB concocted and disseminated the forgeries that claimed that HIV had been "manufactured" by American biological warfare specialists. However, confronted with the official American protest and New Thinking policy Gorbachev announced in July 1987, the KGB abruptly stopped this campaign at the height of its success.[54]

However, this did not halt Soviet active measures altogether. Instead, as pointed out by Chebrikov, the KGB foreign intelligence supported the Soviet leadership's campaign for "the world without nuclear weapons" with "effective active measures."[55] According to another KGB official, the KGB operatives needed to keep an eye on "big politics," taking into account "not only internal processes but also the foreign policy of the Soviet Union, the need to create in the eyes of the foreign public a real, well-to-do image of our country."[56]

Further, Chebrikov reported that "perestroika already had and continue[d] to have a favourable, for the Soviet Union, impact on mass public organizations, liberal-bourgeois, moderate and even part of the conservative political forces of the capitalist states." He argued that *demokratizatsiya* (democratization) of the Soviet society would thus demonstrate "its attractiveness in the world" and increase "the number of our friends," hence "our chance for work," creating "new operational positions" in intelligence work.[57]

"Humanism" Campaign

One of the effective measures for gaining international prestige was the "humanism" campaign with Soviet political prisoners. At the internal meeting, Chebrikov referred to the release of a famous Soviet dissident and Nobel Peace Prize winner Andrei Sakharov at the end of 1986, who had been arrested since 1980 for his public condemnation of the Soviet intervention in Afghanistan. The KGB chief assessed that "the cancellation of Sakharov's administrative expulsion caused a positive political response." Soviet authorities further prepared the procedures for the early release of other dissidents "on the conditions of their official statements about the termination of hostile activities."[58]

A half-year later, Chebrikov further stressed that the pardon of dissidents "politically justified itself fully." First, this measure "significantly muted the Western propaganda campaign" on the "political prisoners in the USSR." Second, this measure "demonstrated the confidence and strength of the socialist system, its humanism." Chebrikov, however, warned that some released dissidents were showing signs to resume anti-Soviet activities. "Unfortunately, a number of [security] organs, despite the instruction given at the beginning of the year [1987], limited their role to registering the hostile activities of such people, they are lagging behind in organizing their deep cultivation." This indicates the KGB's tactics was to cultivate released dissidents to make further use of them, possibly, as agents.[59] The KGB chairman further reminded Chekists that "the release of people convicted for anti-Soviet activities is not identical with their rehabilitation," and that there was no change "in the political assessment of the actions of hostile elements who went abroad at different times."[60] Chebrikov's internal narrative suggests that the release of dissidents was merely a tactical gesture to impress the West.

In the last years of the Soviet Union, Vladimir Kryuchkov, who succeeded the chairmanship from Chebrikov, stated that foreign intelligence officers both in the Center (Moscow) and abroad "underestimate[d] the importance and the role of measures designed to promote influence" and ordered the KGB's foreign intelligence training school to launch new "specialist courses in active measures."[61]

After the collapse of the USSR, the practice of active measures continued to be in place along with disinformation in the KGB successor's agencies.[62] According to Sergei Tretyakov, a former Russian intelligence officer, who defected to the US in the early 2000s, despite the assurance by the Russian Foreign Intelligence Service (the SVR), the successor of the KGB's First Chief Directorate (foreign intelligence), to its American colleagues that it would no longer conduct active measures, the SVR continued to do the same disinformation tasks with the same staff, only renaming the Service "A" to Department "MS"(*meropriyatiya sodeistviya*: support measures).[63]

Internal Evaluations of the Active Measures

Active measures operations were not omnipotent; some were exposed. The KGB was well-aware of inherent risks and warned that compromised operations may not only complicate the KGB's operational circumstances but also have grave ramifications for the USSR's relations with target states. The KGB, therefore, required Chekists to exercise utmost caution—look before you leap. Simultaneously, they were reminded of the opposite—he who hesitates is lost. Extreme vigilance did not allow Chekists to recruit new assets abroad.[64]

The effectiveness of active measures operations was evaluated quantitatively and qualitatively. Measuring success in terms of the number of bogus articles published by a recruited journalist or the number of statements made by a coopted parliamentarian was relatively objective. Rather complicated was measuring the impact of sophisticated operations such as raising the peace movement in Europe—an operation built on the intellectual enterprises and grievances of local leftists. The success of such activities, not solely attributed to a KGB handler, was evaluated subjectively by a KGB residency head or an active measures group leader.[65] Either way, though, the communist leaders paid greater attention to the long-term cumulative effects of a combination of multiple operations rather than the immediate effect of a single operation.[66]

The results of routine operations were reported to the Service "A," which compiled them systematically with "the number and date of the task, time, place, form and channel of implementation (in encrypted form), the results of the event and the reaction of the target of influence, personal numbers of employees who participated in its implementation." The clippings of newspapers, magazines, books, brochures and other printed materials published abroad in the course of active measures were sent to the Service "A".[67]

A future opening of the archive of the Service "A," though least likely in contemporary Russian settings, would shed light on the massive scale and content of Cold War active measures.

Conclusion

As discussed in this chapter, Cold War active measures were shaped through interactions between foreign intelligence and counterintelligence. In other words, internal and external threats, as well as domestic and foreign concerns and opportunities, were often conflated in the modus operandi of Chekists. The counterintelligence nature of the Soviet security and intelligence agencies, whose raison d'être was to protect state security, or the regime, affected fundamental principles and practices of active measures. These historical aspects have had lingering effects on post-Soviet Russian intelligence.

Although external counterintelligence formally falls under the scope of the Russian Foreign Intelligence Service, which inherited the Directorate "K," the counterintelligence agency, i.e., the Federal Security Service (FSB), also plays a crucial role. According to the Law on the FSB, one of the organization's duties is "to carry out … measures to ensure the security of institutions and citizens of the Russian Federation beyond its borders" (Article 12, paragraph "л"). More importantly, the FSB is authorized to "infiltrate special services and organizations of foreign states" (Article 13, paragraph "в"), whereas such a function is not envisaged in the Law on Foreign Intelligence, which governs the activities of the SVR.

Similarly, whereas intelligence from the territory (Directorate "RT") was a responsibility of the KGB foreign intelligence, post-Soviet Russia reinstated it as a prerogative of the FSB by the 2003 revision of the Law on the FSB (Article 13, paragraph "в¹"). For the intelligence activities from the territory of the Russian Federation, the FSB regional directorates took over the KGB regional directorates' first departments, which had been tasked with recruiting foreign travellers in the USSR.[68]

Further research into the Soviet intelligence activities during the Cold War, as well as a comparison of operational circumstances in the past and the present, may reveal the contours of active measures used by successor intelligence agencies in Russia today.[69]

Notes

1 Rid, *Active Measures: The Secret History of Disinformation and Political Warfare*, 10.
2 Kux, "Soviet Active Measures and Disinformation", 19. In the category of active measures, some experts further include kidnappings and assassinations by the "Wet Affairs" Department of the KGB's First Chief Directorate, the activities of spetsnaz and "paramilitary dimension" such as support to foreign guerrilla and insurgency groups in the Third World. See Dziak, "Soviet Intelligence and Security Services in the 1980's: The Paramilitary Dimension"; Knight, *The KGB*, 284–85.
3 Kux, "Soviet Active Measures and Disinformation", 19.

4 Vladimirov and Bondarenko, *Politicheskaya razvedka*, 94. This and other KGB training manuals cited in this chapter was published by TheInterpreter. "KGB Training Manual Revealed".
5 Shultz and Godson, *Dezinformatsia*, 38.
6 Shultz and Godson, 18.
7 Kux, "Soviet Active Measures and Disinformation", 20–22.
8 For the KGB's coordination with other communist intelligence agencies, see e.g., the testimony of Ladislav Bittman, a defector of the Czechoslovak StB. Ladislav Bittman, *The KGB and Soviet Disinformation: An Insider's View* (Washington: Pergamon-Brassey's, 1985); Shultz and Godson, *Dezinformatsia: Active Measures in Soviet Strategy*, 15.
9 Shultz and Godson, *Dezinformatsia*, 15.
10 Shultz and Godson, 14–15. For the political and bureaucratic obstacles for interagency coordination for the US covert actions, see also Godson, "Covert Action: An Introduction", 5.
11 *Istoriya sovetskikh organov*, 3–4.
12 Dziak, *Chekisty: A History of the KGB*; Waller, *Soviet Empire*.
13 Sherr, "The New Russian Intelligence Empire", 13.
14 Sherr, *Soviet Power: The Continuing Challenge*, 152.
15 Vladimirov and Bondarenko, *Politicheskaya razvedka*, 86–88.
16 "Soviet Active Measures in the "Post-Cold War" Era 1988–1991".
17 *Osnovnye napravleniya i ob'ekty*, 54.
18 *Osnovnye napravleniya i ob'ekty*, 75–76.
19 Shultz and Godson, *Dezinformatsia*, 133.
20 Barron, *KGB Today*, 175. For example, KGB defector Stanislav Levchenko testified that most of the agents he handled in Tokyo had believed that he was a Soviet correspondent, but not a KGB officer. Shultz and Godson, *Dezinformatsia*, 176.
21 *Rabota s agenturoi*, 28–32. Usually an intelligence officer's talk to persuade a foreigner into a Soviet agent had a strong psychological effect on the target of recruitment; "consenting to intelligence cooperation means taking a known risk, entrusting his security to intelligence." The recruiter convinced the target that Soviet intelligence was responsible for ensuring the agent's safety. See *Konspiratsiya v razvedyvatel'noi rabote*, 23.
22 Vladimirov and Bondarenko, *Politicheskaya razvedka*, 45,89.
23 Mitrokhin, "Doveritelnyye svyazi vliyaniya".
24 *Verbovka agentury*, 104.
25 *Osnovnye napravleniya i ob'ekty*, 55. However, this does not preclude the existence of agents who combine both roles, i.e., intelligence collection and influence operation.
26 Shultz and Godson, *Dezinformatsia*, 133.
27 For example, Prime minister Aleksei Kosygin ordered the release of Japanese fishermen arrested by the USSR border guards on charges of violation of Soviet territorial waters at the end of the visit of the parliamentary delegation headed by Japanese politician Hirohide Ishida, Soviet agent of influence. Ishida was cited by Japanese media describing the release as evidence of the Soviets' reciprocation and his personal relationship with the Kremlin. The Soviets continue to detain Japanese fishermen and soon other captives were supplied. See Barron, *KGB Today*, 79.
28 Vladimirov and Bondarenko, *Politicheskaya razvedka*, 44.
29 Chebrikov, "Ob itogakh ianvarskogo (1987g.) plenuma", 19–20.
30 Kux, "Soviet Active Measures and Disinformation", 20.
31 Hosaka, "Repeating History", 437.
32 *Osnovnye napravleniya i ob'ekty*, 93.
33 Vladimirov and Bondarenko, *Politicheskaya razvedka*, 8.
34 Barron, *KGB Today*, 446.
35 *Osnovnye napravleniya i ob'ekty*, 93–94. See also Myagkov, *Inside the KGB*, 175. In 1967, the KGB instructed the counterintelligence services to "take active measures for discovering and foiling enemy schemes, and so on."
36 Kalugin, "The KGB Has Not Changed Its Principle", 106.
37 In terms of its aggressiveness, this "active offensive method" resembles, though not identically, "counterespionage" operations by the US intelligence. See Johnson, "Introduction", 10.
38 *Osnovnye napravleniya i ob'ekty*, 96–98.
39 *Osnovnye napravleniya i ob'ekty*, 98.
40 *Osnovnye napravleniya i ob'ekty*, 99.

41 *Osnovnye napravleniya i ob'ekty*, 99–100.
42 Vladimirov and Bondarenko, *Politicheskaya razvedka*, 9–10.
43 Vladimirov and Bondarenko, 23.
44 Vladimirov and Bondarenko, 23.
45 Vladimirov and Bondarenko, 86.
46 Vladimirov and Bondarenko, 92–93.
47 Vladimirov and Bondarenko, 93. The concept has been preserved in contemporary Russian intelli-
    gence. Chekists call a set of active measures operations "Complex of measures."
48 Baigarin, 126–28.
49 Bittman, *The KGB and Soviet Disinformation: An Insider's View*, 95–96.
50 Bukovsky, *Judgment in Moscow: Soviet Crimes and Western Complicity*, chap. 6.1 "Acceleration".
51 For Soviet "peace offensive," see Zubok, "Why Did the Cold War End", 349.
52 Chebrikov, "Ob itogakh ianvarskogo (1987g.) plenuma", 19–20.
53 Chebrikov, "Ob itogakh iyun'skogo (1987 goda) plenuma", 11.
54 Andrew and Mitrokhin, *The Sword and the Shield*, 244–45.
55 Chebrikov, "Ob itogakh iyun'skogo (1987 goda) plenuma", 12. In the early and mid-1980s, the KGB
    launched a complex of measures to exploit Western anti-nuclear movements to prevent NATO from
    installing the Pershing II missiles in Europe. Andrew and Mitrokhin, *The Sword and the Shield*, 484. For
    the KGB's involvement in the "nuclear winter" theory, see Earley, *Comrade J: The Untold Secrets of
    Russia's Master Spy in America After the End of the Cold War*, 169–77.
56 Grishanin, "Perestraivat' i uglubliat' kontrol' [Rebuild and deepen control]", 18.
57 Chebrikov, "Ob itogakh iyun'skogo (1987 goda) plenuma", 11–12.
58 Chebrikov, "Ob itogakh ianvarskogo (1987g.) plenuma", 19.
59 Recruitment of prisoners for the purposes of infiltrating them into nationalist and sectarian groups was
    widespread practice in the post-Stalin KGB. Hosaka, "Repeating History", 439.
60 Chebrikov, "Ob itogakh iyun'skogo (1987 goda) plenuma", 16.
61 Andrew and Mitrokhin, *The Sword and the Shield*, 245.
62 Waller, *Soviet Empire*, 136.
63 Earley, *Comrade J: The Untold Secrets of Russia's Master Spy in America After the End of the Cold War*, 195.
64 For example, see Baigarin, 127.
65 Shultz and Godson, *Dezinformatsia*, 183–84.
66 Shultz and Godson, 173.
67 Vladimirov and Bondarenko, *Politicheskaya razvedka*, 86, 95–96. For the active measures statistics
    reported by the Rome residency in 1975, see endnote 172 in Andrew and Mitrochin, *The Sword and
    the Shield*, 659.
68 Soldatov, "Neizvestnaya razvedka".
69 For such attempts, see e.g., Riehle, *Russian Intelligence*; Hosaka, "Putin's Counterintelligence State".

Bibliography

Andrew, Christopher and Vasili Mitrokhin. *The Sword and the Shield: The Mitrokhin Archive and the Secret
    History of the KGB*. New York: Basic Books, 1999.
Baigarin, A., A. *Sbornik KGB SSSR* 49–50 (1971).
Barron, John. *KGB Today: The Hidden Hand*. London: Hodder and Stoughton, 1984.
Bittman, Ladislav. *The KGB and Soviet Disinformation: An Insider's View*. Washington: Pergamon-Brassey's,
    1985.
Bukovsky, Vladimir. *Judgment in Moscow: Soviet Crimes and Western Complicity*. California: Ninth of
    November Press, 2019.
Chebrikov, V. "Ob itogakh ianvarskogo (1987g.) plenuma TSK KPSS i zadachakh partiinoi organizatsii
    KGB SSSR, vytekaiushchikh iz ego reshenii. Doklad chlena Politbiuro TSK KPSS, Predsedatelia KGB
    SSSR tovarishcha Chebrikova V. M. na sobranii partiinogo aktiva tsentral'nogo apparata Komiteta
    gosbezopasnosti SSSR 14 fevralia 1987 goda" [On the results of the January (1987) plenary session of
    the Central Committee of the CPSU and the tasks of the party organization of the KGB of the USSR
    arising from its decisions. Report of a member of the Politburo of the Central Committee of the
    CPSU, Chairman of the KGB of the USSR, Comrade Chebrikov V.M. at a meeting of the party
    activists of the central apparatus of the USSR State Security Committee on February 14, 1987]. *Sbornik
    KGB SSSR* 116 (1987).

Chebrikov, V. "Ob itogakh iiun'skogo (1987 goda) plenuma TSK KPSS i zadachakh organov i voisk komiteta gosudarstvennoi bezopasnosti SSSR po uglubleniyu perestroiki operativno-sluzhebnoi deiatel'nosti - Doklad chlena Politbiuro TSK KPSS, Predsedatelya KGB SSSR V. M. Chebrikova na soveshchanii rukovodiashchego sostava, sekretarei partkomov i partbiuro podrazdelenii tsentral'nogo apparata KGB SSSR 8 iiulia 1987 goda" [On the results of the June (1987) plenum of the Central Committee of the CPSU and the tasks of the organs and troops of the USSR State Security Committee to deepen perestroika of operational and service activities—Report of the member of the Politburo of the Central Committee of the CPSU, Chairman of the KGB of the USSR V.M. Chebrikov at a meeting of the leadership, secretaries of party committees and party bureau divisions of the central apparatus of the KGB of the USSR on July 8, 1987]. *Sbornik KGB SSSR* 118 (1987).

Dziak, John J. *Chekisty: A History of the KGB*. Lexington: Lexington Book, 1988.

Dziak, John J. "Soviet Intelligence and Security Services in the 1980's: The Paramilitary Dimension". In *Intelligence Requirements for the 1980's: Counter-Intelligence*, edited by Roy Godson. Washington, D.C: National Strategy Information Center, Inc., 1980.

Earley, Pete. *Comrade J: The Untold Secrets of Russia's Master Spy in America After the End of the Cold War*. New York: Berkley Books, 2007.

Godson, Roy. "Covert Action: An Introduction". In *Intelligence Requirements for the 1980's: Covert Action*, edited by Roy Godson. Washington, D.C: National Strategy Information Center, Inc., 1981.

Grishanin, G. "Perestraivat' i uglubliat' control" [Rebuild and deepen control]'. *Sbornik KGB SSSR*, no. 128 (1989).

Hosaka, Sanshiro. "Repeating History: Soviet Offensive Counterintelligence Active Measures". *International Journal of Intelligence and CounterIntelligence* 35, no. 3 (3 July 2022): 429–458. https://doi.org/10.1080/08850607.2020.1822100.

Hosaka, Sanshiro. "Putin's Counterintelligence State: The FSB's Penetration of State and Society and Its Implications for Post-February 24 Russia". Tallinn: International Centre for Defence and Security / Estonian Foreign Policy Institute, December 2022. https://icds.ee/en/putins-counterintelligence-state/.

Istoriya sovetskikh organov gosudarstvennoi bezopasnosti [History of Soviet State Security Organs]. Moscow: Vysshaya krasnoznamennaya shkola komiteta gosudarstvennoy bezopasnosti pri Sovete ministrov SSSR imeni F. E. Dzerzhinskogo, 1977.

Johnson, Loch K. "Introduction". In *Handbook of Intelligence Studies*, edited by Loch K. Johnson, Abingdon. London: Routledge, 2007.

Kalugin, Oleg. "The KGB Has Not Changed Its Principle". In *Perils of Perestroika: Viewpoints From the Soviet Press, 1989–1991*, edited by Isaac J. Tarasulo, 105–110. Wilmington, Delaware: Scholarly Resources Incorporated, 1992 Original was published in Komsomolskaya Pravda, June 20, 1990.

The Interpreter. "KGB Training Manual Revealed", 1 November 2018. https://www.interpretermag.com/kgb-training-manuals-revealed/.

Knight, Amy. *The KGB: Police and Politics in the Soviet Union*. Revised. Boston: Unwin Hyman, 1990.

Konspiratsiya v razvedyvatel'noi rabote na territorii sovetskogo soyuza s pozitsii vedomstv prikrytiya. Analiticheskiy obzor [Tradecraft in Intelligence Work on the Territory of the Soviet Union from the Position of Cover Departments. Analytical Review], 1988. https://www.4freerussia.org/wp-content/uploads/sites/3/2020/01/%D0%9A%D0%BE%D0%BD%D1%81%D0%BF%D0%B8%D1%80%D0%B0D1%86%D0%B8%D1%8F-%D1%80%D0%B0%D0%B7%D0%B2%D0%B5%D0%B4%D1%8B%D0%B2%D0%B0%D1%82%D0%B5%D0%BB%D1%8C%D0%BD%D0%BE%D0%B9-%D1%80D0%B0%D0%B1%D0%BE%D1%82%D0%B5.pdf.

Kux, Dennis. "Soviet Active Measures and Disinformation: Overview and Assessment". *The US Army War College Quarterly: Parameters* 15, no. 1 (4 July 1985).

Mitrokhin, Vasili. "Doveritelnyye svyazi vliyaniya—confidential contacts of influence". In *KGB Lexicon: The Soviet Intelligence Officers Handbook*. Abingdon: Routledge, 2002.

Myagkov, Aleksei. *Inside the KGB*. New York: Ballantine Books, 1945.

Osnovnye napravleniya i ob"ekty razvedyvatel'noi raboty za granitsei [Fundamental Directions and Targets of Intelligence Work Outside the Country], 1970. https://drive.google.com/file/d/1dWB8ak89PeNiQppP7IYNqc7FIO6TZmGM/view.

Rabota s agenturoi [Work with Agents' Networks], 1970. https://drive.google.com/file/d/1myLMhCtnOioNn8a4XrrIayXOkX-GFfi4/view.

Rid, Thomas. *Active Measures: The Secret History of Disinformation and Political Warfare*. Farrar, Straus and Giroux, 2020.

Riehle, Kevin P. *Russian Intelligence: A Case-Based Study of Russian Services and Missions Past and Present*. Bethesda: National Intelligence Press (NI Press), 2022.

Sherr, James. *Soviet Power: The Continuing Challenge*. Palgrave Macmillan UK, 1987. https://doi.org/10.1007/978-1-349-08524-8.

Sherr, James. "The New Russian Intelligence Empire". *Problems of Post-Communism* 42, no. 6 (1995): 11–17.

Shultz, Richard, and Roy Godson. *Dezinformatsia: Active Measures in Soviet Strategy*. Potomac Books, 1984.

Soldatov, Andrei. "Neizvestnaya razvedka [Unknown intelligence]". agentura.ru, 17 January 2022. https://agentura.ru/investigations/neizvestnaja-razvedka/.

"Soviet Active Measures in the 'Post-Cold War' Era 1988–1991". Washington, D.C: United States Information Agency, 1992. http://intellit.muskingum.edu/russia_folder/pcw_era/.

Verbovka agentury [Recruitment of Agents], 1964. https://drive.google.com/file/d/1yPeDz2x-nL2HCXglsSq7pf0YMDIFxsN0/view.

Vladimirov, V.M., and Yu A. Bondarenko. *Politicheskaya razvedka s territorii SSSR [Political Espionage from USSR Territory]*. Moscow: Krasnoznamennyy institut KGB SSSR imeni YU, 1989. https://drive.google.com/file/d/1XxulRAgcXO4XIFW-YrNhi57uzAx3dGwB/view.

Waller, Michael J. *Secret Empire: The KGB in Russia Today*. Boulder: Westview Press, 1994.

Zubok, Vladislav. "Why Did the Cold War End in 1989? Explanations of 'The Turn'". In *Reviewing the Cold War: Approaches, Interpretations, Theory*, edited by Odd Arne Westad. London: Routledge, 2013 original [2000].

5

HISTORICAL DISINFORMATION PRACTICES

LEARNING FROM THE RUSSIANS

Randolph H. Pherson, Deanna Labriny and Abby DiOrio

Examples dating from 1992 are presented showing that the Soviet Union and Russia have employed a wide range of disinformation techniques. Three examples are provided to illustrate Russia's evolution from the simple use of propaganda to the sophisticated manipulation of social media.

A detailed case study illustrates how Russia used the Five D's to deflect blame for the shoot down of MH17 aircraft in Ukraine in 2014. The chapter then explores how this same approach is reflected in subsequent Russian efforts to sow discord in the 2016 US Presidential election, inflame anti-immigration concerns in Germany, disrupt the 2016 Brexit vote in the UK, and interfere with the 2017 presidential balloting in France.

INTRODUCTION

Partisan political actors and social manipulators are increasingly using social media platforms to reshape popular perceptions for partisan political or social purposes. Such actions are rendering democratic processes more vulnerable and inhibiting constructive social dialogue. These attacks on democratic countries, liberal institutions, electoral processes and social norms have come from a variety of sources, ranging from teenage entrepreneurs in Macedonia to Russia's robust disinformation campaign accompanying its invasion of Ukraine. With the growth of social media, the Kremlin has proven particularly adept at providing misleading information that confirms consumers' biases and further hardens mental mindsets to benefit the perpetrators.

As more people migrate their media consumption to online and social media, they are being forced to contend with dubious information from a wide range of sources, resulting in growing arguments over supposed facts. This has contributed to our society becoming more polarized. As a result, the quality of political and social discourse has suffered.

Disinformation has become so widespread that major news outlets and politicians have increasingly fallen for and parroted false information. In 2019, for example, several US Republican senators echoed a Kremlin talking point that Ukraine—and not Russia—hacked the Democratic National Committee's server and interfered in the 2016 elections. The US Intelligence Community and several White House security officials repeatedly stated that this

DOI: 10.4324/9781003190363-8

claim was wholly unsubstantiated.[1] In 2022, Russian President Vladimir Putin's aggressive disinformation convinced most Russians that Ukraine had initiated hostilities against Russians and that the conflict was not a war but a "special military operation."[2]

Social media platforms have proven amenable hosts to Digital Disinformation. Network algorithms facilitate the micro-targeting of users by propagating vivid content tailored to their personal interests. Social media companies have embraced the embedded use of these algorithms in their social media platforms, which have proven biased toward sensational content because it is profitable. Unfortunately, this commercial preference for user engagement is easily exploited by those with intention to inflame and distort.[3] The potency of this form of micro-targeting is enhanced by the fact that users tend to be "cognitively lazy." They usually fail to take measures to validate any skepticism of information populating their newsfeeds. This leaves them vulnerable to believing incorrect information that supports their pre-conceived viewpoints.

Defining the Phenomenon

The phenomenon of Digital Disinformation goes by many monikers. The phrase **Fake News** often appears in the public domain because its use by the former US President to describe news reporting critical of his administration obscured its meaning in public discourse. Experts in the field increasingly recognize that an agreed-upon lexicon is needed both to clarify the debate and to lay the ground for prescribing effective solutions.[4]

A simple way to distinguish between forms of Digital Disinformation is to focus on the motives of the perpetrators.[5] For example:

- **"Entrepreneurial News"** or **"Fraud News"** is usually generated by an individual to mislead a reader for personal or financial gain; the purpose is to attract the viewer to ads and thereby generate revenue.
- **"Agenda-driven News"** or **"False News"** is purposely intended to mislead the reader, most often for partisan political or social purposes. The objective is to provide information that affirms the reader's biases and further hardens mental mindsets. Usually, the product or report includes text and/or images that are a mix of correct and incorrect information.

A better schema for categorizing misleading information is to determine whether the message is true or false and whether the disseminator of the information intends to do harm or does not.[6]

- **Unintentional Misinformation** is usually spread by people with honest intentions because it supports their world view and they do not know or really care if the report is factually incorrect. Misinformation can also include the **mislabelling of data** or when **satire** is taken too seriously.
- **Mal-information** is the deliberate publication of private information for personal or corporate advantage that does not serve the public interest. A good example is **revenge porn**. It can include the deliberate misrepresentation of times and dates or a change of context.
- **Disinformation or Deception** is an intentional act by an adversary or a competitor to influence the decisions or actions of the recipient to the advantage of the deceiver. It is **fabricated or deliberately manipulated content** that can take both narrative and visual forms. **Propaganda** falls into this category. Types of digital disinformation include:

a   **Computational Propaganda** which is the use of algorithms, automation and human curation to purposefully distribute misleading information over social media networks.[7]

b   **Image Manipulation** which is the distribution of pictures with the intention of altering one's perception of that person, object or event. The images can simply be unflattering but true representation or altered with the intent of denigrating the subject or falsely associating the subject with something negative.

c   **Deep Fakes** which are videos that have been digitally altered through artificial intelligence techniques. Laypeople now can plug a photograph or video clip into prewritten code and produce an extremely realistic, life-like false image or video. Deep Fakes are inherently hard to detect and, so far, society is largely ill-equipped to deal with them.

d   **Active Measures** including propaganda, influence operations, or perception management campaigns conducted by nation states, most notably Russia. Active Measures are intended to manipulate the perceptions or actions of individual decision makers, the public, and governments to influence elections and the broader course of world events.

For purposes of this discussion, this paper will use the terms **Propaganda** and **Digital Disinformation,** which encompasses most of these forms. Both involve the purposeful propagation of incorrect or misleading information over publications, broadcasts and social media platforms to manipulate and manage popular perceptions in a way that advantages the political and social agendas of the perpetrator.

This phrase Digital Disinformation has been adopted by the International Association for Media and Communication Research (IAMCR) as a preferred term for describing "Fake News." At its February 2018 Colloquium in Paris, IAMCR noted that Digital Disinformation touches many aspects of our lives, including the politics of climate change, globalization, feminism, health, science and many other concerns. It posits that Digital Disinformation threatens the integrity of knowledge and scientific reasoning.[8]

RUSSIA'S METHODOLOGY

In this chapter, we will review the methodology Russia has employed in implementing robust propaganda and disinformation campaigns. The Russian methodology can be described as **Sowing the Five Ds: Dismiss, Distract, Deflect, Distort** and **Distrust**.[9]

The chapter presents examples dating from 1992 showing how the Soviet Union and Russia have employed a wide range of techniques to promote false narratives including propaganda, TV broadcasts, social media postings and even false-flag operations. Three examples are provided to illustrate Russia's evolution from the simple use of propaganda to the sophisticated manipulation of social media by extending the "Frozen Conflict" in Transnistria; covering up the 2010 Plane Crash in Smolensk that killed the Polish Prime Minister; and efforts to buttress the Assad regime in Syria since 2011.

A detailed case study illustrates how Russia used the Five Ds to deflect blame for the shoot down of MH17 aircraft in Ukraine in 2014. The chapter then explores how this same approach is reflected in subsequent Russian efforts to sow discord in the 2016 US Presidential election, inflame anti-immigration concerns in Germany, disrupt the 2016 Brexit vote in the UK, and interfere with the 2017 presidential balloting in France.

EXTENDING THE FROZEN CONFLICT IN TRANSNISTRIA

*Key themes of the campaign have been to use propaganda to sow **distrust** and **distort** the intentions of leadership, while **distracting** attention to their motives through alleged false-flag operations.*

In March 1992, military hostilities broke out in Transnistria, a sliver of land on the east bank of the River Dniester in Moldova (see Figure 5.1). It lasted until July 1992 when a peace agreement was signed and a Commission on Security and Cooperation in Europe (OSCE) Mission was established to oversee a peaceful but unending "frozen conflict" that left 1,500 Russian troops occupying the territory. Six hundred troops were designated as "peacekeepers" and the remainder occupied a large Russian military base, often referred to as Russia's Kaliningrad in the south.[10]

During the war, some Transnistrians feared losing their Russian identity to "Romanianization." Since the late 1880s, Russian-language publications had detailed stories of Slavs "succumbing" to Romanian influence, all designed to shock the reader. According to these accounts, individuals lost not only their knowledge of the Russian or Ukrainian language, but their identity as a Slav and the Slavic way of life. The 1992 war and Moldovan nationalist policies resurrected these old fears among Transnistria's Slavs.[11]

Figure 5.1. Transnistrian Territories Sandwiched between Moldova and Ukraine

Following the 1992 conflict, Russia exploited the Moldovan-Transnistrian "frozen conflict," hindering efforts to reach a peaceful settlement. Russian state media broadcast anti-Moldovan propaganda to Transnistrian TVs and radios. Russia maintained its military presence in the Transnistria, despite promising to remove its so-called "peacekeepers" in 2002.[12]

More recently, in the wake of the Russian invasion of Ukraine, Transnistria's Interior Ministry and the radio antennae that broadcast Russian propaganda to Ukraine were blasted by an unknown assailant. Those familiar with past Russian tactics suggested that the Kremlin was laying the ground with a false flag operation for intervening in Transnistria. Those fears were heightened when Ukrainian intelligence published what appeared to be a Transnistrian newspaper dated May 2, 2022 that referenced attacks that had not yet occurred and appealed for Russian intervention.[13]

When Kremlin loyalist, President Igor Dodon, lost his November 2020 re-election bid and was replaced by the pro-Western candidate, Maia Sandu, the Kremlin propaganda machine launched a campaign of threats and disinformation attacking the new president and her country. Events took a more serious turn in May 2022 when Russia moved from propaganda to open discussion of military action, projecting a similar pattern to what was seen before Putin ordered his all-out war in Ukraine.[14]

A disinformation campaign began on social media, followed shortly after by unsubstantiated allegations that Romanian troops had infiltrated Moldova's 6,000-strong army, donned its uniforms, and would soon start a war. This "insider information" was revealed by Igor Girkin, aka Igor Strelkov, a Russian intelligence officer, who played a key role in the annexation of Crimea; occupation operations in Donbas, Ukraine; and the downing of flight MH17 over Ukraine in 2014 that killed 298 crew and passengers (a role described in the MH17 case study that follows).[15]

Russian media also blamed Ukraine for a series of bombings in Transnistria. The ineffectiveness of the attacks and the choice of targets, however, makes it unlikely Kyiv ordered the strikes. For example, Transnistrian authorities claimed that a drone from "the Ukrainian side" dropped two grenades on a parking lot used by the Russian peacekeeping forces. The crude bombing only shattered the windshield of a Russian transport truck. With Moldova and Ukraine ruled out as suspects, commentators speculated that the Transnistrians or Russians were staging false-flag attacks.[16]

COVERING UP THE PLANE CRASH IN SMOLENSK, RUSSIA

*Russian efforts to **dismiss** allegations of its complicity in the downing of the aircraft carrying the Polish President required it to **distort** or refuse to publish evidence about the crash and **distract** the public with false alternative explanations.*

On April 10, 2010, a Polish military TU-154M (similar to a Boeing 727) crashed in the city of Smolensk, Russia, killing all the crew and passengers aboard. The deaths included the Polish President and First Lady, the last Polish President in exile, the Chief of the General Staff, Commanders-in-Chief, the Chairman of the Polish National Bank and several Members of Parliament. The airplane was equipped with all the necessary electronic navigation and instrumentation to land safely, even in foul weather. The instrumentation was the latest and best.[17]

The Smolensk airport was fogged in, but the airport had not been declared closed. As the TU-154 approached the airport, the tower suggested the pilot divert to Moscow. The pilot said he would make one attempt to land, and, if that failed, he would look for another airport. At two kilometers from the runway, the airplane was on track. At one kilometer, the plane was suddenly 40 to 60 meters to the left of the centerline, 2.5 meters above the ground, below the glide path, and reported to be traveling 280 K/hr, with throttles applied to abort the landing. The aircraft's reported speed of 280 K/hr was suspicious because it is twice the speed of a normal approach on landing. Despite trying to apply full power to abort the landing, the plane crashed, striking trees, flipping over, and landing well short of the runway.[18]

One hypothesis was that the crash was caused by Russian air controllers who used "meaconing" (the interception and rebroadcast of navigation signals) to down the plane. Until the time of the crash, the controllers reassured the crew that they were "on the course and on the approach lane." Subsequent reporting revealed that the plane was not on course during the entire approach look-and-see procedure.[19] A senior US intelligence official noted that "the Russians have a history of manipulating aviation navigational beacon systems to lure planes off course, while telling them different information from the air control tower."[20]

The Polish government decided to leave the investigation solely in Russian hands. It refused to seek help from NATO. Most of the evidence gathered was left on Russian soil and never returned to Poland, including confidential military and security codes belonging to NATO's armies, "black boxes," and other flight recording devices.[21] The Russians delayed for weeks before returning less sensitive items and kept all items of intelligence value.[22]

In the days following the tragedy, a major disinformation campaign was launched both in Poland and Russian to **distort, dismiss, distract and deflect** from the facts of the case.

- **Distortion:** Within minutes of the crash, the Russian government, with the acquiescence of the Polish government said that the crash was due to "pilot error, lack of training and poor communication skills (i.e., that the pilot could not speak Russian)."[23] Subsequent investigations determined that the pilot, fluent in Russian, had no problem communicating with the Smolensk control tower.[24] He had successfully landed at Smolensk airport only days before the crash and was one of the most experienced pilots of the regiment, with over a thousand hours flying a Tu-154.[25]
- **Dismissal:** In an alleged effort to dismiss potentially incriminating evidence, between 8:20 and 9:14 Warsaw time the airport mysteriously had no outside communication, no power supply, and no eyewitnesses were present.[26]
- **Distraction:** Edmund Klich, chief of Polish investigating committee later revealed to the Polish parliament that the Russian traffic controller who talked with the pilot of the presidential plane had disappeared. The Russians said that he retired.[27]
- **Distortion:** False claims were made of four abortive landing attempts but, in fact, only one look-and-see attempt was made.[28]
- **Dismissal:** Hundreds of examples were documented of evidence being destroyed or overlooked and never considered for examination.[29]
- **Distortion:** Erroneous information was distributed at the time of the crash. For example, the official time given by the Premier of the Russian Federation Ministers is over a dozen minutes later than it was finally set. In addition, it was falsely reported that the aircraft's left wing hit a tree with a 40 cm (16") diameter trunk.[30]
- **Distraction:** Russia claimed that the crash was caused by a fuel explosion that precipitated an emergency landing into a swampy forest ground.[31]
- **Distortion** and **Dismissal:** Shortly after the funerals took place, Russia sent coffins back to Poland with anonymous human remains inside, claiming that thorough DNA examinations had been performed. But these unidentified remains, released for burial, had never been inspected, and no post-mortem examination data was released.[32]

- **Deflection.** The entire landing navigation, performed by Russian airfield ground personnel (the flight control tower), misled the crew, falsely confirming their positions, which led to the crash. Based an evaluation of the flight controller's recordings, officials confirmed the deception.[33]

Following the crash, independent Polish citizens commenced their own inquiries. Their efforts established that the Russian transcript of the "final cockpit recordings" showed signs of editing and splicing (**Distortion**). They also noted that the report delivered by the Russian Aircraft Accidents Investigation Committee contained neither technical data nor official source documentation of the inspection. (**Dismissal**).

BUTTRESSING THE ASSAD REGIME IN SYRIA

Moscow has invested heavily in propaganda and Digital Disinformation campaigns to **distort** *the reason for conflict in Syria,* **dismiss** *and* **deflect** *blame for Russia's actions, and sow* **distrust** *in the international community.*

Russia's efforts to support Assad date back to the outbreak of popular protests against the Assad regime in 2011 and extend to this day. Moscow's disinformation strategy seeks to create an alternate narrative about Syrian military activities to influence Western policies toward Damascus and sow confusion and **distrust** in Western societies. In 2015, the US State Department estimated that Moscow spent over $1.4 billion per year on propaganda in addition to funding think tanks to promote its narrative and general diplomatic efforts.[34]

A key theme of Russian President Vladimir Putin's disinformation campaign has been to repeatedly refer to the war in Syria as a war on terrorism. In early 2018, following a Syrian government offensive that killed hundreds of people in Eastern Ghouta in southern Syria, Russian Foreign Minister Sergei Lavrov reiterated in an address to the UN Human Rights Council in Geneva that Moscow will continue to help Assad's government defeat the "terrorist threat." By **dismissing** and denying the massive human toll of over 500,000 people killed, most by the Syrian regime, and another 100,000 forcibly disappeared, Russia and Syria constantly **deflected** attention, refocusing it on the regime's enemies.[35]

Both the Russian Air Force and Assad regime have adopted tactics to displace people in opposition areas both to weaken areas of possible opposition and create an international refugee crisis that caused severe political and social problems for neighboring and European countries. By 2018, displacement had produced 5.6 million Syrian refugees, with estimates that another 6.7 million have been internally displaced. Russian media has capitalized on this phenomenon by propagating a propaganda campaign targeting the threat presented by Syrian refugees to Europe. It has sowed **distrust** by spreading false stories such as the fabricated rape of Russian nationals by refugee gangs in Germany and reports of sex attacks in Cologne on New Year's Eve 2015–2016.[36]

Russia's RT television outlets and Sputnik News wire services have created a prolific chatter of false narratives. Its Internet Research Agency (IRA) produces a sophisticated blend of state sponsored propaganda, coupled with social media disinformation targeting specific population groups. This process works via **distorted** fake accounts and social bots that spread Moscow's propaganda to advance Russian political interests.[37]

One example is Russia's launch of a trolling campaign following the April 13, 2018 US-led strikes against Syrian regime targets that were associated with a chemical weapons attack on Douma. On April 14, Pentagon Chief Spokeswoman Dana W. White claimed there had been "a 2,000 percent increase in Russian trolls in the last 24 hours." While the number may have been an exaggeration, many accounts were identified as trolls along with a noticeable a crossover between Iranian and Russian content.[38]

Russian disinformation over the years has buttressed the Assad regime and made Russia the main power broker in Syria. And the disinformation campaign continues to this day. In a 2022 report, the Institute for Strategic Dialogue (ISD) identified a network of 20 conspiracy theorists, media outlets, and organizations, backed by a coordinated Russian campaign—all propagating propaganda.[39]

The ISD report documents that three principal false narratives have been promoted by this network of conspiracy theorists. They sought to:

- **Distort** the role of the White Helmets—a volunteer organization working to evacuate people in Syria.
- **Dismiss** and deny facts about Syria's use of chemical weapons.
- Sow **distrust** of the findings of the world's foremost chemical weapons watchdog.[40]

The report notes that the White Helmets were also targeted following the chemical attack on Khan Sheikhoun in 2017. That attack killed 92 people, a third of them children. A UN unit later concluded Russian efforts to **dismiss** and **deflect** accountability were unsupported, stating that there were "reasonable grounds to believe that Syrian forces dropped a bomb dispersing sarin" on the town in Idlib province. White Helmets volunteers were identified as the most frequently attacked targets with more than 21,000 tweets designed to discredit the group or encourage attacks against their first responders.[41]

The ISD research revealed that some 47,000 disinformation tweets were sent by the core of 28 conspiracy theorists over seven years from 2015 to 2021. Of these, 19,000 were original posts, and they were re-tweeted more than 671,000 times.

AVOIDING BLAME FOR THE SHOOTDOWN OF MH17 IN UKRAINE

*Russian employed all "five D's —**Dismiss, Distract, Deflect, Distort**, and **Distrust**—in seeking to avoid accountability for the shootdown of MH17 over Ukraine.*

On July 17, 2014, Malaysian Airlines Flight 17 (MH17)—*en route* from Amsterdam to Kuala Lumpur—was thirty miles from the Russia-Ukraine border flying at 33,000 feet when it was shot down by a missile with a Russian-made 9N314M warhead launched from a Russian Buk missile system (see Figure 5.2).[42] The Boeing 777 carried 283 passengers and 15 crew members. On board were six persons traveling to an international conference on AIDS in Melbourne, Australia, several distinguished scientists, a young rower from Indiana University, a pioneer in aerospace engineering, an international reporter who covered elections in Ukraine,[43] and 80 children—all of whom perished.[44]

Responsibility for investigating a crash is usually assigned to the state within which an incident occurs, according to the International Civil Aviation Organization. Ukraine initiated the probe but asked the Dutch Safety Board (DSB) to head the investigation because most of the victims were from the Netherlands.[45] The DSB formed a Joint Investigation Team (JIT) composed of representatives from the countries most impacted by the tragedy—the Netherlands, Ukraine, Malaysia, Australia, and Belgium.[46]

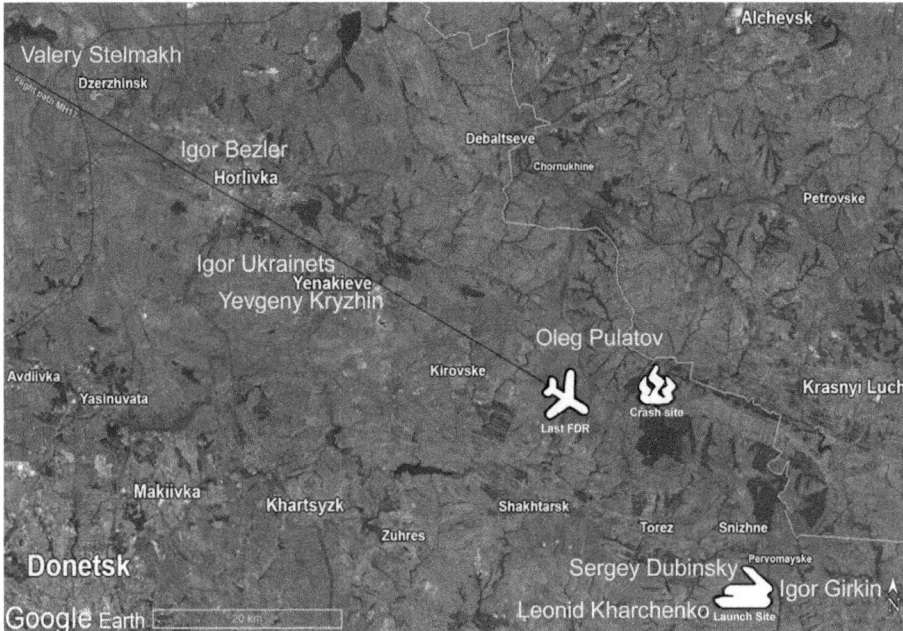

Figure 5.2. Flight Path and Last Flight Data Recorder Point of MH17, Buk Missile Site, and Main Crash

Significant evidence was found linking Russia to the downing of MH17. In October 2015, the DSB announced that its fifteen-month investigation confirmed that the missile was a Russian-made Buk missile.[47] In June 2016, the JIT published a photo of a part from a Buk missile found at the crash site. On September 28, 2016, the JIT released its report, which revealed the origin and site of the missile launch, confirmed the weapon type, and presented alternative hypotheses. JIT cited witness testimony, satellite imagery, photographs and intercepted phone conversations as its basis for confirming that the Buk missile launching platform was manufactured in Russia, transported from Russia to the Ukraine, and returned to Russia an hour after the plane was shot down.[48] The report identified the launch-site location as an agricultural field near Pervomaiskyi, Ukraine, which was controlled by Russian-backed fighters at the time of the incident.[49] The report further elaborated on the JIT statement earlier that year regarding the type of weapon and the methodology used to identify the missile series and system.[50]

The team explored alternative scenarios regarding the source and type of weapon. It ruled out speculation that the downing of the aircraft was a terrorist attack or caused by an explosion originating from inside the plane based on forensic evidence revealed in the investigation. For example, particles of the unique type of glass used for the cockpit windows in a Boeing 777 were found in the bodies of the cockpit crew during the postmortem investigation of the victims. Given that the glass shards were found inside the victims' bodies, the plane had to have been penetrated from the outside.[51]

Another hypothesis was that a military aircraft shot down the plane.[52] This scenario was ruled out by evidence collected from radar data, witness testimonies, and forensic data that indicated no other planes were in the air in eastern Ukraine at the time MH17 was struck down. According to a report issued by the research organization Bellingcat, Russian military veteran Igor "Strelkov" Girkin placed a post on his social media site within minutes of the shootdown saying, "we shot down a Ukrainian military plane." When it became apparent, however, that the plane was a commercial airliner, the post was quickly taken down.[53]

On May 24, 2018, the JIT held a press conference in Utrecht, Netherlands, revealing an image of the Buk TELAR rocket found at the site of the MH17 crash that showed the serial number printed on the rocket. At the conference, JIT announced that the missile belonged to the 53rd Anti-Aircraft Missile brigade, a Russian brigade based in the western Russian city of Kursk.[54] As a result of this evidence, both the Netherlands and Australia announced they were holding Russia accountable for shooting down the plane.

The JIT compiled a list of one hundred potential suspects who could be held responsible for the disaster. In June 2019, the JIT released the names of four men who it alleged were involved in transporting the missile from Russia to Ukraine and declared them primary suspects in the downing of MH17.[55] International arrest warrants issued by the JIT charged Igor Girkin, Sergey Dubinsky, Oleg Pulatov, and Leonid Kharchenko with the murder of 298 passengers and crew members. Girkin, Dubinsky, and Pulatov have served in Russian intelligence or have a history of employment with the Russian military. Kharchenko is a Ukrainian national who lacks a military background, unlike his co-conspirators. He, however, was serving at the time as the commander of a combat unit in eastern Ukraine.

Russia's Aggressive Counternarrative

Russian officials initially responded to the international community's accusations that Russia was responsible for shooting down MH17 by holding a press conference on July 21, 2014. At the press conference, Russia's Defense Ministry presented radar data that depicted another aircraft flying in the vicinity of MH17 moments before the plane was shot down.[56] The Russian Union of Engineers claimed that a thorough examination of the wreckage showed that the plane was shot down by heat-seeking air-to-air missiles.[57] The Russian media furthered this narrative by crediting the testimony of an unverified source alleged to be a Spanish air traffic controller in Kiev, Ukraine. The supposed witness claimed that MH17 was followed by two Ukrainian fighter jets. The testimony was later discredited.[58]

In August 2015, the Russian tabloid Komsomolskaya Pravda released an audio recording of two men impersonating US Central Intelligence Agency (CIA) "agents" who were conspiring to shoot down MH17.[59] The tape contains several linguistic inaccuracies such as improper syntax, awkward lexical and prosodic stresses, and other blatant conversational abnormalities, indicating that the men recorded are not native English speakers. The source of the video and the identities of the men in the recording were not provided by the tabloid, strongly suggesting that the Russian government was using one of its media outlets to spread propaganda and disinformation.

When the JIT revealed that the missile was Russian made and the launch platform had been transported from Russia to Ukraine and then back to Russia an hour after the crash, Russian officials admitted that the Buk missile system was manufactured by Russia in 1986.[60] They claimed, however, that the missile had been delivered to a Russian military unit in Ukraine and was never returned to Russia.

In September 2018, Moscow repudiated the international investigators' claims that assigned responsibility for the downing of the aircraft to Russian separatists.[61] At a press conference held by Russia's Ministry of Defense, officials asserted that the evidence of Russia's involvement in the crash was fraudulent. According to the Russian officials, the videos presented by international investigators showing the missile being transported from Russia to Ukraine were fake. They also claimed that Russia had evidence that the videos were falsified.[62]

Russia tried to deflect blame to Ukraine by insisting it had evidence of Ukraine's liability for the crash. Russia claimed to be in possession of an audio recording of a Ukrainian soldier professing to have shot down the plane. Moreover, JIT revealed in June 2019 that a Russian operative was sharing conspiracies on his Facebook profile that the Buk missile launcher was Ukrainian and that the downing was part of a deliberate false flag attack orchestrated by Ukraine, the United States, and Royal Dutch Shell.[63]

An analysis of Russia's response to allegations that it was responsible for the MH17 shootdown reveals a robust program of denial and deception. The Russians tried to reframe the narrative by fabricating a story, deflecting blame, and using details that could not be verified by an external investigation team or, at the very least, would be difficult and time-consuming to disprove. Russia's disinformation methodology can best be described as the **Sowing the Five Ds: Dismiss, Distract, Deflect, Distort, and Distrust**.[64]

- Russia tried to **dismiss** the allegations and deflect blame by refocusing attention and claiming it had evidence of Ukraine's involvement.
- The press conferences held two years after the accusations and four years after the crash served to **distract** the international media and the public from other more recent or ongoing crimes.
- Russian officials sought to **deflect** the international media and the public from other more recent or ongoing crimes.
- Russian officials attempted to **distort** the facts by stating that the missile launcher was never returned to Russia despite video evidence of the missile launcher's transport that had been released two years earlier.
- Russian propagandists tried to instill doubt and **distrust** in the investigation by dismissing the video footage as fake and propagating other Digital Disinformation.

Separatists' Disinformation Strategies

This disinformation strategy was also reflected in the methods used by members of the four separatist groups associated with the downing of MH17: the Donetsk People's Republic (DNR); the DNR's military intelligence unit; the Main Intelligence Administration (GRU) DNR; the Bezler Group and its Minyor Unit; and the Vostok Brigade (see Figure 5.3).[65]

The following Sensitive but Unclassified (SBU) phone intercept indicated that the Bezler group was involved:

NAEMNIK: Nikolaevich ...
BEZLER: Yes, Naemnik.
NAEMNIK: A birdman [misspeaks] ... a birdie is flying towards you.
BEZLER: Is a birdie flying towards us?
NAEMNIK: Yes ... [just] one, for now ...
BEZLER: A reconnaissance [aircraft] or a big one?
NAEMNIK: Can't see behind the clouds ... [it's flying] too high ...
BEZLER: I see ... roger ... report upwards.[66]

Even though the identities of these men were confirmed by international investigators, Bezler continued to deny any involvement with the downing of MH17. The Russian news

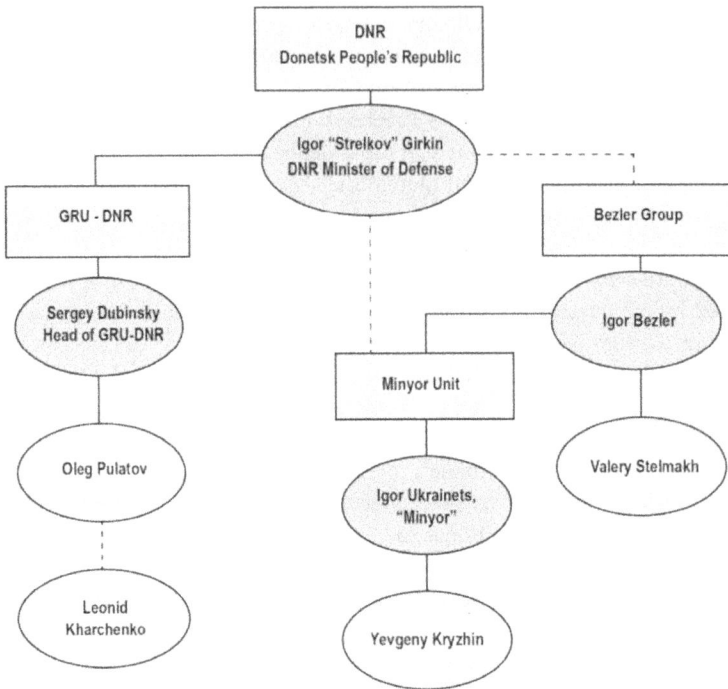

Figure 5.3. Association of Key Actors

agency RIA Novosti leaked a conversation involving Bezler, who defended himself using elements of the Five D's disinformation strategy. Bezler:

- **Dismisses** the link between the conversation and MH17 by claiming that the plane referenced in the phone intercept was a different plane.
- **Distorts** the facts by stating, "The Boeing fell in the area of Snizhne. There are 100 [kilo]meters between [Yenakieve and Snizhne], I don't have weapons capable of downing planes at such a distance."[67] The actual distance between the two cities is 43 kilometers.
- **Deflects** attention away from his group—saying he did not have the weapons capabilities—in an effort to distract from his connections with equipped militant groups and his role in spotting the plane.

Although the Bezler group may not have been involved with the actual downing of the plane, it was involved with the transportation of the Buk missile as evident from the video recordings later published in the JIT report in 2016.[68] In response to the video, Bezler employed the same disinformation strategy and dismissed the video in a post on his social media account: "I looked through the video, and at 01:31 I had to laugh and from then on I watched it as an animated fiction movie."[69]

The day after the downing of MH17, a member of the Minyor unit, Yevgeny Kryzhin, posted "five points" on his LiveJournal blog attesting to the Minyor Unit's innocence. In his post, Kryzhin:

- **Dismisses** allegations against his unit. Kryzhin asserts that the audio recordings of conversations between Minyor members are fake.
- **Distracts** from the Minyor unit's role in the crash by stating that the members of the Minyor group, heard on the SBU recording, were nowhere near the vicinity of the crash at the time.
- **Distorts** the truth. He claims that the group did not have the weapons capabilities required to perform this action.[70]
- **Deflects** and sows **distrust**. Kryzhin accuses the Ukranians of making fake audio recordings, addressing Ukraine directly saying, "So Ukry [Ukrainians] f*** off with your fakes."[71]
- **Deflects** and **distorts**. When confronted by a *Ukrainian LiveJournal* user, stating that the separatists have Buks. Kryzhin replied, "Only Strelkov has Buks."[72] The unit may not have had the weapons capabilities independently; however, they were working with the DNR, which did have the weapons capability. Furthermore, the Minyor unit was involved with the transport of the Buk.

The JIT countered this strategy by presenting persuasive contrary evidence, specifically by (1) showing that the missile that brought down the plane belonged to the 53rd Russian military brigade and (2) providing video showing that the missile was transported from Russia to Ukraine prior to the attack.

Targeting the Domestic Population

The weaponization of "internet-based social media platforms" as an information warfare method first emerged in the early 2000s. Russia used it initially to shape domestic public opinion and suppress political opposition.[73] Following the downing of MH17, the Russian government used Digital Disinformation both as a defensive tactic to counter the international community's accusations and to mold popular views of the incident.

Russia's first efforts to deflect blame for the downing of MH17 were largely directed internally, using social media and state media outlets to spread disinformation. Russian separatists were heard blaming each other on social media and blaming the Ukrainians on the LiveJournal networking site and in TV interviews on Russia Today (RT). Some separatists even took credit for downing the plane on their own social media accounts. The release of an unconvincing tape of Russian agents impersonating CIA agents by a Russian tabloid almost certainly was directed toward the Russian population because major linguistic imperfections were allowed to appear in the audio.

The Russian government insisted that the perpetrators were independent separatists, even though the key players involved in the downing of the plane had ties to the Russian military and the GRU. Disassociating the perpetrators of the shootdown from Russian military forces was central to a disinformation campaign designed to **deflect** blame, preserve domestic support for the Russian military efforts, and promote Russian nationalism.

CRANKING UP DIGITAL DISINFORMATION CAPABILITIES

Liberal democracies have been targeted by Russian/Soviet active measures and disinformation campaigns since the early 20th century.[74] Over the past decade, the Kremlin has increasingly leveraged social media to interfere in democratic elections and civic discourse by means of political and ideological manipulation of democracies. In Europe, Russian disinformation has worked its way into the social media feeds of voters in Germany, France, the Netherlands, the United Kingdom, and many other countries.

Russian state media outlets, for example, were used to inflame anti-immigrant and anti-refugee sentiment in Western Europe.[75] As stated in the SSCI report,

> We are in the midst of a world-wide internet-based assault on democracy. The Oxford Internet Institute has tracked armies of volunteers and "bots," or automated profiles, as they move propaganda across Facebook and Twitter in efforts to undermine trust in democracy or to elect favored candidates in the Philippines, India, France, the Netherlands, the United Kingdom, and elsewhere.[76]

PROMOTING AN ANTI-IMMIGRATION CAMPAIGN IN GERMANY

*Russian propaganda and disinformation tactics have been used to sow **distrust** of German institutions and migrants traveling to Europe while **distracting** attention to Russian shortcomings.*

In Germany, anti-immigrant rhetoric has been aimed at Muslims and refugees from Syria and other parts of the Middle East/North Africa region who began flooding into Europe beginning in 2015. Roughly 5 million (or 25 percent) of the world's refugee population are Syrian refugees.[77] Germany has accepted over 1 million Syrian refugees.[78]

Russia quickly sought to take advantage of the rising anti-immigration sentiments spurred by this influx. A news broadcast on Russia's main news channel falsely claimed that migrants were raping children in Germany. They said a young Russian-German girl named Lisa was kidnapped and raped by Syrian migrants to promote **distrust** of migrants. The segment continued with a blurry video that depicted an assumed Syrian migrant boasting about raping her.[79] It was later discovered that this video had been uploaded to YouTube six years earlier.

German police confirmed that the girl had been reported missing but eventually discovered that she had been at a friend's house.[80] The Russian foreign minister held a press conference directed toward a Russian audience where he discussed migration problems and referred to the girl in the story as "our girl Lisa."[81] In addition, Russian TV (RT) produced several segments focusing on the fear and concerns that Germany's Russian speakers had toward Syrian refugees, further provoking **distrust** of migrants. According to *German Deutsche Welle*, women who were filmed in these segments were later determined to be paid television actors.[82]

In addition to targeting Germans with anti-immigration content, Russia sponsored and promoted far-right and white-nationalist parties in Germany. Russia's weaponization of immigration in Germany was part of their larger disinformation strategy to sow **distrust** of Germany's democratic processes. Anxiety about immigration provided the fuel behind the rise of white nationalism and right-wing parties in Europe.[83]

Germany continues to be one of Putin's major targets in the Russian Information War. Russia has launched *RT Deutsch*, a German-language version of the Russian state-sponsored propaganda machine.[84] *RT Deutsch* presents the same narrative style and propaganda material that has notoriously been used by the Russian government in Kremlin disinformation campaigns to **distort** the news and **distract** people from reporting unfavorable to Russia.

TARGETING THE 2016 US PRESIDENTIAL ELECTIONS

*Russia is using increasingly sophisticated digital disinformation techniques to sow **distrust** in US institutions, social norms, and elective processes; **distort** facts to spur domestic conflict and **deflect** blame for interfering in the 2016 Presidential election.*

Over the past decade, the Russians have increasingly used disinformation methods, particularly the weaponization of social media, as an offensive tactic in their information warfare operations directed toward the United States. Technological advancements in artificial intelligence (AI), advertising, and the rapidly growing global reliance on social media networks and internet news sources were exploited aggressively. Russian Digital Disinformation campaigns reflecting the Five-D's strategy to **Dismiss, Distract, Deflect, Distort,** and **Distrust** have become increasingly sophisticated and destructive.

Several years before the 2016 US presidential election, Reuters reported that a Russian think tank controlled by the Kremlin, the Russian Institute for Strategic Studies (RISS), generated an elaborate plan to influence the US election campaign. The RISS reportedly recommended that the Kremlin launch a propaganda campaign via social media and that their "social media propaganda effort turn toward undermining faith in the American electoral system by spreading false stories of voter fraud."[85] It recommended that Russian state-backed global news outlets encourage US voters to elect a president "who would take a softer line toward Russia."[86]

Leading up to the US presidential election in 2016, Russian operatives affiliated with the Internet Research Agency (IRA) used American social media platforms to spread disinformation and divisive propaganda.

In December 2018, the Senate Select Committee on Intelligence released two studies on Russian efforts to interfere in the 2016 Presidential election campaign and followed the reports with another in October 2019.

- The first stated that Russian disinformation campaigns targeted societal fractures "to divide Americans by race, religion, and ideology," actively "eroding trust in our democratic institutions."[87] US Senator Mark Warner, the panel's senior Democrat, described the effort to "undermine and manipulate our democracy" as "much more comprehensive, calculating, and widespread than previously revealed."[88]
- The second report stated that "Russia-based actors, at the direction of the Russian government, effectuated a sustained campaign of information warfare against the United States aimed at influencing how this nation's citizens think about themselves, their government, and their fellow Americans."[89]

The studies describe a multi-pronged Russian approach on Facebook, Twitter, Instagram, and YouTube to suppress the vote through the use of malicious misdirection such as "text to vote" scams; candidate support redirection such as voting for a third party; and depressing turnout by texting "stay at home, your vote does not matter."[90]

- During the 2016 election campaign, Russia's social media campaign reached 126 million people on Facebook, 10.4 million tweets on Twitter, 20 million postings on Instagram, and uploaded more than 1,000 videos on YouTube.[91]
- Instagram posts by Russia's Internet Research Agency (IRA) received 187 million engagements, and Facebook posts had over 76 million engagements.[92]

- In the final days leading up to the 2016 US Presidential election, junk news was shared in the battleground state of Michigan just as widely as professional news.[93]

Russia's government has denied involvement with the IRA's interference in the 2016 US election; however, the SSCI found "significant Kremlin support, authorization and direction of the IRA's operations and goals."[94] Furthermore, a Russian oligarch known to be a close comrade of President Putin, Yevgeny Prigozhin, funneled millions of dollars to IRA operations devoted to engaging with US citizens on social media in 2016.[95]

Just as the Russian government attempted to disassociate itself from the Russian militants who shot down MH17, the Kremlin has continuously dismissed any blame for or associations with the IRA's operations and the actions of the Russian operatives that impeded US democratic processes by exploiting and manipulating American civic discourse.

Russian trolls targeted Americans by interacting with US citizens through various sophisticated disinformation methods. The goal of this divisive propaganda was to intensify racial, social, and political cleavages within the US population.

Highly skilled Russian disinformationists programmed and deployed bots that repeatedly posted socially divisive and politically polarizing content on US social media sites. They posted inflammatory and provocative comments to engage US citizens and created and shared distorted content on these platforms. The Russian operatives created fake social media accounts that targeted Americans based on sociopolitical demographic factors to influence public opinion on "hot-button" political topics such as race and immigration.[96]

Sixty-six percent of all Facebook advertising content created by the IRA included topics and phrases pertaining to race.[97] The majority of racially divisive content created by the IRA was targeted at African Americans.[98] The IRA created Facebook pages and Instagram, Twitter, and YouTube content geared toward sowing **distrust** in African American audiences.[99]

Seeking in particular to target African Americans living in metropolitan areas, Russian trolls used locational targeting methods. Blacktivist, one of the IRA's Facebook pages, received 11.2 million engagements, and half of its top ten Instagram accounts included racial content that was targeted toward African American audiences.[100]

The 2019 Senate report "found that no single group of Americans was targeted by IRA information operatives more than African Americans. A separate report analyzed the content shared by trolls and found that conservative trolls focused content on "refugees, terrorism, and Islam; while liberal trolls talk more about school shootings and the police."

Russia's use of social media to sow **distrust** and undermine democracy appears to have continued. A CNN investigation discovered that in late 2019 Russia set up a troll factory in a residential community one hour outside of Accra, Ghana. At least 16 employees worked at the site, which was raided and shut down by the police in February 2020. The operators, presenting themselves as the non-profit, Eliminating Barriers for the Liberation of Africa (EBLA), set up hundreds of accounts on Facebook, Twitter and Instagram. The plan was to stir up social unrest in the United States, focusing on alleged racism and police brutality. CNN further reported that a similar trolling operation was established in Nigeria and more may be ramping up across Africa.[101]

Targeting of US Presidential Candidates

A major component of Russia's information warfare strategy to interfere with the 2016 US election and democratic processes was the creation and dissemination of disinformation about the 2016 US presidential candidates. Disinformation was dispatched through social media by

Russian trolls about the US presidential candidates, the Democratic National Committee (DNC), and other US politicians. The Senate Select Committee on Intelligence found that IRA trolls also targeted and disparaged Republican candidates in the primaries including Jeb Bush, Marco Rubio, and Ted Cruz. The trolls' goal was to discredit and **dismiss** any candidate with views and interests counter to those of the Russian government.

On July 22, 2016, three days before the Democratic national convention, WikiLeaks released 19,252 DNC emails.[102] This release and the persistent stream of continuing leaks became a major **distraction** for Hillary Clinton's campaign. Russian GRU hackers, operating under the front "Guccifer 2.0," sent WikiLeaks thousands of DNC emails that they had retrieved by hacking into the DNC's email server.[103]

According to the FBI investigation report, Russian hackers with financial ties to the IRA had roamed through the DNC's network for seven months prior to the release of the hacked messages.[104] In addition, the Russian operatives hacked into Clinton campaign chairman John Podesta's email and delivered hundreds of his emails to WikiLeaks.[105]

During the critical final months of the campaign, Buzzfeed reported that the twenty top-performing false election stories from hoax sites and hyper-partisan blogs generated 8,711,000 shares, reactions, and comments on Facebook.[106] Efforts to sow **distrust** in US political institutions accelerated. According to the SSCI report, Hillary Clinton was especially targeted by Russian Digital Disinformation campaigns to "undermine public faith in the US democratic process, denigrate Secretary Clinton, and harm her electability and potential presidency."[107]

Reinforcing these efforts were other social media posts designed to **deflect**. Following the WikiLeaks release of hacked materials, one fake news story that emerged, commonly known as "Pizza-gate," asserted that one of Podesta's leaked emails revealed records of a child-sex trafficking ring that was, according to this rumor, run by Hillary Clinton out of the basement of a pizza parlor in Washington DC.[108]

The story, which originated on Twitter, was quickly picked up by rightwing entertainment programs such as InfoWars and Breitbart. The story was shared about 1.4 million times by more than 250,000 accounts in the first five weeks following its release.[109] In addition, more than 3,000 accounts tweeted about Pizza-gate five times or more. Experts have concluded that automated bots were largely behind the dissemination of this false narrative.

In response to the false story, in December 2016 a man in North Carolina drove to the pizza parlor in Washington, DC, armed with an AR-15 semiautomatic rifle, a .38 handgun, and a folding knife.[110] He entered the pizza shop determined to "free the abused children he was convinced were being held in the nonexistent basement."[111]

A Twitter account operated by Russian trolls in St. Petersburg used another form of **distraction** by posing as the Tennessee Republican Party with the name @TEN_GOP. It posted a manipulated version of one of the Podesta emails on Twitter that attempted to incriminate Hillary Clinton for the Islamist-militant attacks on the US consulate in Benghazi that resulted in the deaths of four US officials.[112]

The Russian news agency Sputnik published an article regurgitating the @TEN_GOP story and used the fake document as the basis for its news article.[113] Sputnik's retrieval and dissemination of this false narrative provides a good example of Russia's asymmetric information warfare operation. Sputnik's aggressive manipulation of the Benghazi event demonstrated how effectively Russia applied the Five Ds to denigrate Hillary Clinton, undermine her credibility, and create overall distrust in the US Government and its processes. The incident was intended to sow **distrust** of Clinton within the American population.

The US House of Representatives formed a select committee to investigate the validity of these accounts relating to the Benghazi incident. After two years of research and hearings and an expenditure of $7 million, the committee released an 800-page report concluding that no evidence could be found indicating that Clinton was culpable regarding the Benghazi attack.[114]

In the final three months leading up to the November election, the number of Digital Disinformation engagements on Facebook pertaining to the election outperformed the number of postings on legitimate news websites.[115] The two most widely shared distortions of information about the two candidates were the claim that Pope Francis had endorsed Donald Trump's campaign and that Hillary Clinton had sold weapons to the ISIS terrorist group in Syria.[116]

SUPPORTING THE 2016 UK PRO-BREXIT CAMPAIGN

*Russia's use of Digital Disinformation to generate **distrust** of immigrants, **dismiss** arguments for remaining in the EU, and **distort** the economic benefits of leaving the EU may have influenced enough voters to tip the scales in support of the vote to leave.*

When the United Kingdom (UK) announced the result of the Brexit vote was to "Leave" the European Union (EU), the world was shocked, similar to the world's reaction to the result of the subsequent US 2016 presidential election. According to polls of UK voters, "Leave voters were motivated mainly by anti-immigration sentiments." They indicated that while younger city dwellers voted to stay, older citizens living in rural areas who were less exposed to immigrants voted to leave.[117]

In the six months leading up to the Brexit vote, Russian media outlets—RT and Sputnik—ran 261 stories that were anti-EU, pro-Brexit, and anti-immigration.[118] Hundreds of Russian bots and trolls from the IRA were posting about Brexit on thousands of Twitter accounts.[119] The objective was to **dismiss** arguments highlighting the unfavorable consequences of leaving the EU, **distract** attention from concrete pros and cons by emphasizing the threat posed by newly arriving immigrants, and **distort** the economic benefits of leaving the EU.

The House of Commons Digital, Culture, Media and Sport Committee in its final report on Disinformation and "Fake News" noted that "The social reach of these anti-EU articles published by the Kremlin-owned channels was 134 million potential impressions, in comparison with a total reach of just 33 million and 11 million potential impressions for all content shared from the Vote Leave website and Leave.EU website, respectively."[120] It was the same anti-immigration, pro-nationalist messaging that the Kremlin-backed Russian media used to permeate German civic discourse and impact the 2016 US election.

In 2017, the Information Commissioner's Office in the UK launched an investigation into Facebook's use of UK citizens' data and its role in the 2016 Brexit referendum and 2017 UK national elections.[121] Facebook's head of cybersecurity announced that Facebook had removed 289 pages, 75 accounts with roughly 790,000 followers, after linking this activity to Sputnik.[122] According to Facebook's findings, these pages frequently posted anti-NATO content and promoted protest movements through Facebook. In addition, Facebook "removed 107 pages, groups, and accounts that were designed to look as if they were run from Ukraine but were part of a network that originated in Russia."[123]

FALLING SHORT IN DISRUPTING THE 2017 FRENCH ELECTION

*Russian efforts to sow **distrust** of presidential candidate Emanuel Macron were less successful because the French had developed an effective counter strategy.*

Months before the final round of the French presidential election in 2017, Russian operatives launched a disinformation campaign against Emmanuel Macron's presidential bid. Two days before the presidential election, these actors released gigabytes of hacked data, including from the Macron campaign team. The incident came to be known as the "Macron Leaks." The data consisted of emails and forgeries created to denigrate Emmanuel Macron.

The objective was to release large quantities of data that would distort Macron's record and sow **distrust** in his leadership skills. Russia's attempt to impact the outcome of the French presidential election and polarize French society, however, largely failed.

French efforts to counter the disinformation attacks were successful due to several factors:

- **Growing Sensitivity.** Previous Russian disinformation campaigns in the United States and the UK helped make the French aware of the growing threats of cyber attacks and disinformation campaigns directed toward transatlantic democracies, the degree of influence these campaigns have had on democratic civic discourse, and the extent to which they are being used to undercut the security of transatlantic partnerships and interests.
- **Timing.** France's 2017 election was a year and a half after the US election, two years after Brexit, and three years after Russian cyber attacks on German intelligence servers. This enabled France to prepare a counterresponse. As expected, the hackers released their disinformation just hours before the electoral silence period. The timing limited the spread of disinformation, and the poor quality of the reporting rendered the attempt highly suspicious.
- **Structural Factors.** The French voting system posed a more challenging target:

  a   The election process is direct; therefore, interference is more obvious.
  b   Two rounds of voting enable voters to shift their decisions.
  c   The French media is dominated by mainstream and critical media sources, and therefore lacks the tabloid-style and alternative websites, common in the UK and the United States.
  d   Critical thinking habits and moderate skepticism are embedded in French culture and society.

- **Active Countermeasures.** Russian disinformation operatives did not anticipate that Macron's campaign staff would react so quickly. Because the quality of their disinformation was deemed irrelevant in their previous campaigns, the Russians assumed their same mistakes would not make any difference. For Russian trolls and hackers, quantity—not thoroughness or the quality of disinformation—is the goal.
- **Cultural Ignorance.** Disinformation was primarily in English and first spread by the American alt-right community. Russian hackers were unaware of ideological differences in French nationalism, especially pertaining to the historical significance of French-language use in media and literature and its association with the French nationalist movement.

Several measures were taken by the French government and the Macron campaign to counter Russian attempts at election interference and disinformation attacks. The French:

- **Learned from history.** The US 2016 election was the pivotal learning point for democracies around the world. After witnessing the impact of Russian disinformation on the US 2016 election, other democracies realized that they too were vulnerable to Russian disinformation.
- **Had governing bodies in place to regulate.** The National Commission for the Control of the Electoral Campaign for the Presidential Election (CNCCEP) and the National Cybersecurity Agency (ANSII) were trusted governmental bodies that were designated to preserve the integrity and public confidence in electoral processes and results.[124] These agencies kept the public, the media and political parties informed when there was a risk of cyber attacks during the campaign.
- **Communicated a firm resolve.** In December 2016, the minister of defense officially announced France's new cyber command operation would consist of 2,600 cyber-warfare combatants. In addition, the minister stated that "France reserves the right to retaliate by any means it deems appropriate … through our cyber arsenal but also by conventional armed means."[125]
- **Took technical preventative measures.** In addition to establishing the cyber command in March 2017, the French government announced that electronic voting would not be available to citizens abroad.
- **Confused the hackers with their own game.** The Macron campaign mounted an aggressive retaliatory response to Russian phishing attempts. It used a strategy known as digital blurring and planted false information to confuse and distract the hackers.

CONCLUSION

In conclusion, Russia has developed over the past decades a robust capability to manipulate popular perceptions. The impact of such campaigns appears to have been greatly accelerated by the recent rapid growth of social media platforms. Russia has also significantly expanded the number of countries targeted by its Digital Disinformation efforts. The French experience and the West's response to Russia's disinformation campaign accompanying its invasion of Ukraine, however, suggest that as more countries come to anticipate such attacks, they will be able to mount more effective defences. They will also be able to develop what could potentially be even more disruptive offensive strategies.

Digital Disinformation emanating from Russia will continue to pose a significant challenge to the West. Indeed, a 2017 US Senate hearing entitled "Disinformation: A Primer in Russian Active Measures and Influence Campaigns" included prescient testimony from several expert witnesses who provided insights into the mechanics of Russian influence operations. The report warned that Russian social media manipulation "has not stopped since the [US 2016] election and continues fomenting chaos amongst the American populace."[126]

As the public and its elected representatives learn more about the degree to which popular perceptions are being manipulated by Russian Digital Disinformation, pressure to better protect individuals—and society writ large—against targeted disinformation campaigns will grow. Debate will focus on who to blame: governments or online service providers such as Facebook, Twitter, Instagram or Tic Tok. Most likely blame will be placed on all. The incentives for implementing corrective strategies, however, are likely to vary considerably.

As more people migrate their media consumption to online and social media, a failure to address the proliferation and adoption of false narratives will undoubtedly have corrosive effects on the West's ability to operate and govern. Credible studies on this matter consistently proclaim that an effective solution require whole-of-society initiatives in which consumers, corporations and governments all play a role in developing societal resilience. Solutions will require not only organizational directives from the top but individual initiatives from the bottom. That has been vividly demonstrated in the Western response to the Russian invasion of Ukraine.

Notes

1 Edmondson, "GOP Senators," *New York Times.*
2 Otis, "Crisis in Ukraine," *CTV News.*
3 Dobrowolski, "Weapons of Mass Distraction?" *Citizenship as a Regime,* 186–207.
4 Wardle and Derakhshan, "Information disorder" *Council of Europe Report 27.*
5 Pherson and Mort Ranta, "Cognitive Bias, Fake News"
6 Wardle and Derakhshan. "Information disorder" *Council of Europe Report 27.*
7 Woolley and Howard, "Automation, Algorithms, and Politics," 9.
8 IAMCR, "Fake News and Digital Disinformation"
9 *PBS News Hour,* "Inside Russia's Propaganda Machine"
10 Zadeskey, "Transnistria: The History Behind the Russian-backed Region," *Origins*
11 Ibid.
12 Ibid.
13 Lautman, Olga, "Darkness on the Dniester," *CEPA*
14 Ibid.
15 Ibid.
16 Zadeskey, "Transnistria: The History Behind the Russian-backed Region," *Origins*
17 Cathcart, "FOR THE LIFE OF LECH KACZYNSKI," *Georgian Journal*
18 Ibid.
19 *Smolensk Crash News Digest,* "Polish Air Crash Disinformation & Active Measures Campaign with Examples"
20 Cathcart, "FOR THE LIFE OF LECH KACZYNSKI," *Georgian Journal*
21 *Smolensk Crash News Digest,* "Polish Air Crash Disinformation & Active Measures Campaign with Examples"
22 Cathcart, "FOR THE LIFE OF LECH KACZYNSKI," *Georgian Journal*
23 *Smolensk Crash News Digest,* "Polish Air Crash Disinformation & Active Measures Campaign with Examples"
24 Cathcart, "FOR THE LIFE OF LECH KACZYNSKI," *Georgian Journal*
25 Ibid.
26 *Smolensk Crash News Digest,* "Polish Air Crash Disinformation & Active Measures Campaign with Examples"
27 Cathcart, "FOR THE LIFE OF LECH KACZYNSKI," *Georgian Journal*
28 *Smolensk Crash News Digest,* "Polish Air Crash Disinformation & Active Measures Campaign with Examples"
29 Ibid.
30 Ibid.
31 Ibid.
32 Ibid.
33 Ibid.
34 Alami, Mona, "Russia's Disinformation Campaign has changed how we see Syria," *SyriaSource,* September 4, 2018. https://www.atlanticcouncil.org/blogs/syriasource/russia-s-disinformation-campaign-has-changed-how-we-see-syria/
35 Ibid.
36 Ibid.
37 Ibid.
38 Ibid.
39 *Asharq Al-Awsat,* "Russian-Backed Campaign Fueling Disinformation Revealed in Syria,"
40 Ibid.
41 Ibid.

42 BBC News "MH17 Ukraine Plane Crash"
43 Patterson, Sanchez, and Stapleton, "MH17 Crash Victims"
44 *BBC News*, "MH17 Ukraine Plane Crash"
45 Ibid.
46 Parket, "JIT: Flight MH17 Shot Down by BUK"
47 *BBC News*, "MH17 Ukraine Plane Crash"
48 *BBC News*, "MH17 Crash: Buk Missile Part Found"
49 Parket, "JIT: Flight MH17 Shot Down by BUK"
50 Ibid.
51 Ibid., animation 2
52 Ibid.
53 Maza, "Russian Propaganda? Moscow Blames Ukraine," *Newsweek*.
54 *BBC News*, "MH17 Ukraine Plane Crash"
55 Ibid.
56 Standish, Reid. "Propaganda Watch Russians Impersonate CIA," *Foreign Policy*.
57 Ibid.
58 Ibid.
59 Ibid.
60 *BBC News*, "MH17 Ukraine Plane Crash"
61 Maza, "Russian Propaganda? Moscow Blames Ukraine," *Newsweek*.
62 Ibid.
63 Van Huis, "A Birdie Is Flying Towards You," *Bellingcat*.
64 *PBS News Hour*, "Inside Russia's Propaganda Machine"
65 Van Huis, "A Birdie Is Flying Towards You," *Bellingcat*.
66 Ibid.
67 Ibid, 19.
68 Ibid.
69 Ibid.
70 Ibid, 11.
71 Ibid, 12.
72 Ibid, 11.
73 US Congress, SSCI, "Russian Active Measures Campaigns"
74 *SSCI*, US Congress, "Russian Active Measures Campaigns"
75 Vaidhyanathan, "The Disinformation Machine," 187–188.
76 Vaidhyanathan, "The Disinformation Machine," 180.
77 Stengel, "Information Wars," 223.
78 Ibid.
79 Ibid, 225.
80 Meister, Stefan. "The 'Lisa Case:' Germany as a target of Russian Disinformation," NATO *Review*.
81 Stengel, "Information Wars"
82 Ibid.
83 Ibid, 208.
84 Ibid, 228.
85 Vaidhyanathan, "The Disinformation Machine," 177.
86 Vaidhyanathan, "The Disinformation Machine," 175–195.
87 DiResta, et al. "The Tactics & Tropes of the Internet Research Agency," *New Knowledge*.
88 *SSCI Press Release*, "New Reports on Internet Research Agency's Social Media Tactics."
89 US Congress, SSCI, "Russian Active Measures Campaigns," 4.
90 DiResta, et al. "The Tactics & Tropes of the Internet Research Agency," *New Knowledge*.
91 Ibid.
92 Ibid.
93 Kollanyi et al. "Bots and Automation over Twitter during the U.S. Election."
94 *SSCI*, US Congress, "Russian Active Measures Campaigns," 5.
95 Stone and Gordon, "Russia-Sponsored Troll Networks Targeting U.S."
96 *SSCI*, US Congress, "Russian Active Measures Campaigns"
97 Ibid.
98 Ibid.

99 Ibid.
100 Ibid.
101 Ward, "Inside a Russian Troll Factory in Ghana," *CNN World*.
102 Stengel, "Information Wars"
103 Ibid.
104 Ibid.
105 Ibid.
106 Silverman, "How Viral Fake Election News Outperformed Real News," *BuzzFeed News*.
107 Ibid.
108 Ibid.
109 Vaidhyanathan, "The Disinformation Machine," 185.
110 Amanda Robb, "Anatomy of a Fake News Scandal," *Rolling Stone*.
111 Ibid.
112 Eichenwald, "Keep an eye on Sputnik" *Twitter*
113 Ibid.
114 Gambino and Smith, "House Benghazi Report Faults Military Not Clinton," *The Guardian*.
115 Silverman, "How Viral Fake Election News Outperformed Real News," *BuzzFeed News*.
116 Ibid.
117 Ibid, 235.
118 *House of Commons*, "Disinformation and Fake News"
119 Stengel, "Information Wars," 225.
120 House of Commons, "Disinformation and Fake News," 70.
121 Vaidhyanathan, "The Disinformation Machine," 180.
122 *House of Commons*, "Disinformation and Fake News," 70.
123 Ibid.
124 Conley and Vilmer, "Successfully Countering Russian Electoral Interference," *CSIS*.
125 Ibid.
126 *SSCI*, US Congress, "Primer in Russian Active Measures and Influence Campaigns"

Bibliography

Alami, Mona, "Russia's Disinformation Campaign has changed how we see Syria," *SyriaSource*, September 4, 2018.

Al-Awsat, Asharq "Russian-Backed Campaign Fueling Disinformation Revealed in Syria," London, June 20, 2022. https://english.aawsat.com/home/article/3713461/russian-backed-campaign-fueling-disinformation-revealed-syria.

BBC News, "MH17 Crash: Big Buk Missile Part Found in Ukraine," June 6, 2016, https://www. bbc.com/news/world-europe-36462853.

BBC News, "MH17 Ukraine Plane Crash: What .We Know," February 26, 2020, https://www.bbc.com/news/world-europe-28357880.

Brown, Chris, "Putin Was 'Good' and Obama Was 'Bad': Former Russian Trolls Reveal Online Work to Create 'Fake News'," *CBC News*, March 7, 2018. https://www.cbc.ca/news/world/russia-trolls-internet-fake-news-1.4562526.

Cathcart, Will, "FOR THE LIFE OF LECH KACZYNSKI," *Georgian Journal*, February 3, 2011. https://georgianjournal.ge/politics/3259-for-the-life-of-lech-kaczynski.html.

Conley, Heather A., and Jean-Baptiste Jeangène Vilmer, "Successfully Countering Russian Electoral Interference," *Center for Strategic and International Studies*, June 21, 2018, https://www.csis.org/analysis/successfully-countering-russian-electoral-interference.

DiResta, Renee, Dr. Kris Shaffer, Becky Ruppel, David Sullivan, Robert Matney, RyanFox, Dr. Jonathan Alright, Ben Johnson, "The Tactics & Tropes of the Internet Research Agency," *New Knowledge*, December 2018.

Dobrowolski, Alexandra. 2018. "'Weapons of Mass Distraction'?" *Citizenship as a Regime*, 186–207. https://doi.org/10.2307/j.ctv2n7qxg.13.

Edmondson, C. December 2019. "GOP Senators, Defending Trump, Embrace Debunked Ukraine Theory," *New York Times*. https://www.nytimes.com/2019/12/03/us/politics/republicans-ukraine-conspiracy-theory.html.

Eichenwald, Kurt. (@kurteichenwald), "1. Before the year is out….the evidentiary link between Wikileaks, Russian hackers, Kremlin disinformation outlet, and Donald Trump. In 2016, I received a phone call from someone in the intelligence world telling me to keep an eye on Sputnik," Twitter, December 31, 2018, 4:21 p.m., https://twitter.com/ kurteichenwald/status/1079850037088342017.

Gambino, Lauren and David Smith, "House Benghazi Report Faults Military Response, Not Clinton, for Deaths," *Guardian*, June 28, 2016, https://www.theguardian.

House of Commons, Digital, Culture, Media and Sport Committee, *Disinformation and 'Fake News': Final Report*, Eighth Report of Session, 2017–2019, February 14, 2019, 70. https://publications.parliament. uk/pa/cm201719/cmselect/cmcumeds/1791/1791.pdf

International Association for Media and Communication Research (IAMCR), https://iamcr.org/clearing house/challenges_of_digital_disinformation.

Kollanyi, Bence, Philip N. Howard, and Samuel C. Woolley. "Bots and Automation over Twitter during the U.S. Election." Data Memo 2016. 4. Oxford, UK: *Project on Computational Propaganda*. www.poli ticalbots.org.

Lautman, Olga, "Darkness on the Dniester," *CEPA*, May 3, 2022, https://cepa.org/darkness-on-the-d niester/.

Maza, Cristina, "Russian Propaganda? Moscow Releases Audio Blaming Ukraine for Downing of MH17 Flight That Killed Almost 300," *Newsweek*, September 17, 2018, https://www.newsweek.com/russia n-propaganda-moscow-releases-audio-blaming-ukraine-downing-mh17-flight-1124371.

Meister, Stefan. "The 'Lisa Case:' Germany as a target of Russian Disinformation," *NATO Review*, 25 July, 2016. https://www.nato.int/docu/review/articles/2016/07/25/the-lisa-case-germany-as-a-target-of-russian-disinformation/index.html.

Otis, Daniel. "Crisis in Ukraine: To Russia, it's a 'special military operation' not 'war'", *CTV News*, March 8, 2022. https://www.ctvnews.ca/world/crisis-in-ukraine-to-russia-it-s-a-special-military-operation-not-war-1. 5811520.

Patterson, Thom, Ray Sanchez, and AnneClaire Stapleton, "MH17 Crash Victims: Athlete, Young Brothers, Family Among Those Killed," *CNN*, July 23, 2014, http://www.cnn. com/2014/07/18/ world/mourning-mh17/index. html.

PBS News Hour, "Inside Russia's Propaganda Machine," July 11, 2017, https://www.youtube. com/wa tch?v=xSIkkza9TVI. This list of the Five D's is inspired by Atlantic Council senior fellow Ben Nimmo's concept of "The Four D's: Dismiss, Distort, Distract, and Dismay." The expanded version drops Dismay and adds Deflect and Distrust.

Parket, Landelijk, "JIT: Flight MH17 Was Shot Down by a BUK Missile from a Farmland Near Pervomaiskyi," *Openbaar Ministerie*, September 28, 2016, http://adam.curry.com/art/1475112492_ bD4NDhxj.html.

Pherson, Randolph H. and Penelope Mort Ranta, "*Cognitive Bias, Fake News, and Structured Analytic Techniques*." Paper presented at the Intelligence and the Knowledge Society (IKS) Conference, Bucharest, Romania, October 17–19, 2018.

Robb, Amanda. "Anatomy of a Fake News Scandal," Rolling Stone, November 16, 2017, https://www. rollingstone.com/politics/politics-news/anatomy-of-a-fake-news-scandal-125877/.

Senate Select Committee on Intelligence (SSCI)Press Release, *New Reports Shed Light on Internet Research Agency's social Media Tactics*, 17 December 2018. https://www. intelligence.senate.gov/press/new-rep orts-shed-light-internet-research-agency%E2%80%99s-social-media-tactics.

Senate Select Committee on Intelligence, US Congress, *Russian Active Measures Campaigns and Interference in the 2016 U.S. Election*, vol. 2, "Russia's Use of Social Media with Additional Views," 116th Congr., 1st sess., 2019, 15, https://www. intelligence.senate.gov/sites/default/files/documents/ Rep ort_ Volume2.pdf

Senate Select Committee on Intelligence, "*Disinformation: A Primer in Russian Active Measures and Influence Campaigns*," Panel I, March 30, 2017. https://www.intelligence.senate.gov/sites/default/files/hearings/ S%20Hrg%20115-40%20Pt%201.pdf.

Silverman, Craig. "This Analysis Shows How Viral Fake Election News Stories Outperformed Real News on Facebook," *BuzzFeed News*, November 16, 2016, https://www.buzzfeednews.com/article/craigsil verman/viral-fake-election-news-outperformed-real-news-on-facebook.

Smolensk Crash News Digest, "Polish Air Crash Disinformation & Active Measures Campaign with Examples: Highlights of desinformation and active-measures campaign in mass media after the crash," April 9, 2022. https://www.smolenskcrashnews.com/polish-air-crash-disinformation.html.

Standish, Reid. "Propaganda Watch: Listen to Two Russians Badly Impersonate CIA Spies to Pin MH17 on U.S," *Foreign Policy*, July 30, 2019, https://foreignpolicy.com/2015/08/12/propaganda-watch-listen-to-two-russians-badly-impersonate-cia-spies-to-pin-mh17-on-u-s/.

Stengel, Richard, *Information Wars: How We Lost the Global Battle Against Disinformation and What We Can Do About It.* Atlantic Monthly Press, 2019.

Stone, Peter, and Greg Gordon, "Russia-Sponsored Troll Networks Targeting the U.S. May Number in the Hundreds," *McClatchy DC*, October 19, 2017, https://www.mcclatchydc.com/news/nation-world/world/article179799311. html.

Vaidhyanathan, Siva, "The Disinformation Machine," in *Anti-Social Media: How Facebook Disconnects US and Undermines Democracy* (Oxford: Oxford University Press, 2018), 177.

Van Huis, Pieter, "'A Birdie Is Flying Towards You': Identifying the Separatists Linked to the Downing of MH17," *Bellingcat*, June 2019, 42, https://www.bellingcat.com/wp-content/ uploads/2019/06/a-birdie-is-flying-towards-you.pdf.

Ward, Clarissa, "Inside a Russian Troll Factory in Ghana," *CNN World*, 12 March, 2020https://www.cnn.com/2020/03/12/world/russia-ghana-troll-farms-2020-ward/index.html.

Wardle, Claire, and Hossein Derakhshan. "Information disorder: Toward an interdisciplinary framework for research and policy making, "*Council of Europe, Report*," 27, 2017.

Woolley, Samuel C., and Philip N. Howard, "Automation, Algorithms, and Politics/Political Communication, Computational Propaganda, and Autonomous Agents—Introduction," *International Journal of Communication*, 10(0), 9.

Zadeskey, William, "Transnistria: The History Behind the Russian-backed Region," *Origins: Current Events in Historical Perspective*, OSU.edu, July, 2022 https://origins.osu.edu/read/transnistria-history-behind-russian-backed-region?language_content_entity=en.

6

INFLUENCE OPERATIONS AND THE ROLE OF INTELLIGENCE

Jan Goldman

Understanding how influencing operations target intelligence must begin with understanding how information can be weaponized. If so, why should it be used, and how should it be used as a weapon of warfare? This is because intelligence is defined by understanding an adversary's capability and intention to use it. More specifically, a threat is an aggressive intention to employ such a capability with an action whose consequences will be detrimental to another country or target.[1] Intelligence is a collection of information derived from the possibility of this threat. Thus, intelligence is "information about information to combat a perceived threat."[2] The danger of influencing operations can come in two forms: physical intrusion and content intrusion. Intelligence agencies must be aware of both since, to be effective, they must complement each other to influence operations. Finally, intelligence operations (i.e., the collection and analysis of information) must understand the potential of an adversary's capabilities. Intelligence, from this viewpoint, is the basis of understanding human actions and, through an understanding of all aspects of the audience an adversary may want to target. This requires the ability to gauge the profess of adversary programs and their capability to determine their effectiveness in providing persuasive communication.

FOREIGN INFLUENCE OPERATIONS

On February 24, 2022, the Russian military invaded Ukraine. Soon after, the Russians began an influencing operation to overwhelm the "information-sphere" as a much larger hybrid warfare strategy.[3] This was not a new tactic but something the Russians did eight years earlier when the Kremlin began interfering with Ukraine after a popular uprising toppled Russian-aligned President Viktor Yanukovych. The Russian campaign to keep President Yanukovych in office became ineffective. Russia's military spy agency, the GRU, launched a covert influence operation (a tactic that would be repeated against the American presidential campaign two years later.) However, the Kremlin's influence campaign "created a slew of fake personas on the social media platforms of Facebook and its Russian equivalent VKontakte (VK)." The personas represent ordinary people from across Ukraine who are disillusioned with opposition protests at Kyiv's central square, called the Maidan.[4]

DOI: 10.4324/9781003190363-9

The campaign was part of an all-out propaganda offensive against the new government in Kyiv and pro-Western demonstrators. Its goal was to influence key decision-makers and the wider public to pave the way for the Russian military action launched on February 27 with armed men seizing the Crimean parliament building.[5] Russia ultimately annexed Crimea. But this time, the Ukrainian internet infrastructure was resilient.

Russia could not use its influence campaign, and it was reduced to several attacks against Ukraine's power grid and ultimately unleashing a wiper tool called "NotPetya."[6] This tool was disguised as ransomware in Ukraine's financial sector in 2017. Since Russia couldn't influence the information sphere, it would shut down its system and end all communications. To understand the influence of a massive cyberattack on the scale of NotPetya, it is essential to realize that this type of operation can shut an entire city down, making intelligence collection almost impossible to gather. As the Ukraine Navy sought to operate along its border, this cyber worm was released in a medium where national borders have no meaning.[7] NotPetya was eating away at Ukraine's computers, including at least four hospitals in Kyiv alone, six power companies, two airports, more than 22 Ukrainian banks, ATMs and card payment systems in retailers and transport, and practically every federal agency.[8] All parts of the government were dead. At least 300 companies were hit, and one senior Ukrainian government official estimated that 10 percent of all computers in the country were wiped.[9]

Meanwhile, Russia sought to influence online conversations by intruding and producing censorship, disinformation, and leveraging misinformation. But Ukraine's intelligence operations were ready and took several steps to limit Russian propaganda's influence by banning Russian online news outlets and social networks. Consequently, the Kremlin turned to different tools, including using Bots to spread disinformation and misinformation. Russia's online propaganda efforts are not exclusive to Ukraine—Facebook released a report in May 2017 detailing Russia's most extensive disinformation campaigns, including Ghana and Mexico.[10]

As a social media platform in Ukraine, it is estimated that 16 million people use Facebook.[11] Russian state media appears committed to spreading a false narrative through this platform about the invasion—referring to it instead as a "military operation" and spreading misleading information about civilian and Russian casualties. It remains brutal, at times, to tell fact from fiction online during the digital fog of war.[12]

The Soviet Union pioneered such influencing operations during the Cold War to covertly influence political dialogue in the West. Well before the Internet, in 1964, the KGB rezidenturas scattered false stories in newspapers, spread rumors, and worked to raise tensions.[13] A KGB front group was able to publish a sprawling conspiracy theory about John F. Kennedy's assassination, which later became the basis for a major movie.[14] Twenty years later, when AIDS spread worldwide, Russian operatives planted a story in a small Indian newspaper claiming that the U.S. government had manufactured the virus at a U.S. military facility.[15]

Before the first missile was launched, nor even before the first leaflet was distributed, the U.S. military was already waging a psychological war through leaflets, radio broadcasts, the Internet, loudspeakers, and even cell phones in Iraq and Afghanistan in the past two decades. It has been widely recognized that offensive influence operations are psychological operations (PSYOP) intended to influence civilians' and troops' emotions, motives and habits through carefully planned persuasion campaigns. Research in social psychology, such as learning, motivation, cognition, culture and persuasion, is essential for success in offensive influence operations. During the 2003 invasion, U.S. military planes dropped over 25 million leaflets over Iraq, encouraging civilians and the military to listen to U.S. radio and dissuading them from fighting for Saddam Hussein's regime.[16]

HISTORICAL INFLUENCE OPERATIONS TARGETING CIVILIANS

Historically, this is not a new problem; influence operations have been used internally during wartime. In 1917 President Woodrow Wilson established the Committee on Public Information (CPI) immediately after the United States entered the ongoing conflict between European states. The goal of this committee was to shape public opinion about World War I. This committee was created by executive order and congressional approval.[17] The Committee on Public Information was established to turn every communication channel and education to promote the war effort through the press, education, and advertising. One aspect of the committee's work was censorship of potentially harmful material. The committee maintained a stranglehold on the kind of material from the war that could be released to the public. The other arm of the Committee was to generate propaganda materials.[18] Creel ran the committee much like an advertising agency and provided a blitz of media materials, like magazine and newspaper articles, posters, still photographs, film reels and radio broadcasts to reach the citizens of the United States. The agency provided speech makers for public events and numerous other resources for communities across the United States.[19] Wilson had previously sought to mobilize public opinion by proposing a clearinghouse for information on government activities and foreign press monitoring.

The committee was ostensibly set up to provide information to the public regarding the war effort, but in reality, it was a propaganda agency. It used several techniques to dehumanize the enemy and promote anti-German sentiment in the United States to encourage people to support the war.[20] During World War II, American forces and their allies committed atrocities against the enemy, while details of the atrocities were detailed and sometimes with unreliable facts. Resistance to the draft, opposition to the war, and anti-war sentiment were suppressed in the United States. In place of taking up arms, those who refused to back the war were shamed through various methods, including distributing white feathers to young men who appeared draft-age to shame them for perceived cowardice.[21]

TARGET ANALYSIS FOR AN INFLUENCE OPERATION

Whether the influence operations are external or internal, any intelligence effort in developing an influencing operation is conducting target analysis to determine audience receptiveness.[22] Will the target be persuaded to use this influence in ways that would aid and be beneficial in a time of war or national security crisis? In addition to providing intelligence about long-standing attitudes toward political, military, economic, and social topics, basic research also acts as a readiness check. Various age cohorts, social classes, and occupational groups have been separated as much as possible from this data.[23] As long as the fears and anxieties are assessed appropriately and in time, they can be exploited as potential targets. We can build thematic influence operations based on the targets' vulnerabilities and attitudes. For these materials to be effective, they must support the campaign's plans and policies. The exposures and attitudes underlying them must result from scientific analysis and evaluation.

Moreover, any analysis of an influence operation should consider its effectiveness in reaching its intended audiences. Defining the goal and measuring the operation's success can provide insight into this perspective. Answering these questions may appear straightforward, but examining propaganda materials in depth can be challenging. In addition to analyzing the themes and logic of the influence operation, one must also look at the psychological implications of the message, including the symbols, language, and other linguistic devices that accompany it. The quality of intelligence is essential for any military operation, but it seems even more crucial in influencing operations. It is critical throughout the spectrum of operations, from the initial planning to the campaign evaluation. Influence operations are not only an abuser of reality. They are also a producer of intelligence.[24]

PROTECTING INFLUENCE OPERATIONS

Influencing operations against intelligence are broken down into at least two categories. Information warfare includes the communications process. There has been a rapid rise in the use of information technologies over the past few decades, and their popularity has grown throughout our society. This has transformed the communications process and the meaning and implications of information warfare. Accordingly, information warfare uses destructive force to destroy information assets and systems against the systems and computers that support four critical infrastructures (including the power grid, communications, financial, and transportation). It is, however, in the national security interests of a country to protect against the threat of computer intrusion posed by an intelligence agency, even on a smaller scale, and that is important in the current discussion about information warfare.[25] Influence operations in propaganda and deception have always existed in warfare.[26] The intentional and strategic use of visual (e.g. gestures, pictures, or written word) or aural (e.g., spoken words) communication to influence the opinions and behavior of a target audience to achieve politico-military ends during times of conflict—an apt pragmatic definition of "propaganda" for our purposes—probably finds its historical roots in the primordial clashes of the Mesolithic and Epipaleolithic periods.[27] From President Harry Truman's declaration on April 20, 1950, that "this is a struggle, above all else, for the minds of men," to President Ronald Reagan's speech to the British Parliament on June 8, 1982, declaring "for the ultimate determinant in the struggle that's now going on in the world will not be bombs and rockets, but a test of wills and ideas […]", the Cold War was seen as fundamentally a "battle of ideas".[28]

National Security Directive 42, which President George H. Bush issued in 1990, had long recognized the vulnerabilities of telecommunications and information processing systems. In 1993, President Clinton issued Executive Order #12864 to address these vulnerabilities. As a result of this document, the Information Infrastructure Task Force (IITF) was established. The ITTF was responsible for exploring "national security, emergency preparedness, security issues, and network protection implications" of the NII, among other things. An investigation was conducted in 1995 by the Security Policy Board, an inter-agency group set up by President Bill Clinton and headed by former FBI Director John Deutch. According to the report, many countries actively worked on information warfare programs.

With a charter to serve the needs of domestic and military security, the NSA established the Information Warfare Technology Centre in 1996. Additionally, President Clinton issued Executive Order #13010, which established a commission to assess the risks associated with eight critical infrastructure elements and recommend ways to mitigate those risks.

However, an information attack is an ability to use information as a weapon rather than in the broader sense of becoming part of the environmental warfare surrounding a threat. This idea goes back to 1995, as recognized by the U.S. Air Force.[29] Opposing intelligence agencies must now see information as a weapon rather than another warfare environment. Information as a weapon has been called the electron as the "ultimate precision-guided weapon."[30] The U.S. Army field manual on information warfare explains the significance of using the information as a weapon for firepower in combat, the generation of destructive force against an enemy's capabilities and will to fight.[31]

Based on a historical case study of the conditions for detecting foreign influence operations in Norway from 1950 to 2021, blind spots can be defined analytically and how they relate to conflict escalation.[32] They depend on three main factors: (1) domestic political and legal frameworks, (2) an adversary's willingness to engage, and (3) the opportunities offered by technology. Those two issues are the subjects of this article, and their discussion is restricted to the following conceptual aspects. Factor one is the primary focus of the article. Blind spots in this context differ from other related variations on intelligence warnings and horizon scanning in the existing literature. Whether it is called a "blindside" or a "strategic shock," the emphasis tends to be on factors that adversely affect the quality of analysis. As such, blind spots represent inattention, a lack of imagination or an insufficient understanding of adversaries' strategic intent or capability.

In contrast, the current concept is based on externally observable factors that aid in pointing out sectors where unknowns may be hidden. Thus applied, blind spots represent inattention, poor imagination, or insufficient knowledge about adversaries' strategic intent or capacity. The present concept, in contrast, is constructed on externally observable factors that help identify sectors in which unknowns may be concealed.

ANALYSIS OF INFLUENCE OPERATIONS

Operation Desert Storm, the U.S. conflict with Iraq in 1990, is considered the first modern-era influence operation integrated fully into warfare based on the number of incidents that occurred. The Department of Defence published on the tenth anniversary of this conflict a set of interviews and the article "Another aspect of Desert Storm Information Warfare" involved the use of electronic tags that, once activated, revealed the computer locations within Iraq and their potential military applications.[33] But the most significant availability to influence an enemy's information is the U.S involvement in the war between Serbia and Kosovo.

Creating target decoys was an art form mastered by the Serbs.[34] The Allies were aware of the limits of overhead surveillance times and placed targets where they could be observed. As the Allied forces were only allowed to bomb specific windows at night, the Japanese used dummy targets and unserviceable vehicles instead of the real targets. The Serbian military used many fake messages and physical decoys to trick the Allies into bombing empty or non-existent targets. After seeing the Iraqi experience with Operation Desert Storm, the Serbs realized that their radio messages were subject to extensive eavesdropping. To cause Air Force aircraft to bomb empty targets or targets with collateral damage, the Serbs broadcast several false radio messages. Serbia used fraudulent radio traffic so extensively that it was difficult to determine if Allied radio intercepts damaged the enemy's forces.

There was open-source intelligence suggesting that the Serbs had agents or sympathizers with cell phones waiting on lawn chairs at the end of the runways at Aviano and other Allied airbases.[35] It was important for Serb intelligence to know the number and type of aircraft being launched when they took off. Using a relatively simple analysis, the Serbs could determine relatively accurately by using flight time calculations based on the types and numbers of aircraft that the Allies would be flying over various target areas. Undoubtedly, this open-source early warning system played a significant role in the Serbian military not being badly damaged during the air raids.

Moreover, there was the knowledge that the Russians sent electronic eavesdropping ships to the area and that the Russians may have shared intelligence with the Serbs.[36]There was no secure communication interoperability between the Allies; therefore, the Allies were forced to rely on non-secure methods, which compromised operational security.[37] It can almost be regarded as an open-source intelligence that military operations communicate in the open. It would be the equivalent of Serb agents directly listening to NATO communications, with only a slight delay time, if Russia or other nations were indeed listening to NATO communications that were not encrypted. One common misconception about influence operations in IO is that it only involves procedures that aim to influence people's minds. This is also a widespread misconception. It is incorrect (by being too limiting) to say that "IO is the name, influence is the game," but this term is frequently used in the psychological operations community.[38]

It involves delivering messages that help change attitudes and behaviors by influencing select groups of people. In contrast, influence operations ignore technical aspects of information operations that help safeguard friendly information and information systems and act against opposing information and information systems. In the military testing community, this mis-perception had an interesting side effect. The more quantitatively oriented community of developers and operational testers felt that IO only applied to influence operations. But in reality, testing new IO systems and procedures is more vital than ever. The U.S. military becomes more vulnerable by relying more on commercial technologies for communication and computation. Testing environments need to consider both IO's offensive and defensive aspects more effectively.

Unfortunately, policymakers lack adequate information about the problem they are trying to solve.[39] Limited and scattered empirical research has been conducted on how influence operations can affect people and society-for example, by altering beliefs, changing voting behavior, or inspiring political violence. Consequently, policymakers cannot identify influence threats, judge whether the problem is improving or worsening, and develop evidence-based solutions. Evidence shows that some influence operations can affect people's beliefs and behavior. Still, fundamental questions remain, like if and how social media-based influence operations differ from traditional forms of influence.

Several studies had shown that influence operations utilizing media such as newspapers, radio, and television before the Internet effectively swayed voters toward a particular political party.[40]

The behavior of individuals exposed to long-term mass media campaigns has also been affected. The study of mass media exposure to Nazi propaganda over many years found that increased risk-taking by German soldiers during World War II was related to targeted infor-mation operations.[41] In particular, targeted information operations have increased political violence during conflict or civil unrest. In another set of experiments, researchers looked at the short-term effects of social media-based influence operations. Studies found that certain operations changed political beliefs and behavior, increased feelings of xenophobia or dis-crimination, and changed attitudes about vaccines and medical information. Researchers have found that political advertising has a persuasive but short-lived effect on citizens. Similar findings have been found for social media operations. Even short-term shifts in the social media activity of prominent actors can have slight, statistically detectable effects on racial violence. There are gaps in the literature, but critical insights are provided. Most of the most pressing questions are answered by empirical research.[42]

INTELLIGENCE PERSPECTIVE: PREPARING FOR INFLUENCE OPERATIONS

Building on carefully observed and evaluated intelligence collection can prevent delays and failure by developing intelligence assessments for decision-makers. This requires a systematic approach to obtaining reliable information in an environment where an adversary's influence operation may be saturated.

Counterintelligence to neutralize an influence operation depends on knowledge of one's capability and sympathetic and realistic awareness of the influencer's capability to target their population. The more one knows about the targeted population; the more one can understand the message's appeal. Intelligence for influencing operations may be defined as a body of knowledge resulting from the collection, evaluation, collation, and interpretation of pertinent information concerning the attitudes, beliefs, sensitivities, and patterns of rational and nonrational behavior that may characterize a group that hopes to influence. This type of data concerning targeted areas susceptible to influence operations includes two purposes for counterintelligence for influence operations. The purposes include a) enabling planners to draft realistic and feasible plans on understanding the vulnerabilities of those seeking to be targeted, b) it provides an ability to assess the effectiveness of an operation.

The first and, therefore, most obvious need for intelligence collection in an influence operation is to satisfy the systematic examination of such aspects as the military, sociological and political conditions that may affect the influence operation. Collecting and assessing such data is to ascertain or pinpoint the possible psychological vulnerabilities attracting this influence campaign. In the conduct of a broad strategic influence campaign by one country against the other, intelligence data requires the imagination and ingenuity of the men and women who are going to use this data; indicative of this type of data for planning purposes would include any data indicating who influences whom why, and by how much in the area or among the people that are being addressed in this campaign.

In planning a strategic influence campaign, it is desirable, if not necessary, that a study be devoted to the media communications commonly used by the targeted people. Intelligence collection must be directed at who gets what type of information under particular conditions over specific media. Likewise, intelligence must be collected on what competing media of information technology is likely to contest one's right to receive a targeted group's undivided or significant attention.

The need for target analysis has been firmly established.[43] However, essential intelligence requirements necessitate adequate target analysis for all foreign information and influence operations. This is not easy to implement satisfactorily. In a time of armed conflict, the knowledge required is difficult to obtain, and which is secured is always suspect, for in the battle to get sufficient raw data.[44]

CONCLUSION

A foreign influence operation aims to achieve a specific effect among a target audience. Various actors employ tactics, techniques, and procedures to influence a target audience's decisions, beliefs, and opinions. Despite this, there has been little nuanced policy discussion of these activities that can support a coherent intelligence strategy. A significant portion of any influence operation fuels fears of how influence operations could undermine the legitimacy of liberal democracies from adversaries. In evaluating such influence campaigns, it is essential to consider their purpose. Influence operations encompass both coordinated physical intrusion and content intrusion. Nevertheless, the intelligence communities are under increasing pressure to counter this threat against malign influence operations. However, understanding these campaigns is essential to take effective action.

In practice, influence operations are neither intrinsically good nor bad, and societies must determine acceptable behavior and responses. Sometimes it isn't easy to understand the motives driving some influence operations. Governments, including the intelligence community, have fewer opportunities to counter-influence operations through existing laws than social media networks. Various policies govern how social media companies use their platforms, which provides opportunities to combat influence operations. Their access to information about how their platform is used and their domain-specific expertise allow them to develop more customized solutions. There are a few ways to counter such influence operations through government-led legal mechanisms. Online activity is largely unregulated, not just due to the relative lack of laws, which can be difficult in these cases. In other words, governments have generally not provided much assistance for private industry actors in identifying which influence operations are unacceptable and should be combated. De facto, industry actors determine these societal boundaries without such guidance. Governments should participate in multi-stakeholder efforts, some of which have been established by think tanks and non-profit organizations, and consider legal approaches that focus on transparency rather than criminalization while guiding industry players.

It is difficult to counter the influence of operations exposed by intelligence either in peacetime or wartime. Innovative influence operators know how to evade existing rules so that their activities and content do not violate known laws and policies. The reports showcase examples of influence operations and rarely explain how and whether such operations violate existing platform policies or rules. However, this is not such a problem, given that many of these offensive influence operations do not directly violate any current guidelines or laws—raising questions about where the boundaries are (and should be) between what is allowed and what is not and, by extension, which should be deciding where those boundaries should be. Despite existing policies applying, these questions persist. Rather than rushing to demand enforcement, stakeholders should clarify what constitutes problematic behavior.

Notes

1 Jan Goldman and Susan Maret, *Intelligence and Information Policy for National Security: Key Terms and Concepts*, Rowman and Littlefield, 2016, 570.
2 Intelligence is a body of information acquired and furnished in response to the perceived requirements of the consumer. Information is developed to prevent an adversary from providing information to influence an opponent's operation. See, Jan Goldman and Susan Maret, Intelligence and Information Policy for National Security: Key Terms and Concepts, Rowman and Littlefield, p.291.
3 Ellen Nakashima, Inside a Russian disinformation campaign in Ukraine in 2014, Washington Post, December 25, 2017, https://www.washingtonpost.com/world/national-security/inside-a-russian-disinformation-campaign-in-ukraine-in-2014/2017/12/25/f55b0408-e71d-11e7-ab50-621fe0588340_story.html. This article is based on a classified GRU report.
4 Ibid.
5 Ibid.
6 Ibid.
7 Andy Greenburg, The Untold Story of NotPetya, the Most Devastating Cyberattack in History, WIRED, August 22, 2018, https://www.wired.com/story/notpetya-cyberattack-ukraine-russia-code-crashed-the-world/
8 Ibid.
9 Ibid.
10 Ibid.
11 TARA LAW, Time, FEBRUARY 28, 2022, https://time.com/6151572/russian-media-ukraine-coverage/
12 Ibid.
13 Laura Reston, How Russia Weaponizes Fake News. *The New Republic*, May 12, 2017, 6–8.
14 The 1991 movie "JFK," Director Oliver Stone and starring Kevin Costner, Gary Oldman, and Jack Lemmon, produced by Warner Brothers
15 Ibid.
16 Timothy Garden, *International Affairs*, Royal Institute of International Affairs, Vol. 79, No. 4 (Jul. 2003), pp. 701–718 (18 pages). According to news reports, to counter the messages, Iraqi officials used

some of their own psychological strategizing. They sought to dissuade Iraqi citizens from picking up leaflets by claiming that they were contaminated with chemical or biological poisons. The officials wore protective suits to reinforce this as they disposed of the leaflets.

17 National Archives, Records of the Committee on Public Information, Record Group 63, 1917–21, https://www.archives.gov/research/guide-fed-records/groups/063.html
18 Ibid., 265; Haidee Wasson, and Lee Grieveson. *Cinema's Military-Industrial Complex*. University of California Press, 2018, 264.
19 Ibid.
20 Ibid.
21 Ibid.
22 Ronald D. McLaurin, ed., Military Propaganda. Psychological Warfare and Operations (New York: Praeger, 1982). Instrumental was Part IV: "Intelligence and Research," which included, among other pertinent articles, the following: Philip P. Katz, "Intelligence for Psychological Operations," and "Exploiting PSYOP Intelligence Sources," 121–54. Also beneficial was the Department of the Army Field Manual (FM) 33–1, Psychological Operations, 1979.
23 Frank L. Goldstein (editor) *Psychological Operations Principles and Case Studies*, Maxwell Air Force Base, Alabama (Air University Press, 1996). https://www.airuniversity.af.edu/Portals/10/AUPress/Books/B_0018_GOLDSTEIN_FINDLEY_PSYCHOLOGICAL_OPERATIONS.pdf
24 Melissa Dittmann, Operation hearts and minds, Psychological operations are becoming a regular part of military strategy, *Monitor on Psychology,* June 2003, Vol 34, No. 7. https://www.apa.org/monitor/jun03/operation
25 Susan Macdonald, (2006), *Propaganda and Information Warfare in the Twenty-First Century: Altered Images and Deception Operations*, Routledge, London—New York.
26 Ibid.
27 James Arquilla, "Thinking about information strategy." In J. Arquilla and D. Borer, eds. *Information Strategy and Warfare: A guide to theory and practice.* Routledge, New York, 2.
28 Harry Truman, "Address on Foreign Policy at a Luncheon of the American Society of Newspaper Editors." Harry S. Truman Library & Museum, 20 April 1950, http://www.trumanlibrary.org/public papers/index.php?pid=715; R. Reagan, "Address to members of the British Parliament," June 8, 1982, (transcript) https://reaganlibrary.archives.gov/archives/speeches/1982/60882a.htm.
29 George Stein, "Information Attack: Information Warfare in 2025," in *White Papers: Power and Influence, vol.3, issue 1* (Maxwell AFB, Ala: Air University Press, 1996), 98
30 John Correll, "Warfare in the Information Age," Air Force Magazine 79, no. 12, 1996, 3.
31 US Army, FM 100–6, Information Operations: Doctrine, Tactics, Techniques, and Procedures, November 2003.
32 Hallvard Notaker (2022) In the blind spot: Influence operations and sub-threshold situational awareness in Norway, *Journal of Strategic Studies*, DOI: 10.1080/01402390.2022.2039634.
33 Allen Patrick, Information Operations Planning, Artech House, 2006. ProQuest Ebook Central, http://ebookcentral.proquest.com/lib/citadel/detail.action?docID=338743.
34 Timothy, L. Thomas, "Kosovo and the Current Myth of Information Superiority," *Parameters*, Spring 2000, http://www.carlisle.army.mil/usawc/parameters/00spring/contents.htm.
35 Sec. Def. William S Cohen and General Henry H. Shelton, "Joint Statement on Kosovo After Action Review," Defence Link News, October 14, 1999, http://www.defenselink.mil/news/Oct1999/b10141999_bt478–99.html.
36 Frederic H., Levien, "Kosovo: An IW Report Card," Journal of Electronic Defence, Vol. 22, No. 8, August 1, 1999.
37 Patrick Allen and Chris Demchak, "An IO Conceptual Model and Applications framework," *Military Operations Research Journal*, Special Issue on IO/IW, Vol. 6, No. 2, 2001.
38 Jon Bateman, Elonnai Hickok, Laura Courchesne, Isra Change, Jacob Shapiro, Measuring the Effects of Influence Operations: Key Findings and Gaps from Empirical Research, Carnegie Endowment for International Peace, 2021 https://carnegieendowment.org/2021/06/28/measuring-effects-of-influence-operations-key-findings-and-gaps-from-empirical-research-pub-84824.
39 Ibid.
40 Leonid Peisakhin and Arturas Rozenas, "Electoral Effects of Biased Media: Russian Television in Ukraine," American Journal of Political Science 62, no. 3 (2018): 535–550; and Stefano DellaVigna and Ethan Kaplan, "The Fox News Effect: Media Bias and Voting," Quarterly Journal of Economics 122, no. 3 (2007): 1187–1234. Peisakhin and Rozenas, "Electoral Effects of Biased Media"; and Jay C.

Kao, "How the Pro-Beijing Media Influences Voters: Evidence from a Field Experiment," the University of Texas at Austin, December 2020, https://www.jaykao.com/uploads/8/0/4/1/80414216/pro-beijing_media_experiment_kao.pdf. In the case of Taiwan, however, a backfire effect was reported if the targeted audience had a preexisting negative view of China or perceived the outlet as associated with the Chinese government.
41 Benjamin Barber IV and Charles Miller, "Propaganda and Combat Motivation: Radio Broadcasts and German Soldiers' Performance in World War II," World Politics 71, no. 3 (2019): 457–502. 6 David Yanagizawa-Drott, "Propaganda and Conflict: Evidence from the Rwandan Genocide," Quarterly Journal of Economics 129, no. 4 (2014): 1947–1994; and Maja Adena, Ruben Enikolopov, Maria Petrova, Veronica Santarosa, and Ekaterina Zhuravskaya, "Radio and the Rise of the Nazis in Prewar Germany," Quarterly Journal of Economics 130, no. 4 (2015): 1885–1939. Aleksandr Fisher, "Demonizing the Enemy: The Influence of Russian State-Sponsored Media on American Audiences," Post-Soviet Affairs 36, no. 4 (2020): 281–296.
42 Christopher A. Bail, Lisa P. Argyle, Taylor W. Brown, John P. Bumpus, Haohan Chen, M. B. Fallin Hunzaker, Jaemin Lee, Marcus Mann, Friedolin Merhout, and Alexander Volfovsky, "Exposure to Opposing Views on Social Media Can Increase Political Polarization," Proceedings of the National Academy of Sciences 115, no. 37 (2018): 9216–9221.
43 Robert Clark, Intelligence Analysis: A Target-Centric Approach, CQ Press, (2003); Eliot A. Jardines, "Testimony of Eliot A. Jardines", Using Open Source Effectively: Hearings before the Subcommittee on Intelligence, Information and Terrorism Risk Assessment, Committee on Homeland Security, United States House of Representatives, 2005, pp. 11–18; Jack Davis, "A Policymaker's Perspective on Intelligence Analysis", Studies in Intelligence, 38 (5), 2007.
44 A good paper on seeking medical data, much less any other type of data, is outlined in Collecting Survey Data during Armed Conflict by William G. Axinn, Dirgha Ghimire, and Nathalie E. Williams, National Library of Medicine: National Center for Biotechnology Information, June 2013, vol. 28, issue 2, 153–171.

Bibliography

Adena, Maja and Ruben Enikolopov, Maria Petrova, Veronica Santarosa, and Ekaterina Zhuravskaya, "Radio and the Rise of the Nazis in Prewar Germany," Quarterly Journal of Economics 130, no. 4 (2015): 1885–1939.

Allen, Patrick and Chris Demchak, "An IO Conceptual Model and Applications framework," Military Operations Research Journal, Special Issue on IO/IW, Vol. 6, No. 2, 2001.

Arquilla, James and D. Borer, eds, "Thinking about information strategy" in Information Strategy and Warfare: A guide to theory and practice. Routledge, New York.

Bail, Christopher A., Lisa P. Argyle, Taylor W. Brown, John P. Bumpus, Haohan Chen, M. B. Fallin Hunzaker, Jaemin Lee, Marcus Mann, Friedolin Merhout, and Alexander Volfovsky, "Exposure to Opposing Views on Social Media Can Increase Political Polarization," Proceedings of the National Academy of Sciences 115, no. 37 (2018).

Barber, Benjamin and Charles Miller, "Propaganda and Combat Motivation: Radio Broadcasts and German Soldiers' Performance in World War II," World Politics 71, no. 3 (2019)

Bateman, Jon and Elonnai Hickok, Laura Courchesne, Isra Change, Jacob Shapiro, Measuring the Effects of Influence Operations: Key Findings and Gaps from Empirical Research, Carnegie Endowment for International Peace, 2021https://carnegieendowment.org/2021/06/28/measuring-effects-of-influence-operations-key-findings-and-gaps-from-empirical-research-pub-84824.

Clark, Robert Intelligence Analysis: A Target-Centric Approach, CQ Press, 2003.

Cohen, William S. and General Henry H. Shelton, "Joint Statement on Kosovo After Action Review," Defence Link News, October 14, 1999, http://www.defenselink.mil/news/Oct1999/b10141999_bt478–499.html.

Correll, John "Warfare in the Information Age," Air Force Magazine 79, no. 12, 1996, 3.

Davis, Jack "A Policymaker's Perspective on Intelligence Analysis", Studies in Intelligence, 38 (5), 2007.

DellaVigna, Stefano and Ethan Kaplan, "The Fox News Effect: Media Bias and Voting," Quarterly Journal of Economics 122, no. 3, 2007.

Dittmann, Melissa, Operation hearts and minds, Psychological operations are becoming a regular part of military strategy, Monitor on Psychology, June2003, Vol 34, No. 7. https://www.apa.org/monitor/jun03/operation.

Fisher, Aleksandr "Demonizing the Enemy: The Influence of Russian State-Sponsored Media on American Audiences," *Post-Soviet Affairs* 36, no. 4 (2020).

Garden, Timothy, International Affairs, *Royal Institute of International Affairs*, Vol. 79, No. 4 (Jul. 2003).

Goldman, Jan and Susan Maret, *Intelligence and Information Policy for National Security: Key Terms and Concepts*, Rowman and Littlefield, 2016.

Goldstein, Frank L. (editor) *Psychological Operations Principles and Case Studies*, Maxwell Air Force Base, Alabama, Air University Press, 1996 at https://www.airuniversity.af.edu/Portals/10/AUPress/Books/B_0018_GOLDSTEIN_FINDLEY_PSYCHOLOGICAL_OPERATIONS.pdf.

Greenburg, Andy, The Untold Story of NotPetya, the Most Devastating Cyberattack in History, *WIRED*, August 22, 2018, https://www.wired.com/story/notpetya-cyberattack-ukraine-russia-code-crashed-the-world/.

Notaker, Hallvard, In the blind spot: Influence operations and sub-threshold situational awareness in Norway, *Journal of Strategic Studies*, 2021.

Jardines, Eliot A. *Testimony of Eliot A. Jardines*, Using Open Source Effectively: Hearings before the Subcommittee on Intelligence, Information and Terrorism Risk Assessment, Committee on Homeland Security, United States House of Representatives, 2005.

Katz, Philip P. "Intelligence for Psychological Operations," and "Exploiting PSYOP Intelligence Sources," 121–54. Law, Tara, *Time*, February 28, 2022, https://time.com/6151572/russian-media-ukraine-coverage/.

Levien, Frederic H., "Kosovo: An IW Report Card," *Journal of Electronic Defence*, Vol. 22, No. 8, August 1, 1999.

Macdonald, Susan, *Propaganda and Information Warfare in the Twenty-First Century: Altered Images and Deception Operations*, Routledge, London—New York, 2006.

McLaurin, Ronald D. ed., *Military Propaganda. Psychological Warfare and Operations* (New York: Praeger, 1982).

Nakashima, Ellen, Inside a Russian disinformation campaign in Ukraine in 2014, *Washington Post*, December 25, 2017, https://www.washingtonpost.com/world/national-security/inside-a-russian-disinformation-campaign-in-ukraine-in-2014/2017/12/25/f55b0408-e71d-11e7-ab50-621fe0588340_story.html.

National Archives, Records of the Committee on Public Information, Record Group 63, 1917–21, https://www.archives.gov/research/guide-fed-records/groups/063.html.

Patrick, Allen, Information Operations Planning, Artech House, 2006. *ProQuest Ebook Central*, http://ebookcentral.proquest.com/lib/citadel/detail.action?docID=338743.

Reagan, Ronald. "Address to members of the British Parliament," June 8, 1982, (transcript) https://reaganlibrary.archives.gov/archives/speeches/1982/60882a.htm.

Peisakhin, Leonid and Rozenas, Arturas "Electoral Effects of Biased Media: Russian Television in Ukraine," *American Journal of Political Science* 62, no. 3 (2018): 535–550.

Reston, Laura, How Russia Weaponizes Fake News. *The New Republic*, May 12, 2017.

Stein, George, "Information Attack: Information Warfare in 2025," in *White Papers: Power and Influence*, vol.3, issue 1, Maxwell AFB, Ala: Air University Press, 1996.

Stone, Oliver, director. JFK, Warner Brothers Films, 1991.

Thomas, Timothy, "Kosovo and the Current Myth of Information Superiority," *Parameters*, Spring 2000, http://www.carlisle.army.mil/usawc/parameters/00spring/contents.htm.

Truman, Harry "Address on Foreign Policy at a Luncheon of the American Society of Newspaper Editors." Harry S. Truman Library & Museum, 20 April1950, http://www.trumanlibrary.org/publicpapers/index.php?pid=715;.

U.S. Army Field Manual (FM) 33–31, Psychological Operations, 1979.

U.S. Army, Field Manual 100–106, Information Operations: Doctrine, Tactics, Techniques, and Procedures, November 2003.

Wasson, Haidee and Lee Grieveson. *Cinema's Military-Industrial Complex*. University of California Press, 2018.

Williams, Nathalie E. *National Library of Medicine: National Center for Biotechnology Information*, June2013, vol. 28, issue 2

Yanagizawa-Drott, David "Propaganda and Conflict: Evidence from the Rwandan Genocide," *Quarterly Journal of Economics* 129, no. 4, 2014.

7

ASYMMETRICAL CONFLICT IN THE INFORMATION DOMAIN— THE CASE OF RUSSIA

Bjorn Palmertz and James Pamment

Introduction

The changes to the global security environment during latter years have introduced an increased use of a rather complex and elusive challenge: influence operations. A term referring to multiple coordinated information influence activities targeting of opinion-formation in illegitimate, though not necessarily illegal ways, by foreign actors or their proxies.[12] These operations originate from state actors incorporated with a wider array of political, economic and military capabilities. They have also been widely utilized by terrorist organizations as an enabler to increased international recruitment and framing their organizational image. An image enhanced by real world attacks in order to spread fear and show the vulnerability of western society. At the same time, home-grown extremist organisations in the west, especially on the far right, have focused on harnessing the opportunities of social media and other internet resources to influence target audiences where they see an opportunity to gain a foothold.

Of the prominent actors mentioned, the Russian leadership offers one of the most vivid and persistent examples of employing a grand strategy diametrically opposed to the west, integrating influence operations with other offensive capabilities. Since the dissolution of the Soviet Union there was hope on the part of many western leaders, experts and companies to welcome Russia as a partner into the European and international fold through trade, collaboration and international organisations. A hope that has seemed less and less likely since 2014, but definitively came to an end on February 24 2022 as Russia launched a full-scale invasion of Ukraine. The signs and warnings of an increasing willingness to oppose and disrupt the European security order and the sovereignty of other states are numerous. They include the 2007 bronze soldier event of Estonia[3], the Russian invasion of Georgia in 2008[4], the 2014 annexation of the Crimean Peninsula and support to separatist groups in eastern Ukraine[5] as well as influence operations directed at the US elections in 2016[6] and French elections in 2017.[7]

But opportunities for influence operations do not only stem from the willingness and capability of an actor, they also rest on vulnerabilities relating to the modern information environment itself and how it impacts internal discourse and interrelationships within the western democracies being targeted. Many countries now face an increasingly fragmented

DOI: 10.4324/9781003190363-10

information environment, where technological development has enabled analytical tools and near instantaneous reach for governments. But these opportunities, due to their low barriers of entry, have also been made available to a much wider range of groups and individuals in society. In addition, many western states, both due to new communication means and domestic socioeconomic challenges, are increasingly confronted by citizens frustrated by 21st century politics. The issue of how the political leadership and government agencies reach and communicate with their population becomes an immediate issue related to societal resilience, national security and the safekeeping of a functioning democratic system.

Origins

The aim to gain a psychological advantage over a perceived opponent is nothing new in governance or conflict. This is especially true in the case of Russia, and there are historical reasons for the strong institutionalisation and development of such methods to benefit strategic aims, especially in war. According to Giles, "Russia's practice of information warfare has ... developed rapidly, while still following key principles that can be traced to Soviet roots. This development has consisted of a series of adaptations following failed information campaigns by Russia, accompanied by successful adoption of the Internet."[8] It is worth bearing in mind that the historical and ideological context of Russia has produced an Information Warfare theory that differs from that of the West. Across a number of Russian definitions in this area of study, two subdivisions can be recognized—that of information-technical relating to software and hardware, and information-psychological relating to the cognitive aspects of human beings. These are not mutually exclusive and do intersect in certain instances, such as in computer-based phishing and hacking operations.[9]

One of the early terms in the Russian context was deception (maskirovka) meaning the art of ensuring success in combat by deceiving the opponent about key aspects such as the condition, readiness, actions and plans of military units. The Russian Army started a military maskirovka school in 1904 which was active until 1929. Named the Higher School of Maskirovka, it laid the groundwork for later Russian doctrines and operational implementation.[10] Some examples are the 1924 Soviet directive for higher commands outlining the underlying methodologies, and the 1929 Field Regulations of the Red Army which stated that "surprise has a stunning effect on the enemy. For this reason all troop operations must be accomplished with the greatest concealment and speed."[11] The methodologies were outlined as 1) secrecy, the utilisation of camouflage terrain and information security when developing plans, 2) demonstrative actions, for example using diversionary manoeuvres and feint attacks, 3) imitation, such as the creation of units aimed at conducting diversion and the conduct of false signalling as well as 4) disinformation, the dissemination of false information and rumours through mass media and other communication channels.[12]

Building on the early work relating to deception and disinformation, another central theory began developing in Russia during the 1960's and 70's, that of reflexive control (refleksivnoe upravlenie). It refers to means of conveying to a partner or an opponent specially prepared information to incline him or her to voluntarily make the predetermined decision desired by the initiator of the action. It is employed not only on the strategic and tactical levels in war but also on the political strategic level domestically as well as in foreign policy. The target is not necessarily limited to key decision-makers but can also consist of larger categories of a population.

In the study of warfare during Soviet times, or in contemporary Russia, perspectives of warfare and peace are not considered to be on opposite ends of a spectrum of escalation.

Interaction with other states occurs within a framework of struggle regardless of whether they are carried out within an alliance, a state of opposition or temporary collaboration. Such provisional partnerships sometimes occur despite considerable underlying differences. This aspect of Russian political and military perspectives has proven to be remarkably resilient in the face of continuous historical change.[13] Reflexive Control, mirroring this, entails a much wider range, and complex combination of measures than what is involved in the delimited use of tools within the concept of deception. It does not only target an opponent's process of intelligence and flow of information, but takes the cultural, emotional and psychological aspects of the opponent's decision making process into account. This makes possible an indirect means to impose your will by the opponent making decisions and conducting operations that are actually inappropriate for them. Sub-methodologies include camouflage, disinformation, encouragement, blackmail by force and the compromising of various officials and officers.

So, in what ways can an opponent be influenced to make decisions that do not serve their own aims? A few proposed by Russian military thinker, Komov, include:

- Distraction, by creating a real or imaginary threat to one of the enemy's most vital locations during the preparatory stages of combat operations, thereby forcing him to reconsider the wisdom of his decisions to operate along this or that axis;
- Overload, by frequently sending the enemy a large amount of conflicting information;
- Paralysis, by creating the perception of a specific threat to a vital interest or weak spot;
- Exhaustion, by compelling the enemy to carry out useless operations, thereby entering combat with reduced resources;
- Deception, by forcing the enemy to reallocate forces to a threatened region during the preparatory stages of combat operations;
- Division, by convincing the enemy that he must operate in opposition to coalition interests;
- Pacification, by leading the enemy to believe that pre-planned operational training is occurring rather than offensive preparations, thus reducing his vigilance;
- Deterrence, by creating the perception of insurmountable superiority;
- Provocation, by forcing the commander to take action advantageous to your side;
- Suggestion, by offering information that affects the enemy legally, morally, ideologically, or in other areas;
- Pressure, by offering information that discredits the government in the eyes of its population.[14]

Another important concept is active measures (aktivnyye meropriyatiya) utilised by the Soviet Union during the cold war. Definitions of active measures can for example be found in the Dictionary of Counterintelligence, issued by the KGB's Felix Dzerzhinsky Higher School in 1972. It is there described as encompassing "acts of counterintelligence making it possible to penetrate the intentions of the enemy, allowing his unwanted steps to be anticipated, to lead the enemy into error, to take the initiative from him, to thwart his actions of sabotage."[15] Its aim was to influence opinions and/or actions of individuals, governments and/or publics conducted through, for example, covert and deception operations, separate but sometimes supported by espionage, counterintelligence, diplomacy and informational activities.[16]

As in earlier related Russian theories, disinformation (dezinformatsiya) was also a sub-category of active measures. Deliberately disseminating false information in order to influence the perceptions of those that are exposed to it. It was used extensively by the Soviet Union during the cold war period. Pro-Soviet publications and state controlled communication channels were used, albeit a preferred method was attempting to inject disinformation into credible western media sources. Some well documented cases targeting the US during the 1980's include claims that the US manufactured the AIDS virus in a military facility at Fort Detrick, was manufacturing an ethnic weapon aimed at killing only non-whites, that the CIA assassinated Swedish Prime Minister Olof Palme and Indian Prime minister Indira Gandhi and that the FBI assassinated Rev. Martin Luther King Jr.[17]

In addition to disinformation several other methods were used, including 1) front organisations—independent groups supporting Soviet policies or policies in line with the strategic aims of the USSR, such as unilateral nuclear disarmament, 2) agents of influence, for example soviet intelligence personnel that infiltrated foreign organisations in order to disseminate specific messages and narratives; local recruits who were cultivated by the USSR as well as individuals that were being discretely supported by the Soviet state without their own knowledge, 3) forgeries where real-world examples include a false embassy report about US plans to overthrow the government in Ghana, and forgeries of diplomatic communication showing US involvement in an attempt to murder the Pope.[18]

Darczewska and Żochowski accentuate that active measures were not discontinued after the collapse of the Soviet Union and the reformation of its security services. The attachment to methodologies that are tried and tested is a pervasive trait of Russian political and strategic culture. This includes

> the legitimation of the regime by transferring internal tensions within Russian society to external enemies; mythologising its own power, the army and the special forces; the "fortress under siege" syndrome; the primacy of psychological and ideological thinking over thinking in terms of political realism; and imposing an ideologised image of the world, and creating a confrontational approach to the international community.

Another concept used in Russia to describe methods of conducting warfare is New Generation Warfare, for example in articles by Chekinov and Bogdanov during 2012 and 2013. The authors highlight that non-military asymmetric methods such as information and psychological warfare will be central to NGW. This raises the barrier for the opponent to engage in hostile action, opens up opportunities to publicly condemn the aggressive nature of the enemy and makes it more difficult for them to build a positive public opinion image.

They describe an escalation to encompass 1) the aggressor using non-military actions in preparation for an upcoming attack on another state. It would be distributed and target the social system of the country based on promoting democracy and respect for human rights. Disinformation would also be used to maintain information security regarding when and how operations take place, both utilising public diplomacy and mass media channels. 2) Information technology will be employed within both influence- and cyber operations to gain an advantage in the conflict ongoing in the information environment. 3) The aggressor may deploy nonlethal, genetically engineered biological weapons that affect the human psyche making it more difficult to ensure cohesion and effective leadership. Undercover operatives can also be introduced to further fuel unrest and fear among the citizens. 4) A military phase of operations begins with reconnaissance and subversive missions conducted under the guise of information operations. Thereafter key infrastructure and societal systems will be targeted. 5) A military kinetic escalation follows, likely beginning with air assets striking key military, industrial and communication infrastructure. 6) Robot controlled systems able to operate independently and Unmanned Aerial Vehicles will probably be used to supplement the ongoing air operations. 7) Not until the initial phases of the conflict come to a close and the political and military goals are achieved, ground units, especially special operations forces, will target what remains to ensure the completion of all critical objectives.[19]

Thomas highlights that the authors several times state that US or Western armies use the techniques under discussion. He also gives some historical perspectives by explaining:

> Of interest is that in the past, during the days of the Soviet Union, Soviet officers described techniques supposedly used in Western armies to fight wars that, in actuality, were descriptions of how the Soviets might or would fight … Thus if this methodology has carried over to Russian times, then the description of Western techniques below could be a way to openly describe to Russian officers how to fight a future war without stating so. Whether this same template is used is unknown.[20]

Linking historical Russian active measures with present day influence operations, Rid describes how the method takes not only facts but also requires a political judgement call and sometimes a collective one to determine their success or failure. Which means that if a targeted society believes disinformation has been successful it contributes to making it so. As the term active measures grew less common during the 1990s and related Russian capabilities transitioned from the KGB to the SVR, the information environment itself changed with the rise of the internet. The ability to proactively analyse target audiences and then disseminate disinformation under the cover of anonymity or fake personas offered new opportunities. In addition, operational adaptation could be implemented within much shorter time frames than during the cold war era. Easy access to online communities harbouring "petri dishes of vicious, divisive ideas" enabled connecting disinformation to a steady supply of conspiracy theories. This has resulted in citizens and groups of targeted societies themselves, knowingly or unknowingly, aiding in the formation and dissemination of disinformation. Rid highlights that the tempting conclusion that crafting and spreading disinformation therefore has become easier is misleading. The nature of today's active measures results in a disintegration. "For the offender, campaigns have become harder to control, harder to contain, harder to steer, harder to manage and harder to assess. For victims, disinformation campaigns have also become more difficult to manage, more difficult to assess in impact, and more difficult to counter."[21]

Key components of Russian influence methodology

During the last decade the Russian government apparatus has shown an increasing willingness to engage in open propagation of its narratives through state funded media outlets such as RT and Sputnik. But simultaneously they have also employed methods that are more difficult to track and attribute such as trolls, bots and building relationships with individuals and organisations in the west that sympathise with their general view of the world.

The review of Russian thinking concerning influence related subjects during the last century highlights that the methodologies of the trade are rarely, if ever, entirely new. What has changed drastically are the opportunities for global data collection, reconnaissance and targeting, as well as the direct-to-user dissemination capability coupled with software tools that provide near instant feedback on reach and content reception. War and conflict have a centric role in Russian thinking, and the institutionalised development of concepts like deception, reflexive control and active measures during the 20th century mirrors this perception of the world. The adaptive mind-set coming out of a view that interaction with other states occur within a framework of struggle is undoubtedly beneficial for the development and deployment of influence operations.

Current Russian influence operations employ a wide range of methodologies and tools, but they all benefit from the large amount of data readily available that makes it possible to, from a distance, learn how a foreign information environment works. The ability to evaluate and realign operations also greatly increases. It is also easier to engage or disrupt actors without revealing your identity and true motivations. The debate, and institutional roles, in western societies has often separated cyber- and influence operations. In reality, Russian and other proficient offensive actors combine these capabilities when needed. This includes employing social engineering methods to have a user open the door to a system so that data can be extracted. It can also involve hacking into and disrupting critical infrastructure systems potentially affecting government credibility or arousing fear among the targeted population. Disseminating information that fuels polarisation and pollutes public debate in another country also serves an important purpose. If a country experiences internal strife and common ground among politicians or the public is hard to achieve they have less energy and focus to expend on containing and countering the strategies and actions of Russia. Hence, four methodological components are especially illustrative. Two of them that focus on the aim and operational design are target audience analysis and the use of strategic narratives, and an additional two related to implementation are digital dissemination of disinformation and cyber influence operations.

The increasing availability of personal information on internet based social platforms and websites has enabled an increasingly wide reaching and detailed target audience analysis for offensive actors. The ability to map personal preferences and living conditions as well as the social networks of potential targets for influence operations offers a potential for improved precision in terms of visual design and message formulation, and can indicate whether a certain target audience is especially receptive to certain types of influence. In addition, since monitoring can occur in real time, the evaluation of how an operation is going enables a greater capability to conclude if an effort should cease, or be redirected. Also, as was seen in Russia's influence operation leading up to the US elections in 2016, knowing how a target audience communicates and forms their digital footprint makes it easier to select target segments, design the operation, its products and choose suitable dissemination channels. Howard et al. states that based on the most active campaigns of Russia's premier online influence operator Internet Research Agency, they clearly sought to "energize conservatives around Trump's campaign and encourage the cynicism of other voters in an attempt to neutralise their vote."[22]

Building on a target audience analysis, the use of strategic narratives as opposed to situation-based messaging offers greater opportunities in the long term. These are stories that connect the implicit with the explicit, not only describing prior events, but incorporating the moral attitude to these events.[23] Hence, they can enable changing the frames of perspective and context for a number of issues among a target audience. Russia utilises narratives to reinforce their own impact and power while promoting constraints on other nation states and organisations they see as limiting such aims. The decline of the West and rise of the rest is regularly included in various statements, for example in the 2013 Foreign Policy Concept—"The ability of the West to dominate world economy and politics continues to diminish. The global power and development potential is now more dispersed and is shifting to the East, primarily to the Asia-Pacific region."[24] Pynnöniemi outlines a number of narrative components, such as the use of specific terms rooted in soviet history and propaganda, especially in connection with The Great Patriotic

War. Such words include fascism, nazi and banderovtsy. The latter stems from WWII Ukrainian nationalist Stepan Bandera, who fought the Soviet Army during the Second World War. The word has a negative connotation for a majority of the Russian population and was used regularly by the Russian media in reports during the Maidan Uprising with reference to Ukrainians in general and pro-Ukrainian activists in particular. On some occasions fascist allegations were used by official channels, when for example the Russian Foreign Ministry in April 2014 referred to rumours in the media about the construction of "fascist concentration camps" in Ukraine. Another set of rhetorical frames are based on political language in order to move or blur the focus of target audiences. Descriptions aiming to portray the West as a provocation to a Russia simply attempting to defend its valid interests were used regularly during the Cold War, and are again a staple in Russian communication, not least related to the conflict in Ukraine. The term Russophobia has also been a regularly used term to describe the foreign policy of certain western countries and the EU as anti-Russian. Vladimir Putin, in one of his speeches, emphasised that "the West's stoking Russophobic sentiment in Ukraine could lead to disaster". The purpose is not only to strengthen the image of an aggressive West internationally, but also has domestic communication benefits for the political leadership. It attaches a certain stigma for Russians that do not accept the view of the Kremlin, and makes it easier to alienate these individuals from the broad base of the citizenry's political life, fuelling a general acceptance of them being treated as outsiders.[25]

Disinformation, the creation, presentation and dissemination of verifiably false content for economic gain or to intentionally deceive the public[26], is very likely as old as communication itself. Relatively new, however, is the reach, speed, potential for concealment and near real time adaptability afforded by the modern information environment through its multitude of platforms and the integration of personal communication devices into our daily lives. Russia has demonstrated this through numerous influence operations, one of the more creative ones applying the sockpuppet technique—imposter accounts managed by someone who hides their real identity or intentions. They can be used to join online communities, introduce disinformation, and in a coordinated way partake in opposing sides of a debate to sow discord.[27] During the U.S. election campaign 2016 this was illustrated through two opposing Facebook pages, "Heart of Texas" and "United Muslims of America" being infiltrated by the Internet Research Agency conducting influence operations on behalf of the Russian state out of an office in St. Petersburg.[28] In a display of the integration between the digital information environment and the real world, the actions of the IRA did not stop at disinformation, at times connected to existing events, but included promoting a protest to "Stop Islamification of Texas" while also advertising a Muslim counter-protest at the same time and place; in front of a mosque in downtown Houston.

With the invasion of Ukraine, and a renewed unity in terms of drawing a line against Russian aggression among many Western nations, certain opportunities regarding disinformation may be on the decline. Facebook, now renamed Meta, has stated that they have increased resources to monitor issues related to the war, while removing a number of Russian influence operations utilising or attempting to hack into social media accounts on Facebook. They have also started restricting access to Russian state-controlled media and removed their opportunity to advertise on their platforms.[29] The United States have also imposed sanctions on individuals connected to Russian intelligence service-directed influence activities designed to destabilise Ukraine.[30] At the same time, Russian reach into developing countries in for example Africa does not seem to be diminishing, but potentially increasing. The memory of western colonisation, the persistent threat of violent extremism and lower barriers of entry into domestic information environments offer a more palatable stage from which the Kremlin can launch influence initiatives. Russian social media accounts in South Africa are for example actively promoting pieces on American biolabs in Ukraine, "Ukrainian war crimes", and western imperialism.[31]

In line with the technological development of how we create, store and consume information Cyber Influence Operations is another methodological development employed by Russia in recent years. As a confluence of influence and cyber operation capabilities it uses communication technology, networked systems and cyberspace to change or influence an audience's choices, ideas, opinions, emotions and motivations.[32] Such combined operations can serve to extract non-public information from computer systems to then release select parts, sometimes interspersed with falsified information, with the aim of discrediting the targeted individual or institution, or to paint them as responsible for acts or views that are not accurate. Systems can also be hacked or targeted by Distributed Denial of Service attacks that limit or halt their functionality, which can fuel an already existing sense of insecurity among people relying on that system, and thereby support the narrative of a concurrent influence operation.[33]

On Friday May 5 2017, two days before the second and final French presidential election round, about nine gigabytes of data extracted from the Emmanuel Macron campaign computer system was leaked.[34] Within 30 minutes a large amount of social media accounts began to disseminate the information. The timing was not a coincidence, just a few hours later the 44 hour discretionary period prohibiting media quoting the candidates close to Election Day was to begin.[35] Cyber security company Trend Micro pointed to long-standing Russian hacker group APT28 who also had targeted the Democratic National Committee prior to the 2016 U.S. elections.[36] In a 2018 high confidence assessment, the UK National Cyber Security Centre described that APT28 are associated with the Russian military intelligence service, GRU.[37] The operation has not been considered successful due to a number of reasons. French institutions and the Macron campaign were prepared and well-coordinated having observed Russian attempts at election influence against the U.S. the previous year. They also exhibited transparency, actively utilising both proactive and rapid reactive strategic communication once the leak occurred. French media, which generally retains relatively high levels of credibility among the population, was also more wary of anonymous data leaks, and employed an amplified level of source criticism. Lack of cultural understanding on the part of Russian operatives also played a role since some fabricated documents inserted into the leaked data were rather amateurish from a French audience perspective. Finally, the timing of the leak close to Election Day was surely chosen with intent, but also limited the ability to spread the hacked information widely.[38]

Building societal resilience and counter influence capability

So, drawing from lessons learned, what can be done to counter influence operations from a democratic nation state perspective? First, it is worth considering that influence operations often utilise a combination of methodologies, aimed at vulnerabilities that vary depending on the target audience and can be difficult to attribute to a specific state or non-state actor, especially in the short term. This points to the importance of preparation and resilience, maintaining reasonable expectations of what can be achieved by responding to influence operations when they are underway. It is also unrealistic to expect a universal list of tasks that would benefit all societies in raising the barrier of entry and operational response capability in this area. What rather can be done is to highlight a number of relevant categories that could serve as a foundation for more case specific action plans.

Influence operations can span over a wide range of methodologies and target areas, sometimes combined for greater effect. At the same time, public as well as private organisations, develop internal cultures and approaches focused on their core mission. This can result in gaps, internal competition and differences in perception between societal entities that can be exploited by an offensive actor. National capability development and coordination is therefore central to be able to identify, analyse and counter threats in this space.[39] It requires building a broad awareness of both relevant threat actors as well as the strengths and weaknesses of the society to be defended, recognizing for example the close relationship between societal resilience in both the cyber and influence threat space.[40] This needs to occur at, and be adapted to, the local, regional and not least national level to ensure that for example reporting of incidents not only is produced, but also can be received, collated and put into an overarching context. In addition, knowledge needs to be continuously incorporated into capability development by conducting regular exercises for key personnel and establishing forums where professionals with different skill sets can get to know each other and find ways to collaborate preemptively and operationally. This also enables each organisation to consider how influence operations relate to their area of responsibility and how countering them can be included into their everyday routines and decision making processes.

International collaboration is another area to consider since influence operations often target several countries simultaneously. Established links between like minded national entities can not only contribute to improving situational awareness, offer quality exchange of experiences through training or joint exercises, but also set the foundation for a coordinated response. This can involve combining complementary capabilities utilising for example strategic communication, diplomacy, or symbolic action through economic, military or cyber related means.

Well-coordinated strategic communication is also an important basis from which to counter influence operations. Considering the dissemination speed of the modern information environment, maintaining a continuous presence in the information environment and combining the use of long term narratives and situation-specific messaging that connects to them is vital. This enables setting a factual based frame around potentially vulnerable issues and prepares a range of key societal actors in case of an escalation, limiting the risk of getting lured into a reactive stance within a malign narrative frame initiated by an opponent.[41] In certain cases this also enables the ability to attribute a symbolic action or an influence operation, i.e. expose the actor behind it, their aims and methodology. This was for example done by the UK in the aftermath of the Skripal poisoning by Russia in 2018, and by the US releasing select intelligence to show Russian plans for influence operations prior to the invasion of Ukraine in 2022. Requirements include maintaining good situational awareness, target audience analysis, a clear view of the division of responsibility between different actors and being able to produce communication adapted from secret intelligence quickly and efficiently. The media environment is no longer dominated by large entities sharing journalistic ethical guidelines which raises the communicative demands on societal actors. A proactive ability to utilise for example social media, a greater knowledge of different audiences and a higher readiness to provide media outlets and the public with statements and relevant materials is therefore a necessity. Archetti also underscores the importance of a say-do consistency: "What matters more in the information age and the era of 'soft power' is not so much messages and ideas themselves, but the consistency between the communication of those messages and ideas and policy action. It is no good having a strategic narrative that is not based on (an easily Google-checked) reality."[42]

The media sector of today offers both challenges and opportunities. The range of outlets is considerably more diversified and fragmented than only a few decades ago. New technology and dissemination methods, coupled with increased competition, commercialization and compressed production time frames has had a considerable effect on traditional media companies.[43] The opportunities for journalists to specialise, produce in depth reporting and maintaining routines relating to fact-checking and source criticism have decreased. These especially add vulnerabilities during rapidly developing situations where there many contradictory pieces of information circulate such as in times of conflict or crisis, something which can be exploited by malign actors.

As the dominance of traditional media sharing journalistic ethical guidelines has faded the communicative demands, expectations of transparency and responsiveness are increasing for key societal actors. This points to an increased importance for government and civil society organisations alike to share knowledge and be available to provide expert commentary or background information when requested. As well as building proactive abilities to utilise social media and having a greater understanding of their different audiences and how to frame narratives and messaging to reach them. Training of journalists and programs supporting media companies following ethical guidelines of reporting can also be implemented, while upholding journalistic independence and a clear view of the different roles of media and for example government agencies.

Ongoing dialogue between government and social media platforms is another related area which is hard to ignore. It has been one of the main tools of election related influence operations by Russia in 2016 and 2017, as well as a number of other offensive state and non-state actors exploiting its potential for anonymity, speed and reach to spread disinformation. This means that counter influence professionals need to understand how the platforms work, their main target audiences and be on the lookout for new methodological developments in that space. An ability to have an exchange with social media companies about serious vulnerabilities and malign actors is also vital, even though certain topics are difficult to resolve between a nation state perspective and that of international corporations in the business of attracting users, commodification of their data and raising advertising revenue. Governments and institutions maintaining oversight, ensuring that platforms act in accordance with their terms of service, are transparent about their failures and act to diminish their recurrence is therefore an important component of a national counter influence strategy.[44]

Government institutions are in today's world not always seen as the most credible or primary source for information, and influence operations are not only a challenge for the state itself. Vulnerabilities such as alienation or radicalization of certain groups in society are therefore highly relevant also from a counter influence and resilience perspective. For example, groups that lack social trust or align with non-democratic ideas may relate more to a foreign actor that they feel is closer to their identity or goals. They are therefore receptive to influence operations, and can in certain cases knowingly, or unknowingly, be used as proxies for malign actors. Collaboration between different government agencies becomes a necessity since such problems can present themselves in many societal sectors. Support for civil society actors and academia is also important since they may have reach and credibility in related areas that government institutions do not possess. This could involve funding research, contributing to education programs, enabling local outreach programs and improving grassroots response capabilities through for example social media.[45]

In closing, the limits and use of legislation in the face of malign influence operations has garnered a lot of debate in western democracies during latter years. Regulation, and sanctions against information dissemination channels is a complex issue in open societies with freedom of speech and the press. Legislation related to the Internet, social media and data protection is also an area in constant motion. Where does a state draw the line between provocative reporting or activism and malign influence campaigns? In a fast moving information environment how can there be a clear differentiation between information and imagery that do not meet standards for source criticism and deliberate disinformation? How often can the sender responsible for a message on social media platforms be ascertained beyond a reasonable doubt? And last but not least, the consequences of any legislative action must be considered with great care to ensure that they do not violate the values of the democratic society they are implemented to protect.

Notes

1 Pamment, J., Nothhaft, H., Agardh-Twetman, H. & Fjällhed, A. (2018) Countering information influence activities: The state of the art, Lund University, p. 8
2 Wanless, A. & Pamment, J. (2019) How Do You Define a Problem Like Influence? Journal of Information Warfare 18.3: pp. 1–14.
3 Schmidt, A. (2013) The Estonian Cyberattacks, in "The fierce domain—conflicts in cyberspace 1986–2012", edited by Jason Healey, Washington, D.C.: Atlantic Council.
4 Vendil Pallin, C. & Westerlund, F. (2009) Russia's war in Georgia: lessons and consequences, Small Wars & Insurgencies, 20:2, pp. 400–424.
5 Nilsson, N. (2021) "De-hybridization and conflict narration: Ukraine's defence against Russian hybrid warfare." in Hybrid Warfare: Security and Asymmetric Conflict in International Relations. By Weissmann, M., Nilsson, N., Palmertz, B. and Thunholm, P. (eds). London: I.B. Tauris, pp. 214–231.
6 Mueller, R. S. (III) (2019) Report on The Investigation into Russian Interference in the 2016 Presidential Election, Volume I of II. Washington D.C.: U.S. Department of Justice.
7 Jeangène Vilmer, J.-B., Escorcia, A., Guillaume, M. and Herrera, J. (2018) Information Manipulation: A Challenge for Our Democracies, report by the Policy Planning Staff (CAPS) of the Ministry for Europe and Foreign Affairs and the Institute for Strategic Research (IRSEM) of the Ministry for the Armed Forces, Paris.
8 Giles, K. (2016) Russia's "New" Tools for Confronting the West: Continuity and Innovation in Moscow's Exercise of Power. Chatham House.
9 Thomas, T. (2011) Recasting the Red Star. Fort Leavenworth, KS: Foreign Military Studies Office.
10 Korotchenko, E. & Plotnikov, N. (2–17–1994) Informatsiia—tozhe oruzhie: O chem nel'zia zabyvat' v rabote s lichnym sostavom. Krasnaia zvezda (Red Star).
11 Glantz, D. (1989) Soviet Military Deception in the Second World War. London: Routledge. Frank Cass.
12 Ulfving, L. (2006) Överraskning och Vilseledning. Svenskt Militärhistoriskt Bibliotek.
13 Chotikul, Diane (1986): The Soviet Theory of Reflexive Control in Historical and Psychocultural Perspective: A Preliminary Study, U.S. Naval Postgraduate School.
14 Komov, S. A. (1997) About Methods and Forms of Conducting Information Warfare, Military Thought, No. 4.
15 Кёнтрразведывательный слёварь (Dictionary of Counterintelligence), 1972, pp. 161–2.
16 Darczewska, J. & Żochowski, P. (2017) Active Measures: Russia's key export. OSW—Centre for Eastern Studies, Point of View no. 64.
17 United States Information Agency (1988) Soviet Active Measures in the Era of Glasnost, Report to Congress.
18 Cull, N. J., Gatov, V., Pomerantsev, P., Applebaum, A. & Shawcross. A. (2017) Soviet Subversion, Disinformation and Propaganda: How the West Fought Against it. London School of Economics and Political Science.
19 Chekinov, S. G. & Bogdanov, S. A. (2013) The Nature and Content of a New–Generation War, Voennaya Mysl (Military Thought), No. 10.
20 Thomas, T. (2015) Russia: Military Strategy Impacting 21st Century Reform and Geopolitics. Fort Leavenworth, KS: Foreign Military Studies Office.
21 Rid, T. (2021) Active Measures—The Secret History of Disinformation & Political Warfare. London: Profile Books, pp. 433–4.
22 Howard, P. N., Ganesh, B., Liotsiou, D., Kelly, J. & Francois, C. (2018) The IRA, social media and Political Polarisation in the United States, 2012–2018. University of Oxford, p. 32.
23 Miskimmon, A., O'Loughlin, B. & Roselle, L. (2013). *Strategic Narratives: Communication Power and the New World Order*, 1st Ed, p. 5. London: Routledge.

24 *Russian Foreign Policy Concept* (2013).
25 Pynnoniemi, K. & Racz, A. (2016). *Fog of Falsehood: Russian Strategy of Deception and the Conflict in Ukraine*. Helsinki, FI: The Finnish Institute of International Affairs, pp. 71–109.
26 Pamment, J. (2020) The EU's Role in Fighting Disinformation: Crafting A Disinformation Framework. Carnegie Endowment for International Peace, p. 3.
27 Swedish Civil Contingencies Agency (2018) Countering information influence activities: A handbook for communicators.
28 Riedl, M. J., Strover, S., Cao, T., Choi, J. R., Limov, B. & Schnell, M.(2021) Reverse-engineering political protest: the Russian Internet Research Agency in the Heart of Texas. Information, Communication & Society.
29 Nix, N. (April 7, 2022) Facebook cracks down on covert influence networks targeting Ukraine. The Washington Post.
30 U.S. Department of State (January 20, 2022) Press statement: Taking Action to Expose and Disrupt Russia's Destabilization Campaign in Ukraine.
31 Sild, E. K. (June 1, 2022) Commentary: War in Ukraine Exposes Russia's Influence in Africa. International Centre for Defence and Security.
32 Cordey, S. (2019) Cyber Influence Operations: An Overview and Comparative Analysis. ETH Zurich, p. 5.
33 Brangetto, P. & Veenendaal, M. (2016) Influence Cyber Operations: The Use of Cyberattacks in Support of Influence Operations, NATO CCD COE.
34 Jeangène Vilmer, J-B. (2018) Successfully Countering Russian Electoral Interference, Center for Strategic & International Studies.
35 France 24 (2017-05-07) French media rules prohibit election coverage over weekend.
36 Trend Micro (2017-04-25) From Espionage to Cyber Propaganda: Pawn Storm's Activities over the Past Two Years.
37 UK National Cyber Security Center (2018-10-04) Reckless campaign of cyber attacks by Russian military intelligence service exposed.
38 Jeangène Vilmer, J-B, Escorcia, A., Guillaume, M. & Herrera, J. (2018) Information Manipulation: A Challenge for Our Democracies, report by the Policy Planning Staff (CAPS) of the Ministry for Europe and Foreign Affairs and the Institute for Strategic Research (IRSEM) of the Ministry for the Armed Forces, pp. 111–115.
39 NATO StratCom COE (2019) Hybrid Threats: A Strategic Communications Perspective, p. 28
40 The Center for European Analysis (2021) The Evolution of Russian Hybrid Warfare, pp. 44–46.
41 Paul, C. & Matthews, M. (2016) The Russian "Firehose of Falsehood" Propaganda Model. RAND Corporation, p. 10.
42 Archetti, C. (2014). *Soft power, new media and diplomacy*. Evidence submitted to the House of Lords select committee on soft power and the UK's influence. Oral and written evidence.
43 Zacharia, J. & Grotto, A. (2020) How to Report Responsibly on Hacks and Disinformation: 10 Guidelines and a Template for Every Newsroom. Stanford University Cyber Policy Center, pp. 1–7.
44 The Alliance for Securing Democracy at The German Marshall Fund of the United States (2019) European Policy Blueprint for Countering Authoritarian Interference in Democracies, p. 28
45 Polyakova, A. & Fried, D. (2019) Democratic Defense Against Disinformation 2.0, p. 3. Atlantic Council.

Bibliography

Archetti, C. (2014). *Soft power, new media and diplomacy*. Evidence submitted to the House of Lords select committee on soft power and the UK's influence. Oral and written evidence.
Brangetto, P. & Veenendaal, M. (2016) *Influence Cyber Operations: The Use of Cyberattacks in Support of Influence Operations*, NATO CCD COE.
Chekinov, S. G. & Bogdanov, S. A. (2013) *The Nature and Content of a New–Generation War*, Voennaya Mysl (Military Thought) No. 10.
Chotikul, Diane (1986): The Soviet Theory of Reflexive Control in Historical and Psychocultural Perspective: A Preliminary Study, U.S. Naval Postgraduate School.
Cordey, S. (2019) *Cyber Influence Operations: An Overview and Comparative Analysis*. ETH Zurich, p. 5.
Cull, N. J., Gatov, V., Pomerantsev, P., Applebaum, A. & Shawcross. A. (2017) *Soviet Subversion, Disinformation and Propaganda: How the West Fought Against it*. London School of Economics and Political Science.

Darczewska, J. & Żochowski, P. (2017) *Active Measures: Russia's key export*. OSW—Centre for Eastern Studies, Point of View no. 64. France 24 (2017-05-07) French media rules prohibit election coverage over weekend.

Giles, K. (2016) *Russia's 'New' Tools for Confronting the West: Continuity and Innovation in Moscow's Exercise of Power*. Chatham House.

Glantz, D. (1989) *Soviet Military Deception in the Second World War*. London: Routledge. Frank Cass.

Howard, P. N., Ganesh, B., Liotsiou, D., Kelly, J. & Francois, C. (2018) *The IRA, Social Media and Political Polarization in the United States, 2012–2018*. University of Oxford, p. 32.

Jeangène Vilmer, J.-B., Escorcia, A., Guillaume, M. and Herrera, J. (2018) *Information Manipulation: A Challenge for Our Democracies*, report by the Policy Planning Staff (CAPS) of the Ministry for Europe and Foreign Affairs and the Institute for Strategic Research (IRSEM) of the Ministry for the Armed Forces, Paris.

Jeangène Vilmer, J.-B., Escorcia, A., Guillaume, M. & Herrera, J. (2018) *Information Manipulation: A Challenge for Our Democracies*, report by the Policy Planning Staff (CAPS) of the Ministry for Europe and Foreign Affairs and the Institute for Strategic Research (IRSEM) of the Ministry for the Armed Forces, pp. 111–115.

Jeangène Vilmer, J.-B. (2018) *Successfully Countering Russian Electoral Interference*, Center for Strategic & International Studies.

Komov, S. A. (1997) *About Methods and Forms of Conducting Information Warfare*, Voennaya Mysl (Military Thought) No. 4.

Korotchenko, E. & Plotnikov, N. (2–17–1994) *Informatsiia—tozhe oruzhie: O chem nel'zia zabyvat' v rabote s lichnym sostavom*. Krasnaia zvezda (Red Star).

Miskimmon, A., O'Loughlin, B. & Roselle, L. (2013). *Strategic Narratives: Communication Power and the New World Order*, 1st Ed, p. 5. London: Routledge.

Mueller, R. S. (III) (2019) *Report On the Investigation into Russian Interference in the 2016 Presidential Election*, Volume I of II. Washington D.C.: U.S. Department of Justice.

NATO StratCom COE (2019) *Hybrid Threats: A Strategic Communications Perspective*, p. 28

Nilsson, N. (2021) "De-hybridization and conflict narration: Ukraine's defence against Russian hybrid warfare." in *Hybrid Warfare: Security and Asymmetric Conflict in International Relations*. By Weissmann, M., Nilsson, N., Palmertz, B. and Thunholm, P. (eds). London: I.B. Tauris, pp. 214–231.

Nix, N. (April 7, 2022) Facebook cracks down on covert influence networks targeting Ukraine. *The Washington Post*.

Pamment, J. (2020) *The EU's Role in Fighting Disinformation: Crafting a Disinformation Framework*. Carnegie Endowment for International Peace, p. 3.

Pamment, J., Nothhaft, H., Agardh-Twetman, H. & Fjällhed, A. (2018) *Countering information influence activities: The state of the art*, Lund University, p. 8

Paul, C. & Matthews, M. (2016) *The Russian "Firehose of Falsehood" Propaganda Model*. RAND Corporation, p. 10.

Polyakova, A. & Fried, D. (2019) *Democratic Defense Against Disinformation 2.0*, p. 3. Atlantic Council.

Pynnoniemi, K. & Racz, A. (2016). *Fog of Falsehood: Russian Strategy of Deception and the Conflict in Ukraine*. Helsinki, FI: The Finnish Institute of International Affairs, pp. 71–109.

Rid, T. (2021) *Active Measures—The Secret History of Disinformation & Political Warfare*. London: Profile Books, pp. 433–434.

Riedl, M. J., Strover, S., Cao, T., Choi, J. R., Limov, B. & Schnell, M. (2021) *Reverse-engineering political protest: the Russian Internet Research Agency in the Heart of Texas*. Information, Communication & Society.

Schmidt, A. (2013) The Estonian Cyberattacks, in *The fierce domain—conflicts in cyberspace 1986–2012*, edited by Jason Healey, Washington, D.C.: Atlantic Council.

Sild, E. K. (June 1, 2022) *Commentary: War in Ukraine Exposes Russia's Influence in Africa*. International Centre for Defence and Security.

Swedish Civil Contingencies Agency (2018) Countering information influence activities: A handbook for communicators.

The Alliance for Securing Democracy at The German Marshall Fund of the United States (2019) *European Policy Blueprint for Countering Authoritarian Interference in Democracies*, p. 28

The Center for European Analysis (2021) *The Evolution of Russian Hybrid Warfare*, pp. 44–46.

Thomas, T. (2011) *Recasting the Red Star*. Fort Leavenworth, KS: Foreign Military Studies Office.

Thomas, T. (2015) *Russia: Military Strategy Impacting 21st Century Reform and Geopolitics*. Fort Leavenworth, KS: Foreign Military Studies Office.

Trend Micro (2017-04-25) *From Espionage to Cyber Propaganda: Pawn Storm's Activities over the Past Two Years*.

U.S. Department of State (January 20, 2022) Press statement: Taking Action to Expose and Disrupt Russia's Destabilization Campaign in Ukraine.

UK National Cyber Security Center (2018-10-04) Reckless campaign of cyber-attacks by Russian military intelligence service exposed.

Ulfving, L. (2006) Överraskning och Vilseledning. Svenskt Militärhistoriskt Bibliotek.

United States Information Agency (1988) *Soviet Active Measures in the Era of Glasnost*, Report to Congress.

Vendil Pallin, C. & Westerlund, F. (2009) *Russia's war in Georgia: lessons and consequences, Small Wars & Insurgencies*, 20:2, pp. 400–424.

Wanless, A. & Pamment, J. (2019) How Do You Define a Problem Like Influence? *Journal of Information Warfare* 18.3: pp. 1–14.

Zacharia, J. & Grotto, A. (2020) *How to Report Responsibly on Hacks and Disinformation: 10 Guidelines and a Template for Every Newsroom*. Stanford University Cyber Policy Center, pp. 1–7.

Кёнтрразведывательный слёварь (Dictionary of Counterintelligence), 1972, pp. 161–162.

PART III

Contemporary Challenges

8

DISINFORMATION: THE JIHADISTS' NEW RELIGION

Jamil Ammar

Introduction

With the introduction of machine learning–based counter-terrorism technology,[1] the "conventional" digital presence of both the so-called Islamic State (IS) and al-Qaida appears to have diminished. Appearances, however, can be deceiving. In their effort to circumvent the removal of their propaganda, al-Qaida, IS, and newly emerged extreme groups are using different disinformation techniques that are enabling them to maintain a determined presence on a wide range of networking sites.

As early as 2013,[2] both al-Qaida and IS showed very high awareness of the importance of disinformation and used all means of communication to debunk any story that contradicted their narrative, going to great lengths to explain and justify their actions while sowing distrust in official accounts. By systematically and consistently describing, discrediting, and—to some extent—undermining official state narratives across platforms, both IS and al-Qaida have successfully managed to damage the credibility of already weak and corrupt political elites in Iraq and Syria and to spread and justify their violence more convincingly. Both groups use a lethally effective combination of fact-checking, ideological literacy, and audience engagement to create a culture of ideological and propaganda loyalty. Capitalizing on the economic interest of social media companies to promote emotional hyper-partisan stories that generate great disinformation; the politicization of human rights and counter-terrorism regulations;[3] and finally the lack of effectively functioning governments, especially in Syria and Iraq, where both authorities are involved in disinformation campaigns,[4] al-Qaida and IS have managed to muddy the water. So much so that, for many people in Syria and Iraq, it is oftentimes hard to agree on a basic set of facts.

As will be demonstrated in this chapter, disinformation raises acute and novel challenges to national security. Material that is manifestly illegal is not the primary concern. This type of speech can be legally taken down by platforms. In this context, falsity is not relevant to determining illegality. The greatest concern is when hate speech and disinformation overlap. In this context, disinformation that does not meet the threshold of illegal speech—such as the mostly lawful but incendiary, divisive, and emotionally charged speech that extreme jihadi groups spread—is often subject to debunking, demotion, or removal, three strategies that are unlikely to have the desired impact.[5]

This chapter aims to develop a clear picture of al-Qaida and IS disinformation strategies. To this end, based on extensive empirical research, three different disinformation techniques used by a range of extremist groups in Syria, Iraq, and Europe will be investigated. The initial goals of the analysis are to examine disinformation techniques used by extreme groups, and then to assess the extent to which moderate and radical groups differ in their disinformation strategies. But first, a few words on methodology.

Data Collection and Methodology

Deciphering al-Qaida and IS disinformation strategies is a stringent task. In the case of this study, it involved collecting cross-platform social media data (open-source data) from 2014 to March 2022. To be clear, the goal here is to examine al-Qaida and IS disinformation strategies, not to investigate the content itself posted by these terrorist groups. All data were systematically collected from multiple platforms and stored by the author without the use of technical instruments. This near real-time cross check enables a better understanding of how content is posted, shared, and consumed across different platforms. The data collected for this chapter support the theoretical discourse analysis.

To build a comprehensive view of religious extremist disinformation strategies and techniques, this chapter investigates the propaganda activities of three distinct disinformation campaigns, each led by a different group: (i) Hay'at Tahrir al-Sham, or the "Organization for the Liberation of the Levant", HTS—an al-Qaida-affiliated group operating in Syria; (ii) the so-called Islamic State in Iraq, referred to in this chapter as IS; (iii) and, finally, an ongoing disinformation campaign against the Swedish service authorities in Europe. While it is possible to identify some of the players involved in the campaign against the Swedish authorities (including known radical and "moderate" groups), the true identity of people actually behind it is still unclear.

Al-Qaida as a Moderate Entity: A New Disinformation Strategy for Public Socialisation

Al-Qaida has evolved considerably over the past decade.[6] Facing intense scrutiny from local population and stiff rivalry in the form of IS, the group has shown an impressive deal of flexibility and strategic maturation. Al-Qaida and its affiliates began their initial deliberations about modifying their media strategy just under a decade ago.[7] Two prominent jihadi scholars, Mohamad Al-Maqdisi and Abdulmonem Mustafa Halimah (often referred to as Altartousi), have initiated discussions around this new media strategy. Both scholars overtly criticise al-Qaida for publicly showcasing atrocities and for its link to international jihadi groups. Both scholars have also advocated for what might be called a new more publicly acceptable media theory.[8] Understanding how this al-Qaida media strategy in Syria has recently changed elucidates the role—and formidable challenges—of future disinformation campaigns.

Al-Qaida New Media Strategy

Hay'at Tahrir al-Sham, or the "Organization for the Liberation of the Levant", HTS is an al-Qaida-affiliated group operating in Syria. It has been systematically working to re-position itself as a moderate entity, developing a newly orchestrated media strategy for public socialisation. According to Al-Maqdisi and Altartousi, the essence of this new media strategy is twofold: Jihad and all other fundamental Islamic principles are and will remain the cause that unifies committed and devout jihadi groups; however, the way to communicate these principles to the public should be more public and media friendly.[9] To this end, among many others, Al-Maqdisi makes three major proposals.

First, a cornerstone of this new approach is to cease showcasing atrocities; that is, the Mujahideen should not record or publish the killing of civilians or the "killing or torturing of hostages." According to Al-Maqdisi, such activities enable the enemies of the Mujahideen to rally support against them.[10] The results of over eight years of examination of HTS's media activities online in Syria suggest that HTS is following this new media approach. The overwhelming majority of channels run by HTS or its supporters in Syria followed in this study publish mainly daily news and other conventional religious content.

Secondly, the Mujahideen ought not to embrace or adopt "names of jihadi groups" to advance the "tactical and political" interests of jihad, such as "al-Qaida." As long as jihad is maintained, it matters less under which flag or organisation a Mujahid fights.[11] While maintaining its jihadist agenda, the HTS publicly severed its alliance with al-Qaida in 2016. However, HTS retains an austere local Salafi-jihadist ideology. For example, in March 2017, as an example, two HTS suicide bombers attacked Damascus, Syria, killing at least seventy-four people, including eight children.[12]

Thirdly, the Mujahideen ought to develop a more "realistic" and workable plan to establish a Caliphate. Outlandish statements and policies that could unify the international community against this objective ought to be avoided, Maqdisi argues. Striking a softer tone with the West and its allies is key. In the case of Syria, as an example, for the time being, Al-Maqdisi suggests that jihadi groups not make their real intention toward Israel public. "Even if we [the Mujahideen] consider Jews our worst enemies, still we currently have a lot in our plate [busy fighting the Syrian regime] … this is not the right time … if we try too hard [fighting Jews and the Syrian regime simultaneously] we might fail."[13] This new propaganda theory, at least in Syria, departs from al-Qaida's global four-part narrative frame, described by Clark McCauley and Sophia Moskalenko.[14] Anonymity towards the West is no longer a priority, at least media wise. HTS however does not always follow this particular suggestion by Al-Maqdisi. Speaking to his supporters in the northern province of Idlib, Syria, the leader of HTS pointed out "God permitting, Jerusalem will be awaiting our arrival", suggesting a global rather than a local jihad agenda.[15]

Al-Maqdisi's suggestions seem to be working. HTS has already established a strong foothold in north Syria, and has been providing essential services to the local community for quite some time. In the same vein, HTS's current focus on recruiting a small, mostly local[16] vanguard of supporters might give it the upper hand over other international jihadi groups such as IS, at least in the short run.[17] For over five years, the head of HTS in Syria, Abu Muhammad al-Jolani, has successfully managed to maintain the territorial integrity of his mini Islamic Emirate in the northern province of Idlib, Syria. Al-Jolani has made his local authoritarian group a useful asset to pressure the Assad regime. In an interview with PBS, James Jeffrey, a special envoy for Syria under former U.S. president Donald Trump, pointed that "The United States is not targeting these people [HTS militia]. The United States is focused on our policy in Syria, which is mainly to put pressure on the Assad regime. So go draw your own conclusions."[18]

This being said, many countries continue to consider HTS, an al-Qaida affiliate.[19] The number of the successful U.S drone strikes in this area seems to suggest that Idlib has become a sanctuary for not only al-Qaida but also some IS leaders and fighters. A score of both al-Qaida and IS leaders has been killed in this region. The recent elimination of two top IS leaders in the northern province of Idlib, including the head of IS, Abu Ibrahim al-Hashimi al-Qurayshi is likely to enforce the United States to re-examine its assessment of the relationship among HTS, al-Qaida, and IS, and to re-consider its unrealistic goal of taming this jihadi group.[20]

It remains to be seen whether the latest Russian invasion of Ukraine and the subsequent increasing offensives by the Syrian government in the northern province of Idlib will bring HTS closer to other international jihadi groups. The Russian invasion of Ukraine presents an opportunity for a new rapprochement between al-Qaida and HTS. For now, evidence suggests that HTS is steadily inching towards becoming a local jihadi organisation.

Disinformation: The New Face of Radicalization Leading to Violence

By contrast with HTS in Syria, the Islamic State in Iraq has adapted a much more sophisticated multi-tiered disinformation strategy.[21] To this, we turn our attention. IS disinformation focuses on the deliberate mischaracterization of real events in coordinated social media campaigns that are intended to shape public perception and interpretation of these events. To influence public understanding and political responses, IS combines two disinformation techniques defined by Martin Innes as "fogging" and "flooding."[22]

"Fogging" basically involves the spreading of multiple interpretations of an event related, in our case, to IS or to its supporters or allies. The essence of these alternative interpretations of real events is to sow seeds of doubt and confusion about the underlying causes of these events. This tactic is also intended to increase hostility towards the political establishment. This disinformation technique is often followed by multiple posts across platforms to ensure that the information space is dominated by a particular confounding message (or, as called by Innes, "flooding"). According to this disinformation strategy, a distorted or deceptive message reinforces and reproduces the perception of mistreatment and unfair targeting of jihadi groups and their supporters.

To ensure the success of its disinformation strategy, IS deploys different, mainly Telegram, channels to share content with its audience. In this context, Telegram's different channels play different roles within the network. Some Telegram channels insure a steady flow of conventional propaganda content online daily. In this chapter, we will refer to this type as "Class A Channels". The titles used for such channels are generic with a clear symbolic but no political, religious, or security connotations. This class of Telegram channels covers political and military events only. To enhance the credibility and trust of this type of channel, all covered events are real. Published military operations are always combined with images of victims and their real names. With a few narrow exceptions, to reveal the true identity of this class of channels (as an IS-leaning channel), terms such as IS, the "Islamic State', "Al Dowla", "jihad", or any key term that could raise a red flag is commonly avoided. In the same vein, graphic images and videos are not posted. Only images that are available widely in many other social media platforms.

In this class of channels, two disinformation strategies are used to construct and transmit disinformation under the guise of journalistic credentials. The first covers IS operations against the Iraqi government and its allies. The second covers security operations against IS personal. Under the first disinformation strategy all successful IS operations are described as:

1   "Car accidents that led to the death" of individuals, often police officers, military personals, Iraqi from the Shia ethnic group (Shia Muslims are a branch of Islam that IS strongly despises), or civilian working for the Iraqi government (all perceived as IS enemies). For example, on May 18, 2021, at 18:05 PM, a representative of IS posted the following:

"Tigris#Al Sharkat, the death of member of the Mobilization Forces … [al-Ḥashd ash-Shaʻbī- Shia group] in a car accident."

2   Kidnaping of individuals or a freak incident such as fire that destroys crops or property. For example, on May 23, 2021, at 05:26 AM, a representative of IS posted the following: "North Baghdad# a fire destroys three orchards in Al Bofaraj-Al Tarmiya."

3   Destruction of power towers or other industrial facilities. For example, on June 09, 2021, at 05:17 AM, a representative of IS posted the following: "Saladin# Tigris, due to the use of explosives, the ministry of electricity announced the destruction of three power towers … ". All these operations are combined with real images.

4   Terrorist attacks: "Breaking News: Terrorists publish images of the terrorist ambush against the Shia army in Dyala"; "Breaking News: Terrorist attack in Tor Ghoz Matto."

Under the second disinformation strategy news related to IS supports is covered by communicating multiple explanations and interpretations of events related to IS followers/supporters with particular focus on "unjustified attacks against innocent victims" or "unfair treatment" by the "secular" political establishment. On May 22, 2021, at 07:48 AM, a representative of IS posted the following: "Dyala# the death of a detainee in Al Naseya prison in Baghdad under torture"; "the Shia Mobilization Force kidnaped a Sunni woman in Mosul."

A second category of channels, namely, Class B Channels, functions similar to Class A, with a few notable differences: In this type of channel, the name of IS, commonly referred to as "Al Dowla", is mentioned few times. IS fighters, however, are conspicuously referred to as "Mahdieen" or "Jaysh al-Mahdī". In this context, it is worth noting that "Jaysh al-Mahdī (JAM)" is one of the most powerful Shia militia accused of contributing heavily to the civil conflict in Iraq.[23] The use of the "Jaysh al-Mahdī" title by two competing militias (IS—Sunni group, and Jaysh al-Mahdī—Shia group) is likely to further complicate the process of countering disinformation online.

While it does not publish graphic content, Class B Channels compete for attention and credibility by being the first to report on the details of insurgency activities to be published on other channels. The integrity of the substantive material advertised on this type of channel is difficult to compromise or question, and is always combined with temporary timed hyperlinks to newly created jihadi channels on which propaganda materials can be freely accessed. For extra clarity, IS propaganda materials are not directly posted on this type of channel, and all hyperlinks are active for a limited time, only to then be de-activated. Some channels in this class attract dozens of thousands of followers per channel. Apart from the hyperlinks, all posted materials are open and available to the public. Notably, the same content published on a Class A Channel is simultaneously re-posted on a Class B Channel in high volumes and frequency, to make it abundantly visible to its audience. Different channels with different names are used to spread the same message in Arabic, English, French, and Kurdish.

A third type of channel, Class C, provides more comprehensive articles to explain IS views, comment on recent jihadi and political events, and—at the same time—debunk stories against IS. Different Telegram channels are used to communicate the same message/article in different languages. No images or videos are posted in this class.

A fourth type, Class D Channels, operates with the sole purpose of inviting users to join these channels so that if/when a major IS channel is deleted, users will be directed to a new one. IS admins do not publish any material on these channels. Still, the number of followers for some of these channels exceeds a few thousands.

Finally, a fifth type, Class E Channels, provides full access to conventional graphic IS images and videos. This type of channel is only accessible via the temporary timed hyperlinks provided by IS admins on other channels such as Class B. Graphic content published on this type of channel is usually accessible for a couple of hours only. All Class E Channels are hidden within a short period of time, and a new channel is created for each major insurgency event such as new group swearing allegiance to IS around the world or the appointment of a new caliph.

Samuel Paty and the "Kidnaping of Muslim Children" in Sweden

After examining al-Qaida and IS disinformation strategies in Syria and Iraq, we move our attention to another ongoing disinformation campaign in Europe. The Swedish social services are reeling under immense pressure to debunk an orchestrated campaign accusing some of its social service officials of systemically and unfairly targeting Muslim families in Sweden and taking their children into care.[24] The Swedish authorities have been battling this rolling global disinformation campaign for some time with little success. This disinformation campaign asserts that Swedish social services routinely kidnap Muslim children in an attempt to Westernise them.[25]

One particular YouTube video, published by Shuounislamiya,[26] compounds conspiracy theories with emotional and incendiary claims such as "kidnapping of Muslim children", "Muslim children are placed with paedophiles and are being raped", " children forced to eat pork", and a child's family described as "Muslim terrorists."[27] Multiple social media accounts with links to violent extremist groups are involved in this scheme, which prompted Swedish authorities to warn of violent threats being made against social services personnel. Some Swedish investigative news sites[28] exposed this disinformation campaign and alerted that the names of individual social service workers had been published.[29]

Truly troubling is that this ongoing global disinformation campaign has not only gained traction with some extreme groups, but also with a much wider audience. Even some academics and public speakers with no link to extreme groups are enchanted by this conspiracy theory and have endorsed the Shuounislamiya's propaganda efforts.[30]

In this context, if a Samuel Paty 2.0 catastrophe is to be averted, the threats against the Swedish social services ought to be taken very seriously.[31] Samuel Paty, a French teacher who believed in the public school's role in assimilating immigrant children into the French society, was beheaded by Abdoullakh Anzorov in a horrifying and shocking crime. Paty's beheading highlights the challenges faced by many racially and ethnically diverse countries in Europe and the struggle to integrate generations of newcomers into their societies.

Abdoullakh Anzorov was allegedly offended by Paty's choice to show cartoons of the Prophet Muhammad in a class on free speech.[32] Abdoullakh was informed and enraged by a cynical social media campaign that led to him to murder Paty. Most of the claims made in that disinformation campaign have since been debunked, and the French schoolgirl whose complaints sparked the online campaign against Paty has admitted to spreading false claims and that she did not, in fact, see the cartoons.[33] The lesson here is chilling: disinformation can and did lead to violence; for this reason, the Swedish authorities' warnings ought to be taken with all gravity.

Three notable complicating factors warrant discussion. The first is that this conspiracy theory is gaining support with very wide audience. It is sometimes difficult to verify the political affiliation of groups active online. Both names and affiliations change regularly. However, it is unrealistic to assume that all who believe and support the disinformation campaign against the Swedish authorities are al-Qaida or IS followers. Clearly, many are not.

People involve in this disinformation milieu are not necessarily violent radicals.[34] A possible explanation could be that some individuals got involved in this conspiracy theory as a symbolic act of brotherhood. The ideological pressure to defend and justify the actions of other Muslims against out-group fuels disinformation and lures some members of the Muslim Community to believe in absurd forms of conspiracy theories. As the perceived sense of alienation or dis-crimination grows, the ideology of radical groups and their conspiracy theories becomes more and more expanded and enriched. In the case of Sweden, as an example, one may interpret this absurd believe in disinformation and conspiracy theories about the kidnapping of Muslim kids as a medium through which some individuals vent their frustration of the childcare system rather than as a support to terrorist groups.

This distinction is relevant because, apart from IS's graphically violent content, very little differentiates how peaceful Salafi groups and designated terrorist groups such as al-Qaida or IS conduct their propaganda campaigns.[35] Both peaceful and violent groups are equally persuasive. Both types of groups have a perceived sense of injustice and adhere to the misguided and clearly untenable view that Muslims are mistreated in the West. For these reasons, both Salafi and radical groups are more than happy to endorse disinformation campaigns that falsely claim to defend the Muslim community at large. In the case of Sweden, to substantiate dubious claims, the opinions and suggestions provided by supporters of this conspiracy draw upon real, individual stories from parents who feel discriminated against.[36] That peaceful online activists who are not known for their extreme views adhere to an austere interpretation of faith steams the process of countering disinformation. Due to this ideological bias, even the wholesale banning of super influencers or disinformation spreaders is unlikely to find success. Focusing the discussion solely on designated terrorist or extreme groups such as IS or al-Qaida does not help either. In this context, the shared ideological base is a prominent influencing factor; not only does it affect the speed at which a disinformation campaign is spread, but it also could impede efforts to debunk such a conspiracy in the first place.

Concluding Remarks

The Jihadist new media strategy intentionally combines both hate speech and disinformation. This new media strategy raises novel challenges to national security. Disinformation that does not meet the threshold of illegal speech—such as IS disinformation strategies used to construct and transmit disinformation under the guise of journalistic credentials or the orchestrated campaign again the Swedish social services, as mere examples — often stays online for an extended period of time.

For propaganda purposes, extremist groups utilise mostly publicly accessible social media platforms to attract the attention of their targeted audience. Regardless of the communication channel used by these extremists, the aim of using publicly available communication channels remains the same; to spread propaganda material widely so that ulti-mately the content will be found and consumed by the targeted audience. Both, al-Qaida's new media strategy to re-position itself as a moderate entity, and the Islamic State's use of news reporting and journalistic credentials to deliberately mischaracterize real events in coordinated social media campaigns that are intended to shape public perception are likely to significantly eliminate the need to use secured communication channels, such as Telegram, to communicate with targeted audiences.

Al-Qaida and IS new media strategies will make the already-challenging task of identifying terrorist speech even harder, as it used to typically contain violent scenes and key terminol-ogy. This is no longer "always" the case. The anonymity feature of social media platforms will also become less relevant, opening the door for wider more publicly open communication tools, thus more clandestine support for terrorists' agenda.

Countering disinformation that leads to violence is an almost insoluble task. Various fundamental political, social, economic, legal, and technological limitations continue to challenge different aspects of this puzzle.[37] A coherent decision-making process relies on the average individual's ability to assess and differentiate fact from fiction. If individuals do not have access to accurate information or are too confused to make sense of the facts on the ground, they are unlikely to be able to draw rational conclusions about the best course of action. In the case of Sweden, a potential policy failure can, in some cases, become a threat to the public or the government itself. If the tragedy of Samul Paty teaches us anything, disinformation can lead to violence.

The challenge of curbing disinformation campaigns is redoubled by the recent Russian invasion of Ukraine and the formation of the so-called international brigade of foreign "volunteers" (a more acceptable term than foreign fighters or Mujahideen) to fight Russia.[38] Facebook and Instagram's recent revision of hate speech rules to permit posts calling for violence against Russians and the death of Putin has not gone unnoticed.[39] These misguided policies have already further fueled jihadi extremists' erroneous perception of being unfairly treated. This discrepancy in the media's representation of foreign "volunteers" and hate speech–related polices are likely to encourage an environment in which disinformation and antipathy towards non-Muslims can be readily believed and easily justified.

Notes

1 Ammar, "Cyber Gremlin," 238–265; House Homeland Security, "Artificial Intelligence and Counterterrorism;" Crime and Justice Research Institute, "The Malicious Use Of Artificial Intelligence," 1–58; The United States Government, "The Global Engagement Center;" Macdonald, Correia and Watkin, "Automation and the Rule of Law,"183–197.
2 Al-Maqdisi, "Alwasaya Alaaliya," 8–10.
3 In both Syria and Iraq counter terrorism regulations are heavily used to suppress political dissidents and to justify crimes against civilians. Human Rights Watch, "Report on Syria"; Human Rights Watch, "Report on Iraq."
4 CIJA, "Syria Disinformation Campaign;" France 24, "Virtual Battlegrounds;" al-Kaisy, "Disinformation in the Iraqi media."
5 Tompros, Crudo, Pfeiffer, and Boghossian, "Social Media-Obsessed World," 66–108; McArdle, "Deception spreads faster than truth;" The Canadian Government, "Changes to the Canadian Human Rights Act;" Ó Fathaigh, Helberger, and Appelman, "Defining disinformation;" Bazelon, "In the age of disinformation;" Berger and Morgan, "The ISIS Twitter Census;" Ammar, "Religious Hate Speech," 97–124; Ammar, "Cyber Gremlin," 238–265; Ammar, and Xu, "Ideology Meets Today's Technology;"
6 Observations are based on eight years examination of al-Qaida activities online.
7 Al-Maqdisi, "Alwasaya Alaaliya," 8–10; Altartousi, "Jabhat al-Nusrah."
8 Ibid.
9 Id.
10 Id.
11 Id. Ammar, "Religious Hate Speech," 97–124.
12 Public Safety Canada.
13 Al-Maqdisi, "Alwasaya Alaaliya," 8–10.
14 These are: Islam is under attack by the United States and its allies; jihadist are defending against this attack and their actions are religiously sanctified; and it is the duty of Muslims to support these actions. McCauley and Moskalenko, "Lone Wolf Terrorists," 70.
15 SITE, "HTS."

16 HTS in Syria still enjoys the support of the formidable Uyghur militant group, the Turkistan Islamic Party in Syria, and to a lesser extend Saudi and Tunisian groups. Most of those fighters, however, have previously defected from IS.
17 CSIS, "Salafi-Jihadist Threat."
18 Jeffrey, "U.S. Ambassador."
19 Public Safety Canada.
20 BBC News, "Qurayshi killed in Syria."
21 With regards to IS disinformation strategy in this section, observations are based on four years examination of IS activities online with particular focus on Syria and Iraq.
22 Innes, "Countering Extremist Mis/Disinformation."
23 CISAC, "Mahdi Army."
24 Doku.nu, "Sweden is Attacked Online."
25 Ibid.
26 Ltaif, "Kids in Sweden?" (m10-m13). As of March 28, 2022, this You Tube channel is followed by 646000 subscribers. This very same video is viewed 156,769 views.
27 Ibid.
28 Id.
29 Id.
30 A very notable example is Eyad Qunaibi, a charismatic speaker with over 66703 Telegram followers and 1.31 million YouTube subscribers. He posted two videos (one was viewed 483,047 times and the other gained 295,509 views as of March 28, 2022) specifically supporting and recommending the Shuounislamiya's coverage of this story in Sweden. Both videos are in Arabic; Qunaibi, "Do Not Flee to Sweden Please;" Qunaibi, "The Kidnapping of Muslim Children."
31 Henley, "Samuel Paty murder."
32 The cartoons in question are related to the Prophet Mohammed which was published in Danish newspaper *Jyllands-Posten*. Portraying the Prophet Mohammed is strictly forbidden in Islam, and for this reason, the cartoons were considered highly offensive by many Muslims around the world.
33 BBC. News, "French schoolgirl admits lying;" Euronews, "Samuel Paty murder."
34 The concept of "radical milieu" was first raised by Malthaner and Waldmann, "The radical milieu," 979–998.
35 Ammar and Xu, *When Jihadi Ideology Meets Social Media*, 46.
36 Ltaif, "Kids in Sweden?" (m10-m13).
37 Ammar, "Cyber Gremlin," 12.
38 Harding, "Fighters flocking to Ukraine."
39 Dwoskin, "Violence against Russian invader."

Bibliography

Al-Maqdisi, M. "Alwasaya Alaaliya Leansar Alsharia Algalyia". Archive.org. Last modified 2013. https://archive.org/details/doctor25250_gmail_20170317_1732/page/n9/mode/1up?view=theater.
Al-Kaisy, A. "Covid-19 and the dangers of disinformation in the Iraqi media." *Ethical Journalism Network*. Accessed March 31, 2022. https://ethicaljournalismnetwork.org/covid-19-and-the-dangers-of-disinformation-in-the-iraqi-media.
Ammar, J. "Cyber Gremlin: Social Networking, Machine Learning and the Global War on Al-Qaida-and IS-Inspired Terrorism." *International Journal of Law and Information Technology*, Issue 3, (Autumn 2019): 238–265. https://doi.org/10.1093/ijlit/eaz006.
Ammar, J. "Counterterrorism and the Commercialization of Religious Hate Speech". In *Novos Estudos Sobre Liberdade Religiosa, Risco de Segurança no Século XXI*, edited by Davide Argiolas, 97–124. Petrony: Aogsto, 2018.
Ammar, J. and Xu, S. *When Jihadi Ideology Meets Social Media*. Cham: Switzerland Palgrave Macmillan, 2018.
Ammar, J. and Xu, S. "Yesterday's Ideology Meets Today's Technology: A Strategic Prevention Framework for Curbing the Use of Social Media by Radical Jihadists, 26 *Journal of Science and Technology* (2016). https://www.albanylawscitech.org/article/19249.
Bazelon, E. "The First Amendment in the age of disinformation." *The New York Times* (October 13, 2020), https://www.nytimes.com/2020/10/13/magazine/free-speech.html.
BBC. News, "Islamic State leader Abu Ibrahim al-Qurayshi killed in Syria, US says,". Last modified February 4, 2022, https://www.bbc.com/news/world-middle-east-60246129.

BBC. News, "Samuel Paty: French schoolgirl admits lying about murdered teacher". Last modified March 9, 2021, https://www.bbc.com/news/world-europe-56325254.

Berger, J. and Morgan, J. "The ISIS Twitter Census: Defining and Describing the Population of ISIS Supporters on Twitter" (2015) Brookings Project on U.S. Relations with the Islamic World Analysis Paper No. 20, https://www.brookings.edu/research/the-isis-twitter-census-defining-and-describing-the-population-of-isis-supporters-on-twitter/.

CIJA. "Statement on the Syria Disinformation Campaign." Accessed March 26, 2021. https://cijaonline.org/news/2021/3/26/cija-statement-on-the-syria-disinformation-campaign.

Doku.nu, "*Sweden is Attacked Online Due to the LVU Law,*" February 2, 2022, https://doku.nu/2022/02/02/sweden-is-attacked-online-due-to-the-lvu-law/.

Dwoskin, E. "Facebook breaks its own rules to allow for some calls to violence against Russian invader," *The Washington Post,* March 10, 2022, https://www.washingtonpost.com/technology/2022/03/10/facebook-violence-russians/.

Euronews, "Samuel Paty murder: One year on, what impact has the teacher's killing had in French schools?" last modified October 2021, https://www.euronews.com/2021/10/15/samuel-paty-murder-one-year-on-what-impact-has-the-teacher-s-killing-had-in-french-schools.

GNET- The Global Network on Extremism and Technology Countering. "Extremist Mis/Disinformation After Terror Attacks." Accessed March 31, 2022. https://gnet-research.org/2021/11/08/fogging-and-flooding-countering-extremist-mis-disinformation-after-terror-attacks/.

Harding, L. "My plan is there is no plan': the foreign fighters flocking to Ukraine", *The Guardian*, March 11, 2022, https://www.theguardian.com/world/2022/mar/11/ukraine-russia-war-foreign-fighters-volunteers.

Halimah, M. "Monakashat Hawael Makalatana Bekosus Al-enthemam Ela Jabhat al-Nusrah". Accessed March 30, 2022. www.abubaseer.bizland.com/. Also available via this link. https://www.enabbaladi.net/archives/31029 (Arabic text).

Henley, J. "Samuel Paty murder: French police raid dozens of Islamist groups," The Guardian.com, October2020, https://www.theguardian.com/world/2020/oct/19/samuel-paty-french-police-raid-dozens-of-islamist-groups.

Human Rights Watch. "*Human Rights Watch Report on Syria 2020.*" Accessed March 30, 2022. https://www.hrw.org/world-report/2021/country-chapters/syria#.

Human Rights Watch. "*Human Rights Watch Report on Iraq 2020.*" Accessed March 30, 2022. https://www.hrw.org/world-report/2021/country-chapters/iraq.

Jeffrey J. "Former U.S. Ambassador to Iraq and Turkey, interview by filmmaker Martin Smith." Accessed March 31, 2022. https://www.pbs.org/wgbh/frontline/interview/james-jeffrey/.

Ltaif, Z. "What is Happening to Kids in Sweden?" Shuounislamiya December 28, 2021, You Tube Video, https://www.youtube.com/watch?v=BIdUUdGzgaQ&t=200s. See in particular m10-m13. As of March 28, 2022, this You Tube channel is followed by 646000 subscribers. This very same video is viewed 156,769 views.

McArdle, E. "Oh, what a tangled web we weave. Deception spreads faster than truth on social media. Who — if anyone — should stop it?," *Harvard Law Bulletin*, Summer 2021, Jul 07, 2021, https://hls.harvard.edu/today/oh-what-a-tangled-web-we-weave/.

McCauley, C. and Moskalenko, S. "Toward a Profile of Lone Wolf Terrorists: What Moves an Individual From Radical Opinion to Radical Action." *Terrorism and Political Violence* Volume 26, Issue 1 (2014): 70. https://doi.org/10.1080/09546553.2014.849916.

Macdonald, S. Giro, S. and Watkin. A. "Regulating Terrorist Content on Social Media: Automation and the Rule of Law", *International Journal of Law in Context*, Issue 2 (June2019):183–197. https://doi.org/10.1017/S1744552319000119.

Malthaner, S. and Waldmann, P. "The Radical Milieu: Conceptualizing the Supportive Social Environment of Terrorist Groups." *Studies in Conflict and Terrorism*, Vol. 37, No. 12 (2014): 979–998. https://doi.org/10.1080/1057610X.2014.962441.

Ó Fathaigh, R., Helberger, N. and Appelman, N. "The perils of legally defining disinformation," *Internet Policy Review*, 10(4), (2021). https://doi.org/10.14763/2021.4.1584.

Stanford University, Center for International Security and Cooperation. "Mahdi Army." Accessed March 31, 2022. https://cisac.fsi.stanford.edu/mappingmilitants/profiles/mahdi-army#highlight_text_16994.

The Center For Strategic and International Studies (CSIS). "*The Evolution of the Salafi-Jihadist Threat.*" Accessed March 31, 2022. https://www.csis.org/anlysis/evolution-salafi-jihadist-threat.

The Government of Canada. Public Safety Canada. *Currently Listed Terrorist Entities*. Public Safety Canada, 2020. https://www.publicsafety.gc.ca/cnt/ntnl-scrt/cntr-trrrsm/lstd-ntts/crrnt-lstd-ntts-en.aspx#29.

The Canadian Government, "*Combatting hate speech and hate crimes: Proposed legislative changes to the Canadian Human Rights Act and the Criminal Code*," https://www.justice.gc.ca/eng/csj-sjc/pl/chshc-lcdch/index.html.

The United States Government. *House Homeland Security: Artificial Intelligence and Counterterrorism: Possibilities and Limitations.* Congress.gov. https://www.congress.gov/event/116th-congress/house-event/LC64673/text?s=1&r=1.

The United State Government. *The Global Engagement Center to Counter the Messaging and Diminish the Influence of International Terrorist Organizations, Including the Islamic State and Al Qaida.* Accessed March 30, 2022, https://2009-2017.state.gov/r/gec/index.htm#:~:text=The%20Global%20Engagement%20Center%20is,Obama%20on%20March%2014%2C%202016.

Tompros, L., Crudo, R., Pfeiffer, A. and Boghossian, R. "The Constitutionality Of Criminalizing False Speech Made On Social Networking Sites In A Post-Alvarez, Social Media-Obsessed World," *Harvard Journal of Law & Technology*, Volume 31, (Number 1 Fall 2017), 66–108.

United Nations Interregional Crime and Justice Research Institute (UNICRI—A Joint Report by UNICRI and UNCCT-). *Algorithms And Terrorism: The Malicious Use Of Artificial Intelligence For Terrorist Purposes.* Accessed March 31. https://www.un.org/counterterrorism/sites/www.un.org.counterterrorism/files/malicious-use-of-ai-uncct-unicri-report-hd.pdf.

Qunaibi, E. A charismatic speaker with over 66703 Telegram followers and 1.31 million You Tube subscribers. He posted two videos (one was viewed 483,047 times and the other gained 295,509 views as of March 28, 2022) specifically supporting and recommending the Shuounislamiya's coverage of this story in Sweden. Both videos are in Arabic; Qunaibi, E. "Do Not Flee to Sweden Please," Feb 12, 2022, You Tube Video, https://www.youtube.com/watch?v=nuguVuKHasc&t=1s;

Qunaibi, E. "The Kidnapping of Muslim Children in the West," January 28, 2022, You Tube Video, https://www.youtube.com/watch?v=ZZz7x8owUlQ&t=459s.

9

DIGITAL DISINFORMATION, ELECTORAL INTERFERENCE, AND SYSTEMIC DISTRUST

Josephine Lukito

Despite the so-called principles of Westphalian sovereignty, countries have historically sought to intervene in other states' affairs to achieve foreign policy goals, employing a multitude of tactics in the hopes that a foreign country will elect a government leader that is sympathetic to the intervening country.[1] However, the efficacy of these tactics is questionable. Coercive tactics, such as military interventions, can be especially costly, ineffective and even prone to backfiring.[2]

One area that has seen a rise in electoral interference from foreign actors, however, is in digital communication systems. Digital disinformation, defined as intentionally false information that is shared between two or more individuals through a computer or digital technology, has been found in many subjects, from health communication[3] and science information[4] to "fake news" about social issues.[5] In the context of political communication, digital disinformation is most often studied in the context of democratic elections. More specifically, scholars, political actors and citizens alike have expressed concerns that foreign actors utilize digital disinformation to interfere in democratic elections, in both established and emerging democracies.[6]

This raises the following questions: why are digital media vulnerable to disinformation? How effective are digital disinformation campaigns at interfering in elections? And what can democracies do to combat digital disinformation?

In this chapter, I seek to answer these questions. To do so, I first highlight the vulnerabilities of the digital media ecology that make it easy for malicious actors to disseminate disinformation. Next, I discuss the tactics of disinformation campaigns during elections, with a particular focus on troll armies, one of the most common tactics employed by foreign actors. Then, I discuss and highlight one well-known case: Russia's Internet Research Agency and their digital disinformation campaign during the 2016 and 2020 U.S. Presidential election. Finally, I end with some recommendations for democracies seeking to reduce digital disinformation during their election periods.

Disinformation in the Digital Era

Though governments have long sought to interfere in other countries' democratic processes, digital media makes disinformation production even easier. There are three key reasons for this. First, digital disinformation is much cheaper to produce compared to other forms of

DOI: 10.4324/9781003190363-12

foreign intervention.[7] Because governments are able to effectively participate in the public spheres of other countries without physically having to go to that country, there are substantially fewer costs associated with producing content to manipulate these public spheres.

Second, digital disinformation exploits the anonymous nature of online accounts and pseudonyms. Because disinformation, as a form of black propaganda, relies on the ability to hide one's identity,[8] the internet is a prime space for state actors to spread disinformation. Disinformation producers may attempt to further obscure their true identity by using virtual private networks (VPNs) to conceal their IP addresses or to give the appearance that their IP address originates from somewhere else. Additionally, the anonymity of online profiles and identities facilitates disinformation at two levels: false personas and false accounts. Even when a disinformation account shares predominantly true information, it is often couched in a false identity sharing fake and politically motivated opinions. To create this identity and exploit online anonymity, disinformation accounts often present as if they are a citizen of the country being targeted.[9]

Third, the ability to utilize automation allows malicious actors to create and spread unparalleled amounts of political disinformation that can target large political groups as well as specific and niche communities. Often, to do this, disinformation is spread using automated accounts, also known as bots.[10] While not all bots are bad,[11] malicious actors rely heavily on bots to amplify computer-mediated disinformation, making these messages appear more popular than they actually are.[12] Bot accounts can "follow" other disinformation accounts managed by real people, making the latter appear more popular. Disinformation actors can also use bots to share (or "retweet") and like disinformation content, increasing the messages' perceived social value. Another strategy that disinformation actors can use to manipulate their follower account is to buy followers from a third party.[13]

Though any disinformation actor can (and do) exploit these considerations, state-sponsored production benefits from an overabundance of resources that allow them to scale up the production of digital disinformation. One person can produce only so much digital disinformation, but a government can hire hundreds of individuals to work together to conduct a digital disinformation campaign, using state-surveillance strategies to inform their content production. It is also worth re-emphasizing that state-sponsored disinformation campaigns serve broader political goals, including framing discourse around national security affairs,[14] coordinating with other information warfare tactics like cyberattacks and hacks,[15] or damaging social media platforms as spaces for public discourse. For example, state-sponsored disinformation campaigns can help disseminate information that is stolen through other cyberwarfare strategies.

Digital Disinformation and Elections

As of 2019, over 44 government agencies employed some sort of digital disinformation campaign.[16] The activities of these campaigns may be especially noticeable during elections, a time when citizens in a democracy may pay closer attention to their media systems to make informed decisions about voting.[17] In other words, political disinformation campaigns—coordinated attempts to spread disinformation—may be especially powerful during times of planned bureaucratic shifts.

However, it is worth noting that, during elections, disinformation campaigns have both short-term *and* long-term goals. In the short-term, these goals focus primarily on maligning a candidate or discouraging voting.[18] For example, during the 2018 Brazilian election, then-candidate Jair Bolsonaro utilized WhatsApp and Twitter to spread disinformation about electronic ballots in hopes that this would help Bolsonaro claim fraud in the case that he lost.[19] Long-term goals, on the other hand, focus on sowing distrust in the political system,[20] which can have lasting and compounding consequences. Even when this disinformation is identified and removed, the presence of it at all may produce distrust in a country's media ecology.

To achieve both their long-term and short-term goals, states utilize a variety of mechanisms, including intentional actors, and unintentional processes that exploit existing flows in the target country's media system.

One of the most popular tactics presently is the use of social media troll armies, sometimes known as troll farms. Troll armies engage in the coordinated production of disinformation messages by humans using social media accounts with false identities; these social media accounts are called "sockpuppets."[21] The disinformation that troll armies produce is a form of *astro-turfing*, a type of disinformation where messages funded by a government are meant to look like they come from regular people or grassroots organizers.[22]

Because countries utilize troll armies to serve different purposes—including trying to make their own country look good, delegitimizing social movements, or framing public conversation about wedge issues—the disinformation messages are as varied as the goals they serve. Notably, the content that troll armies (and other disinformation actors) spread does not inherently have to be false—computer-mediated disinformation includes both the dissemination of false information and the use of false identities to advocate for insincere opinions.

The relationship between governments and their troll armies varies greatly. Some countries simply outsource the production of disinformation to a private company (like the United States' Operation Earnest Voice, which is maintained by the private company Ntrepid).[23] Alternatively, political operatives can play a more active role in the development of messages (like the 50 Cent Party, which is managed by Chinese authorities).[24] And yet others are somewhere in the middle, functioning as a somewhat independent group but with affiliations to government politicians (like Russia's Internet Research Agency).[25]

Another tactic, as I have noted, is the use of bots, automated accounts that can help amplify disinformation messages. While they are easier to identify compared to the "organic" production of troll armies, bots have the advantage of speed—it is much faster for a computer to retweet 1,000 messages compared to a human. And perhaps most importantly, disinformation campaigns will use bots and sockpuppets in tandem to spread disinformation, with sockpuppets creating the disinformation content and bots amplifying that content by falsely liking and sharing it.[26]

These intentional disinformation actors also exploit the anonymous nature of the digital ecosystem to encourage citizens of their target country to share disinformation as misinformation, making it even more difficult to remove the source of these falsities. To do so, disinformation campaigns will exploit societal cleavages, including (but not limited to) differences in socioeconomic status, ethnicity, religion and gender identities/sexual preference.[27] When citizens are confronted with misinformation that confirms their prior beliefs, they may be more likely to share it, effectively making citizens unwitting spreaders of disinformation.[28]

In addition to utilizing unwitting citizens to spread false information, disinformation campaigns will also target news organizations, politicians and other public figures to "trade [information] up the chain,"[29] a term first coined by Ryan Holliday to refer to the process of amplifying false information so that it is picked up by larger and larger media outlets until it reaches mainstream acknowledgement. During an election, when news organizations are hungry for political information, they may be especially vulnerable to helping trade disinformation up the chain. As a result, they too become unwitting spreaders of false information.

Knowing the key role that news organizations and public figures play as information providers, foreign disinformation actors are likely to target these organizations, sometimes by presenting themselves as news organizations[30] or by at-mentioning ("@mentioning") news organizations' Twitter accounts.[31]

It is worth noting that these information tactics do not operate in isolation—typically, disinformation actors (particularly those backed by foreign states) will use a variety of malicious tactics to help inform and produce digital disinformation. One such tactic is to combine disinformation with surveillance, as Russia did in 2016,[32] to help produce more salacious disinformation. Additionally, foreign actors have also combined disinformation with mal-information, which is true and often sensitive information that is stolen or exaggerated to make a political opponent look bad.[33]

To understand these tactics in action, I now turn to a well-known digital disinformation campaign as an exemplar: Russian disinformation during, and in between, the 2016 and 2020 U.S. Presidential election.

Case Study: Russian Disinformation in the United States

Perhaps the best-known modern case of foreign influence is Russia's Internet Research Agency (IRA). The IRA targeted elections and voting periods around the world, including the 2016 and 2020 U.S. Presidential election, the 2016 Brexit vote, and the 2021 Ugandan general election.

Prior to infiltrating the U.S. digital media ecology, the IRA was already active within Russia and post-Soviet countries, including Ukraine, where Russia sought to build support for their annexation of Crimea.[34] Even in these early stages of Russian digital disinformation, it was clear that these disinformation tactics were informed in great part by disinformation strategies employed by the Soviet Union including the use of active measures[35] to distract citizens from important news stories, distort information, dismiss claims that made Russia appear as the aggressor, and dismay audiences with inflammatory political statements.[36]

Though difficult to define because of its scope, active measures broadly refer to a range of tactics employed by Russia (formerly, the Soviet Union) to subvert and undermine their opposition.[37] A key characteristic of active measures is its persistence and repetition—active measures rely on a multitude of strategies to produce a torrent of misleading messages that, over time, wear down the opponent to the point that the opponent cannot distinguish fact from fiction. While developed long before the internet, the digital media ecosystem is particularly optimal for active measure tactics when one considers how troll activity and bot amplification can be combined.

To achieve their short-term goal of getting Donald Trump elected in the 2016 and 2020 election,[38] and their long-term goal of destabilizing the U.S. media ecosystem, the Internet Research Agency produced a vast amount of disinformation, across multiple social media platforms, in advertisements, and in tandem with Russian-owned media organizations such as RT (formerly, "Russia Today").[39]

From mid-2015 to 2017, Internet Research Agency specialists produced an unprecedented amount of online political disinformation in English, not only on the 2016 Presidential Election but on a range of salient political issues. During that time, the Internet Research Agency produced or retweeted 1,886,919 Twitter messages, commented on or produced 12,603 Reddit posts, and purchased 3,126 Facebook advertisements,[40] costing slightly over 100,000 USD.[41] Despite focusing on election periods, IRA disinformation during this time actually increased after the 2016 U.S. Presidential election,[42] highlighting Russia's continued interest in targeting the United States long after the election.

Russian trolls hired at that time were expected to produce at least "80 comments and 20 shares a day,"[43] covering a range of political and non-political topics. In their political content, IRA disinformation agents focused on salient and controversial wedge issues, including sanctuary cities, LGBTQ+ rights, Syria, and police brutality. However, Russian trolls were also expected to produce non-political content to create a semblance of an "authentic" perceived personas.[44] As a result, many Russian IRA accounts during this time also shared inspirational quotes and played "hashtag" games, encouraging interaction between themselves and their followers.

To exploit contentious issues within the United States, the Russian IRA engaged in asymmetric targeting and produced targeted forms of disinformation for different U.S. audiences. This was done by carefully crafting thousands of separate accounts posing as Americans with differing false identities. One study found that, on Twitter, IRA accounts could be grouped into five categories: "right troll, left troll, news feed, hashtag games and fearmonger."[45]

Two groups of accounts deserve additional exploration. First, research suggests that conservative individuals were more likely to be exposed to Russian disinformation.[46] One reason why conservative may have been especially vulnerable to IRA disinformation may be because of the density of the conservative social media network, which the IRA exploited to great effect.[47] Notably, many conservative news organizations unwittingly quoted a Russian troll during the 2016 U.S. Presidential election,[48] and being quoted helped IRA trolls like @TEN_GOP grow their followership.[49]

The second group that Russia heavily targeted was Black Americans.[50] By impersonating Black Americans, Russian IRA trolls engaging in digital blackfacing produced the greatest amount of engagement for Russian IRA Twitter content.[51] On Facebook, IRA accounts impersonating Black activists also created and promoted false rallies and protests.

A key reason why these two groups were targeted could be because both conservatives and Black Americans communicate online through large and well-developed digital networks.[52] In other words, the digital networks that help build these communities were the very same networks that Russian IRA accounts exploited to spread disinformation during the 2016 U.S. Presidential election. In the case of the former (conservatives), IRA trolls encouraged inflammatory remarks, including promoting a "#HeterosexualPride-Day" hashtag and using a portmanteau of "rape" and "refugee." In the case of the latter (Black Americans), IRA trolls impersonated activists to organize fake events and create misleading portrayals of social justice movements. Another reason why Russian IRA trolls targeted conservatives and Black Americans is because both were essential voting groups in the 2016 U.S. Presidential election.

To strategically produce this disinformation, members of the Internet Research Agency traveled to the United States on June 2014 to gain information about partisan issues, ongoing social movements and contentious social issues.[53] The IRA began producing disinformation as early as 2014, but amplified their disinformation in 2015 and 2016 as more people were paying attention to the election. These tactics reveal the long-term strategic planning of Russia's 2016 disinformation campaign.

Importantly, IRA accounts also helped amplify the leaked emails of John Podesta, whose account was hacked by Kremlin-linked Russian cyber espionage team Fancy Bear.[54] Previous work suggests that this leak contributed to Donald Trump's victory over Hillary Clinton in the 2016 election.[55] Following Wikileak's release of these emails in October and November 2016, the Internet Research Agency was quick to share false, exaggerated or misleading information, including claims that Clinton was an alcoholic[56] or claims that Clinton's campaign manager had taken money from a foreign agent.[57] Using variations of the hashtag #PodestaEmails, Internet Research Agency trolls repeatedly amplified and sometimes misconstrued the emails, highlighting ways in which Russia's other election interference strategies strategically coincided with their digital disinformation campaign.

Despite attempting to remove Russian trolls from multiple digital platforms in 2017 and 2018,[58] their presence persisted on mainstream social media, where they continued to exploit these platforms' algorithms to spread inflammatory political disinformation.[59] Unlike 2016, where Russia was both a producer and amplifier of disinformation, in 2020, Russian bots were more likely to amplify disinformation produced by U.S. citizens.[60] Perhaps because Russian disinformation in 2016 was also marked by notable linguistic errors,[61] the Internet Research Agency also hired unsuspecting U.S. journalists to produce stories in 2020 through the shell website "Peace Data."[62] These developments highlight the cat-and-mouse nature of combating disinformation—for every tactic that is identified and mitigated, foreign actors will device other strategies that make it even harder to detect election-targeted digital disinformation.

Did these strategies work? In truth, there is substantial disagreement regarding the effectiveness of Russian disinformation. Some scholars claim that the IRA's messages were effective in persuading citizens to vote for Donald Trump (or effective in dissuading citizens from voting for Biden); Ruck et al. make such a claim, pointing to a temporal relationship between IRA Twitter activity and election polls.[63] Focusing specifically on the month before the election, Jamieson argues that Podesta's email leak specifically contributed to Trump's success,[64] suggesting that it is not digital disinformation per se, but the broader active measures campaign (which includes hacking and mal-information) that affected the election.

However, other scholars note that the IRA was not producing disinformation that was unique to what U.S. citizens were producing.[65] Researchers making this argument also importantly distinguish between exposure and persuasion: just because a person saw an IRA message does not necessarily mean they were persuaded by it.[66] These studies suggest that claims of the effectiveness of IRA disinformation are overblown, partly because it is unlikely a few disinformation messages shifted the election in one direction or another and partly because the IRA was not creating something new in the digital media system—it was simply exploiting the polarization that had already existed.

But both of these arguments focus only on the short-term goal of the IRA's digital disinformation campaign (i.e., to get Donald Trump elected in 2016 and 2020). And, perhaps, it is in the long-term goal of sowing distrust in the U.S. electoral and media system where we see the most harm of digital disinformation. To put it bluntly: IRA disinformation may not have caused a substantial number of individuals to change their vote, but it likely contributed to distrust in the process of the 2016 U.S. Presidential election. By focusing repeatedly on Russian attempts to interfere with the election, regardless of whether Russia's attempts were actually successful, actors in the U.S. media system (citizens, journalists, public figures and politicians) inadvertently played into the goals of the IRA's active measures tactics. And the distrust culled after the 2016 U.S. Presidential election likely opened the door to distrusting the 2020 U.S. Presidential election, a distrust that then–President Donald Trump exploited in the lead up to election day, and after.[67]

Conclusion

Owing to the cheapness and ease by which foreign actors can produce digital disinformation during elections, it is unlikely that democracies will be able to fully rid themselves of this harm, during elections and outside of them.[68] However, there are things that democracies, and the companies and citizens within them, can do to stop the spread or effectiveness of digital disinformation campaigns.

First, it is important to recognize why digital disinformation campaigns from foreign actors are effective at all: as other scholars have noted, these campaigns exploit existing societal disagreements, including (but not limited to) polarization, racial tensions and other social and economic issues. When political systems fail their citizenry, foreign actors will be there to provide alternative rationales and solutions that exacerbate these societal cleavages, particularly during elections. In other words, digital disinformation is not only a problem unto itself—its success is a consequence of broader political issues.

Second, governments and digital media corporations cannot combat digital disinformation campaigns apart from each other. Recent scholars have instead advocated for co-regulation;[69] for example, social media companies using AI to remove disinformation could themselves be overseen by a transparent auditing process that includes government entities, scholarly experts and ethical AI activists. Corporations and governments can also work together to fund research on the development of more accurate tools to identify and remove troll farms and bots.

Finally, the effectiveness of digital disinformation campaigns during elections hinges on distrust—distrust in the media system (because there is disinformation) and distrust in the political system (because people believe elections have been interfered with by foreign actors). For tools to combat digital disinformation to work, they need to be trustworthy. And for these tools to be trustworthy, they need to be transparent in application, uniformly applied and clear in scope. Ambiguous terms such as "coordinated inauthentic behavior," do society no good if there is no clear definition for what this means, or how a social media platform determines what is "coordinated inauthentic behavior" or not.[70]

While these solutions may seem daunting in the face of repeated attempts to manipulate and build distrust within our media ecosystem during elections, it is worth noting that collective efforts by governments, companies, publics, and citizens can do much to benefit society. Active measures (and, by extension, digital disinformation) have been described as "drops of water falling on a stone: five minutes, ten minutes, fifteen minutes, one hour, one day, nothing happens, but five years, ten years, fifteen years – you've worn a hole in the stone."[71] If this is the case, our solutions must be equally persistent and systematic. Stone by stone, we can build a dam to stop the flow of digital disinformation.

Notes

1 Corstange, Daniel, and Nikolay Marinov. "Taking sides in other people's elections: The polarizing effect of foreign intervention." *American Journal of Political Science* 56, no. 3 (2012): 656–657; Dixon, Paul. "'Hearts and minds'? British counter-insurgency from Malaya to Iraq." *Journal of Strategic Studies* 32, no. 3 (2009): 355.

2 Dell, Melissa, and Pablo Querubin. "Nation building through foreign intervention: Evidence from discontinuities in military strategies." *The Quarterly Journal of Economics* 133, no. 2 (2018): 764.

3 Nguyen, An, and Daniel Catalan. "Digital mis/disinformation and public engagement with health and science controversies: Fresh perspectives from Covid-19." *Media and Communication* 8, no. 2 (2020): 323–328.

4 Lewandowsky, Stephan. "Climate change disinformation and how to combat it." *Annual Review of Public Health* 42, no. 1 (2021): 1–21.

5 Lee, Taeku, and Christian Hosam. "Fake news is real: the significance and sources of disbelief in mainstream media in Trump's America." In *Sociological Forum*, vol. 35 (2020): 996–1018.

6 Parahita, Gilang Desti. "Voters (dis)-believing digital political disinformation in gubernatorial election of DKI Jakarta 2016–2017." *Jurnal Ilmu Sosial dan Ilmu Politik* 22, no. 2 (2018): 127–143; Schia, Niels Nagelhus and Lars Gjesvik. "Hacking democracy: managing influence campaigns and disinformation in the digital age." *Journal of Cyber Policy* 5, no. 3 (2020): 413–428.

7 Bennett, Huw. "'Words are cheaper than bullets': Britain's psychological warfare in the Middle East, 1945–60." *Intelligence and National Security* 34, no. 7 (2019): 925.

8 Becker, Howard. "The nature and consequences of black propaganda." *American Sociological Review* 14, no. 2 (1949): 221.

9 Xia, Yiping, Josephine Lukito, Yini Zhang, Chris Wells, Sang Jung Kim, and Chau Tong. "Disinformation, performed: Self-presentation of a Russian IRA account on Twitter." *Information, Communication & Society* 22, no. 11 (2019): 1660–1661.

10 Starbird, Kate. "Disinformation's spread: bots, trolls and all of us." *Nature* 571, no. 7766 (2019): 449–450.

11 Tsvetkova, Milena, Ruth García-Gavilanes, Luciano Floridi, and Taha Yasseri. "Even good bots fight: The case of Wikipedia." *PloS one* 12, no. 2 (2017): e0171774.

12 Beskow, David M., and Kathleen M. Carley. "Characterization and comparison of Russian and Chinese disinformation campaigns." In *Disinformation, misinformation, and fake news in social media*, pp. 63–81. Springer, Cham, 2020; Santini, Rose Marie, Debora Salles, Giulia Tucci, Fernando Ferreira, and Felipe Grael. "Making up audience: Media bots and the falsification of the public sphere." *Communication Studies* 71, no. 3 (2020): 466–487.

13 Dawson, Andrew, and Martin Innes. "How Russia's internet research agency built its disinformation campaign." *The Political Quarterly* 90, no. 2 (2019): 247–249.

14 Gallacher, John D., Vlad Barash, Philip N. Howard, and John Kelly. "Junk news on military affairs and national security: Social media disinformation campaigns against us military personnel and veterans." *arXiv preprint arXiv:1802.03572* (2018).

15 Loui, Ronald, and Will Hope. "Information Warfare Amplified by Cyberwarfare and Hacking the National Knowledge Infrastructure." In *2017 IEEE 15th Intl Conf on Dependable, Autonomic and Secure Computing, 15th Intl Conf on Pervasive Intelligence and Computing, 3rd Intl Conf on Big Data Intelligence and Computing and Cyber Science and Technology Congress (DASC/PiCom/DataCom/CyberSciTech)*, (2017): 280–283.

16 Bradshaw, Samantha, and Philip N. Howard. "The global disinformation order: 2019 global inventory of organised social media manipulation." (2019):15.

17 Mitchelstein, Eugenia, and Pablo J. Boczkowski. "Online news consumption research: An assessment of past work and an agenda for the future." *New media & society* 12, no. 7 (2010): 1085–1102.

18 Ekdale, Brian, and Melissa Tully. "African elections as a testing ground: Comparing coverage of cambridge analytica in Nigerian and Kenyan Newspapers." *African Journalism Studies* 40, no. 4 (2019): 28–29.

19 Recuero, Raquel, Felipe Soares, and Otávio Vinhas. "Discursive strategies for disinformation on WhatsApp and Twitter during the 2018 Brazilian presidential election." *First Monday* (2021).

20 Mueller, Robert S. *The Mueller Report.* e-artnow, 2019.

21 Freelon, Deen, Michael Bossetta, Chris Wells, Josephine Lukito, Yiping Xia, and Kirsten Adams. "Black trolls matter: Racial and ideological asymmetries in social media disinformation." *Social Science Computer Review* (2020): 561.

22 Keller, Franziska B., David Schoch, Sebastian Stier, and JungHwan Yang. "Political astroturfing on Twitter: How to coordinate a disinformation campaign." *Political Communication* 37, no. 2 (2020): 256–280.

23 Imamverdiyev, Yadigar N. "Social Media and Security Concerns." *İTP Jurnalı* (2016):20.

24 Yang, Xiaofeng, Qian Yang, and Christo Wilson. "Penny for your thoughts: Searching for the 50 cent party on sina weibo." In *Proceedings of the International AAAI Conference on Web and Social Media*, vol. 9, no. 1, pp. 694–697. 2015.

25 Rodriguez, Manuel. "Disinformation Operations Aimed at (Democratic) Elections in the Context of Public International Law: The Conduct of the Internet Research Agency During the 2016 US Presidential Election." *International Journal of Legal Information* 47, no. 3 (2019): 160.

26 Starbird, 2019.

27 Wigell, Mikael. "Hybrid interference as a wedge strategy: a theory of external interference in liberal democracy." *International Affairs* 95, no. 2 (2019): 255–275.

28 Zhou, Yanmengqian, and Lijiang Shen. "Confirmation Bias and the Persistence of Misinformation on Climate Change." *Communication Research* vol. 4, no. 4(2021):515–516

29 Krafft, P. M., & Donovan, J. (2020). Disinformation by design: The use of evidence collages and platform filtering in a media manipulation campaign. *Political Communication, 37*(2), 195; Holiday, R. (2013). *Trust me, I'm lying: confessions of a media manipulator.* Penguin.

30 Jensen, Michael. "Russian trolls and fake news: Information or identity logics?." *Journal of International Affairs* 71, no. 1.5 (2018): 115–124

31 Magelinski, Thomas, Lynnette Ng, and Kathleen Carley. "A Synchronized Action Framework for Detection of Coordination on Social Media." *Journal of Online Trust and Safety* 1, no. 2 (2022):1–24.

32 Lee, David. "The tactics of a Russian troll farm" *The Guardian*.https://www.bbc.com/news/technology-43093390

33 Wardle, Claire. "Information disorder: Toward an interdisciplinary framework for research and policy making." (2017). https://rm.coe.int/information-disorder-toward-an-interdisciplinary-framework-for-researc/168076277c; Tenove, C., & Tworek, H. (2019). Online Disinformation and Harmful Speech: Dangers for Democratic Participation and Possible Policy Responses. *Journal of Parliamentary & Political Law, 13*, 215–232.

34 Doroshenko, Larissa, and Josephine Lukito. "Trollfare: Russia's Disinformation Campaign During Military Conflict in Ukraine." *International Journal of Communication* 15 (2021): 4666.

35 Gioe, David V., Richard Lovering, and Tyler Pachesny. "The soviet legacy of Russian active measures: new vodka from old stills?" *International Journal of Intelligence and Counter Intelligence* 33 (2020): 514–539.

36 Richey, Mason. "Contemporary Russian revisionism: understanding the Kremlin's hybrid warfare and the strategic and tactical deployment of disinformation." *Asia Europe Journal* 16, no. 1 (2018): 101–113.

37 Abrams, Steve. "Beyond propaganda: Soviet active measures in Putin's Russia." *Connections* 15, no. 1 (2016): 7.

38 Bastos, Marco, and Johan Farkas. "'Donald Trump is my President!': The internet research agency propaganda machine." *Social Media+ Society* 5, no. 3 (2019): 2056305119865466; National Intelligence Council. "Foreign Threats to the 2020 US Federal Elections" https://www.dni.gov/files/ODNI/documents/assessments/ICA-declass-16MAR21.pdf

39 Lukito, Josephine. "Coordinating a multi-platform disinformation campaign: Internet Research Agency activity on three US social media platforms, 2015 to 2017." *Political Communication* 37, no. 2 (2020): 238–255

40 Ibid.

41 Spangher, Alexander, Gireeja Ranade, Besmira Nushi, Adam Fourney, and Eric Horvitz. "Characterizing Search-Engine Traffic to Internet Research Agency Web Properties." In *Proceedings of The Web Conference 2020*, pp. 2253–2263. 2020.

42 Lukito, 2020.

43 MacFarquhar, N. (2018, February 18). Inside the Russian Troll Factory: Zombies and a Breakneck Pace. *The New York Times*. https://www.nytimes.com/2018/02/18/world/europe/russia-troll-factory.html

44 Xia et al., 2019

45 Linvill, Darren L., and Patrick L. Warren. "Troll factories: Manufacturing specialized disinformation on Twitter." *Political Communication* 37, no. 4 (2020): 447–467.

46 Hjorth, Frederik, and Rebecca Adler-Nissen. "Ideological asymmetry in the reach of pro-Russian digital disinformation to United States audiences." *Journal of Communication* 69, no. 2 (2019): 168–192.

47 Bastos & Farkas, 2019.

48 Lukito, Josephine, Jiyoun Suk, Yini Zhang, Larissa Doroshenko, Sang Jung Kim, Min-Hsin Su, Yiping Xia, Deen Freelon, and Chris Wells. "The wolves in sheep's clothing: How Russia's Internet Research Agency tweets appeared in US news as vox populi." *The International Journal of Press/Politics* 25 (2020): 196–216.

49 Zhang, Yini, Josephine Lukito, Min-Hsin Su, Jiyoun Suk, Yiping Xia, Sang Jung Kim, Larissa Doroshenko, and Chris Wells. "Assembling the networks and audiences of disinformation: How successful Russian IRA Twitter accounts built their followings, 2015–2017." *Journal of Communication* 71, no. 2 (2021): 305–331.

50 Freelon, Deen, and Tetyana Lokot. "Russian Twitter disinformation campaigns reach across the American political spectrum." *Harvard Kennedy School Misinformation Review* 1 (2020).

51 Freelon et al., 2020.

52 For more on the conservative network, see Crosset, Valentine, Samuel Tanner, and Aurélie Campana. "Researching far right groups on Twitter: Methodological challenges 2.0." *New media & society* 21, no. 4 (2019): 939–961.; For more on Black Twitter, see Clark, Meredith. "Black Twitter: Building connection through cultural conversation." *Hashtag publics: The power and politics of discursive networks* (2015): 205–218.

53 Mueller, 2019.

54 Franceschi-Bicchierai, Lorenzo. "How Hackers Broke Into John Podesta and Colin Powell's Gmail Accounts" *Vice*. https://www.vice.com/en/article/mg7xjb/how-hackers-broke-into-john-podesta-and-colin-powells-gmail-accounts (2016, October 20).

55 Jamieson, Kathleen Hall. *Cyberwar: how Russian hackers and trolls helped elect a president: what we don't, can't, and do know.* Oxford University Press, 2020

56 Russian IRA account, @Pamela_Moore13, tweeted "Hillary Clinton is an alcoholic. No wonder she has health issues. #WorldMentalHealthDay #HillarysHealth #PodestaEmails" on October 10, 2016 which was retweeted 304 times.

57 Russian IRA account, @TEN_GOP, twice tweeted this claim on October 17, 2016. The first tweet, with the text, "I would only do this for political reasons (i.e. to make Soros happy)' ~ Robby Mook, Hillary's camp manager" was retweeted 2,245 times and the second tweet, with the text "Hillary's camp manager accepting money from foreign agents" was retweeted 2,269 times. Both are in reference to an October 16, 2016 email dump shared on Wikileaks in which Clinton's campaign manager (Robby Mook) discussed taking money from "foreign registered agents" (i.e., lobbyists). Subsequent emails reveal that Clinton was not privy to this conversation and it is unclear whether the campaign took the money. For more see https://www.cbsnews.com/news/wikileaks-emails-show-clinton-camp-chatter-on-foreign-lobbying-health-taxes/

58 Exposing Russia's Effort to Sow Discord Online: The Internet Research Agency and Advertisements. https://intelligence.house.gov/social-media-content/

59 Hao, Karen. "Troll farms reached 140 million Americans a month on Facebook before 2020 election, internal report shows" *MIT Technology Review*. https://www.technologyreview.com/2021/09/16/1035851/facebook-troll-farms-report-us-2020-election/ (2021, September 16)

60 Howard, Brad. "How the 2020 election war on bots and trolls differs from 2016" *CNBC*. https://www.cnbc.com/2020/10/27/how-the-2020-election-war-on-bots-and-trolls-differs-from-2016.html (2020, October 27)

61 Suk, Jiyoun, Josephine Lukito, Min-Hsin Su, Sang Jung Kim, Chau Tong, Zhongkai Sun, and Prathusha Sarma. "Do I sound American? How message attributes of IRA disinformation relate to Twitter engagement." *Computational Communication Research* (Forthcoming).

62 Collier, Kevin and Ken Dilanian. Russian internet trolls hired U.S. journalists to push their news website, Facebook says. *NBC*. https://www.nbcnews.com/tech/tech-news/russian-internet-trolls-hired-u-s-journalists-push-their-news-n1239000 (2020, September 1)

63 Ruck, Damian J., Natalie M. Rice, Joshua Borycz, and R. Alexander Bentley. "Internet Research Agency Twitter activity predicted 2016 US election polls." *First Monday* (2019)

64 Jamieson, 2020.

65 Berghel, Hal. "Oh, what a tangled web: Russian hacking, fake news, and the 2016 US presidential election." *Computer* 50, no. 9 (2017): 87–91.

66 Bail, Christopher A., Brian Guay, Emily Maloney, Aidan Combs, D. Sunshine Hillygus, Friedolin Merhout, Deen Freelon, and Alexander Volfovsky. "Assessing the Russian Internet Research Agency's impact on the political attitudes and behaviors of American Twitter users in late 2017." *Proceedings of the national academy of sciences* 117, no. 1 (2020): 243–250.

67 Inskeep, Steve. Timeline: What Trump told supporters for months before they attacked. *NPR*. https://www.npr.org/2021/02/08/965342252/timeline-what-trump-told-supporters-for-months-before-they-attacked (2021, February 8).

68 Van Raemdonck, Nathalie, and Trisha Meyer. "Why Disinformation is Here to Stay. A Socio-technical Analysis of Disinformation as a Hybrid Threat." In *Addressing Hybrid Threats: European Law and Policies*. Edward Elgar, 2022

69 Marsden, Chris, Trisha Meyer, and Ian Brown. "Platform values and democratic elections: How can the law regulate digital disinformation?." *Computer Law & Security Review* 36 (2020): 1–18.

70 McGregor, Shannon. "What even is 'coordinated inauthentic behavior' on Platforms?" *Wired*. https://www.wired.com/story/what-even-is-coordinated-inauthentic-behavior-on-platforms/ (2020, September 17).

71 "Dennis Kux, Former Head of the Active Measures Working Group, 1984" as quoted in Abrams, 2016

Bibliography

Abrams, Steve. "Beyond propaganda: Soviet active measures in Putin's Russia." *Connections* 15, no. 1 (2016): 5–31.

Bail, Christopher A., Brian Guay, Emily Maloney, Aidan Combs, D. Sunshine Hillygus, Friedolin Merhout, Deen Freelon, and Alexander Volfovsky. "Assessing the Russian Internet Research Agency's impact on the political attitudes and behaviors of American Twitter users in late 2017." *Proceedings of the National Academy of Sciences* 117, no. 1 (2020): 243–250.

Bastos, Marco, and Johan Farkas. ""Donald Trump is my President!": The internet research agency propaganda machine." *Social Media+ Society* 5, no. 3 (2019): 1–13.

Becker, Howard. "The nature and consequences of black propaganda." *American Sociological Review* 14, no. 2 (1949): 221–235.

Bennett, Huw. "'Words are cheaper than bullets': Britain's psychological warfare in the Middle East, 1945–60." *Intelligence and National Security* 34, no. 7 (2019): 925–944.

Berghel, Hal. "Oh, what a tangled web: Russian hacking, fake news, and the 2016 US presidential election." *Computer* 50, no. 9 (2017): 87–91.

Beskow, David M., and Kathleen M. Carley. "Characterization and comparison of Russian and Chinese disinformation campaigns." In *Disinformation, misinformation, and fake news in social media*, pp. 63–81. Springer, Cham, 2020.

Bradshaw, Samantha, and Philip N. Howard. "*The global disinformation order: 2019 global inventory of organised social media manipulation.*" (2019).

Clark, Meredith. "Black Twitter: Building connection through cultural conversation." *Hashtag publics: The power and politics of discursive networks* (2015): 205–218.

Collier, Kevin and Ken Dilanian. Russian internet trolls hired U.S. journalists to push their news website, Facebook says. *NBC*. https://www.nbcnews.com/tech/tech-news/russian-internet-trolls-hired-u-s-journalists-push-their-news-n1239000 (2020, September 1)

Crosset, Valentine, Samuel Tanner, and Aurélie Campana. "Researching far right groups on Twitter: Methodological challenges 2.0." *New Media & Society* 21, no. 4 (2019): 939–961.

Corstange, Daniel, and Nikolay Marinov. "Taking sides in other people's elections: The polarizing effect of foreign intervention." *American Journal of Political Science* 56, no. 3 (2012): 655–670.

Dawson, Andrew, and Martin Innes. "How Russia's internet research agency built its disinformation campaign." *The Political Quarterly* 90, no. 2 (2019): 245–256.

Dell, Melissa, and Pablo Querubin. "Nation building through foreign intervention: Evidence from discontinuities in military strategies." *The Quarterly Journal of Economics* 133, no. 2 (2018): 701–764.

Dixon, Paul. "'Hearts and minds'? British counter-insurgency from Malaya to Iraq." *Journal of Strategic Studies* 32, no. 3 (2009): 353–381.

Doroshenko, Larissa, and Josephine Lukito. "Trollfare: Russia's Disinformation Campaign During Military Conflict in Ukraine." *International Journal of Communication* 15 (2021): 4662–4689.

Ekdale, Brian, and Melissa Tully. "African elections as a testing ground: Comparing coverage of cambridge analytica in Nigerian and Kenyan Newspapers." *African Journalism Studies* 40, no. 4 (2019): 27–43.

Franceschi-Bicchierai, Lorenzo. "How Hackers Broke Into John Podesta and Colin Powell's Gmail Accounts" *Vice*. https://www.vice.com/en/article/mg7xjb/how-hackers-broke-into-john-podesta-and-colin-powells-gmail-accounts (2016, October 20).

Freelon, Deen, Michael Bossetta, Chris Wells, Josephine Lukito, Yiping Xia, and Kirsten Adams. "Black trolls matter: Racial and ideological asymmetries in social media disinformation." *Social Science Computer Review* (2020): 560–578.

Freelon, Deen, and Tetyana Lokot. "Russian Twitter disinformation campaigns reach across the American political spectrum." *Harvard Kennedy School Misinformation Review* 1 (2020).

Gallacher, John D., Vlad Barash, Philip N. Howard, and John Kelly. "Junk news on military affairs and national security: Social media disinformation campaigns against us military personnel and veterans." *arXiv preprint arXiv:1802.03572* (2018).

Gioe, David V., Richard Lovering, and Tyler Pachesny. "The soviet legacy of Russian active measures: new vodka from old stills?" *International Journal of Intelligence and Counter Intelligence* 33 (2020): 514–539.

Hao, Karen. "Troll farms reached 140 million Americans a month on Facebook before 2020 election, internal report shows" *MIT Technology Review*. https://www.technologyreview.com/2021/09/16/1035851/facebook-troll-farms-report-us-2020-election/ (2021, September 16)

Hjorth, Frederik, and Rebecca Adler-Nissen. "Ideological asymmetry in the reach of pro-Russian digital disinformation to United States audiences." *Journal of Communication* 69, no. 2 (2019): 168–192.

Holiday, Ryan. *Trust me, I'm lying: confessions of a media manipulator*. Penguin, 2013.

Howard, Brad. "How the 2020 election war on bots and trolls differs from 2016" *CNBC*. https://www.cnbc.com/2020/10/27/how-the-2020-election-war-on-bots-and-trolls-differs-from-2016.html (2020, October 27)

Imamverdiyev, Yadigar N. "Social Media and Security Concerns." *İTP Jurnalı* (2016):18–23.

Inskeep, Steve. Timeline: What Trump told supporters for months before they attacked. *NPR*. https://www.npr.org/2021/02/08/965342252/timeline-what-trump-told-supporters-for-months-before-they-attacked (2021, February 8).

Jamieson, Kathleen Hall. *Cyberwar: how Russian hackers and trolls helped elect a president: what we don't, can't, and do know.* Oxford University Press, 2022.

Jensen, Michael. "Russian trolls and fake news: Information or identity logics?." *Journal of International Affairs* 71, no. 1.5 (2018): 115–124.

Keller, Franziska B., David Schoch, Sebastian Stier, and JungHwan Yang. "Political astroturfing on Twitter: How to coordinate a disinformation campaign." *Political Communication* 37, no. 2 (2020): 256–280.

Krafft, P. M., and Joan Donovan. "Disinformation by design: The use of evidence collages and platform filtering in a media manipulation campaign." *Political Communication* 37, no. 2 (2020): 194–214.

Lee, David. "The tactics of a Russian troll farm" *The Guardian.* https://www.bbc.com/news/technology-43093390.

Lee, Taeku, and Christian Hosam. "Fake news is real: the significance and sources of disbelief in mainstream media in Trump's America." In *Sociological Forum*, vol. 35 (2020): 996–1018.

Lewandowsky, Stephan. "Climate change disinformation and how to combat it." *Annual Review of Public Health* 42, no. 1 (2021): 1–21.

Linvill, Darren L., and Patrick L. Warren. "Troll factories: Manufacturing specialized disinformation on Twitter." *Political Communication* 37, no. 4 (2020): 447–467.

Loui, Ronald, and Will Hope. "Information Warfare Amplified by Cyberwarfare and Hacking the National Knowledge Infrastructure." In *2017 IEEE 15th Intl Conf on Dependable, Autonomic and Secure Computing, 15th Intl Conf on Pervasive Intelligence and Computing, 3rd Intl Conf on Big Data Intelligence and Computing and Cyber Science and Technology Congress (DASC/PiCom/DataCom/CyberSciTech)*, (2017): 280–283.

Lukito, Josephine. "Coordinating a multi-platform disinformation campaign: Internet Research Agency activity on three US social media platforms, 2015 to 2017." *Political Communication* 37, no. 2 (2020): 238–255.

Lukito, Josephine, Jiyoun Suk, Yini Zhang, Larissa Doroshenko, Sang Jung Kim, Min-Hsin Su, Yiping Xia, Deen Freelon, and Chris Wells. "The wolves in sheep's clothing: How Russia's Internet Research Agency tweets appeared in US news as vox populi." *The International Journal of Press/Politics* 25 (2020): 196–216.

MacFarquhar, Neil. Inside the Russian Troll Factory: Zombies and a Breakneck Pace. *The New York Times.* https://www.nytimes.com/2018/02/18/world/europe/russia-troll-factory.html (2018, February 18).

Magelinski, Thomas, Lynnette Ng, and Kathleen Carley. "A Synchronized Action Framework for Detection of Coordination on Social Media." *Journal of Online Trust and Safety* 1, no. 2 (2022).

Marsden, Chris, Trisha Meyer, and Ian Brown. "Platform values and democratic elections: How can the law regulate digital disinformation?" *Computer Law & Security Review* 36 (2020): 1–18.

McGregor, Shannon. "What even is 'coordinated inauthentic behavior' on Platforms?" *Wired.* https://www.wired.com/story/what-even-is-coordinated-inauthentic-behavior-on-platforms/ (2020, September 17).

Mitchelstein, Eugenia, and Pablo J. Boczkowski. "Online news consumption research: An assessment of past work and an agenda for the future." *New media & society* 12, no. 7 (2010): 1085–1102.

Mueller, Robert S. *The Mueller Report.* (2019)

National Intelligence Council. "*Foreign Threats to the 2020 US Federal Elections*" https://www.dni.gov/files/ODNI/documents/assessments/ICA-declass-16MAR21.pdf.

Nguyen, An, and Daniel Catalan. "Digital mis/disinformation and public engagement with health and science controversies: Fresh perspectives from Covid-19." *Media and Communication* 8, no. 2 (2020): 323–328.

Parahita, Gilang Desti. "Voters (dis)-believing digital political disinformation in gubernatorial election of DKI Jakarta 2016–2017." *Jurnal Ilmu Sosial dan Ilmu Politik* 22, no. 2 (2018): 127–143.

Recuero, Raquel, Felipe Soares, and Otávio Vinhas. "Discursive strategies for disinformation on WhatsApp and Twitter during the 2018 Brazilian presidential election." *First Monday* (2021).

Richey, Mason. "Contemporary Russian revisionism: understanding the Kremlin's hybrid warfare and the strategic and tactical deployment of disinformation." *Asia Europe Journal* 16, no. 1 (2018): 101–113.

Rodriguez, Manuel. "Disinformation Operations Aimed at (Democratic) Elections in the Context of Public International Law: The Conduct of the Internet Research Agency During the 2016 US Presidential Election." *International Journal of Legal Information* 47, no. 3 (2019): 149–197.

Ruck, Damian J., Natalie M. Rice, Joshua Borycz, and R. Alexander Bentley. "Internet Research Agency Twitter activity predicted 2016 US election polls." *First Monday* (2019)

Santini, Rose Marie, Debora Salles, Giulia Tucci, Fernando Ferreira, and Felipe Grael. "Making up audience: Media bots and the falsification of the public sphere." *Communication Studies* 71, no. 3 (2020): 466–487.

Schia, Niels Nagelhus, and Lars Gjesvik. "Hacking democracy: managing influence campaigns and disinformation in the digital age." *Journal of Cyber Policy* 5, no. 3 (2020): 413–428.

Spangher, Alexander, Gireeja Ranade, Besmira Nushi, Adam Fourney, and Eric Horvitz. "Characterizing Search-Engine Traffic to Internet Research Agency Web Properties." In *Proceedings of The Web Conference 2020*, pp. 2253–2263. 2020.

Starbird, Kate. "Disinformation's spread: bots, trolls and all of us." *Nature* 571, no. 7766 (2019): 449–450.

Suk, Jiyoun, Josephine Lukito, Min-Hsin Su, Sang Jung Kim, Chau Tong, Zhongkai Sun, and Prathusha Sarma. "Do I sound American? How message attributes of IRA disinformation relate to Twitter engagement." *Computational Communication Research*, 4, no. 2 (2022): 590–628.

Tenove, C., & Tw-orek, H. (2019). Online Disinformation and Harmful Speech: Dangers for Democratic Participation and Possible Policy Responses. *Journal of Parliamentary & Political Law*, 13, 215–232.

Tsvetkova, Milena, Ruth García-Gavilanes, Luciano Floridi, and Taha Yasseri. "Even good bots fight: The case of Wikipedia." *PloS one* 12, no. 2 (2017): e0171774.

Van Raemdonck, Nathalie and Trisha Meyer. "Why Disinformation is Here to Stay. A Socio-technical Analysis of Disinformation as a Hybrid Threat." In *Addressing Hybrid Threats: European Law and Policies*. Edward Elgar, 2022.

Wardle, Claire. "*Information disorder: Toward an interdisciplinary framework for research and policy making.*" (2017). https://rm.coe.int/information-disorder-toward-an-interdisciplinary-framework-for-researc/168076277c.

Wigell, Mikael. "Hybrid interference as a wedge strategy: a theory of external interference in liberal democracy." *International Affairs* 95, no. 2 (2019): 255–275.

Xia, Yiping, Josephine Lukito, Yini Zhang, Chris Wells, Sang Jung Kim, and Chau Tong. "Disinformation, performed: Self-presentation of a Russian IRA account on Twitter." *Information, Communication & Society* 22, no. 11 (2019): 1646–1664.

Yang, Xiaofeng, Qian Yang, and Christo Wilson. "Penny for your thoughts: Searching for the 50 cent party on sina weibo." In *Proceedings of the International AAAI Conference on Web and Social Media*, vol. 9, no. 1, pp. 694–697. 2015.

Zhang, Yini, Josephine Lukito, Min-Hsin Su, Jiyoun Suk, Yiping Xia, Sang Jung Kim, Larissa Doroshenko, and Chris Wells. "Assembling the networks and audiences of disinformation: How successful Russian IRA Twitter accounts built their followings, 2015–2017." *Journal of Communication* 71, no. 2 (2021): 305–331.

Zhou, Yanmengqian, and Lijiang Shen. "Confirmation Bias and the Persistence of Misinformation on Climate Change." *Communication Research* vol. 4, no. 4 (2021): 500–523.

10

A PERCEPTION MANAGEMENT TAKE ON PROPAGANDA AS POLITICAL WARFARE

Adrian Tudorache

Introduction: the *rediscovery* of political warfare

As the 2017 U.S. National Security Strategy recognized the reality of a world defined by great power competition, experts pointed out to the need for a refocus on political warfare as a distinct instrument of statecraft. There were various reasons for such a call. For instance, it was justified through the "catastrophic"[1] consequences of direct conventional or nuclear war between great powers, and, therefore generated the need for an alternative way of competition. Other reasons to be able to wage political warfare were to substantially increase U.S. influence abroad and to consolidate politically the military gains, for instance in certain regions such as the Middle East.[2] However, the need for a political warfare capability was also connected to the lack of political deliverables despite tactical military successes, reflecting "fundamental flaws in [U.S.] approach to modern warfare."[3]

Increasing U.S. decision-makers' awareness for political warfare was deemed necessary as competitors like Russia and China have continuously relied on it in attempts to exploit Western societies' transparency and openness. Notably, Russia relied on hybrid and information warfare in performing the annexation of Crimea in 2014, with some success in complicating the response of the international community; furthermore, Russia interfered in elections in Western societies, such as in the U.S. presidential elections in 2016, in the French presidential elections in 2017, and continued, according to director of national intelligence Dan Coates, to aggressively engage to "manipulate social media and to spread propaganda focused on hot-button issues that are intended to exacerbate socio-political divisions" in relations to 2018 [U.S.] mid-term elections, despite Kremlin's denials of such acts[4].

In 1948, George F. Kennan outlined the need for organized political warfare to allow the United States to effectively compete with the Soviet Union while rejecting a clear-cut distinction between war and peace:

> Political warfare is the logical application of Clausewitz's doctrine in time of peace. In broadest definition, political warfare is the employment of all the means at a nation's command, short of war, to achieve its national objectives. Such operations are both overt and covert. They range from such overt actions as political alliances,

economic measures (as ERP – the Marshall Plan), and 'white' propaganda to such covert operations as clandestine support of "friendly" foreign elements, 'black' psychological warfare and even encouragement of underground resistance in hostile states.[5]

In Clausewitz's conceptualization, even war would pass as political warfare: "War is a mere continuation of policy by other means"[6]. Furthermore, political warfare today would include the use of wide range of tools, from the digital and cyber domains to weaponization of energy and would aim for a whole-of-society approach in both offensive and defensive actions.

As key elements of political warfare, propaganda and disinformation have raised significant challenges in international affairs and domestic settings, as addressing them proved problematic: the tools to distinguish and counter them properly have taken long to develop and to adapt to the pace of the threat. This chapter aims to explore how propaganda is used as a tool of political warfare, to expand national influence abroad, to improve domestic approval ratings, and to support the realization of foreign policy and fulfilment of strategic objectives. It will also examine applications of propaganda with impact on foreign audiences, which include both decision-makers, as well as the general public. This remains particularly important for democratic societies where elected officials remain accountable to public opinion, as well as for the foreign policy conduct of nations in the international arena.

Such exploration would be seen from the perspective of perception management, defined by U.S. Department of Defense as

> actions to convey and/or deny selected information and indicators to foreign audiences to influence their emotions, motives, and objective reasoning as well as to intelligence systems and leaders at all levels to influence official estimates, ultimately resulting in foreign behaviours and official actions favourable to the originator's objectives. In various ways, perception management combines truth projection, operations security, cover and deception, and psychological operations.[7]

The definition originates in the military sphere; however, the concept should be intrinsic to statecraft at large as it relates to decision-making, to targeting professional audiences, as well as the general public or "the human dimension." Shaping the information space through perception management includes sub-elements of public affairs, public diplomacy, psychological operations, deception and covert action.[8]

The abundance of concepts related to "shaping the information space" or "waging political warfare" is to be noted, as well as the close interactions or overlaps between those concepts, and their variable definitions. Thus, the terminology also includes, *inter alia,* hybrid warfare, grey zone, active measures, disinformation, and reflexive control (approximately the Russian version of perception management). Measuring the results of such actions may, however, remain the most challenging aspect due to the difficulty to assess the direct causality between such efforts and changes in foreign policy decision-making.

A framework of analysis regarding political warfare and propaganda

Kennan's definition of political warfare faces a critique related to the use of "warfare" by experts making a distinction between military actions and other types of actions, of non-military nature that are needed for effectively engage in competition. According to such views, hybrid threats would frame better a competitor's action, defined as "any adversary that simultaneously employs a tailored mix of conventional weapons, irregular tactics, terrorism, and criminal behaviour in the same time and battlespace to obtain their political objectives," for instance with specific reference to Russia's actions in Georgia in 2008 and in Ukraine in 2014.[9] Also referring to Putin's war against Ukraine in 2014, Anne Applebaum used the term "masked warfare", as well as "the new war of words", asking for more "human intelligence" and for preparing to face this kind of challenge.[10]

Galeotti seems more inclined to use Kennan's definition, by relating it to what Russia does in practice; thus, Russia itself equates its campaign against the West with "war", though those actions are actually "short of war" and kept below that threshold. In relation to political warfare, Galeotti quotes Lewis Galantiere, Head of Propaganda Policy at Radio Free Europe, who in 1952 characterized it as

> the sum of the activities in which a government engages for the attainment of its objectives without unleashing armed warfare. But it is a description which applies to none of those activities when each of them is carried independently of the others. In that case, they become 'mere' diplomacy, intelligence, propaganda, economic negotiation, armament production, and so on. The essence of political warfare is that it is planned and the means employed to carry it on are coordinated.[11]

Galeotti points out that political warfare, in the Russian view, is essentially "undeclared and unacknowledged […] a new policy of diplomacy with a military mindset where politics, spycraft and propaganda are virtually indistinguishable."[12]

Russian experts speak of non-linear warfare, with the Russian Chief of General Staff Valery Gerasimov pointing out to the "blurring the lines between states of war and peace," and the increased role of non-military means to achieve strategic and political goals, while recognizing their effectiveness in "exceeding the power of force of weapons."[13] Such effectiveness is attained by a "correlation on nonmilitary and military measures 4 to 1" and by targeting vulnerabilities of the adversary, its population and by a tailor-made approach to each situation, thus rejecting a template-type response.[14] In an analysis of Gerasimov's perspective, Schnaufer points out to the fact that non-linear warfare aims at the social and political structure of the enemy, has limited planning to allow exploitation of opportunities, and relies on "propaganda, political and social agitators, and cyberattacks," without excluding kinetic action.[15] Its overall objectives are to weaken the adversary socially, politically, and military, respectively diminishing its political will, its military capabilities, and potential response to foreign interference.[16]

In a manner quite close to perception management, the Soviet reflexive control is defined "as a means of conveying to a partner or an opponent specially prepared information to incline him to voluntarily make the predetermined decision desired by the initiator of the action."[17] The theory of reflexive control has been developed during the Cold War, aiming to target: (1) the decision process or the system of decision making; (2) individuals responsible for decision making; (3) cultural complex within which the decision is embedded.[18] However, a Soviet scholar in reflexive control highlights that feeding the adversary with a flow of information favourable to the initiator leads the adversary to a decision that is not optimal to him, but predetermined by the initiator.[19] The application of Russian propaganda and information warfare over the past two decades has proven quite consistent with the Soviet reflexive control tactics and reflected their defining characteristic of accomplishing obfuscation or a so-called 4D approach —dismiss, distort, distract, dismay.[20] While such Russian use of Soviet reflexive control reflects also the Russian military system "aversion to innovation," analysts also point out to the tactics' limited effectiveness.[21] Such limitations may come from the acceleration and speed which are defining factors for today's information space, and the inherent difficulty of any bureaucratic system to keep up with the pace of its evolution.

One of the essential components of political warfare is the so-called active measures, a concept of Soviet origin as well. Richard Shultz equates active measures with non-kinetic political warfare, which includes overt media tactics, disinformation and fake news, election operations in the Facebook Age, and hybrid force multipliers, which are essential to Russian hybrid warfare.[22]

From a historical point of view, Thomas Rid distinguishes four waves of political warfare/active measures/disinformation in the modern era: (1) the interwar period, with the development of the radio and influence operations characterized by innovation, conspiracy, twists, "a weapon of the weak"; (2) after World War II, in the form of "political warfare" (West) or disinformation (Eastern bloc) performed by intelligence agencies to subvert the adversary in a professional way; (3) Late 1970s, active measures became sophisticated, resourced and managed by an effective bureaucracy to attain global impact; (4) mid-2010s, "high-tempo, low-skilled, remote, and disjointed", as well as less measured.[23] Rid also distinguished the following features of active measures: (1) methodical output of large bureaucracies; (2) include an element of disinformation; (3) have purpose, most common to weaken the adversary; (4) pragmatic, not perfect; (5) attacks against liberal order to erode the foundation of open societies; (6) accelerated exponentially by the digital revolution. He also noted the challenges these tactics face today given the facilitating medium of application, respectively on a large scale to target whole societies: "the internet […] made active measures less measured: harder to control, harder to steer, and harder to isolate engineered effects", thus more dangerous.[24]

Turning to propaganda, I note the overlaps and interconnections the concept has with active measures, disinformation, and political warfare. Historically, the term comes from Latin, where it means to propagate, but originates in the Catholic Church creation in 1600 of the *Sacra Congregatio de Propaganda Fide*, with the objective to promote faith to New World and to oppose Protestantism. For some experts, this was the beginning when the term lost its neutrality; its neutrality was fully lost when it was associated with Nazi Germany, Communist propaganda of Soviet Union, and Maoist China.[25]

J. Ellul defines propaganda as "a set of methods employed by an organized group that wants to bring about the active or passive participation in its actions of a mass of individuals, psychologically unified through psychological manipulations and incorporated into an organization."[26] In his view, propaganda engages in psychological action, respectively it attempts to modify opinions,[27] and it needs to impact the collective to be effective, not just the individual.[28] In addition, propaganda involves *re-education* or *brainwashing*, "transforming the adversary into an ally," and public and human relations, which aim to make the individual to conform[29]. It exists only in relation with technology as well as the state,[30] which adds to the understanding that a bureaucracy and systematic planning are needed for using propaganda.

Furthermore, in an analysis of Ellul's work, Van Vleets underlines that propaganda does psychological warfare with the aim to make the target audience doubt and mistrust its knowledge and actions; in turn, this leads to dependence on the propagandist, and less critical and analytical thinking of the target audience.[31] Propaganda presents inaccurate versions of the reality and, in this sense, it misrepresents symbols, it also "appeals to emotions and prejudices, bypassing rational thought", and is effective when "it is embedded in underlying attitudes or behaviours."[32]

Welch adds that its objective is to "convince people to think and act in a particular way and for a particular persuasive purpose."[33] An important aspect concerning propaganda remains its ethics. Despite misconceptions, experts agree on the neutrality of propaganda, either good or bad depending on the purpose it serves.[34]

Based on the above analysis of concepts, some underlying reflections can be drawn regarding propaganda as political warfare:

- It is a tool of statecraft with the purpose to influence the target, foreign or domestic, and attain political objectives and strategic goals. Thus, persuasion is the essential feature and not communication of the truth;
- Propaganda is human or population centric, depending on the subsequent objectives: influence decision-making, sow distrust within the society, weaken governance and explore vulnerabilities of a liberal democratic society; feeding the information that corresponds to the target's attitude, behaviour, and expectations leads it to predetermine decision-making or form an opinion favourable to the initiator;
- It is boosted by technology, without which it cannot function and be effective; the digital age and social media accelerate the propaganda's effects, though the recent advancements in these areas make it harder to control and command, though more dangerous; measuring its impact remains a challenge;
- It requires a certain level of planning and coordination, as well as integration with other tools – such as technology, cyber etc — to achieve political objectives;
- Its application evolves continuously;
- Coping with the propaganda's challenge requires constant awareness and nimble adaptation to its development, as well as a whole-of-government/society approach;
- It is characterized by pragmatism not perfectionism and its ethics is dependent on the purpose it serves.
- Staying ahead of the adversary's actions, and thus controlling the information space, is essential in successfully employing propaganda as political warfare; this also requires a significant level of flexibility and adaptation, as the adversary can itself adapt and take the initiative.

Russian propaganda machine: what adaptation, winning, and losing the information space looks like

Over the past 15 years, the Russian propaganda machine provided analysts with troves of activities that deserve research. Such activities are all part of the Russian leaders' self-described "war against the West," even if the West meant the democratic aspirations of Georgia and Ukraine, or the Western societies proper. They demonstrate strengths, weaknesses, adaptation, development, and over-reach, as well as ineffectiveness resulting from a gross misalignment of political goals with stated narratives.

The U.S. Department of State's Global Engagement Centre, whose core mission listed on the official website as of October 1, 2022 is "to direct, lead, synchronize, integrate, and coordinate efforts of the Federal Government to recognize, understand, expose, and counter foreign state and non-state propaganda and disinformation efforts aimed at undermining or influencing the policies, security, or stability of the United States, its allies, and partner nations," presents the five main pillars of Russian disinformation and propaganda in a 2020 report: (1) official government communications; (2) state-funded global messaging (media for both foreign and domestic publics, based in and outside Russia, and cultural institutions); (3)

proxy sources (conscious and unconscious proliferators of Russian narratives or other outlets); (4) weaponization of social media (to weaken societies including by sowing distrust in institutions); (5) cyber-enabled disinformation. While the official communications are visible and attributed to Russia, as one moves towards the last pillar the connection to Russia fades until it is denied[35]. However, this kind of organization and integration was not always the norm: drawing from the Soviet legacy of the Cold War, it was built over the years, from the announcement of the information security doctrine of 2000 to the lessons learned of the Russian invasion of Georgia in 2008, and the establishment of an integrated system that can wage political warfare with propaganda, active measures, and hybrid actions at its core.

The Russian invasion of Georgia in 2008 was a military conflict which combined conventional military action with cyber and information warfare on both sides. The initial Russian narrative focused on the "genocide" (a common theme in Russian propaganda) allegedly perpetrated by the Georgians against South Ossetians, as well as on the rightfulness of Russia's war. Initially, Georgia managed to get its message out, thus having the upper hand in the information warfare; later, Russia managed to perform better and convince the domestic audience which recorded widespread approval as it increasingly controlled the information space—using a military spokesperson, bringing Russian journalists, and presenting the Russian military advances and "Georgian atrocities."[36]

Opinions vary in relation to the effectiveness of Russia's use of information warfare in this conflict. Thomas points out to Russian experts who argued that the 2008 war demonstrated Russia's inability to advance its goals in the international information arena or its incapacity to convince the world of its narrative; they underlined that fulfilment of such objectives required a comprehensive integrated system to manage the information space.[37] Such a system should be comprised of bureaucratic institutions and management positions: a council on public diplomacy headed by the prime minister, a presidential advisor for information and propaganda activities, state foreign affairs media company, state internet company, information crisis action centre, information countermeasure centre, NGOs and training centre for personnel conducting information warfare ("information troops" consisting of diplomats, experts, journalists, publicists, web designers, hackers etc.).[38] These proposals reflect the consensus regarding propaganda as a planned activity, employed systematically and in a coordinated fashion by an effective bureaucracy.

Some positive assessments regarding Russia's actions point out to the combination of cyber activities designed to cripple the adversary's command and control infrastructure and the use of "mass information armies" to interact directly with the people via the internet, unlike the Georgian approach who preferred addressing the leaders.[39]

Despite Russian propaganda's limited effectiveness in the international arena in 2008, the invasion did not trigger any condemnation of Russian invasion (with notable exceptions from the Baltic states and Poland). There were no UN Security Council Resolutions to condemn it; Russia itself proposed its own resolutions against Georgia. Russian recognition of the independence of Abkhazia and South Ossetia was indeed condemned which probably triggered Russian not to proceed with annexation, as it happened in 2014 with Crimea. Nonetheless, Russia has had *de facto* control over the two provinces, and their symbolic value was not comparable to Crimea's. Furthermore, NATO's 2010 Strategic Concept presents the Euro-Atlantic security environment as being at peace, with no mention of the invasion or of Russia's conduct; it also depicts NATO-Russia cooperation as being of strategic importance.[40]

In Russia's use of propaganda in this conflict, one can notice: the scope to influence foreign and domestic targets; actions performed with media and technology support (including social media to focus directly on the targeted population); the adaptation to the environment and gaining the upper hand later in the conflict despite initial failure.

In 2014, in response to the Ukrainian protests to President Yanukovich changing course from the population's expected closer integration with the European Union, and his subsequent abandonment of presidency, Russia used a complex hybrid warfare toolbox to gain control of and annex Crimea. Unlike 2008, Russia information warfare machine was better fit to act in support of the Russian military actions in Crimea and in Eastern Ukraine.

At least two entities deserve more attention in this regard, including during and in the aftermath of the 2014 events: RT and the Internet Research Agency. According to Margarita Simonyan, Russian state media company RT's chief editor, the search for alternative sources of information—"the other side of the truth"—represented a fertile ground for the RT which can feed it. To this end, RT acted to grow its audience, foreign and domestic, and defined its purpose in a manner like that of a defence ministry. However, RT's approach in many critical moments, including Russia's annexation of Crimea, proved its use as an information weapon and that journalistic standards and values did not define the organization.[41] The Internet Research Agency, "the troll factory" in Sankt Petersburg, has the goal to influence foreign public opinion along the lines of Russian government political goals. To this end it relies heavily on the social media and the transparency existing in democracies. Unlike 2008, both entities appear to be part of the plan mentioned above for Russia to adapt, to better manage the information space and to increase the effectiveness of its political warfare.

While real Russian strategic goals of 2014 cannot be easily discerned, those might have included: keeping Ukraine out of the West, securing a land corridor in the South of Ukraine, thus linking Russia with Crimea and Transnistria, and limiting the West's response to Russia's actions. Propaganda, denial, and deception came into play. First, Russian covert actions in Crimea and subsequent control of the peninsula ("the little green men") were denied by Russian leadership who claimed that those were done by local militia, despite the need for advanced command and control to perform such operations (later, the Russian President admitted involvement of Russian forces in taking Crimea, but justified the actions with no loss of life). Second, the information space was flooded with news that depicted, as before in Georgia, Ukrainian atrocities, and genocide, thus putting the blame for triggering Russian actions on Ukraine itself. An escalation of Russian rhetorical aggression was a novelty when compared with stricter Soviet propaganda tactics, for instance, "turning America in radioactive ash."[42] Third, Russian domestic approval rating of Russian leadership was very high, especially with regard to the annexation of Crimea. An indication in that regard was, for instance, the volunteerism among the Russians to fight in Eastern Ukraine, a significant feat compared to the 2022 rejection of mobilization. Fourth, attempts to bring to fruition the concept of Novorossiya with protests sparking in Odessa and other parts of Ukraine were not successful, though the concept itself was in line with justification of Russian actions. Fifth, such justification also relies heavily on Putin's thinking that "Ukraine is not even a country," allegedly expressed at the NATO Summit in Bucharest in 2008 (which basically denies the existence of Ukraine despite the 1994 Budapest Memorandum on Security Assurances), and his narrative that Russians and Ukrainians are one people, which he later expressed.[43]

Compared to 2008, the effectiveness of Russian political warfare has significantly improved. Surely, Russia was sanctioned internationally, lost the UN General Assembly vote regarding its annexation of Crimea (100 countries supported UNGA Resolution 68/262 to 11 against and 58 abstentions), which was not recognized by most of world nations. However, Russia managed to delay and weaken the international response through denying its involvement in the conflict or later justification of the annexation, and played the aversion of Western leaders to involve themselves in a conflict. Thus, "Russia's information operations have provided support for the policies of inaction."[44] Furthermore, they increased the domestic approval rating (after the annexation of Crimea, approval rating reached 89% as reported by polls). Deniability of Russian military presence triggered a difficulty to attribute the actions to Russia and, despite its incapacity to fulfil all objectives (for instance, securing a corridor to Transnistria), allowed Russia to demonstrate its military superiority in the conflict and play the role of mediator along France and Germany in the Minsk process.

Within NATO, there was progress at the 2014 Summit to recognize Russia's challenges though without designating its conduct as threat. Also, NATO initiated a refocus on collective defence, which included limited adaptation and assurance measures, with the stated objective to deter aggression of NATO territory. Analysts point out though: "Were the West determined to resist Russia's destruction of the Ukrainian state, the dissimulation and confusion Putin has spread would have much less effect."[45]

Russian aggression against Ukraine in 2022 showed a different story. Like in other instances, Russia amassed a significant number of forces at the Ukrainian border, simultaneously denying that their purpose was to invade Ukraine. The Russian narrative before the invasion continued to refer to the NATO's expansion into Russian sphere of interests, to the illegitimate leadership in Kyiv, as well as to "nazification" in Ukraine. The Russian proposed agreements for the security of Russia, NATO, and the U.S., basically advanced to be rejected, were designed to signal a fake Russian openness to dialogue.

Once the invasion began, Russia portrayed it as a special operation, designed to de-nazify and to demilitarize Ukraine, an inconsistent objective, given the Jewish origin of the Ukrainian President and the fact that Ukrainian military posed no threat to Russia. Russian losses on the battlefield could not be explained by propaganda stating that everything goes according to the plan, even when Kyiv was out of the reach of Russian military. Strategic goals changed subsequently with a Russian refocus on the south of Ukraine; internationally, as grain exports from Ukraine were stopped, impacting African countries, Russia attempted to blame the West and the NATO expansion once again for triggering the war, with narratives targeting the "Global South."

Unlike 2014, Western declassified intelligence regarding Russia's intentions put Moscow on the defensive early on, respectively since November 2021. There were many international outreaches to Moscow, asking for withdrawal of Russian troops, even though Moscow continued to deny any offensive plans. Russian narratives regarding NATO (specifically the promise not to expand, that it is aggressive and aims to encircle Russia) were debunked. Furthermore, President Putin's narrative of the Russian and the Ukrainians being one people[46] supported the idea of a quick victory, which may have led to the expectation that Ukrainians would welcome the Russian armed forces with open arms or at least not fight with the determination they have, proved an essential limitation of propaganda: when propaganda, not the reality, feeds the initiator's plans, their realization is doomed to failure.

After the Russian attack, Ukrainian propaganda gained the initiative in the information space quickly. President Zelensky famous quote, "The fight is here, I need ammunition, not a ride," considered "one of the most-cited lines of the Russian invasion of Ukraine," had as source an unnamed senior American intelligence official.[47] Zelensky also made it personal, stating that he was Russia's first target and his family was the second one.[48] He personally appeared in videos and photos in Kyiv, alongside his team, to show proof he stands firm. Memes with him portrayed as *Captain Ukraine* flourished over the Internet,[49] unlike President Putin portrayed in his office, far away from his top generals. Zelensky also addressed Western Parliaments via videoconferences to garner support, his messages including visuals of Russian armed forces' atrocities, targeting of civilians and cities, and, on the other hand, portraying Ukrainians' determination to resist and fight back.

Videos of Ukrainian armed forces destroying Russian capabilities contributed to challenging the myth of Russian invincibility or superiority. News of Russian generals killed on the frontline were significant as well, as they proved the passivity of a very centralized command and control system and added to the symbolic value of Ukrainian success. Other memes or videos showed that the fight engages all Ukraine, including civilians that did their part, for instance a Russian tank being pulled by a tractor driven by a Ukrainian farmer.[50] The obscene, but brave rejection of the defenders of the Snake Island of the Russian flagship Moskva's ultimatum was a rallying cry. Memes of Russian armed forces returning home with loot covered in blood also debunked the idea of the unity of the two peoples as advocated by Putin. Furthermore, a significant presence of international journalists in Kyiv and the regular updates of the Ukrainian military contributed to the Ukrainian pre-eminence in the information sphere[51].

Such actions allowed Ukraine to take control of the information space, to fight potential fatigue in providing support by the West, to ask for accountability for Russian atrocities and war crimes, while focusing on the fact that no other support (with reference to Western boots on the ground) except weapons was needed.

Russian attempts for adaptation of its propaganda to the Ukrainian information counteroffensive have included a more aggressive rhetoric, invoking the nuclear weapons for defending Russian territory (with reference also to the newly annexed territories after sham-referenda). However, in the domestic area, propaganda failed to get the expected support, with the efforts of partial mobilization being delayed and Russian men fleeing the country. More dissenting voices inside Russia challenged the official narrative that the special operation goes according to the plan.

While the current Russian-Ukrainian war is not yet over, the information war proved far more challenging for Russia in the international arena. Its political warfare failed on many fronts:

- Russia was condemned through a UN General Assembly Resolution with 141 yes-votes against 5 no-votes, 35 abstentions and 12 absent.[52]
- NATO, European Union, and their members were essentially supportive of Ukraine and stood united, including by gradually adopting harsh sanctions.
- Sweden and Finland applied for NATO membership, thus renouncing their longstanding neutrality.
- NATO Strategic Concept of 2022 recognized Russia as a threat as well as the reality that the Euro-Atlantic area is not at peace due to Russia's violation of "norms and principles that contributed to a stable and predictable European security order."[53]
- Against the expectation of fuelling disunity in the West, Russia encountered coherence and unity and the West finally recognized the importance of having a more realist approach regarding Russia.

Conclusion

Political warfare proves to be a much-needed capability for effective statecraft in both international and domestic settings, either to advance foreign policy and strategic objectives or to cope with an adversary's similar actions. Propaganda remains a principal tool of political warfare whose aim is to persuade both leaders and societies, and ultimately to shape the decision-making process in a way favourable to the initiator. Its success relies heavily on consistency of stated objectives, careful planning and coordination and use of technology and social media. As seen in the example of Russian propaganda related to the invasion of

Ukraine in 2014, concealment of Russia's real objectives and denial of its presence in Crimea and in Eastern Ukraine has played to its advantage. It could have claimed "victory" any time, after the annexation of Crimea, or after securing a land corridor covering the whole Ukrainian South, had that succeeded. Russian political warfare checked most of the boxes in 2014: it evolved after 2008 invasion of Georgia, with new institutions, planned actions, and better coordinated means.

However, changes of objectives raise challenges for propaganda which may lead to a loss of control over the narratives. During Russian aggression against Ukraine of 2022, the stated objectives have not been consistent in themselves and across time, and the narratives neither fit, nor supported the reality on the ground. Thus, continuously stating that "everything goes to plan" backfired in the information space. Moreover, the initiator planned and set objectives based on its propaganda narrative; belief in propaganda misled the strategic goals and the in-field action, which constituted fatal errors.

Flexibility and adaptation are key features of propaganda needed to maintain control of the information space, with technology and social media as essential enablers. Getting the right message out in the right time with an adequate level of legitimacy contributes to a significant impact, as evidenced by Ukraine's taking over of the information space.

Ultimately, an adequate and balanced use of propaganda, bearing in mind the ethics of its purpose and consistency of its message, overlaps to a very large extent the concept of strategic communication, which might be more acceptable to audiences. Nonetheless, propaganda remains essential for the success of political warfare.

Notes

1 Jones, "Return of Political Warfare," 1.
2 Boot and Doran, "Political Warfare," 1–3.
3 Cleveland et al., "An American Way," 1.
4 Coats, "Remarks," 2.
5 Kennan, "Political Warfare," 1.
6 Von Clausewitz, Book 1, Chapter 1.
7 Department of Defense, Dictionary, 415.
8 Dearth, "Shaping the 'Information Space'," 1–2.
9 Hoffman, "Not-So-New Warfare."
10 Applebaum, "Putin's New Kind of War."
11 Galeotti, *Russian Political War*, 53.
12 Ibid., 55.
13 Gerasimov, "Foresight," 24.
14 Ibid., 28–9.
15 Schnaufer, "Redefining Hybrid Warfare," 21.
16 Ibid., 23.
17 Thomas, "Russia's Reflexive Control," 237.
18 Chotikul, The Soviet Reflexive Control, 48.
19 Reid, "Reflexive Control," 294.
20 Emerson, "Exposing Russian Disinformation".
21 Snegovaya, "Putin's information war," 12.
22 Shultz, "Russia's Hybrid Warfare," Presentation.
23 Rid, Active Measures, p. 6–7.
24 Ibid., 9–14.
25 Bolsover, "Computational Propaganda," 123.
26 Ellul, *Propaganda,* 61.
27 Ibid., xiii.
28 Ibid. 28.

29 Ibid., xiii.
30 Ibid., 28
31 Van Vleet, *Dialectical Theology and Jacques Ellul*, 130.
32 Bolsover, "Computational Propaganda," 124–5.
33 Welch, "A Brief History of Propaganda," 21.
34 Ibid., 32; Stengel, *Information Wars*, 290
35 U.S. Department of State, GEC Report, 8–9.
36 Donovan, "Russian Operational Art," 21.
37 Thomas, "Russian Information Warfare," 279.
38 Ibid., 279–82.
39 Ibid., 283–5.
40 NATO, *Strategic Concept 2010*, 10; 29.
41 Atlantic Council's Disinformation Lab, *Russia Today's role as an "information weapon"*.
42 Snegovaya, "Putin's information war," 15.
43 Putin, "Historical Unity Russians Ukrainians".
44 Ibid., 21.
45 Ibid., 21.
46 Hill and Stent, "The World Putin Wants," 109.
47 Kessler, "Zelensky's Famous Quote".
48 Shuster, "Zelensky Defended Ukraine."
49 Mtn, "Captain Ukraine."
50 The Economic Times, "farmer steals Russian tank."
51 Åslund, "Putin Losing Information War."
52 United Nations, "General Assembly Demands Withdrawal."
53 NATO, *Strategic Concept 2022*, 3.

Bibliography

Applebaum, Anne. "Putin's New Kind of War," *Slate*. April 16, 2014. https://slate.com/news-and-politics/2014/04/vladimir-putins-new-war-in-ukraine-the-kremlin-is-reinventing-how-russia-invades.html.

Åslund, Anders. "Why Vladimir Putin is losing the information war to Ukraine," Atlantic Council, March 6, 2022. https://www.atlanticcouncil.org/blogs/ukrainealert/why-vladimir-putin-is-losing-the-information-war-to-ukraine/.

Atlantic Council's Disinformation Lab. Question That: RT's Military Mission. Assessing Russia Today's role as an "information weapon,"January 8, 2018. https://medium.com/dfrlab/question-that-rts-military-mission-4c4bd9f72c88.

Bolsover, Gillian. "Social Media, Computational Propaganda, and Control in China and Beyond." In *The World Information War: Western Resilience, Campaigning, and Cognitive Effects*, edited by Clack, Timothy, and Johnson, Robert. London: Routledge, 2021.

Boot, Max, and Michael Doran. "Political Warfare." Council on Foreign Relations, 2013. http://www.jstor.org/stable/resrep05693.

Chotikul, Diane. *The Soviet Theory of Reflexive Control in Historical and Psychocultural Perspective: A Preliminary Study*, Monterey, 1986. https://apps.dtic.mil/sti/pdfs/ADA170613.pdf.

Cleveland, Charles T., Ryan Crocker, Daniel Egel, Andrew M. Liepman, and David Maxwell. "*An American Way of Political Warfare: A Proposal*." RAND Corporation, 2018. http://www.jstor.org/stable/resrep19897.

Coats, Dan. Comments at Hudson Institute's "Dialogues on American Foreign Policy and World Affairs" series, July 13, 2018. https://s3.amazonaws.com/media.hudson.org/files/publications/CoatsFINAL.pdf.

Dearth, D. H. "Shaping the 'Information Space.'" *Journal of Information Warfare* 1, no. 3 (2002): 1–15. https://www.jstor.org/stable/26504099.

Department of Defense. Dictionary of Military and Associated Terms. https://web.archive.org/web/20091108082044/http://www.dtic.mil/doctrine/jel/new_pubs/jp1_02.pdf. 2001.

Donovan, George T. "Russian Operational Art in the Russo-Georgian War of 2008", U.S. Army War College, 2009. https://apps.dtic.mil/dtic/tr/fulltext/u2/a500627.pdf.

Ellul, Jacques. *Propaganda: The Formation of Men's Attitudes*. New York: Random House, 1965.

Emerson, John B. "*Exposing Russian Disinformation*", Atlantic Council, June 29, 2015, https://www.atlanticcouncil.org/blogs/ukrainealert/exposing-russian-disinformation/.

Galeotti, Mark. *Russian Political War: Moving Beyond the Hybrid*, London: Routledge, 2019 (kindle edition)

Gerasimov, Valery. "The Value of Science Is in the Foresight", *Military Review*, January-February2016. https://www.armyupress.army.mil/portals/7/military-review/archives/english/militaryreview_20160228_art008.pdf.

Hill, Fiona and Stent, Angela. "The World Putin Wants," *Foreign Affairs*, Volume 101, No. 5, September/October2022.

Hoffman, Frank. "On Not-So-New Warfare: Political Warfare vs. Hybrid Threats," War on the Rocks, 28 July2014. https://warontherocks.com/2014/07/on-not-so-new-warfare-political-warfare-vs-hybrid-threats/.

Jones, Seth G. "The Return of Political Warfare," Center for Strategic and International Studies, February 2, 2018. https://csis-website-prod.s3.amazonaws.com/s3fs-public/publication/180202_Jones_ReturnPoliticalWarfare_Web.pdf.

Kennan, George F. "The Inauguration of Organized Political Warfare,", April 30, 1948. History and Public Policy Program Digital Archive, National Archives and Records Administration, Record Group 59, Entry A1 558-B, Policy Planning Staff/Council, Subject Files, 1947–1962, Box 28. Obtained by Brendan Chrzanowski. https://digitalarchive.wilsoncenter.org/document/208714.

Kessler, Glenn. "Zelensky's famous quote of 'need ammo, not a ride' not easily confirmed, " *The Washington Post*, March 6, 2022. https://www.washingtonpost.com/politics/2022/03/06/zelenskys-famous-quote-need-ammo-not-ride-not-easily-confirmed/.

Mtn, Bogdan. "Volodymyr Zelensky—The Ukrainian Captain." February 27, 2022. https://medium.com/new-writers-welcome/the-ukrainian-captain-volodymyr-zelensky-3e022c99ab09.

North Atlantic Treaty Organization. *Active Engagement, Modern Defense. Strategic Concept for the Defence and Security of the Members of the North Atlantic Treaty Organization*. Lisbon, 2010. https://www.nato.int/nato_static_fl2014/assets/pdf/pdf_publications/20120214_strategic-concept-2010-eng.pdf.

North Atlantic Treaty Organization. *NATO 2022 Strategic Concept*. Madrid, 2022. https://www.nato.int/nato_static_fl2014/assets/pdf/2022/6/pdf/290622-strategic-concept.pdf.

Putin, Vladimir. "On The Historical Unity of Russians and Ukrainians," July 12, 2021. http://en.kremlin.ru/events/president/news/66181.

Reid, Clifford. "Reflexive Control in Soviet Military Planning." In *Soviet Strategic Deception*, edited by Brian Daily and Patrick Parker, 293–311. Lexington, Mass.: Lexington Books, 1987.

Rid, Thomas. *Active Measures: The Secret History of Disinformation and Political Warfare*. New York: Farrar, Straus and Giroux, 2020.

Schnaufer, Tad A. "Redefining Hybrid Warfare: Russia's Non-Linear War against the West." *Journal of Strategic Security* 10, no. 1 (2017): 17–31. http://www.jstor.org/stable/26466892.

Shultz, Richard. "Russia's Hybrid Warfare: A Cold War Legacy System adapted to the Facebook Age". Presentation. January 8, 2020, London. Fletcher School of Law and Diplomacy, GMAP.

Shuster, Simon. "How Volodymyr Zelensky Defended Ukraine and United the World," *Time*, March 2, 2022. https://time.com/6154139/volodymyr-zelensky-ukraine-profile-russia/.

Snegovaya, Maria. "Putin's Information Warfare in Ukraine: Soviet Origins of Russia's Hybrid Warfare." Institute for the Study of War, 2015. http://www.jstor.org/stable/resrep07921.1.

Stengel, Richard. *Information Wars: How We Lost the Global Battle Against Disinformation And What We Can Do About It*. New York: Grove Press, 2019.

The Economic Times. "Epic! Ukrainian farmer 'steals' Russian tank using his tractor" March 1, 2022. https://www.facebook.com/EconomicTimes/videos/4898702183529152/.

Thomas, Timothy L. "Russia's Reflexive Control Theory and the Military," *Journal of Slavic Military Studies* 17: 237–256, 2004, doi:10.1080/13518040490450529, https://www.rit.edu/~w-cmmc/literature/Thomas_2004.pdf.

Thomas, Timothy L. "Russian Information Warfare Theory: The Consequences of August 2008." In *The Russian Military Today and Tomorrow: Essays in memory of Mary Fitzgerald*. Edited by Blank, Stephen, and Weitz, Richard. Strategic Studies Institute, US Army War College, 2010. https://www.jstor.org/stable/pdf/resrep12110.8.pdf..

U.S. Department of State. *GEC Special Report: Pillars of Russia's Disinformation and Propaganda Ecosystem*. August2020. https://www.state.gov/wp-content/uploads/2020/08/Pillars-of-Russia%E2%80%99s-Disinformation-and-Propaganda-Ecosystem_08-04-20.pdf.

United Nations. "General Assembly resolution demands end to Russian offensive in Ukraine." March 2, 2022. https://news.un.org/en/story/2022/03/1113152.

Van Vleet, Jacob E. *Dialectical Theology and Jacques Ellul*, 1517 Media: Fortress Press, 2014. www.jstor.org/stable/j.ctt9m0tt6.9.

Von Clausewitz, Carl. *On War*. London, 1873, https://www.clausewitz.com/readings/OnWar1873/BK1ch01.html#a.

Welch, David. "A Brief History of Propaganda: A much maligned and misunderstood word." In *The World Information War: Western Resilience, Campaigning, and Cognitive Effects*, edited by Clack, Timothy, and Johnson, Robert. London: Routledge, 2021.

11

THE USE AND ABUSE OF HISTORY BY RUSSIAN EMBASSIES ON TWITTER

THE CASE OF THE BALTIC STATES

Corneliu Bjola and Ilan Manor

I. INTRODUCTION

On the eve of the Russian invasion in Ukraine, President Putin stated that the goal of the military campaign was nothing less than "the demilitarisation and denazification of Ukraine"[1]. While his statement might have surprised some opinion leaders in Western countries, his discourse merely reflected an entrenched narrative, which Russian authorities had been actively promoting through their digital channels for several years, centred on the idea that Ukraine possessed no right to statehood and sovereignty.[2] Putin's views were subsequently reinforced by the former Russian President, Dmitry Medvedev, who declared that "Ukrainian identity is one big fake and the goal of the de-Nazification is to change how Ukrainians perceive their identity".[3] In the same vein, an article published by Russian state-owned news agency RIA-Novosti on April 5, 2022, repeated Putin's claim that "Ukrainians are an artificial anti-Russian construct", it proclaimed that "Ukraine's political elite must be eliminated," and it declared that ordinary Ukrainians were "passive Nazis" who "must experience all the horrors of war and absorb the experience as a historical lesson and atonement for their guilt."[4]

The Russian attempt to discursively annihilate Ukraine's state- and nation-hood highlights the immense power that narratives may have in shaping the image of countries, for better, but especially for worse. In all fairness, national images have always had important diplomatic implications, as the extensive literature on soft power keeps reminding us.[5] Nations that are viewed as adhering to international norms or are perceived to be guided by a moral compass are less likely to encounter resistance to their foreign policies, can more easily form coalitions in multilateral forums and importantly, and are more likely to secure support from allies in times of crises as the case of Ukraine currently demonstrates. By contrast, countries with a negative image such as North Korea, Iran, Syria, Venezuela and now Russia, following its brutal invasion of Ukraine, are likely to face serious difficulties in convincing the members of the international community to collaborate with them.

DOI: 10.4324/9781003190363-14

The arrival of digital platforms has complicated, however, the task of national image management. While previously, national images and reputations have been directly connected to countries' policies and their patterns of international behaviour, social media has now made possible to break the link between them by creating "alternative realities" in which a country's image may have little or no connection to its policies. Bots, trolls, and fake news may be utilised in order to promote a negative image of the targeted nation, all with the goal of discrediting and isolating it diplomatically.[6] Doctored images and fictitious historical accounts go viral online thus challenging the targeted nation's ability to pursue or defend its foreign policy goals. The narrative demolition of Ukraine's image attempted by Russia in the past decade has served exactly this purpose: to isolate Ukraine from its European and transatlantic partners and thus to make it easier to be absorbed in what Kremlin insisted to define as its geopolitical "sphere of influence".

While the case of Ukraine is significant in view of the ongoing humanitarian tragedy and the broader geopolitical implications of Russia's aggression for the European and international security, the situation of other countries in the region that have been subjected to similar discreditation campaigns is also important to examine for two reasons. First, it allows us to assess the broader role of historical narratives in Russian disinformation campaigns, and second, it helps reveal the scope of Russian geopolitical ambitions in the region and whether they cover not only prospective but also existing NATO members such as the Baltic states. To this end, the goal of this study is to examine how Russian diplomats use media channels to frame the second World War (WWII) and its legacy in general, and of the Baltic countries' role in WWII in particular. The study does this by analysing the individual frames, or arguments promoted by Russia on social media, while also examining how these frames coalesce to form a narrative or story through which the past illuminates the present. Our analysis shows that Russia's "denazification's narrative" is not unique to Ukraine. It has been also deployed, in a slightly different version, against the Baltic states to undermine their internal cohesion and to fracture their relationship with its European and NATO allies.

The chapter will develop this argument in three steps: the first section will introduce the literature on narratives and discuss its relevance for the study of international politics. The second part will outline the research design and methodology of the paper and analyse two interrelated historical narratives that Russia has deployed strategically against the Baltic states: the first one concerning the 75th anniversary of the end of WWII and the second involving the lessons drawn by President Putin about WWII. The third section will discuss the findings of the study and review their theoretical implications for the study of digital disinformation. The chapter will conclude with a short conclusion summarising the main argument and outlining a few directions for further research.

II. NARRATIVES—PAST AND PRESENT

The concept of narrative has started to gain epistemic authority in social sciences rather recently in an attempt to add more analytical clarity to studies examining the structural role of language and discourse in shaping social action. Unlike the notion of discourse, which is considered to represent the "raw material" of communication that actors plot into a narrative, the latter is viewed as a "particular structure made up of actors and events, plot and time, and setting and space".[7] As Hagström and Gustafsson point out in their comprehensive review of the literature on narrative power, the epistemic shift from discourse to narrative has made possible to better theorise about the capacity of actors to produce effects by disseminating strategic narratives, as well as about the capacity of narratives to enable and constrain actors in the first place.[8] Narratives differ, of course, in their ability to generate effects and constrain actors, so it is important to understand what makes them powerful, how do they change and eventually decline, and by extension, how do they contribute to influencing matters of war and peace in international politics.

In basic terms, narrative analysis rests on identifying the story that actors create to influence the media's coverage of international, national, or issue-specific events as well as the public's understanding of these events. When used in a foreign policy context, narratives have a strategic goal: they serve to rally support for a nation's foreign policies while reducing the complexity of geo-political processes. For this reason, narratives often identify a problem, a protagonist, a stage, a course of action and it typically ends with a prescriptive evaluation. The problem represents the particular challenge the protagonist is expected to solve by countering the actions of various opponents, usually within the context of geopolitical difficult yet morally enlightening circumstances. Narratives also summon the past to the present as the past can serve as a template through which the present can be understood. For instance, the Cold War was a narrative used by Western and USSR-aligned governments to explain the state of the world before 1989. Within this narrative, there were different frames (military, economic, scientific, cultural etc.), which depicted the zero-sum competition between the two superpowers. The Russian aggression against Ukraine is now increasingly perceived as ushering in a new Cold War between United States and its western allies against Russia and its authoritarian supporters, especially China.[9] Within this narrative, zero-sum frames are used again to comment on specific events (the G20 summit, the UN Security Council resolutions) or specific policies (energy security, military assistance, international development).

As Roselle, Miskimmon and O'Loughlin explain in their seminal study, narratives could be analysed from three different perspectives: formation, projection, and reception.[10] Formation refers to the domestic and international conditions that may encourage or constrain actors to strategically articulate their narratives around certain frames as opposed to others. Russia's aggressive use of disinformation narratives is viewed, for instance, as a strategic instrument, within its repertoire of hybrid warfare measures, to counter its declining geopolitical influence, as well as to mask its domestic weaknesses.[11] Projection involves how narratives connect to each other horizontally or vertically and how the flow of narratives travels through the social and traditional media landscape. The projection of a narrative is facilitated by the existence of other narratives that help permeate, structure and prime knowledge on a specific topic. Russia's victimisation meta-narrative against the West[12] draws power from lower-level narratives that portray the European Union or NATO as Washington's 'puppets', working together to deny Russia's alleged rightful place in the global order, all while seeking to promote a decadent, "Russophobic", "aggressive" and even "fascist" international agenda.[13] Finally, reception denotes how the audience makes sense, embraces, or by case rejects the proposed narratives. While there is no universal formula for explaining narrative success or failure, studies shows that the quality of the public sphere of a society can make a big difference in building resilience against hostile, disinformation-driven narratives.[14]

An important subgenre of the narrative literature deals with historical aspects of past conflicts, diplomatic events, and geopolitical turning points. As Freel and Bilali explain, historical narratives are social representations of the past that serve key social functions: to provide a sense of temporal continuity with past struggles, contribute to the formation and reinforcement of group identities, act as a symbolic resource that can be mobilised in the service of political aims and most importantly, to serve as interpretive frames for understanding the present.[15] Historical narratives are made of curated collections of events so the process of selection of what is included and excluded from the narrative plays a critical role for understanding how the shared memory of the past if formed. Chatterje-Doody[16] has found, for instance, that while Kremlin's standard historical narrative has evolved over time it has largely stayed centred on a few core themes: stability and continuity in political power, Russia's long history and uniqueness, and Russia's international identity as great power with a special role to provide regional leadership. The legacy of WWII plays a central role in the construction of this historical narrative. As Stallard points out, Putin has transformed the memory of the Soviet victory in WWII into a national religion, positioned himself as the heir to that legacy, and has framed contemporary threats against Russia and Russians everywhere in the same dark colours. There are signs that the abuse of history through the WWII narrative is being embraced by other autocracies such as China, for exactly the same reasons, to boost the legitimacy of the regime.[17]

III. NARRATIVE ANALYSIS

Drawing on the previous discussion, this study seeks to examine the use of historical narratives by Russia in support of its foreign policy in the Baltic region. To this end, the study has focused on two time periods. The first period included tweets that marked the 75th anniversary of Russia's victory over Nazi Germany in 2020. In total, 3,376 tweets, published between April and August of 2020, were analysed. The second time period included tweets posted by 24 official Russian channels[18] following the publication of President Putin's article in the *National Interest* in June 2020.[19] In total, the second period included 2,515 tweets published between June and July of 2020. Alongside the actual tweets, and the dates they were posted, the rates of engagement for each tweet were also calculated (e.g., number of retweets and likes garnered by each tweet). Content and frame analysis were then employed to compress the dataset of 5,891 tweets assembled from the two time periods to include only those tweets that dealt with WWII and the Baltic States' role in WWII. In total, we have found 2,500 tweets to have high relevance for our empirical analysis.

In the first stage of analysis, all tweets published over a seven-day period were collated and grouped based on subject matter. For instance, a large proportion of tweets dealt with Russia's call for an objective, historical analysis of WWII. Another group of tweets dealt with WWII's impact on Russia's foreign policy. Once all tweets had been grouped by subject matter, they were categorised, using Entman's definition of framing,[20] into those that offered a problem definition, causal attribution, moral evaluation and suggested remedy. This led to the mapping of several frames including, for example, a frame dedicated to the post-war world order that was shaped by the USSR. Another frame dealt with Eastern European nations accused by Kremlin of banning the Russian language. Lastly, the core argument made in each frame was extracted and grouped together leading to the identification of a narrative that Kremlin would promote to explain the legacy of WWII. Two narratives were thus identified. The first is that of 'Russia Remembers' and the second is 'The Lessons of WWII'.

3.1 The 75th anniversary of the end of WWII

The first narrative, disseminated on the 75th anniversary of Russia's victory over Nazi Germany, is that of 'Russia Remembers'. This narrative is comprised of eleven frames or arguments.

The first frame focused on the preservation of historic monuments to the USSR's victory over Nazi Germany. This frame contrasted Russia, who cherished the memory of WWII through monuments, with other nations who allegedly desecrated such monuments and in the process, betrayed the memory of Red Army soldiers that had fought valiantly against Nazi Germany.

This second frame of 'We Remember' highlighted the various means through which Russia remembered the sacrifices made by Red Army soldiers during WWII. This included honouring battles and units as well as holding tributes to Red Army soldiers across the world, including in Israel.

The third frame emphasised the fact that Russia, the US and the UK were allies during WWII. Tweets in this frame demonstrated, or 'proved' that the USSR mourned the loss of President Roosevelt and that the USSR and the Allies shared a common foe. The frame also recognized British soldiers who were honoured by the USSR for their role in WWII.

The fourth frame dealt with preserving the memory and lessons of WWII for younger generations. Tweets in this frame suggested that younger generations must learn what state-egotism, xenophobia and hate could lead to. One way in which the memory of WWII had been preserved was through the UN-system which sought to end state-egotism by creating a mechanism for resolving crises peacefully.

The fifth frame offered evidence of the Red Army's long struggle against Nazi Germany. This included bulletins published by the USSR's Embassy in Washington as well as black and white images documenting battles, regiments and soldiers. Several tweets in this frame focused on female soldiers echoing the supposed social cohesion of the Red Army where men and women fought and died together.

The sixth frame detailed the heavy price that the USSR paid in its struggle against Nazi Germany. This included the loss of 26 million citizens, as well as large numbers of casualties among the Red Army. This price was personified through the siege of Leningrad and the hardship that Soviet citizens endured during the siege.

The seventh frame focused on 'Faces of Victory', or individual soldiers whose heroism and self-sacrifice facilitated the victory over Nazi Germany. Tweets comprising this frame introduced heroes from across the Soviet Union while once again emphasizing the role that women played in the victory over Nazi Germany. Importantly, these tweets humanised WWII and 'put a face' on Soviet soldiers.

The eighth frame argued that Russia liberated Eastern Europe from the yoke of Nazism. Tweets in this frame celebrated the liberation of areas in Slovakia, Belarus, Poland and the Baltics while demonstrating that Russia would still celebrate these liberations today.

The ninth frame focused on audience engagement. Russian Embassies and institutions invited digital publics to partake in the celebration of the victory in WWII. Digital publics were invited to online art exhibits, to hear songs celebrating the USSR and to watch movies documenting the Red Army's great victory.

The tenth frame emphasized that Ukraine was an integral part of the USSR's victory over Nazi Germany and that Ukrainian soldiers were integral to the Red Army. The eleventh and final frame focused on the Baltic States which were accused of seeking to obscure the memory of WWII be it by prohibiting the display of USSR medals or by hiding the fact that Baltic States collaborated with Nazi Germany.

Taken together, these frames create a narrative of 'Russia Remembers'. This is made most evident in the frequent use of the hashtag 'We Remember'. Yet this hashtag is not meant as a statement but, rather, as a question posed to followers: 'Russia remembers the trials of WWII and the heavy price paid to rid the world of Nazism. Do you?' Imbued within this narrative is the assertion that those who forget history are doomed to repeat it. It is for this reason that Russia cherishes the memory of WWII.

Notably, two actors are depicted within this narrative. The first is the Red Army, which is portrayed as a people's army, one made up of ordinary men and women who were willing to sacrifice their lives to defeat Nazism. Second, the Baltic States are portrayed as trying to white-wash history by preventing the commemoration of WWII or by masking their collaboration with the Nazi Germany. This narrative thus creates a moral dichotomy between a 'good' Russia that cherishes the memory of WWII, and the 'immoral' Baltics who refuse to face objective historical evidence of their collaboration with the Nazis. It is by associating the Baltic States with the most abhorrent regime in history that Russia's narrative includes a clear moral evaluation.

As can be seen in *Figure 11.1*, the four most prevalent frames disseminated by official Russian Twitter channels were 'We Were Allies', 'Historical Evidence', 'Faces of Bravery' and 'Russia Liberated the East'. The two least prevalent frames were 'We Remember' and 'Lessons of WWII'. This analysis demonstrates that the Russian narrative has in fact focused on making a claim to historical truth while emphasising the depiction of the Red Army and the objective, historical fact that Russia liberated Eastern Europe from the yoke of Nazism. In light of these frames, the moral dichotomy between Russia and the Baltics becomes even more pronounced as the Baltic States emerge as 'ungrateful'. After being liberated by Russia, the Baltics deny Russia their "gratitude" and recognition.

Figure 11.2 above offers an analysis of engagement rates with Russian frames. As can be seen, the three frames to obtain the highest average number of re-tweets were 'Faces of Bravery'. 'Historical Evidence' and 'We Were Allies'. The two frames to obtain the lowest average number of re-tweets were 'Baltics Whitewash History' and 'Audience Engagement'. It is important to note that many images used in these frames were black and white photos. The lack of colour adds a historical dimension to photographs as they supposedly document a past that can no longer be approached. For instance, one can no longer approach WWII as most veterans are either dead or living in old age homes. It is thus through historical images that the past can be revisited and brought to the present. Moreover, black and white photos are similar to the one would expect to find in an archive. They thus could be used to 'prove' that certain states acted in certain ways. By using black and white photos Russian diplomats thus tried to imply that Russia's interpretation of WWII rested on "objective" historical accuracy.

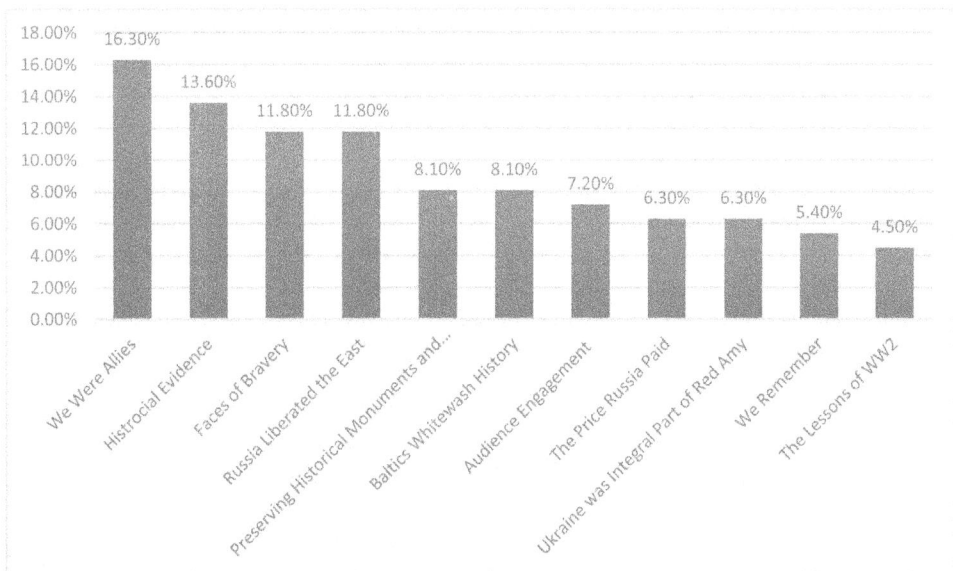

Figure 11.1 Prevalence of Russian Frames (in %)
Source: Author's Creation

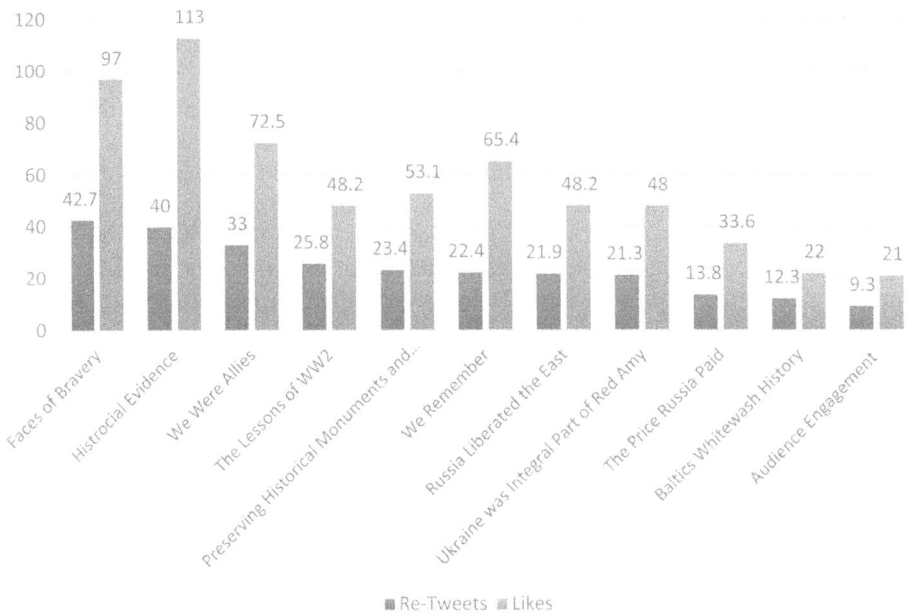

Figure 11.2 Average Engagement Rates of Russian Frames
Source: Author's Creation

3.2 President Putin's views on the lessons of WWII

The second narrative emerges from President Putin's article in the *National Interest* and it refers to 'The Lessons of WWII'. Notably, in his article, President Putin made 10 central arguments:

1 WWII was not caused by the USSR, or USSR non-aggression pact. WWII was the result of the Munich Betrayal. The USSR was the last country to sign a non-aggression pact with Nazi Germany.
2 Russia honours the memory of WWII by cherishing patriots, preserving monuments, and remembering the aid offered by the Allies.
3 The USSR met all its obligations to the Allies and refused to join a Nazi assault on Allies.
4 The USSR carried the brunt of the fighting in WWII. By the time Allies invaded Europe, the Nazis were already defeated on the Eastern front as the USSR liberated cities across Europe.
5 WWII shapes Russian foreign policy which rests on collaborative approaches to shared challenges.
6 Russia helped shape the post-War world order. As such, it remains committed to the mechanisms of the United Nations (UN).
7 Discussions on WWII should be based on objective, historical accuracy. Thus, Russia opened its WWII archives and other countries should follow suit.
8 Other nations cherish the lessons of the Cold War, instead of the lessons of WWII.
9 Some countries glorify the Nazi past.

10 Baltic States seek to re-write history which leads to punitive measures against Russian minorities. Such is the case with legislation banning Russian language and Russian television stations.

Nine of the ten arguments made in Putin's article were echoed in frames shared on Twitter.

The first frame focused on the Munich Betrayal. Tweets in this frame dealt with the blame for WWII, by arguing that the UK and France betrayed Czechoslovakia and that the USSR was the last nation to sign a non-aggression pact with Germany. Moreover, the frame argued that the UK and France deliberately misled the USSR by drawing out negotiations on a possible pact.

The second frame dealt with the memory of WWII indicating that Russia would never forget the War or its lessons. Tweets comprising this frame dealt with preserving war monuments, honouring war veterans, and remembering the help offered by the Allies in WWII.

The third frame suggested that WWII shaped Russia's foreign policy. In the wake of the war, Russia realised the need for collaborative state action. This is the presumed essence of Russian foreign policy, be it on global matters, such as cybersecurity, or regional issues such as maintaining peace in the Arctic. In this frame, Russia demonstrates how the lessons of WWII shape its present-day actions, 'proving' that it refuses to forget the lessons of WWII.

The fourth frame dealt with documented, historical truth. Tweets in this frame referred to Russia's decision to open its WWII archives so that the USSR's role in WWII could be fully understood. Tweets in this frame referenced Russian museums, art exhibits and archives where the 'truth' could be found, alongside historic documents that attested to the USSR's important role during WWII. Tweets also urged other nations to open their archives so that the historic truth could be retrieved.

The fifth frame dealt with the nature of the Red Army. The Red Army was comprised of many peoples and many nations that came together to defeat Nazism. The Red Army's shared heritage manifested not only in the victory over Nazi Germany but also in the Russian language.

The sixth frame argued that the USSR bore the brunt of the fighting during WWII. It was thanks to the heroism of Red Army soldiers that the USSR was able to carry this heavy load. Evidence of this was provided by the numbers of soldiers and citizens that Russia had lost in WWII which far outnumbered the fatalities of the other Allies.

The seventh frame focused on the post-War world order which the USSR helped shape and that Russia still cherishes today. Tweets in the frame emphasised Russian support of the UN peacekeeping operations, abiding by UN resolutions on Libya and Sudan and using the Security Council as a mechanism to resolve global crises through dialogue and diplomacy.

The eighth frame focused on nations that preferred to cherish the spirit of the Cold War as opposed to the spirit of WWII. The Netherlands and the US were both identified as nations that viewed the world through the competitive prism of the Cold War. The same was true of NATO, a Cold War institution that allegedly still refused to bridge the remaining dividing lines across Europe.

The ninth frame focused on those nations that glorified the Nazi past. Such is the case with the alleged "ultra-nationalists" who "staged a coup" in Kiev; the Czech authorities and nations where WWII veterans received the same level of esteem as Nazi collaborators. This frame set the tone for the tenth and final frame which asserted that the Baltics States were attempting to re-write history. Tweets in this frame offered historical evidence that the Baltics were not conquered by the USSR, that Baltic Ambassadors used false historical documents to besmirch the USSR and that Baltic States denied their history by banning USSR symbols and prohibiting the use of the Russian language in schools and on television.

Taken together, these frames, which may be seen in *Figure 11.3*, create the narrative of the "Lesson of WWII". This narrative summons the past to make sense of the present. It suggests that Russia's present-day policies are informed by the lessons of WWII. Indeed, just as the USSR took part in shaping the post-War period, so present-day Russia allegedly remains committed to this world order as manifest in its support of the UN, maintaining European stability, while also seeking to negotiate a settlement to the Arctic territories. Russia is also committed to "objective history", unlike other nations that presumably have yet to open their WWII archives.

As can be seen in *Figure 11.3*, the three most prevalent frames disseminated on Twitter were 'Baltics Re-Write History', 'Objective Historical Accuracy' and 'Lessons of WWII Shape Russia's Policies'. The least prevalent frame was 'Other Nations Can't Forget Lessons of the Cold War'. This analysis demonstrates that following President Putin's article, official Russian channels extensively tweeted about the Baltics States criticising their policies toward Russian minorities as well as their attempts to allegedly glorify the Nazi period. The high prevalence of the 'Objective Historical Accuracy' frame may be explained by the fact that Russia has a credibility deficit. After being widely perceived as interfering in the Brexit referendum and the 2016 US elections, and in annexing Crimea by stealth, Russian authorities are less likely to be seen as trustworthy sources by global audiences. To overcome this deficit, Russian accounts insist on using historical materials, black and white photographs, and archival references.

Figure 11.4 below offers an analysis of the engagement rates of the Russian frames. As it can be seen, the three frames to obtain the highest average number of re-tweets were 'The Brunt of Fighting Fell on the USSR'. 'The Munich Betrayal' and 'Red Army as Russia's Pride'. The two frames to obtain the lowest average number of re-tweets were 'Baltics Re-Write History' and 'Objective Historical Accuracy'. The rates of engagement suggest that users following Russian accounts expressed interest in learning more about the human and

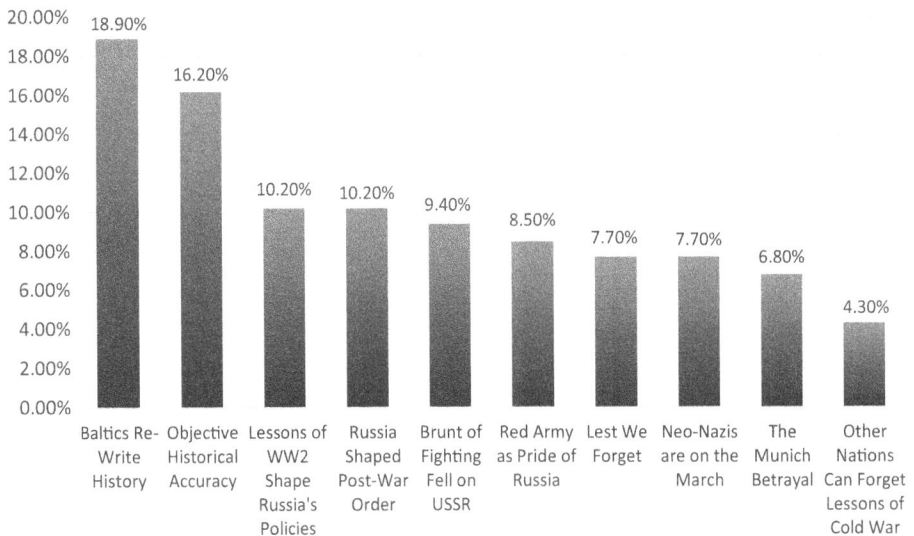

Figure 11.3 Prevalence of Russian Frames (in %)
Source: Author's Creation

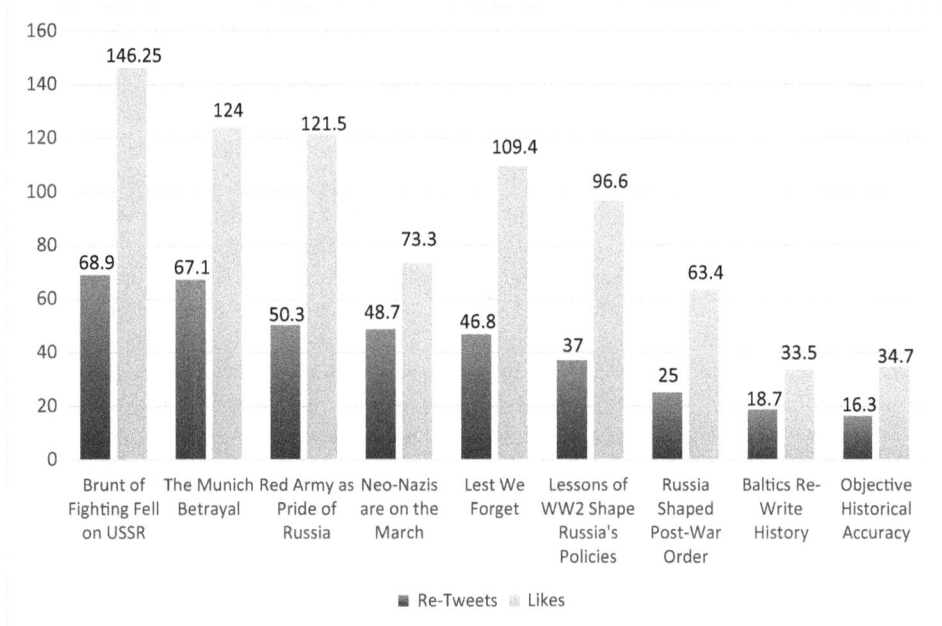

Figure 11.4 Average Engagement Rates of Russian Frames
Source: Author's Creation

military sacrifices suffered by the USSR during WWII, but they found much less credible Russian accusations about Baltic states being involved in historical revisionism. They also seem worried of that fact that Russian calls for revisiting the historical events from a more "objective" perspective may rather constitute an attempt to politicise the history of WWII even more.

IV. DISCUSSION

This study has found that Russia has disseminated two narratives regarding WWII, and Baltic states' role in WWII. The first narrative of 'Russia Remembers' was disseminated on the 75th anniversary of Russia's victory over Nazi Germany. The second narrative of 'Lessons of WWII' was disseminated following President Putin's article in the *National Interest*. Both narratives share certain characteristics. In both narratives, Russia sought to validate its depiction of WWII by relying on carefully selected historical facts. In both narratives, the past was summoned to make sense of the present and in both narratives, Russia framed Baltic states as countries that allegedly refuse to learn from history. In the first narrative, Baltics states were depicted as 'immoral' due to their collaboration with Nazi Germany. In the second narrative, Baltic states were depicted as countries cultivating Russo-phobia and historical revisionism. Finally, in both narratives Russia depicted the USSR as an ethnically diverse collective in which peoples from different nations all came together to defeat evil.

The quantitative analysis found that the 'Lessons of WWII' narrative was more focused on the Baltic states. Indeed, the 'Baltics Re-Write History' was the most prevalent frame in this narrative. However, the 'Baltics Whitewash History' and 'Baltics Re-Write History' frames received a low number of re-tweets. This indicates that Russia's success in besmirching that national image of Baltic States was limited. That said, the frame of 'Historical Evidence' received high levels of re-tweets suggesting that Russia may have been able to restore some of its credibility by making a claim to historical accuracy. It should be noted that both Russian narratives were promoted on a global scale. Tweets comprising both narratives were shared by Russian Embassies in Australia, Belarus, Brunei, Canada. Estonia. Israel, Japan, Pakistan., Sri-Lanka, the UK and more. The narrative was also disseminated by Russian missions to multilateral forums including the UN, the OSCE and the EU.

These findings echo the conclusions of a NATO Stratcom report discussing how Russia uses WWII when criticising Baltic states.[21] In one instance from 2019, Kremlin-linked digital accounts propagated falsities regarding the European Parliament resolution denouncing the Molotov–Ribbentrop Pact. In the wake of the resolution, pro-Kremlin accounts argued that the Baltic and EU states have long sought to re-write history and obscure Latvia and Lithuania's collaboration with the Nazi regime. Another account argued that the resolution was but one more attempt to undermine the USSR's contribution to ending WWII. A third article highlighted Lithuania's alleged glorification of Nazi soldiers calling them accomplices who took part in the Holocaust. Finally, another account lamented the fact that the EU chose to focus on the Molotov–Ribbentrop Pact, and not the 1938 appeasement of Hitler which led him to invade Poland and start WWII. During that same year, another issue came to dominate Kremlin-linked digital accounts, such as allegations that German troops had used a tank to desecrate a Jewish cemetery in Lithuania.

These are prime examples of how the past is summoned to the present by Russia. In line with the theoretical considerations discussed in the second section, Russia justifies its policies through normative and value-based arguments. The structure of Russian narratives thus centres on four main complements: the *problem* (the Baltic states allegedly ignoring the lessons of WWII), the *protagonist* (Russia reminding the Baltic states about their duty not to let the demons of the past to resurface again), the *stage* (the cultural sphere of historical archives, exhibitions, and monuments) and the *course of action* (Baltic states should adopt Russian's interpretation of WWII lessons). In essence, in Russian tweets, WWII lives on in the sense that the lessons of WWII still shape Russian foreign policies via a moral dichotomy that distinguishes between those who remember and abide by the lessons of WWII, and those who have forgotten these lessons.

It is also important to note, as suggested by Roselle *et. al.*, that the formation of the Russian narratives is the result of specific domestic conditions. The focus on WWII historical narratives stems from a structural weakness of the Russian political system. In the absence of a positive narrative about the contemporary political, economic, and social situation in Russia, the Kremlin has been trying to relive more positive moments from its past (e.g., the defeat of Nazi Germany). The projection of these narratives is also facilitated by ongoing discussions about geopolitical transformations in the context of the "rise of China" and the return of great powers' politics. In the same way in which it contributed to the formation of the international order at the end of WWII, Russia should play, the argument goes, an important role in shaping the global order in the current context as well. There are signs, however, that this narrative strategy has been overused and its effectiveness has dwindled. The tepid reactions of the online public to the Russian frames accusing Baltic states of historical revisionism (see Figures 11.3 and 11.4) call into question the viability of the strategy in the Baltic region. More importantly, the war of aggression against Ukraine which Russia has tried to justify using a similar narrative about Ukraine forgetting the lesson of history and embracing Nazi ideology has removed any possible doubts about the imperialist objectives of Russia's use of historical narratives.

V. CONCLUSION

The past plays an important role in national image management as nations that are perceived to have moral blemishes may face present day challenges, which in turn may limit their ability to foster international partnerships. In this study, we have argued that Russia has learned to strategically deploy historical narratives to tarnish the image of target countries in an attempt to provoke internal tensions and to fracture their diplomatic relations with other countries. Drawing on a corpus of 2500 tweets posted by Russian digital channels during the 75th anniversary of the end of WWII, the study has shown that Russia has tried to re-frame the causes, outcomes, and lessons of WWII and to use the resulting narratives for two purposes: to advance its claims to geopolitical status, and to undermine the international reputation of the Baltic states as responsible members of the international community. Theoretically, the study contributes to the literature on digital diplomacy by explaining the strategic role of historical narratives as an instrument of Russian foreign policy. Empirically, it contextualises the importance of historical narratives in support of Russian geopolitical and imperialist ambitions towards not only prospective, but also existing NATO members such as the Baltic states. Further research should examine the conditions that can enable democratic countries to more effectively counter hostile historical narratives in the online medium. Research should particularly focus on better understanding the conditions facilitating the formation, projection, and reception of hostile historical narratives and to design digital strategies that can reduce their effectiveness.

Notes

1 Putin (2022)
2 Perrigo (2022), "How Putin's Denial of Ukraine's Statehood Rewrites History," Time, Feb 22, 2022, https://time.com/6150046/ukraine-statehood-russia-history-putin/
3 Hirsch (2022)
4 Sergeytsev (2022)
5 Nye (2004), Mor (2012), Wilson (2008).
6 Aro (2016), Bennett and Livingston (2018), MacFarquhar (2016).
7 Miskimmon et al. (2014) 9, 256.
8 Hagström and Gustafsson (2019), 400.
9 Applebaum (2022).
10 Roselle et al. (2014), 78–79.
11 Suchkov, (2021), Elswah & Howard (2021).
12 Miskimmon & O'Loughlin (2017).
13 Tyushka (2021).
14 Bjola & Papadakis (2020); Nothhaft et al. (2018).
15 Freel & Bilali (2022), 205.
16 Chatterje-Doody (2014).
17 Stallard (2022).
18 These included: The Russian Government Account, the Russian MFA and Ministry of Defence, the Russian Embassies in Australia, Belarus, Brunei, Canada, Estonia, Indonesia, Ireland, Israel, Japan, Pakistan, Senegal, Sri Lanka, US, UK, Russian Consulate in Edinburgh, as well as the Russian Missions to the UN in New York and Geneva, Russian Mission to the OSCE, EU and Vienna.
19 Putin (2020).
20 Entman (1993).
21 Rodríguez (2020)

Bibliography

Applebaum, Anne. "There is No Liberal World Oder," *The Atlantic*, Mar 31, 2022, Atalantic, https://www.theatlantic.com/magazine/archive/2022/05/autocracy-could-destroy-democracy-russia-ukraine/629363/.

Aro, Jessikka. "The Cyberspace War: Propaganda and Trolling as Warfare Tools." *European View* 15, no. 1 (June 16, 2016): 121–132. https://doi.org/10.1007/s12290-016-0395-5.

Bennett, W Lance, and Steven Livingston. "The Disinformation Order: Disruptive Communication and the Decline of Democratic Institutions." *European Journal of Communication* 33, no. 2 (April 2, 2018): 122–139. https://doi.org/10.1177/0267323118760317.

Bjola, C., and K. Papadakis. "Digital Propaganda, Counterpublics and the Disruption of the Public Sphere: The Finnish Approach to Building Digital Resilience." *Cambridge Review of International Affairs* 33, no. 5 (2020). https://doi.org/10.1080/09557571.2019.1704221.

Chatterje-Doody, P. N. "Harnessing History: Narratives, Identity and Perceptions of Russia's Post-Soviet Role." *Politics* 34, no. 2 (2014): 126–137. https://doi.org/10.1111/1467-9256.12026.

Elswah, Mona, and Philip N. Howard. "'Anything That Causes Chaos': The Organizational Behavior of Russia Today (RT)." *Journal of Communication* 70, no. 5 (2021): 623–645. https://doi.org/10.1093/JOC/JQAA027.

Freel, Samuel H., and Rezarta Bilali. "Putting the Past into Action: How Historical Narratives Shape Participation in Collective Action." *European Journal of Social Psychology* 52, no. 1 (February 1, 2022): 204–222. https://doi.org/10.1002/EJSP.2813.

Hagström, Linus, and Karl Gustafsson. "Narrative Power: How Storytelling Shapes East Asian International Politics." Https://Doi.Org/10.1080/09557571.2019.162349832, no. 4 (July 4, 2019): 387–406. https://doi.org/10.1080/09557571.2019.1623498.

Hirsch, Francine. "'De-Ukrainization' is genocide—Biden was right to sound the alarm," *The Hill*, April 4, 2022https://thehill.com/opinion/international/3267060-de-ukrainization-is-genocide-biden-was-right-to-sound-the-alarm/.

MacFarquhar, Neil. "A Powerful Russian Weapon: The Spread of False Stories." *New York Times*, 2016. https://www.nytimes.com/2016/08/29/world/europe/russia-sweden-disinformation.html.

Miskimmon, Alister, and Ben O'Loughlin. "Russia's Narratives of Global Order: Great Power Legacies in a Polycentric World." *Politics and Governance* 5, no. 3 (September 29, 2017): 111–120. https://doi.org/10.17645/PAG.V5I3.1017.

Miskimmon, Alister, Ben O'Loughlin, and Laura Roselle. *Strategic Narratives. Strategic Narratives: Communication Power and the New World Order*. Routledge, 2014. https://doi.org/10.4324/9781315871264.

Mor, Ben D. "Credibility Talk in Public Diplomacy." *Review of International Studies*. Cambridge University Press, 2012. https://doi.org/10.2307/41485555.

Nothhaft, Howard, James Pamment, Henrik Agardh-Twetman, and Alicia Fjallhed. "Information Influence in Western Societies: A Model of Systemic Vulnerabilities." In *Countering Online Propaganda and Extremism: The Dark Side of Digital Diplomacy*, edited by Corneliu Bjola and James Pamment, 28–43. Abingdon and New York: Routledge, 2018.

Nye, Joseph S. *Soft Power: The Means to Success in World Politics*. 1st ed. New York: Public Affairs, 2004.

Perrigo, Billy. "How Putin's Denial of Ukraine's Statehood Rewrites History," *Time*, Feb 22, 2022, https://time.com/6150046/ukraine-statehood-russia-history-putin/.

Putin, Vladimir. "The Real Lessons of the 75th Anniversary of World War," *The National Interest*, June 16, 2020, https://nationalinterest.org/feature/vladimir-putin-real-lessons-75th-anniversary-world-war-ii-162982.

Putin, Vladimir. "Putin's declaration of war on Ukraine," Feb 24, 2022, *The Spectator*. https://www.spectator.co.uk/article/full-text-putin-s-declaration-of-war-on-ukraine.

Roselle, Laura, Alister Miskimmon, and Ben O'Loughlin. "Strategic Narrative: A New Means to Understand Soft Power." *Media, War & Conflict* 7, no. 1 (April 24, 2014): 70–84. https://doi.org/10.1177/1750635213516696.

Sergeytsev, Timofey. "What should Russia do with Ukraine?" April 3, 2022, *RIA Novosti*, https://ria.ru/20220403/ukraina-1781469605.html.

Suchkov, Maxim A. "Whose Hybrid Warfare? How 'the Hybrid Warfare' Concept Shapes Russian Discourse, Military, and Political Practice." *Small Wars and Insurgencies* 32, no. 3 (2021): 415–440. https://doi.org/10.1080/09592318.2021.1887434.

Stallard, Katie. "Putin isn't the only Autocrat Misusing History," *The Atlantic*, April 30, 2022, https://www.theatlantic.com/ideas/archive/2022/04/russia-propaganda-influences-north-korea-china/629627/.

Tyushka, Andriy. "Weaponizing Narrative: Russia Contesting Europe's Liberal Identity, Power and Hegemony." https://Doi.Org/10.1080/14782804.2021.1883561 30, no. 1 (2021): 115–135. https://doi.org/10.1080/14782804.2021.1883561.

UNFCCC. *"Establishment of an Ad Hoc Working Group on the Durban Platform for Enhanced Action,"* 2011. http://unfccc.int/files/meetings/durban_nov_2011/decisions/application/pdf/cop17_durbanplatform.pdf.

Wilson, E J. "Hard Power, Soft Power, Smart Power." *Annals of the American Academy of Political and Social Science* 616 (2008): 110–124. https://doi.org/Doi10.1177/0002716207312618.

12

USING MIMETIC AND RHETORICAL THEORY TO CRITIQUE A DISINFORMATION CONSENSUS

THE PROBLEM OF "PERSONAL VIGILANCE"

Hamilton Bean and Bryan C. Taylor

Introduction: Socially Mediated Disinformation and the Problem of Personal Vigilance

Disinformation expert Renée DiResta defines disinformation as "information that is spread with the deliberate intent to influence and deceive someone."[1] This definition is widespread and echoed in countless U.S. government reports, scholarly books (including this volume) and academic and popular articles. For example, a 2019 consensus report from the U.S. Department of Homeland Security declared, "One untrue meme or contrived story may be a single thread in a broader operation seeking to influence a target population through methods that violate democratic values, societal norms and, in some jurisdictions, the law."[2] Here, commentators generally agree that disinformation can be a powerful weapon used to erode or subvert liberal-democratic norms. Its effects can include undermining public trust, thwarting rational discourse and decision-making, intensifying social and political divisions, radicalising audiences and instigating violent confrontation.[3]

In this chapter, we address the phenomenon of "socially mediated disinformation" (SMD). By socially mediated, we mean disinformation created for and circulated within social media environments (e.g., Facebook, Instagram, TikTok, Twitter and YouTube). SMD gained considerable public attention following the Russian Internet Research Agency's (IRA) attempt to influence the 2016 U.S. presidential election.[4] Following Donald Trump's unexpected victory, concerned stakeholders attempted to identify the form, extent and influence of the IRA's SMD activities.[5] Researchers have since highlighted how much of the IRA's SMD downplayed or omitted factual claims and instead emphasised textual and visual elements to reinforce American's pre-existing social, cultural, and political identities.[6] In this chapter, we focus on the identity constructing, reinforcing and affirming qualities of SMD (elaborated below).

DOI: 10.4324/9781003190363-15

While researchers have usefully described how IRA SMD induced social media engagement (likes, comments, and shares), they have generally not problematised its communicative dimensions—or their social and political consequences.[7] Here, what counts as evidence of "consequences" remains unclear. For example, a consensus report from the National Academy of Sciences found *no measurable impact* of IRA disinformation on changes in Twitter users' political beliefs,[8] a finding that raises a question about whether current instruments can properly assess SMD induced behaviour. Nevertheless, most commentators maintain as an article of faith that the dangers of disinformation are intensifying.[9] In 2021, for example, the U.S. Office of the Director of National Intelligence (ODNI) authorised the creation of the Foreign Malign Influence Center to help consolidate institutional efforts to combat foreign disinformation.

As a key plank in the current, broad-based U.S. response to disinformation, citizens are called upon to patriotically defend the integrity of the nation's social media environment. The DHS's 2019 consensus report mentioned above, for example, advocated the "expansion of media literacy programs to build societal resilience in the face of disinformation campaigns."[10] The DHS claimed that "media literacy could be framed as a patriotic choice in defence of democracy."[11] Similarly, a March 2, 2020, joint statement from eight U.S. federal agencies underscored this personal vigilance approach (hereafter "PVA"), declaring, "A well-informed and vigilant republic is the best defence against disinformation."[12] Officials, scholars and commentators have advanced PVA in innumerable guides such as "How to Fight Lies, Tricks, and Chaos Online"[13] and in reports from research organisations such as Data & Society. A flood of recent studies have explored how to improve citizens' (especially journalists') disinformation detection abilities.[14]

PVA presumes that citizens can recognise the form and content of disinformation as a means of limiting its amplification. Recent studies suggest, however, that developing such recognition capabilities is immensely difficult for social media users. In a study that investigated how perceived social presence influences fact-checking, for example, researchers found that "social contexts may impede fact-checking by, at least in part, lowering people's guards in an almost instinctual fashion."[15] In other words, social media users are less likely to fact-check statements when engaging with others on platforms such as Facebook or Twitter. Another study found that empirically measurable message factors (i.e. source, inconsistencies, subjectivity, sensationalism and manipulated images) were less important for users' evaluations of disinformation than users' individual differences. In other words, one's *opinion* towards the news topic, rather than the message itself, determined whether people believed the news and shared it online.[16] Given such challenges, it is unsurprising that DiResta has remarked, "When people ask what they can do personally [to combat disinformation], really, the very dissatisfying answer is 'think before you share'."[17]

We sympathise deeply with scholars, security professionals, and elected officials wrestling with this dilemma. However, we argue that PVA overestimates citizens' abilities to perceive SMD and downplays the uniquely rhetorical appeal of identity affirming messages. Here, we diverge from post-positivist interventions such as PVA, which hold that disinformation is an *a priori*, objective phenomenon. Specifically, PVA exemplifies post-positivist efforts to counter disinformation in seeking to identify its unique features, and in training human and non-human moderators to flag and block their manifestations on social media platforms. Related programs following this approach include fact-checking, citizen engagement that reveals and undermines misleading or deceptive information and the cultivation of machine learning systems.[18] Arguably, this "scientistic" approach to disinformation research (and national security research in general) underwrites the PVA consensus—one that often "yields quantitative data, yet questionable conclusions."[19]

Our critique of the PVA consensus instead draws upon a tradition of communication theorising that maintains that "meaning" is not "objective;" rather, it is a "local, situated, and transient accomplishment."[20] This tradition, drawn from mimetic and rhetorical theory (introduced in the next section below), complicates current calls for citizens to "think" before they "share." This is partly because presumed distinctions between "inauthentic" and "authentic" messages are often unclear, irrelevant, or subordinated to other user values and motives. Viewed from this non-foundationalist tradition, the goal of achieving reliable and pre-dictable citizen detection of disinformation is a chimera. This is partly because ontologically speaking, "disinformation" is not a pre-existing object that is passively received by its audience. Alternately, disinformation is a form of rhetorical "calling"—one that evokes an active response from its audience. In this response, audiences subsequently *constitute* that communication (i.e., as relevant, meaningful, and consequential), in and through their creation and repetition of the *identities that it encourages*. In revealing disinformation's ontological instability, mimetic and rhetorical theory suggest that stakeholders' assertions of the deleterious consequences of identity-affirming disinformation may be incomplete.

To support this argument, in this chapter, we highlight the ambiguity and unpredictability of communication due to creative practices of mimesis.[21] In its concern for identity and subjectivity, mimetic and rhetorical theory centre the question of *what type of being* is made (in)secure vis-à-vis online (dis)information and (in)authentic exchange—and also *how*. This reflection subsequently questions how stakeholders might reassess the dangers of identity affirming messages if "mimesis" served as their underlying theory of communication. In what follows, we highlight the value of mimetic and rhetorical theory in understanding key features of today's disinformation landscape. Using an example of IRA SMD directed at U.S. military veterans and their supporters, as well as a non-SMD message aimed at the same audience, we compare the application of mimetic and rhetorical theory to PVA. Before proceeding, we confirm that we are *not* arguing that the dan-gers of disinformation are exaggerated. Rather, we aim to demonstrate the benefits of mimetic and rhetorical theory for improving institutional and public understanding of the relative dangers of identity affirming messages, as well as the significant challenge of creating effective PVA-based interventions.

Innovative Perspectives: Mimetic and Rhetorical Theory

Historically, the term "mimesis" denotes an ancient and sprawling intellectual conversation, conducted among philosophers and theorists across the fine arts, humanities, and social sci-ences.[22] Generally, this conversation has been devoted to classifying "real" and "fake" phe-nomena, understanding their distinction, and exploring its implications. Traced to Plato's expression of disdain towards the artistic derivation of ideal forms, mimetic theory has evolved over the centuries to encompass concerns related to particular kinds of human creativity. Specifically, these concerns involve conditions and practices of *imitation*—including those of *adaptation, assimilation, hybridity, masquerade, mimicry, passing, possession* and *repetition*.

Historically, cultural response to these imitative phenomena has ranged from suspicious monitoring and discipline, to cynical appropriation and artistic appreciation. Why this ambivalence? Because these practices are prolific, and they create an evocative—yet fre-quently problematic—world of *copies, doubles, models*, and *translations*. These products manifest in various—and often uncanny—forms, ranging from *androids, avatars* and *clones*, to *reenact-ments, replicas, simulations* and *twins*. What these phenomena all share is the potential for excessive, unauthorised, and boundary-defying agency. That agency is controversial because it frequently interrogates and subverts conventional premises of identity and reality (e.g., autonomy, objectivity, etc.) that underwrite the authority and legitimacy of societal institutions. In modernist and realist paradigms, subsequently, mimetic actors are often moralised as inauthentic rogues who employ dubious means such as deceit and insinuation to access and corrupt a society's cherished assets. In critical paradigms such as post-colonialism, however, these phenomena are often viewed with sympathy (if not enthusiasm), due to their capacity for disrupting and transforming (e.g., through the induction of desire) relations of meaning and power that presume to govern the vital circulation of identity and difference.

Inevitably, mimetic theory intersects with two central concerns of this volume: *mediation* and *deception*. Concerning the former, the conservative tradition of mimetic theory noted above aspires to regulate the potential of mediation to *distort* the originality of its objects. Here, the concern is that mediation will create insufficient and corrupt forms that compromise the desirable functioning of their audiences. In this tradition, elites (e.g., state officials) often demand that artistic and journalistic actors ensure valid (i.e., accountable and factual) correspondence between their images, performances and narratives, and their presumably "real" sources—a demand which coexists in generative tension with the aesthetic use by these actors of stylistic and figurative devices such as metaphor. The products of professional mediation are then assessed contractually, according to ideals of "reflection" and "authenticity" that protect audiences against the circulation of unmarked fakery (e.g., forgeries).[23]

Alternately, postmodern theorising of mediation (e.g., as contingent realism) emphasises its productivity, rather than deficiency. In these traditions, media are liberated from conservative regimes of rigid accountability to ostensible origins and essences (i.e., as *re-presentation*). Instead, they are endorsed for their capacity to *create* novel meanings and pleasures that challenge and enlighten audiences. These effects are not created through a work's correspondence with an *a priori* reality, but through its *representation* of alternate, plausible, and progressive worlds. Here, the aesthetics of play and resistance guide mediation as a potential source of human flourishing (arising from the recovery of repressed knowledge). That goal is pursued by deconstructing regimes (e.g., of copyright protection) that prematurely foreclose and punitively enforce distinctions between the real and the fake. How, asks mimetic critique, do these regimes perpetuate unequal power relations and stifle creativity?

Our discussion has already acknowledged that the concern of *deception* follows closely on conceptions of mimesis and mediation. While this association is ancient,[24] modern political scholars have often avoided one of its implications: that even democratic governments practise imitative manipulation (e.g., through propaganda).[25] One relevant body of work, however, involves late nineteenth-century theorising among western social scientists that associated imitation with the tropes of *contagion* and *suggestion*.[26] In this literature, mimesis augmented the power of those latter concepts for theorising ambient disorientation created by modern conditions such as industrialization and urbanisation. This disorientation both threatened the integrity of traditional subjectivities (i.e., by producing alienation and agitation), and stimulated novel forms of association among diverse groups (e.g., established and recent immigrants). Here, the idealised development of liberal-democratic society (e.g., manifest in communal trust, accountable institutions, and an active, informed citizenry) was shadowed by dissonant phenomena such as entrepreneurial deceivers (e.g., frauds), unruly crowds, and impulsive—if not dangerous—displays of affect.

One persistent theme in this inquiry involved the mysterious appeal and volatile circulation of non-rational influence producing imitative outcomes. This influence was manifest in both interpersonal interactions (e.g., hypnotic suggestion) and public life (e.g., frenzy created by rumourmongering). Here, imitation was moralised as *viral* circulation, in which a pathogenic agent penetrated and overwhelmed its subject, establishing that subject not merely as a host of undesirable influence, but (reflexively) as the embodied agent of its ongoing reproduction. In this doubling of mimetic influence, naïve and vulnerable audiences might *themselves* become media of further deception.

Digital Mimesis and Socially Mediated Disinformation

As media evolve, their changing logics and affordances affect their potential for activating the mimesis-deception relationship.[27] Indeed, this current handbook exists largely because of urgent consequences arising from the related innovation of *digital media*. Here, readers are already familiar with fundamental changes in existing systems of analog and mass media,[28] including:

- Digitalization permits the malleable convergence of previously discrete media types;
- Audiences are considered active and participatory, rather than passive;
- Users create, modify and/or "spread" content in multi-linear fashion (i.e., compared to uni-linear broadcasting);
- Users may access and interact with content across multiple platforms;
- Facilitated by ambient connectivity, users may utilise both fixed and portable interfaces; and finally,
- Users' production and circulation of content fosters their network ties (if not communal membership).

In this chapter, we are concerned with conditions of imitation and deception that distinguish digital media. As influences on user agency and experience, these conditions arise from key qualities of digital communication, which include its being *anonymous, pervasive* (i.e., atmospheric), *ephemeral, fragmented* (i.e., indexical), *continuous* (e.g., high-volume and high-velocity), *algorithmic* (i.e., remotely strategized and automated), *generic* (i.e., conventionally composed and normatively interpreted), and *searchable* (i.e., categorised and archived). Variously configured in digital systems, these qualities interact to foster communicative synergy between imitation and deception. Recent scholarship has identified numerous examples of this synergy, including the following:

- Digital content is inherently *reproducible*—a condition which subverts conventional expectations for its authorship, uniqueness, and originality;[29]
- Algorithms are driven by fundamentally reductive logics, which reduce the diversity of digital data by *constructing patterns and standardizing knowledge*;[30]
- As a substitute for authentic and accountable identities, online participants routinely employ *iconic avatars*;[31]
- Human and/or non-human users of *fake accounts* (e.g., bots) engage in the *artificial performance* of socially—and politically-significant communication (e.g., astroturfing);[32]
- Political campaigns and social movements strategically deploy (if not weaponize) familiar digital forms (e.g., memes), whose ideological appeal and capacity for limited modification typically serve to *perpetuate existing modes of user community* (e.g., homophily). This process may also assume ritualistic properties that *stimulate users' inclination towards envy and competition*;[33]
- Digital audiences frequently engage in superficial, motivated, and impulsive reception of media content. They are more likely, as a result, to *prematurely endorse and distribute* that content—creating problematic outcomes such as "hashtagged contagion";[34]
- Contemporary movements display conflicting impulses to *both stigmatise and normalise digital fakeness* (e.g., satire) as a playful critique of official politics;[35]

- Partisan groups have created online sources *claiming the status* of professional "news" production, increasing the volume and circulation rate of dubious information;[36]
- Violent extremists have successfully engaged in online *radicalization and recruitment* that appear to "convert" unsuspecting audiences,[37] and finally:
- These developments (e.g., identity, deception, virtual reality, etc.) *are themselves represented* in mainstream media genres (e.g., science fiction television) using *(meta-) mimetic capabilities* (e.g., computer-generated imagery);[38]

This inventory establishes that current struggles around online disinformation do not merely involve the immediate practices of producing, circulating, and consuming deceptive content. Instead, these struggles unfold in a larger cultural and historical context—one involving the ongoing assertion, deployment, and challenge by related groups of *mimetic discourses*. That is, in defining and configuring relevant standards of authenticity and fakery, these discourses depict those immediate activities (and their products) as relatively *authorised and legitimate* (e.g., innocent appropriation, virtuous deceit *of* adversaries, etc.) versus *unauthorised and illegitimate* (e.g., hacker spoofing, malign deceit *by* adversaries, etc.).

Finally, our case study in this chapter depicts imitation and deception occurring on a specific type of digital media—*social media*. Here again, readers are doubtless familiar with the fundamental properties of social media, including users' creation and posting of personal profiles, their production and circulation (e.g., forwarding) of content, their interaction with other users' produced and circulated content (e.g., liking), and their subsequent cultivation of network ties.[39] Further, most of the examples provided above of imitation/deception in digital media (generally) also pertain to social media—particularly, their capacity to stimulate aggressive rivalry and scapegoating.[40] However, it is important to remember that some affordances of specific social media platforms (e.g., the user sign-up process; content-editing features; content-creation norms, etc.) may influence users' subsequent participation in processes of imitation and deception.[41]

Rhetorical Theory: First and Second Personae

Our goal in this chapter is to develop an association between the bodies of mimetic and rhetorical theory that helps to explain the process by which disinformation constitutes the subjectivity of social media users. In developing this explanation, we diverge from two conventional models. The first is a *mechanistic* model of mediation, which emphasises the motivated design by remote sources of "messages," their contingent transmission through physical channels, and their subsequent decoding by receivers (i.e., audiences) to yield "information" that may be objectively assessed as relatively true or false. Our second rejected model is a modernist (i.e., post-positivist) view of user identity that emphasises the pre-existing, essential status of that self (i.e., as "a unique individual"), empirically manifest in cognitive traits and behavioural habits. Alternately, we view disinformation as *a form of rhetoric* that is—however disturbing and destructive—also *productive*, in that it ideologically interpellates (i.e. hails and sutures) the potential subjectivity of users. It does so, specifically, through advancing *discourses* (i.e., narratives) that provide perpetually incomplete, unstable and questing subjects with plausible and gratifying (even if illusory) visions of their Self, their Other(s), and their World. Such discourses provide users with symbolic resources for understanding and configuring those phenomena to create relatively coherent experience, and engage in relatively consistent and meaningful interaction. In the rhetorical shaping of social media users' agency, we find an important engine of its ethics and politics.[42]

Here, subsequently, we also diverge from conventional models that prematurely measure and analyse users' "reception" of mediated disinformation as an effectual cause. We believe such models err in presuming the prior existence of a formed user, who wilfully exerts their original agency to cognitively process and behaviourally respond to mediated disinformation. Alternately, we are concerned with the rhetorical capacities of disinformation to shape the constitution of users' identity as a contingent (and impermanent) orientation of consciousness. Crucially, we argue that *it is through these rhetorically constituted orientations that users conceptualise, experience, and deploy their agency.* In this sense, the persuasive qualities of mediated disinformation are not so much *pre-existing empirical features requiring measurement,* as they are *emergent discursive technologies requiring interpretation.* How, we ask, does the rhetoric of disinformation alternately enable and constrain users' forms of self-understanding, such that their response constitutes (for them) a meaningful act?

Here, we establish the connection between disinformation and mimesis by drawing upon rhetorical conceptions of *persona.*[43] Briefly, contemporary rhetorical theory rejects the conventional image of communicative performance (e.g., public address) as-if an "emission" of a rhetor's existing identity. Alternately, it emphasises that rhetors must engage in artful, multi-layered design of their performances. This is because rhetors can only partly and temporarily control how audiences draw upon their existing experience and knowledge (e.g., of the rhetor's prior performances, of relevant cultural and generic norms, etc.) in attributing the rhetor's apparent identity (e.g., as consistent or inconsistent with their expectations).

As a result, rhetors draw upon their available symbolic resources to craft a particular, desired identity—one whose benefits will ideally accrue to the embodied rhetor presumed by the audience to author the discourse that they experience. In this process, rhetors shape audience impressions of *who* (i.e., what type of social, moral and/or political person) *they are speaking as.* This concept is known as "the first persona."[44]

But this achievement is only half the story. Typically, rhetors also seek to align that first persona with the self-experienced identities of their audiences (i.e., as a condition for securing their assent to a presented proposal). As a result, rhetors continually *characterise* those audiences throughout their performances—implicitly and explicitly evoking their history and culture. In this effort of hailing, rhetors encourage audience members to recognise themselves *as* particular kinds of persons (i.e., virtuous, free, entitled, etc.), predisposed to respond to the rhetor's discourse in a particular manner. This concept is known as "the second persona." It is significant partly because the rhetor's induction of alignment inevitably *excludes, minimises,* and/or *negates alternative self-understandings* that audiences might otherwise adopt in their response (a concept known as "the third persona"[45]). Finally, rhetors undertake this effort at alignment so that audiences will not merely *agree* with the rhetor's claims (i.e., as attitudinal congruence), but will socially and politically *identify* with the symbolic world that is evoked through the rhetor's first persona. If this process is successful, rhetors and audiences achieve *consubstantiality*—experiencing themselves as similar, united—and perhaps identical—subjects who inhabit and enact a shared symbolic world.[46]

The implications here for understanding the configuration of imitation and deception by SMD are clear. Our claim, specifically, is that the effectiveness of SMD is partly achieved through a two-stage rhetorical process. First, the SMD text simultaneously evokes both a deceptive first persona, and a strategic second persona. Second, the SMD text rhetorically evokes audience desire to achieve *alignment and affiliation* (i.e., imitation) between these two personae. This meta-communicative process, we believe, represents an important precondition for the favourable reception of SMD by users, one that is achieved not so much *because of what SMD says* (i.e., its apparent content), but *because of who and what it implies its communicators are (or should be).*

Recently, rhetorical scholars of persona have reassessed the relevance of this concept for understanding digital (as opposed to oral or written) communication. They emphasise that, in responding to continuous streams of provocative rhetorical fragments, projections of first personae by online audiences have become increasingly impulsive, holistic, and inductive.[47] This condition may facilitate "bad actors," who strategically manipulate rhetorical cues that resonate with audience members' cultural knowledge concerning authentic and trustworthy first personae, thus encouraging their misrecognition of those personae.[48] It is important to note here that, while these performances encourage audiences to accept the claims of SMD campaigns, their success is not guaranteed. Promoters of SMD, for example, must manage over time to maintain first personae that appear both *plausible* and *consistent* for their audiences—perhaps even achieving "micro-celebrity" status. Managing the tension between these two goals can be challenging, however, because SMD campaigns simultaneously desire that these accounts achieve expectations for productivity—for example, imposed quotas for frequent posting. Here, we recall that the rhetors behind SMD accounts are playing with complex cultural knowledge required for the competent simulation of familiar identities—and those identities are often associated with foreign societies. As a result, each new posting creates a fresh opportunity for audience members to experience either consistency or inconsistency with an account's existing cultivation of first and second *personae*. Finally, the complexity of this challenge only increases when we appreciate that SMD designers must simultaneously modulate their rhetorical appeals (i.e., as "winks") to implicitly engage a "fourth" (i.e., unacknowledged but collusive) audience persona in the form of *platform algorithms*.[49] This is because the successful elevation of SMD depends on stimulating the recognition of that non-human audience (e.g., through using popular keywords), while simultaneously suppressing recognition of this very performance by SMD's human audience. For all these reasons, we may appreciate the uniquely *mimetic and rhetorical contingency* of SMD's effectiveness.

Application: Comparing Two Models for Analysing the Influence Exerted by SMD

In this section, we compare applications of post-positivist PVA and mimetic/rhetorical theory by assessing two examples of SMD. Our first example is derived from a corpus of 3,393 IRA Facebook and Instagram advertisements, collected for a 2018 U.S. Senate Intelligence Committee report on Russian interference in the 2016 U.S. presidential campaign.[50] Among other genres, that corpus contained approximately 158 *veteran-themed* postings that stimulate audience feelings of pride, sympathy, frustration, and despair surrounding U.S. public (mis-) treatment of U.S. veterans. These postings included appeals to both right- and left-leaning audiences via a spectrum of images and appeals, including "veterans [no possessive] lives matter," "wounded warrior," "support our wounded veterans," and "treat veterans better." Presumably, the IRA advanced such messages to entice U.S. social media users to like and follow IRA social media accounts, thereby exposing these users to future IRA "payload" SMD that contained more explicitly partisan sentiments. As a result, it is useful to examine how these messages represented and configured particular identities of American "civilian" and "military" subjects, as a means of cultivating audience reception.

Visually, the picture in the Instagram post provided in the link https://archive.mith.umd.edu/irads/items/show/9031.html is a photo-realistic painting that depicts a lone white, male soldier against a murky, dark brown background. The soldier appears to be a young adult in his twenties. His hair is buzzed short, and he sits in a chair facing the viewer. His nationality is indicated by adjacent placement of the phrase "Veterans ★US★" in smaller, pink font. The style of his camouflage-pattern combat uniform appears current. That uniform is mostly devoid of service-related markings, but an American flag patch is barely visible on his right shoulder. The soldier appears to be suffering an agony of inauthenticity. In his left hand, he holds a literal mask depicting his own stoically masculine expression; in his right, he holds an assault rifle by its barrel. The soldier's positioning of this mask reveals a searing "behind" image of his face, whose features are convulsing. Verbally, the banner of this posting (in large white font) queries the audience's knowledge concerning PTSD levels among U.S. military veterans. It claims that more of them have died since 1999 from (implied, PTSD-related) suicide than in combat. As a result, the presence of the soldier's weapon seems ominous indeed. Combined, these words and images evoke the ongoing tragedy of many American soldiers who have returned from deployments in the U.S.-led War on Terror to lives marked by isolation, pain and abandonment.

Now compare Image 1 to Image 2, taken from a January 2020 posting to a Facebook page belonging to the non-profit group "Veterans for Natural Rights." A copy of the post is available at this link: https://digital.auraria.edu/work/ns/017d39d4-58bd-48ea-8410-40f6e273d36e. The first author had earlier "liked" this group because his advisee (a combat veteran) was affiliated with and had recommended it. Image 2 drew the first author's attention because it was surprisingly similar to examples of IRA disinformation he had been reviewing for another project. Image 2 is also similar to hundreds of other images included on the Veterans for Natural Rights page. This group claims to advocate for "a community of support, awareness, and education to veterans and civilians alike [,] while promoting natural treatment, personal rights, individual liberty, and overall well-being through activities like seminars, public screenings and community gatherings." The group reports its founding in 2014; it has no known connection with IRA campaigns.

In that posted image, viewers see a cartoon-style depiction of a crying woman and child, standing in the green grass of what appears to be a U.S. military cemetery (i.e., American flags wave on poles in the background). The towheaded boy is wearing sneakers, shorts, and a t-shirt; he holds a bunch of flowers. The maternal figure embraces him from behind. She wears a cap whose lettering suggests the commemoration of a U.S. military unit—it is not clear whether it belongs to her, or to a deceased veteran they are visiting. Her sleeveless dress reveals at least one tattooed arm. The two figures stand before a grouping of white crosses that distinguish veterans' grave-markings. The crosses recede from the figures and the viewer towards the horizon; each is labelled with a culturally- (if not pathologically-) marked emotional state (e.g., "fear," "depression," etc.). The number "22" is written on the foremost cross, a reference to the estimated number of veterans who take their own lives each day. The ironic dialogue depicted between the mother and child invokes (and implicitly laments/condemns) the "war at home" that has "caused" these (implied, unnecessary and unjust) veteran deaths. In its bottom right-hand corner, this image bears the signature of its implied artist. Above it, the apparent identity of the account holder is indicated by the group's title, and an avatar of a grey owl. In addition to providing the date (January 8, 2020), the poster uses distinctively vernacular, populist discourse (e.g., "gonna"; "whole n'other"; "itching for," "has its pick," etc.) to warn their audience of further casualties resulting from continued U.S. military adventurism (e.g., war with Iran).

Viewed through the lens of PVA, both images use similar emotional imagery/commentary. As a result, they do not provide convenient candidates for disinformation remedies that emphasise detectable distinctions between veracity and deception. To be sure, both figures are identity-affirming messages that console grieving survivors, and aim to stimulate feelings of sadness, sympathy, and (perhaps) resolve for action. Nonetheless, PVA's post-positivist commitments maintain that viewers are somehow able to determine that the first image referenced is (likely) proffered by a pro-Russian actor, while the second image is not. Subsequently, PVA holds that engaging with the second image (e.g., through "likes," "shares," or "comments") is a benign act requiring little reflection or hesitation, while engagement with image 1 is fraught with moral peril and anti-patriotic implications. Here, we do not mean to dismiss these possibilities outright. However, the strong thematic similarity between these two figures indicates not only the creative mimesis employed by IRA actors (i.e., in recognizing and reproducing exploitable tropes that shape U.S. public opinion), but also the challenges that face SMD audiences in making these kinds of attributions. Consider, for example, the frequent lag between official recognition and identification of disinformation accounts on social media platforms (to say nothing of the disdain expressed towards that very regulation by some SMD audiences). As a result, it is not immediately clear from PVA protocols how SMD audiences might recognise that the features of Image 1 (but not Image 2) are cause for checking its provenance and validity.

Using mimetic and rhetorical theory to analyse these texts, however, yields an interpretation that is both more subtle and complicated. Image 1, for example, employs a *meta*-mimetic strategy by *taking mimesis as its very object*: the depicted soldier can no longer maintain his pretence of simulating stoicism—his mask can no longer contain the alleged truth of his "inner" pain. As a result, the ethos of this move accrues to the epistemological and moral authority of the first persona associated with this text's creators and posters—they presume to *see, know* and *reveal* this falsehood. So established, this authority hails the audience to assume a second persona that is subordinate in nature—the position of witness to both a revelation and an urgent warning.

Image 2 produces much the same effect, although its evocation of first persona utilises both greater pathos (e.g., in depicting the grieving mother and child) and dialogism (e.g., the simultaneous, supplemental voice of the poster). Here, that persona laces the normal conventions of epideictic rhetoric (i.e., that would honour a fallen defender in a virtuous cause) with the bitterness of experience. It subsequently invites the audience to assume a second persona marked by *ironic, doubled mourning*—both of the soldier's deaths and of the cruel indifference of the American public.

Implications for Designing and Conducting SMD Interventions

What are the implications of these brief readings? Two are apparent. First, they establish that disinformation traffics in the skilful imitation of existing cultural rhetoric. As a result, we do well to educate online audiences about rhetorical concepts such as persona, so that they may become more adept at understanding the symbolic means by which inauthenticity may be both cultivated and detected. Second, these readings give us pause concerning potentially premature conclusions regarding the dubious effects of identified disinformation—particularly compared with online discourse that is not so classified. Here, we refer to the growing practice among both external adversaries and domestic extremists of targeting U.S. veterans and serving military (as well as those who identify with them) with disinformation intended to mobilise visceral resentment towards their government and fellow citizens.[51] It is undeniable that the cynical exploitation of veteran-related symbolism in these campaigns distorts what would ideally be more rational and open public deliberation concerning the causes, consequences, and lessons of America's War on Terror. While conceding U.S. failure to adequately care for its veterans, then, and without slandering the author or audiences of Image 2, we gently note the rhetorical resonance between these two texts, and their shared alignment with Russian interests concerning U.S. post-GWOT Global War on Terror foreign policy (e.g., non-interference in Russia's strategic and military alliance with Iran). At the very least, this claim suggests that valid classification of apparent disinformation may be more complex—and require greater consideration of relative effects—than we typically assume.

Conclusion: The Challenges of Advancing a Melancholy Argument

In this chapter, we have argued that people spread SMD not because a message is either inherently true or false, but (largely) because they seek to affirm and perform their social, cultural, and political identities. As a result, mimetic and rhetorical theory tarnish the appeal of PVA as a key plank of U.S. counter-SMD strategy. This is a melancholy conclusion for us to draw because it implies that ordinary citizens may continue to face acute challenges when confronting identity affirming online discourse in the future. Here, PVA requires patriotic citizens to "think" before they "share"—that is, carefully consider the source, content, and intent of the social media messages they engage. Yet, our application of mimetic and rhetorical theory above indicates the futility of attempting to distinguish (i.e., definitively) between "authentic" and "inauthentic" messages from analysing only their empirical features. Here, we resonate with the recent complaint by Okros and Jensen that "holistic analysis that takes seriously culture and politics has been set aside in favour of an approach that yields excellent intercoder reliability."[52]

In that spirit, our chapter makes a theoretical case for reconsidering the PVA consensus. To be sure, we do not eschew here the responsibility of considering its practical implications. Our sense of what is "practical," however, differs considerably from its conventional, instrumental denotations (e.g., as policy goal). Instead, we believe that our argument implicates nothing less than a whole-of-society effort to promote the convergence of democratic ideals with the benefits of therapeutic reflection and moral vigilance.[53] That is, our argument implies that social media users would benefit from engaging in online discourse with the following questions firmly in mind: Who might I desire to become, that I may be more likely to injure others and myself? How might I recognise the familiar voice of the deceiver, who calls me to adopt that self-understanding as a prelude to performing that (self-)injury?

Notes

1 DiResta, "Online Misinformation and Disinformation," para. 5.
2 U.S. Department of Homeland Security, *Combatting Targeted*, 4.
3 Ivan et al., "Whole of Society."
4 DiResta et al., "Tactics & Tropes."
5 Office of the Director of National Intelligence, *Background to "Assessing Russian.*
6 Linvill and Warren, "That Uplifting Tweet."
7 U.S. Department of Homeland Security, *Combatting Targeted*; DiResta et al., "Tactics & Tropes"; Lawson, "What If Russian Disinformation"; Tucker et al., "Social Media."
8 Bail et al., "Assessing the Russian Internet."
9 Foer, "Putin Is Well On."
10 U.S. Department of Homeland Security, *Combatting Targeted*, 3.
11 Ibid.
12 FBI National Press Office, "Joint Statement," para. 5.
13 Robertson, "How to Fight Lies."
14 Benkler et al., *Network Propaganda*; Ireton & Posetti, *Journalism, Fake News.*
15 Jun et al., "Perceived Social Presence," para. 51.
16 Schaewitz et al., "When Is Disinformation."
17 World Affairs, "Renee DiResta: Media."
18 Bode and Vraga, "Swiss Cheese Model"; Lie et al., "Multi-Modal Classification."

19 Okros and Jensen, "Which Gap?" 3.
20 Taylor and Munoz, "Meaning," 2.
21 Taylor, "Imitation (In) Security."
22 Lawtoo, "Mimetic Condition"; Taylor, "Imitation (In) Security."
23 Horbyk et al., "Fake News."
24 Lester and Yambor, "Visual Deception."
25 Miller and Robinson, "Propaganda, Politics and Deception."
26 Borch, "Imitative, Contagious and Suggestible."
27 Ong, "From Mimesis to Irony."
28 Chayko, *Superconnected*.
29 For example, see Becker, "Desiring Fakes."
30 Bowman, "Of Algorithms and Mimesis."
31 Ibrahim, *Digital Icons*.
32 Keller, et al., "Political Astroturfing on Twitter"; Santini, et al., "Bots as Online Impersonators."
33 Brighi, "Charlie Hebdo."
34 Mitchell and Münch, "#Contagion."
35 Ferrari, "Sincerely Fake"; Taylor, "Defending the State."
36 Reilly, "F for Fake."
37 Singer and Brooking, *Likewar*.
38 Lawtoo, "Black Mirrors."
39 Humphreys, *Social Media: Enduring Principles*.
40 Shullenberger, "Mimesis and Violence."
41 Zulli and Zulli, "Extending the Internet Meme."
42 Charland, "Constructive Rhetoric"; Debrix, *Language, Agency*.
43 Jasinski, "Persona."
44 Black, "Second Persona."
45 Wander, "Third Persona."
46 Burke, *A Rhetoric of Motives*.
47 Kor, "Commenting Persona."
48 Xia, et al., "Disinformation, Performed."
49 Gibbons, "Persona 4.0."
50 DiResta et al., "Tactics & Tropes."
51 Barash, "Hijacking Our Heroes."
52 Okros and Jensen, "Which Gap?" 3.
53 Ivan et al., "Whole of Society."

Bibliography

Bail, Christopher A., BrianGuay, EmilyMaloney, AidenCombs, D. SunshineHillygus, FriedolinMerhout, DeenFreelon and Alexander Volfovsky. "Assessing the Russian Internet Research Agency's Impact on the Political Attitudes and Behaviors of American Twitter Users in Late 2017." Proceedings of the National Academy of Sciences USA 117, no. 1 (2020): 243–250. doi:10.1073/pnas.1906420116.

Barash, Vlad. "Hijacking our Heroes: Exploiting Veterans Through Disinformation on Social Media. Hearing of the House Committee on Veterans' Affairs." U.S. Congress, November 13, 2019. https://www.congress.gov/116/meeting/house/110183/witnesses/HHRG-116-VR00-Wstate-BarashV-20191113.pdf.

Becker, Daniel. "Desiring Fakes: AI, Avatars, and the Body of Fake Information in Digital Art." In *Faking, Forging, Counterfeiting: Discredited Practices at the Margins of Mimesis*, edited by Daniel Becker, Annalisa Fischer, and Yola Schmitz, 199–222. Transcript Verlag, 2018.

Benkler, Yochai, Robert Faris, and Hal Roberts. *Network Propaganda: Manipulation, Disinformation, and Radicalization in American Politics*. Oxford: Oxford University Press, 2018.

Black, Edwin. "The Second Persona." *Quarterly Journal of Speech* 56 (1970): 109–119. doi:10.1080/00335637009382992.

Bode, Leticia, and Emily Vraga. "The Swiss Cheese Model for Mitigating Online Misinformation." *Bulletin of the Atomic Scientists*, 77, no. 3 (2021): 129–133. doi:10.1080/00963402.2021.1912170.

Borch, Christian. "The Imitative, Contagious, and Suggestible Roots of Modern Society." In *Imitation, Contagion, Suggestion: On Mimesis and Society*, edited by Christian Borch, 3–34. New York: Routledge, 2019.

Bowman, Jonathan. "Of Algorithms and Mimesis—GAFA, Digital Personalization, and Freedom as Nondomination." *Constellations* 28, no. 2 (2021): 159–175. doi:10.1111/1467–8675.12483.

Brighi, Elisabetta. "'Charlie Hebdo' and the Two Sides of Imitation." In *Imitation, Contagion, Suggestion: On Mimesis and Society*, edited by Christian Borch, 126–140. New York: Routledge, 2019.

Burke, Kenneth. *A Rhetoric of Motives*. Berkeley, CA: University of California Press, 1969.

Charland, Maurice. "Constitutive Rhetoric: The Case of the Peuple Quebecois." *Quarterly Journal of Speech* 73, no 2 (1987): 133–150. doi:10.1080/00335638709383799.

Chayko, Mary. *Superconnected: The Internet, Digital Media, and Techno-Social Life*. Thousand Oaks, CA: SAGE Publications, 2020.

Debrix, Francois. *Language, Agency, and Politics in a Constructed World*. Armonk, NY: M. E. Sharpe, 2003.

DiResta Renée, Kris Shaffer, Becky Ruppel, David Sullivan, Robert Matney, Ryan Fox, Jonathan Albright, and Ben Johnson. "The Tactics & Tropes of the Internet Research Agency." *New Knowledge*, 2018. ttps://disinformationreport.blob.core.windows.net/disinformation-report/NewKnowledgeDisinformation-Report-Whitepaper.pdf.

DiResta Renée. "Online Misinformation and Disinformation in the Age of COVID-19: A Conversation with Disinformation Expert Renée DiResta." UNICEF, July 9, 2020. https://www.unicef.org/globalinsight/stories/online-misinformation-and-disinformation-age-covid-19.

FBI National Press Office. "Joint Statement from DOS, DOJ, DOD, DHS, ODNI, FBI, NSA, and CISA on Preparations for Super Tuesday." FBI, March 2, 2020. https://www.fbi.gov/news/pressrel/press-releases/joint-statement-from-dos-doj-dod-dhs-odni-fbi-nsa-and-cisa-on-preparations-for-super-Tuesday.

Ferrari, Elisabetta. "Sincerely Fake: Exploring User-Generated Political Fakes and Networked Publics." *Social Media + Society* 6, no. 4 (2020): 1–111. doi:10.1177/2056305120963824.

Foer, Franklin. "Putin is Well on His Way to Stealing the Next Election." *The Atlantic*, June 2020. https://www.theatlantic.com/magazine/archive/2020/06/putinamericandemocracy/610570.

Gibbons, Michelle G. "Persona 4.0." *Quarterly Journal of Speech* 107, no. 1 (2021): 49–72. doi:10.1080/00335630.2020.1863454.

Horbyk, Roman, Isabel Löfgren, Yana Prymachenko, and Cheryll Soriano. "Fake News as Meta-Mimesis: Imitative Genres and Storytelling in the Philippines, Brazil, Russia and Ukraine." *Popular Inquiry* 1 (2021): 30–54. http://urn.fi/URN:NBN:fi:aalto-2021112410434.

Humphreys, Ashlee. *Social Media: Enduring Principles*. Oxford: Oxford University Press, 2016.

Ibrahim, Yasmin. *Digital Icons: Memes, Martyrs and Avatars*. New York: Routledge, 2020.

Ireton, Cherilyn, and Julie Posetti. *Journalism, Fake News & Disinformation: Handbook for Journalism Education and Training*. Paris: UNESCO Publishing, 2018.

Ivan, Christina, Irena Chiru, and Ruben Arcos. "A Whole of Society Intelligence Approach: Critical Reassessment of the Tools and Means Used to Counter Information Warfare in the Digital Age." *Intelligence and National Security* 36, no. 4 (2021): 495–511. doi:10.1080/02684527.2021.1893072.

Jasinski, James. "Persona." In *Sourcebook on Rhetoric: Key Concepts in Contemporary Rhetorical Studies*, edited by James Jasinski, 429–432. Thousand Oaks, CA: SAGE Publications, Inc., 2001.

Jun, Youjung, Rachel Meng, and Gita Venkataramani Johar. "Perceived Social Presence Reduces Fact-Checking." *PNAS* 114, no. 23 (2017): 5976–5981. doi:10.1073/pnas.1700175114.

Keller, Franziska B., David Schoch, Sebastian Stier, and JungHwan Yang. "Political Astroturfing on Twitter: How to Coordinate a Disinformation Campaign." *Political Communication* 37, no. 2 (2020): 256–280. doi:10.1080/10584609.2019.1661888.

Kor, Ryan. "The Commenting Persona: Reception Theory and the Digital Rhetorical Audience." *Journal of Media Research* 11, no. 1 (2018): 55–70. doi:10.24193/jmr.30.4.

Lawson, Sean. "What If Russian Disinformation Isn't as Effective as We Thought?" *Forbes*, December 6, 2019. https://www.forbes.com/sites/seanlawson/2019/12/06/what-if-russian-disinformation-isnt-as-effective-as-we-thought/#a3fc5164e8b6.

Lawtoo, Nidesh. "Black Mirrors: Reflecting (on) Hypermimesis." *Philosophy Today* 65 no. 3 (2021): 523–547. doi:10.5840/philtoday2021517406.

Lawtoo, Nidesh. "The Mimetic Condition: Theory and Concepts." *CounterText* 8 no. 1 (forthcoming).

Lester, Paul Martin, and Marjorie Yambor. "Visual Deception: From Camo to Cameron." In *The Palgrave Handbook of Deceptive Communication*, edited by Tony Docan-Morgan, 857–875. Palgrave Macmillan, 2019. doi:10.1007/978-3-319-96334-1_44.

Lie, Tobby, Haadi Jafarian, Stephen Hartnett, Hamilton Bean, and Farnoush Banaei-Kashani. "Multi-Modal Classification for Polarization Intent Detection in Social Media." *GITHUB*, June 20, 2020. https://github.com/tobby-lie/Russian-Disinformation-Project.

Linvill, Darren, and Patrick Warren. "That Uplifting Tweet You Just Shared? A Russian Troll Sent It." *Rolling Stone*, November2019. ttps://tigerprints.clemson.edu/communication_pubs/18/.

Miller, David, and Piers Robinson. "Propaganda, Politics and Deception." In *The Palgrave Handbook of Deceptive Communication*, edited by Tony Docan-Morgan, 969–988. Palgrave Macmillan, 2019. doi:10.1007/978-3-319-96334-1_50.

Mitchell, Peta, and Felix Victor Münch. "#Contagion." In *Imitation, Contagion, Suggestion: On Mimesis and Society*, edited by Christian Borch, 107–125. Routledge, 2019.

Office of the Director of National Intelligence. Background to "Assessing Russian Activities and Intentions in Recent US Elections": The Analytic Process and Cyber Incident Attribution. *ODNI*, January 6, 2017. https://www.intelligence.senate.gov/sites/default/files/documents/ICA_2017_01.pdf.

Okros, Alan, and Rebecca Jensen. "Which Gap?–What Bridge?" *Armed Forces & Society*. (2022). doi:10.1177/0095327X211035820.

Ong, Walter J. "From Mimesis to Irony: The Distancing of Voice." *The Bulletin of the Midwest Modern Language Association* 9, no. 1/2 (1976): 1–24. https://www.jstor.org/stable/1314783.

Reilly, Ian. "F for Fake: Propaganda! Hoaxing! Hacking! Partisanship! and Activism! in the Fake News Ecology." *The Journal of American Culture* 41, no. 2 (2018): 139–152. doi:10.1111/jacc.12834.

Robertson, Adi. "How to Fight Lies, Tricks, and Chaos Online." *The Verge*, December 3, 2019. https://www.theverge.com/21276897/fake-news-facebook-twitter-misinformation-lies fact-check-how-to-internet-guide.

Santini, Marie., Debora Gomes Salles, Charbelly Estrella, Carlos Eduardo Barros, and Daniela Orofino. "Bots as Online Impersonators: Automated Manipulators and their Different Roles on Social Media." *The International Review of Information Ethics* 30, no. 1 (2021): 1–12. doi:10.29173/irie402.

Schaewitz, Leonie, Jan P. Kluck, Lukas Klösters, and Nicole C. Krämer. "When is Disinformation (In) Credible? Experimental Findings on Message Characteristics and Individual Differences." *Mass Communication and Society* (2020): 484–509. doi:10.1080/15205436.2020.1716983.

Shullenberger, Geoff. "Mimesis and Violence, Part 1: Peter Thiel's French Connection." *Cyborgology*, August 2, 2016. https://thesocietypages.org/cyborgology/2016/08/02/mimesis-and-violence-part-1-peter-thiels-french-connection.

Singer, P. W., and Emerson T. Brooking. *Likewar*. Boston, MA: Houghton Mifflin Harcourt, 2018.

Taylor, Bryan C. "Imitation (In) Security: Cultivating Mimetic Theory to Critique the Media/Security Nexus." *Communication Theory* 27 (2017): 48–69. doi:10.1111/comt.12104.

Taylor, Bryan C. "Defending the State from Digital Deceit: The Reflexive Securitization of Deepfake." *Critical Studies in Media Communication* 38, no. 1 (2021): 1–17. doi:10.1080/15295036.2020.1833058.

Taylor, Bryan C., and Ricardo V. Munoz. "Meaning." *The International Encyclopedia of Communication Theory and Philosophy* (2016): 1–6. doi:10.1002/9781118766804.wbiect168.

Tucker, Joshua A., Andrew Guess, Pablo Barberá, Cristian Vaccari, Alexandra Siegel, Sergey Sanovich, Denis Stukal, and Brendan Nyhan. "Social Media, Political Polarization, and Political Disinformation: A Review of the Scientific Literature." *SSRN*. (2018). doi:10.2139/ssrn.3144139.

U.S. Department of Homeland Security. *Combatting Targeted Disinformation Campaigns: A Whole-of-Society Issue*. DHS, October2019. https://www.dhs.gov/sites/default/files/publications/ia/ia_combatting-targeted-disinformation-campaigns.pdf.

Wander, Philip. "The Third Persona: An Ideological Turn in Rhetorical Theory." *Central States Speech Journal* 35 (1984): 197–216. doi:10.1080/10510978409368190.

World Affairs. "Renee DiResta: Media and Disinformation: What's at Stake for Democracy?" YouTube Video, 1:00:06. June 27, 2019. https://youtu.be/7FahDJqYCbY.

Xia, Yiping, JosephineLukito, YiniZhang, ChrisWells, Sang JungKim, and Chau Tong. "Disinformation, Performed: Self-Presentation of a Russian IRA Account on Twitter." *Information, Communication & Society* 22 no. 11 (2019): 1646–1664. doi:10.1080/1369118X.2019.1621921.

Zulli, Diana, and David James Zulli. "Extending the Internet Meme: Conceptualizing Technological Mimesis and Imitation Publics on the TikTok Platform." *New Media & Society* (2020): doi:10.1177/1461444820983603.

13

DEEPFAKE DISINFORMATION

HOW DIGITAL DECEPTION AND SYNTHETIC MEDIA THREATEN NATIONAL SECURITY

Agnes E. Venema

Deepfakes Introduced

The digital revolution creates a number of unique opportunities and threats with regard to disinformation. While increased computing power may allow us to swiftly recognise social media bots spreading disinformation,[1] that same increased computing power lies at the foundation of this national security threat version 2.0. This chapter specifically focuses on threats emanating from the creation of audio-visual forgeries and fake documentary evidence by computer systems employing Artificial Intelligence/Machine Learning (AI/ML) techniques; outputs that are collectively known as "synthetic media" and are often colloquially referred to as "deepfakes"[2].[3]

Disinformation, misinformation, and mal-information are sometimes used interchangeably[4] but to help understand the nature of the threats deepfakes pose, accurate definitions are necessary. At the core of the distinction is intent, coupled with whether the information is factually real or false.[5] Mal-information is based on true information, but taken out of context in order to inflict harm. Misinformation is false information but does not carry an intent to cause harm. Disinformation is both factually false and intends to cause harm. In the interest of national security, it is vital to combat deepfake disinformation because its highly realistic nature can easily lead deepfakes to quickly spread without people realizing its content is synthetically created. It is worth noting that, particularly when it comes to disinformation, the distinction between national and international operations, and therefore security, is increasingly vague.[6]

The reason this particular threat is a relatively recent phenomenon is because increased computing power and increased interconnectivity through the internet has resulted in people progressively relying on internet sources, including social media, for their consumption of news and information.[7] Furthermore, increased interconnectivity and computing power also mean that computers are rapidly becoming capable of taking over work that humans painstakingly used to do. In particular, the developments in computer science related to Artificial Intelligence (AI) and Machine Learning (ML) have aided this process. These two factors combined pose a

DOI: 10.4324/9781003190363-16

threat unlike any we have seen before in the domain of disinformation because disinformation used in hybrid threat scenarios or foreign influence campaigns can now increasingly be automated, both in terms of content creation and in terms of targeting and dissemination.

The computer science of deepfakes

AI and ML are terms used to describe a process whereby a computer imitates human intelligence. The interchangeability of these terms remains contested even within the field of computer science; some have opined that ML is the means to the end state of AI, whereas others see them as markedly different processes.[8] Furthermore, some systems employ what is called deep learning—a subcategory of AI/ML. This chapter will not discuss the merits of these different schools of thought. Instead, the working definition of AI/ML that is adopted describes the process whereby the computer system is not simply following a human written script (code) of: if A, then B, but it figures out this path itself. This process is sometimes referred to as a black box since humans cannot (yet) examine such systems and figure out their line of reasoning.

While the foundations for this field of computer science have been laid decades ago, the use of AI/ML technology in a sustained manner is a relatively recent phenomenon. Chatbots, chat programs emulating a human by using natural language processing (NLP), are one such use that has found widespread application in only the last decade. There are several reasons for these recent fast-paced developments. Firstly, computing power of computers has increased rapidly. Secondly, the data needed to train AI/ML models, including images and voice recordings, have become more widely available and accessible, in no small part due to our own social media habits, including the sharing of images and video. Some of those pictures have been scraped and used in disinformation campaigns in the past.[9] Thirdly, and related to the previous point, consumers increasingly use multiple internet-connected devices daily to consume and produce media and organise themselves online. As Baker puts it

> the digitalisation of social life has made public space more dynamic with public life temporally and spatially contingent on a range of mediated communication practices—including mobile smart phones, instant messenger applications, online social networks—and the public deriving their status from collective identity, rather than as co-inhabitants of a shared geographical locale.[10]

The latter is of particular importance to disinformation as a threat to national security, as its potential spread is no longer confined to the geographical area within the reach of an individual or group. Indeed, disinformation can now easily be spread from abroad, fuelling the belief systems of online communities and subsequently lead to in-person congregations of members of a predominantly online community, as we saw during the attack on the US Capitol on 6 January 2021. The potential for organisation around belief systems fuelled by disinformation, and therefore its impact, has vastly increased compared to the analogue era.

Increasing computing power has also given rise to powerful GANs: Generative Adversarial Networks. GANs employ deep learning methods and can learn patterns from data without human intervention—so-called unsupervised learning.[11] A generative model can then, based on the patterns it learns from a dataset, generate output that is similar to the input. For example, if the model was given pictures of swans, it will learn what a swan looks like and can then generate an image of a swan that is so good that it might have well been part of the dataset that it learned from. If two such models are put together in a networked environment, one of the models can be tasked to create new examples based on the dataset (the generator), while the other one (the discriminator) will try to distinguish real images from the database from artificially images created by the generator.[12] This is how GANs can be used to create deepfakes.[13] But much like black swan events, a GAN is often unlikely to deal well with an outlier, such as a picture of a black swan in a database of white swans.

Nevertheless, AI/ML has made impressive progress over the last few years as illustrated by the ability of AI/ML systems to beat humans in playing strategic thinking games. These games are also increasing in complexity: whereas in 1997 chess world champion Garry Kasparov was beaten by computer system DeepBlue,[14] by 2015 the AI/ML system AlphaGo beat a human champion at the complex strategy game Go,[15] and the same company announced in 2019 that its successor AlphaStar beat professional computer gamers at the real-time strategy game of StarCraft.[16] What is important about AlphaStar's win is the fact that it demonstrates the capability of the AI/ML to adapt in real-time.

While certainly not all AI/ML machines are of AlphaStar quality, we are seeing tremendous improvement in ever shorter periods of time. This also means that those working in national security need to constantly be aware of the changing parameters in this field of computer science to understand the possibilities and have adequate foresight. The next section discusses in more detail how these computer systems can be used to generate audio-visual forgeries and fake documentary evidence.

II Taxonomy of Fakes

Manual Forgeries

In order to contextualise the speed and ingenuity with which disinformation is spreading, it serves to take a step back to look at some of the history of manipulation for political gain. Manual forgeries are at least as old as the print press. Cases of forged documentation can be found throughout history, ranging from falsified signatures and manipulated pictures, to forged military documents.[17]

US President Abraham Lincoln used visual forgeries to counter rumours spread by his political opponents about his appearance, as photography was not yet in widespread use for news reporting during his presidential campaign. Lincoln earned the title of "the first American president to fashion his self-image through photography"[18] by employing the services of a photographer who not only instructed Lincoln on his pose, but also altered the images.[19] After his assassination, Lincoln became the subject of one of the first recorded "face swaps"; a photo of the head of the late president was superimposed onto an engraving of John C. Calhoun[20] to satisfy the demand for a different style of photographs of Lincoln.

A lot changed since the Lincoln face swap. Most notably, the use of digital editing techniques and tools, such as Photoshop, made it easier and cheaper to digitally alter material. These techniques do not use AI/ML but instead are manually executed. A notable example of such "Photoshopping" in the security space has been the digital alteration of images released of a 2008 Iranian missile test. The image is of four missiles being launched, yet one of the launched missiles appears to be edited into the image later, either because it did not launch in sync with the others or not at all.[21] Information on the success of such missile tests can provide useful information to Iran's adversaries about military capabilities and development of new missiles, so it is understandable why they would attempt to cover up a partially failed launch.

AI/ML Alterations

The digital revolution not only gave rise to digital alteration tools and techniques, it also saw the dawn of computer-generated and altered imagery. This section discusses in broad terms the types of audio-visual forgeries and fake documentary evidence that can be created as output by computer systems. This field, however, develops incredibly fast, which is why this section is a snapshot of current capabilities that undoubtedly will be surpassed within a matter of years, if not sooner.

As briefly mentioned above, natural language processing (NLP) is a branch of AI/ML which concerns itself with the understanding of text and speech by a machine.[22] NPL systems can also be used to create text, and preliminary research by the Center for Security and Emerging Technology showed that at least one of the systems currently on the market can create convincing contents "that fit a viewpoint, manipulate narratives with a slant, seed new conspiracy narratives, and draft divisive posts" based on minimal input.[23] In short, this particular system is capable of creating disinformation if it is provided with similar content as input. If coupled with autonomously operating dissemination channels, such as bot farms or avatars-on-demand,[24] the threat to national security becomes more obvious. The combination of an NPL system used to create disinformation content with autonomous dissemination tools means that no human intervention is necessary to pump large volumes of disinformation into the online ecosystem. The larger the volume of disinformation messages, and the larger the amount of autonomous dissemination sharing that disinformation, the higher the chances of one of those messages going viral. Even without a viral hit, a constant flow of disinformation may muddy the waters enough for people to start doubting the truth and official accounts of events, which is one of the end goals of disinformation.

While disinformation spread through text is dangerous, people are inclined to believe what they *see* even more, as will be discussed later in this chapter. The saying "a picture paints a thousand words" was already being questioned in light of Photoshop, but it takes a new turn now that video can be artificially altered as well and lifelike synthetic faces can be computer generated.

While deepfake technology has been around for longer, it gained the attention of the wider public in 2017 when the digital manipulation technique took off on internet platform Reddit. The vast majority of the early deepfakes entailed sexual exploitation of famous women whose faces were superimposed onto pornographic footage.[25] While sexual gratification led to this type of image-based sexual abuse, it was made possible because famous women had many pictures of themselves online that could be scraped as system input. However, as the developments around deepfakes move fast, fewer images as input are now needed to render realistic enough results. Furthermore, superimposition of the face onto pornographic footage has now made way for deepfaking nude bodies onto images of dressed women.[26]

The two most important deepfakes techniques that alter existing audio-visual footage are superimposition and morphing. Superimposition is the technique described above. It takes the face or body of one person and seamlessly imposes it onto existing footage. Morphing is another technique used to create deepfakes. Here the image of a person is not superimposed, but rather existing footage is manipulated by altering it to match the facial expressions or movements of a source. The source is never imposed onto existing footage, but their movements are copied by the person in the existing footage, meaning that if the source jumps, the target in the original footage also jumps. The result is a re-enactment of an off-camera source using existing footage of the target. While these processes focus on video, a similar process can be used to deepfake audio; provide a deep learning system with existing audio of a person to create new audio that sounds like the original person.

At the time of writing, multiple examples of these techniques can be found online, some for entertainment, others aimed at raising awareness. For example, the UK's retired football star David Beckham featured in a malaria awareness campaign. A deepfake of Beckham was used so that the campaign could be broadcasted in 9 different languages—languages that Beckham does not speak.[27] Politicians have also been the subject of (satirical) deepfakes. Italian satirical TV show *Striscia la notizia* broadcasted a deepfake of former Italian Prime Minister Matteo Renzi in 2019 in which he was seen swearing at his colleagues and making vulgar gestures.[28] Criticism followed, and the TV show released a "behind the scenes" clip of how the video was made.[29] In 2021, a deepfake of Dutch Prime Minister Rutte was made by news outlet *De Correspondent*. In it, deepfake Rutte tells the public his advisors discouraged him from addressing the public and telling the "honest truth" about the climate crisis, but that he is now speaking out anyway because the citizens deserve the truth.[30] Opponents of these deepfakes, such as former director of the Transatlantic Commission on Election Integrity Eileen Donahoe,[31] have invoked existing disinformation in the political arena as one of the reasons to stop creating deepfakes of politicians, but the uptake remains slow. Legislation regarding the (political) use of deepfakes is haphazardly being implemented, and references made in legal documents can sometimes be myopic in nature, such as the provisions covering the weaponization of deepfakes by foreign actors in the US National Defense Authorization Act for Fiscal Year 2020.[32]

The Beckham example also highlights another trend: the creation of digital clones or avatars, including as a means of alleviating workload. While at the time of writing this is legally uncharted territory, we have seen actor Bruce Willis use a clone of himself in commercials[33] and professional services network company EY announced that several of their partners use digital avatars of themselves in a bid to become more digitally approachable and, for example, pitch in the language of the client.[34] The concern about the use of clones and avatars becoming mainstream is that the question as to whether one is speaking to the real person becomes a question of trust. Furthermore, if such an avatar is hacked, it opens the door to fraud (at the time of writing there are two known cases of fraud by deepfake voice clone),[35] can lead to the stealing of industrial secrets (including those with national security implications, such as dual-use goods), and are a counter-intelligence concern. Therefore, depending on what type of data this hacked avatar has access to, their existence has the potential to seriously threaten national security.

While the use of avatars and deepfake impersonation may seem farfetched, several parliaments in Europe were briefly thought to have become a victim of exactly such a scheme in the spring of 2021. News reports came out that at least three national parliaments or members of parliaments had spoken to what they thought was Russian opposition leader Navalny's chief of staff, Leonid Volkov.[36] It turned out to be a prankster in make-up wearing a bathrobe, but alarm bells were raised that it could have been a deepfake operated by adversaries. Nevertheless, the concern did not translate into widely shared awareness on this issue: in June of 2022 several European mayors had calls with a deepfake pretending to be the mayor of Kiev.[37] This scenario demonstrates the harm deepfake avatars can do when the identity behind the avatar cannot be verified. Furthermore, especially with the increase of remote working, this incident underscores the necessity of digital situational awareness and the need for practicing basic cyber security protocols.

In addition to deepfakes, a phenomenon named "shallow fakes" deserves mention, as this technique also uses original audio-visual footage. In contrast to deepfakes, shallow fakes do not alter the content of the video in any manner but are manipulated by adjusting the speed of certain segments of the videos, or by selectively cutting the footage. This technique has made people look either overly aggressive, or slow and perhaps under the influence.

Therefore, such footage might be classified as mal-information rather than disinformation. Some researchers classify all audio-visual mal-information as shallow fakes[38] but in my opinion this classification should be reserved for those videos that are original in content but have been manipulated, rather than for those that have been mis-identified or misattributed with malicious intent.

US Speaker of the House Nancy Pelosi became the victim of a shallow fake in 2019. A video of her was released and shared on social media claiming she was drunk while delivering remarks. In fact, the video had been slowed down so that it appeared she slurred her words, creating mal-information by shallow fake. While these types of manipulations are easier to refute, due to the original video footage being available, there is an inherent risk in shallow fakes going viral on social media as part of a disinformation campaign.

Personas

While the previous examples of deepfakes have been alterations of existing footage, the same techniques can be used to create completely new personas, or "sock puppets". The word personas is used to avoid confusion with existing persons, although a persona might represent an existing person. Newly created personas in audio-visual material are already being used and have garnered much interest. For example, deepfake newsreaders have made their debut.[39] The generation of imagery of a persona deserves attention because personas can be given any identity and credentials and can therefore be presented as a credible source or expert. With every refresh of the website, thispersondoesnotexist.com generates completely artificially created portrait photos of people who, as the name suggests, do not exist in real life. The faces this website creates have already been used as profile avatars by instigators of disinformation campaigns[40] as they are less easy to detect than the abuse of a real person's picture, for example by a reverse image search.

A recent example of the use of a fake persona to give credence to disinformation was during US President Biden's 2020 election campaign. His son, Hunter Biden, was supposedly involved in shady business deals and proof of his wrongdoings was lifted off his computer that he brought in for repair. The "evidence" turned out to be fabricated and the repairman Martin Aspen, the witness in this case, turned out to be a fake persona who even had his own picture.[41] Such images of personas can also be used to give more credibility to honeypots looking to connect with potential targets through social media, including LinkedIn and dating apps. While most social media platforms prohibit the use of fake profiles, they are difficult to detect and accounts can be made, bought, or rented through "influence-for-hire" firms[42] to use these fake personas to spread and give credence to disinformation.

Voice clones and auditory deepfakes

Audio-visual deepfakes are not the only cause for concern. The same techniques used to create them can also be used to create voice clones or auditory deepfakes. Voice clones is a technique that can be somewhat compared to the morphing technique described in the previous section. By training a system on the audio data of a certain person, for example a president or a famous athlete, output can be generated that sounds eerily similar to the original source the data was trained on. The risk for security compromising and fraudulent phishing calls is enormous if this technique is successfully used in real time. At the time of writing it was estimated that most attackers use text-to-speech tools that are mostly considered to be too slow in order to sound like a natural conversation on the phone—the lag between respondents is generally speaking too long.[43]

In 2019 the BBC reported that security firm Symantec has identified three cases of attempted phishing that were likely to have been executed using an auditory deepfake, all claiming to be top-placed officials requesting money transfers.[44] It is likely that that number has far increased since, but proving such fraud in a court of law may be tricky, as is discussed in section III. Fraud is not the only use case discovered until now. In 2020 UK, *The Telegraph* published an article on a custody case turned sour in which one of the parties appears to have made use of a manipulated voice recording, possibly an auditory deepfake, to paint a negative picture of the opposing parent.[45] Such personal attacks are a par for the course when it comes to malicious deepfakes, as the next section discusses.

III Security Implications of Deepfakes

Since disinformation is not new, what is it about deepfakes that makes them so nefarious? *Seeing is believing* and *a picture paints a thousand words* are both frequently used idioms and there is a truth to them. Research suggests that vision, out of all the sensory data humans process, is considered to be the most reliable.[46] However, since photography and later Photoshop emerged, we have had to adjust our way of thinking about the provenance of what we see. If before the ethics of airbrushing models on the covers of magazines was the predominant topic of discussion regarding altered media, we are now entering an era in which the very trust in what we see is being challenged and, with that, perhaps our very notion of the truth.

Secondly, the developments are moving so fast that it is difficult to maintain up to speed with the technical capabilities of creating audio-visual forgeries. In some cases the metadata of a file may prove manipulation, but particularly files that have been shared and re-shared online will have that data missing. Furthermore, once a method to detect deepfakes is known, the models that create them improve as well, incorporating the knowledge about the detection method in their creation of better deepfakes. That means that artefacts of manipulation will continue to become more difficult to spot.

The evolution of deepfakes described above benefits from an example to illustrate the point. Much of the training data for the computer models that created deepfakes was imagery scraped from the internet. This means that models trained on a highly curated set of images. For example, celebrities posing on the red carpet would mostly be flattering and awkward pictures of them with eyes closed would not be uploaded. This resulted in a distorted idea of what the human face looks like, so the first deepfake videos looked unnatural because the computer models had not quite figured out how often people blink their eyes.[47] As these detection methods for deepfakes become known, the models adapt and the deepfakes become harder to identify.[48] This cat-and-mouse game continues today and makes researchers and governmental agencies, such as intelligence agencies and law enforcement, weary of sharing their methods for detection lest they unwittingly aid the enemy.

The national security implications of deepfakes manifest themselves on multiple levels. Often there is a knock-on effect where the effect on one level impacts national security on another level. Therefore, these categories should be seen as intersecting and mutually enforcing. Furthermore, as stated earlier, the hard lines between international and national security are increasingly blurring, particularly in the field of disinformation as the next sections will also illustrate. The effects of deepfakes can be examined at the following levels: international, national, sub-group and personal.

International level

At the international level, deepfakes can threaten the survival and existence of states and state systems, as well the relations between states. Where relations between states are precarious and honest brokers are attempting to mend the peace, a well-timed and believable deepfake of one of the members of the negotiation team or a deepfake social media post can undermine the trust in the entire process. Particularly in high-stakes negotiations with low levels of trust to begin with, it may be hard if not impossible to convince counterparts in the negotiation that a deepfake is, in fact, fake.

Furthermore, a deepfake may be released as part of a "false-flag" operation—an operation to hide the true intent of the actors. In a border dispute, one party to a conflict may, for example, release a deepfake video of them being attacked or an auditory deepfake "recording" discussing plans for an imminent attack, so that their own actual attack can be framed as a pre-emptive strike or self-defence. While the idea may seem far-fetched, false-flag operations have been carried out throughout history, including by Nazi Germany to justify invading Poland.[49]

National level

Similarly, at the national level, a strategically released deepfake or a series of audio-visual forgeries, perhaps reinforced by more traditional means of spreading disinformation, can lead to kicking the hornet's nest. The idea is that an already fragile situation, like a low-level national conflict is further destabilized through disinformation by deepfake. Similarly, the risk of a false flag operation is present in this scenario.

In 2021 in Myanmar, for example, traditional disinformation that was spread through social media platforms (specifically Meta's *Facebook*) led to, what some call, a genocidal campaign[50] against the Rohingya Muslim minority. Such instigated violence is not new and also happens through analogue dissemination channels; disinformation broadcasted by radio undermined an already fragile coexistence between the two major populations in Rwanda, eventually culminating in the 1994 genocide. Of course, such disinformation campaigns can also affect international conflicts, including some of the so-called frozen conflicts in the former Soviet Union, as well as in long-standing disputed territories such as Cyprus, Kashmir, Palestine, and Tibet.

It is important to reiterate that deepfakes do not have to touch upon conflicts and, as discussed below, internal strife or social tension to impact national security. An auditory deepfake of the president of the national bank of a country warning about the imminent collapse of the national currency can become a self-fulfilling prophecy as people react to such news by pulling their money out of their bank accounts. The scarcity that follows may result in widescale riots or looting that can constitute a risk to national security.

Sub-group level

Deepfakes intended to derail national processes or pit groups against one another will often make use of existing tensions within society. Like more traditional disinformation, these deepfakes will exploit sentiments that are already present and make them bubble to the surface. At the sub-group level, deepfakes can be made to target and discredit certain sub-groups of society as a form of audio-visual dog whistles. This can include certain professional groups, such as journalists or virologists, can include members of the LGBTIQA+ or migrant communities, or can be used in support of conspiracy theories that vilify government authorities at large. Disinformation by deepfake in this last category may be used less to elicit a response that triggers a wider community reaction, but as a tool to verify a belief system that a particular sub-group holds. Such deepfakes may be made in support of conspiracy theories and provide its believers with the "proof" their view of reality is in fact true. Furthermore, as certain conspiracy theories become more widespread and prominent figures advocate for them, the sub-groups believing such theories may attempt to (violently) put a stop to the perceived wrong they or society at large is experiencing, as was demonstrated during the attack on the US Capitol on 6 January 2021.

Consider, for example, a deepfake of the CEO of a major news outlet confessing that they have been lying to the public for years. Such a deepfake could erode the public's trust in the independent media, a cornerstone of democratic societies, as well as reinforce the belief system of certain sub-groups in society that the media is part of a (government) conspiracy. The social contract starts to unravel and the core pillars upon which democracy rests start to crumble, potentially leading to civil strife. Similarly, a deepfake claiming a particularly large bank is going bankrupt may lead to a bank run which can go hand-in-hand with social disorder, including hoarding and looting.

Personal level

Finally, at the personal level we see character assassinations happening by both shallow- and deepfakes. The slowed down video of Nancy Pelosi was intended to make her less credible. The deepfake "witness" that was produced to provide "evidence" during President Biden's 2020 election campaign was intended to sway voter opinion. The risk to national security that manifests itself through deepfakes that attack specific people can be sub-divided into two categories: attacks on the person and attacks on the office.

Disinformation by deepfake designed to attack the person can be aimed at discrediting a person to the point where they are no longer able to perform their job. Rana Ayyub is an Indian journalist who became the victim of a pornographic deepfake campaign in 2018 that was coupled with doxing—the publishing of her personal information online.[51] Ayyub self-censored her online interaction trying to avoid the barrage of negative attention. Whomever instigated this deepfake attack successfully managed to make Ayyub quit her profession, at least temporarily.

Attacks on an office are personal attacks that are executed in order to remove someone from the function they are performing, including political office. In Malaysia in 2019, a deepfake confession to sodomy implicating a minister resulted in the arrest of the alleged culprit in a country where homosexuality is illegal.[52] What this case illustrates is the real-life consequences these types of attacks can have, even if the quality of the deepfake is mediocre at best.

As both these examples demonstrate, the position of certain groups of people makes them more vulnerable to a deepfake intended to harm them. As described earlier, women have faced the brunt of exploitative deepfakes and due to the still prevalent purity myth, the social repercussions for women being depicted in lewd deepfakes is often very high. The LGBTIQA+ community faces similar social risks and in countries where homosexuality is illegal, implied or real homosexuality may result in a criminal charge which in certain countries is punishable by death. In addition, extortion, or sextortion deepfakes may pray on members from these communities, and their fears of reprisals once the deepfake is made public. This form of blackmail could lead to a host of compromised behaviours that can harm national security interests, including vote buying, the sharing of classified information, or the provision of access to restricted areas.

It is therefore important to reiterate that online disinformation has offline consequences and can make its way into the criminal justice system. When the "proof" is a deepfake, it will be up to the defence to prove that evidence is not real, putting the criminal justice system on its head.[53] Whereas certain defendants or criminal justice systems may be able to finance the painstaking process of investigating an alleged deepfake, without guarantees as to its outcome, such scrutiny is not afforded to everyone even in countries with well-intended criminal justice systems rather than kangaroo courts.

Liar's Dividend

The opposite is also true; if deepfakes can be made of anyone and anything, how do we know that things that did actually happen are true and real? This dilemma was coined the Liar's Dividend by Chesney and Citron in their leading article on deepfakes.[54] The idea is that doubt will be cast on unfavourable real events by those implied in those events. The world was privy to an example of this tactic when former US President Donald Trump claimed that the audio footage taped for the TV show *Access Hollywood*, in which he said demeaning things about women, was fake, while the show's host confirmed its authenticity.[55] However, the fact that doubt has been created already works in the favour of the accused.

The consequences of such doubt became clear in 2019 in Gabon. President Ali Bongo of Gabon had not been seen in public and had allegedly been transported to a different country for medical treatment after a stroke. When he appeared on television for his New Year's address, he looked markedly different, fuelling rumours that the video was not real. A military coup ensued because military leaders believed the televised address was a deepfake, proving that the actual President was incapacitated.[56] The video turned out to be of the real President Bongo and the coup failed. The fact that it happened at all, however, illustrates the dangers of the Liar's Dividend to national security.

IV The Future Is Now

Fundamentally, deepfakes that are designed as a tool to spread disinformation harm national security by attacking the fabric of society: trust in people, processes, and institutions. Such deepfakes pull at the thread of the social contract, potentially unravelling it quicker than any government can sew it back together. Democratic processes, such as elections and referendums, can be unduly influenced by deepfakes that spread disinformation, and institutions founded on trust, such as the stock market or the banking system, are at risk of collapse if concerted efforts are made to undermine them. Furthermore, as the Liar's Dividend demonstrates, the mere existence of a possible deepfake is enough to confuse people.

Furthermore, deepfakes do not have to be good to do harm. Particularly when the deepfake is a form of social engineering that plays into people's emotions, people are quick to share deepfake disinformation especially when it confirms their belief system.[57] In addition, if extortion is the aim of the deepfake quality may not matter, because the fear it strikes into a person may be enough of a motivator.

What then, if anything, can be done to arm ourselves against the adverse effects of deepfakes in the future? At the time of writing, we see a focus on societal resilience, a number of technical solutions, and legislation being suggested as possible defences against deepfakes. The section below briefly discusses their merits.

Rage Against the AI/ML

Alarmingly, people are generally not very good at recognising deepfakes, even when they have been told a dataset they watch contains deepfakes.[58] This is important, because in the context of national security there has been a focus on increasing resilience of the general population against disinformation. For example, panellists at the 2021 *deepfakes, trust & international security innovations dialogue*, held under the auspices of the United Nations Institute for Disarmament Research (UNIDIR), grappled with the question of "how to build a resilient, trustworthy digital ecosystem" but offered few concrete solutions,[59] while the US Department of Homeland Security (DHS) suggest to increase the resilience of the general population by "engineering the environment", a strategy that includes educating both the general public and the media workforce.[60] The question then remains whether this is a viable strategy or if we are perhaps asking too much of the general population to arm themselves against such a sophisticated tool as deepfakes when their ability to recognise them remains low even when educated about them.

People may, in fact, become "useful idiots" of a foreign power without having the slightest idea they are involved, by spreading disinformation further, providing credence to it, and by amplifying the confirmation bias present in echo chambers. After all, "Uncovering the "fake" in fake news does not necessarily convince the recipients of its forgery".[61] The opposite is also true: the more awareness raising and resilience advocacy is undertaken as a strategy to increase a society's defence against deepfakes, the more people will be aware that deepfakes exist. This in turn can play into the hands of those who wish to abuse the Liar's Dividend and sow doubt as to whether the facts are really true.

The knowledge about the existence of deepfakes, however, should be used to our advantage. Those duped by early adaptors of the technique stood little chance, but approximately five years after deepfakes became mainstream we must practice digital hygiene to guard ourselves against malicious deepfakes. For individuals, corporations, and government officials this means vetting those who you have conversations with, be they via telephone, voice-over IP, or one of the many video call systems currently in operation. Ideally this should happen prior to the meeting but mitigating efforts can be put in place during a meeting as well, such as those requiring real-time verification. These systems may also not be completely unhackable, but at the very least they add another layer of protection and may deter some from trying gain access. It needs to be recognised, however, that this is a rapidly changing field and any efforts that may now provide a layer of protection may become obsolete sooner rather than later.

Resilience is not the only tool against deepfakes. Technologically speaking, detecting deepfakes is an ongoing cat-and-mouse game where the creators are generally one step ahead. Some initiatives, such as the *Content Authenticity Initiative* and the company TruePic, are using technology to establish the provenance of imagery, including blockchain. The question is if this is an attainable solution for all internet users if the services are not provided for free. Furthermore, this solution would need wide uptake to be effective. We also need to ask ourselves the question if these initiatives are successful, does that mean that anything lacking provenance will be disregarded as fake by default?

Detection tools could be useful to curb a deepfake going viral on a social media platform, but that would also require a change in algorithms of social media platforms pushing viral content that can harm society. Furthermore, if the deepfake is used to mislead officials or as a false flag, debunking the deepfake may come too late, especially if the deepfake is part of a coordinated attack where potential verification mechanisms have also been compromised.

Legislators also play a role in determining what is acceptable and what deepfake content should fall foul of the law. That often does not mean creating completely new statutes but amending existing laws to update them to include deepfaked materials. For example, criminal codes dealing with fraud, sexual abuse, or impersonation can all be updated to include the use of the deepfake as a tool to commit such an act.

As this chapter illustrated, the fact that deepfakes pose a danger to national security is far from hypothetical. Further case-studies in a conflict situation presented itself during the writing of this chapter's conclusions, when in February 2022 Ukraine became the target of a full-scale armed invasion. While there had been fears that Russia would use footage of staged incidents to justify attack, within days first reports about deepfaked footage surfaced. Two claims stand out: firstly, misinformation and disinformation surrounded the incident at Snake Island, where Ukrainian border guards told a Russian warship "go f★ck yourself". This act was believed to result in the certain demise of the guards, while some suggested Ukrainian forces would not allow their survival to get in the way of a heroic story.[62] Secondly, in an alleged deepfake video, Chechen leader Ramzan Kadyrov, expresses trepidation about sending his troops.[63] In fact, Kadyrov—a known ally of Moscow—has sent reinforcements to fight in Ukraine alongside Russia, although the numbers of troops appear massively inflated, as per Russian propaganda channels.[64] Both incidents seem to fall within the realm of psychological warfare, trying to affect morale and public opinion.

The future looks grim. In a world where disinformation spreads almost at the speed of light, the idea that the marketplace of ideas would bring about enlightenment as competing ideas reveal the truth is wishful thinking. The simple solution may seem to prohibit deepfakes, perhaps inspired by Karl Popper's *paradox of tolerance* idea.[65] Rawlsian philosophers and human rights advocates alike, however, caution that too much intolerance brings about injustice. The balance is a precarious one.

Until we find a consensus, ethically, legally, and politically, as to the acceptance of deepfakes, knowing that they are used as tools of disinformation, we accept the risk of mob violence over a deepfake, as we have seen violence erupt over WhatsApp disinformation in India.[66] As the attack on the US Capitol demonstrated, dog whistles can be effective in whipping up sentiments leading to deadly clashes. A strategically placed deepfake, even when debunked quickly, can lead to escalations that will be impossible to contain, particularly if those in charge instigate the disinformation and have little interest in defusing tensions.

Notes

1 Yang et al., "Arming the Public with Artificial Intelligence to Counter Social Bots," 50–51, 55–56.
2 Some authors prefer to spell the term as "deep fake". That term is a portmanteau of deep learning and fake.
3 Huijstee et al., *Tackling Deepfakes in European Policy*, 2.
4 Bayer et al., "Disinformation and Propaganda—Impact on the Functioning of the Rule of Law in the EU and Its Member States," 24–25.
5 Wardle and Derakhshan, "Information Disorder—Toward an Interdisciplinary Framework for Research and Policymaking," 5.
6 Sedova et al., "AI and the Future of Disinformation Campaigns," December 2021, 1.
7 Shearer, "More than Eight-in-Ten Americans Get News from Digital Devices."
8 Russell and Norvig, *Artificial Intelligence*, 32. Russel and Norvig state "In the public eye, there is sometimes confusion between the terms 'artificial intelligence' and 'machine learning'. Machine learning is a subfield of AI that studies the ability to improve performance based on experience. Some AI systems use machine learning methods to achieve competence, but some do not." In the computer science debate, however, there are some that add further nuance that statement by claiming that

machine learning is a subfield of both AI and statistics, such as Skansi (Skansi, *Introduction to Deep Learning*, v.). Furthermore, this is not a debate that merely takes place in the public eye. Scott William Burk and Gary Miner claim that "most practitioners consider AI to be a subset of machine learning." (*It's All Analytics! The Foundations of AI, Big Data, and Data Science Landscape for Professionals in Healthcare, Business, and Government*, 1 Edition (New York: Routledge, 2020), 140). A recent publication of the U.S. National Security Agency (NSA) went as far as to say that machine learning is often described as artificial intelligence or deep learning, conflating the terms even further McCloskey and Mountain, "Guest Editor's Column.".

9   Sedova et al., "AI and the Future of Disinformation Campaigns," December 2021, 21.
10 Baker, "From the Criminal Crowd to the "Mediated Crowd,"" 44.
11 Russell and Norvig, *Artificial Intelligence*, 1206.
12 Russell and Norvig, 1433–34.
13 Russell and Norvig, 1674–75.
14 Russell and Norvig, 84.
15 "AlphaGo."
16 *DeepMind StarCraft II Demonstration.*
17 Sedova et al., "AI and the Future of Disinformation Campaigns," December 2021, 22.
18 Forbes, "Stefan Lorant," 91.
19 Widmer, "Lincoln Captured!"
20 "Abraham Lincoln."
21 Hadhazy, "Is That Iranian Missile Photo a Fake?"
22 Burk and Miner, *It's All Analytics! The Foundations of AI, Big Data, and Data Science Landscape for Professionals in Healthcare, Business, and Government*, 145–46.
23 Sedova et al., "AI and the Future of Disinformation Campaigns," December 2021, 24–25.
24 Sedova et al., 22.
25 Ajder et al., "The State of Deepfakes: Landscape, Threats, and Impact," 1–2.
26 The apps or bots providing these "services" typically do not work on men Hao, "A Deepfake Bot Is Being Used to 'Undress' Underage Girls.".
27 "How We Made David Beckham Speak 9 Languages."
28 "Deepfake Video of Former Italian PM Matteo Renzi Sparks Debate in Italy," 24.
29 "Come Nasce Un Deepfake? Il Dietro Le Quinte Della Rubrica di Striscia."
30 Mommers, Nuyens, and De Wit, "Beste Mark Rutte, Zo Klink Je Als Je #klimaatleiderschap Toont."
31 Parkin, "The Rise of the Deepfake and the Threat to Democracy."
32 Inhofe, National Defense Authorization Act for Fiscal Year 2020.
33 Humphries, "Bruce Willis Deepfake to Star in Russian TV Ads."
34 Simonite, "Deepfakes Are Now Making Business Pitches."
35 Damiani, "A Voice Deepfake Was Used to Scam A CEO Out Of $243,000"; Brewster, "Fraudsters Cloned Company Director's Voice In $35 Million Bank Heist, Police Find."
36 Roth, "European MPs Targeted by Deepfake Video Calls Imitating Russian Opposition"; Venema, "Opinie: Je kunt dus doodleuk als deepfake met de Kamer vergaderen, en dat is géén grap."
37 Oltermann, "European Politicians Duped into Deepfake Video Calls with Mayor of Kyiv."
38 Johnson, "Deepfakes Are Solvable—but Don't Forget That 'Shallowfakes' Are Already Pervasive."
39 Glover, "China Deploys AI News Anchor."
40 Sedova et al., "AI and the Future of Disinformation Campaigns," December 2021, 20.
41 Stokel-Walker, "How to Spot a Deepfake, According to Experts Who Clocked the Fake Persona behind the Hunter Biden Dossier."
42 Sedova et al., "AI and the Future of Disinformation Campaigns," December 2021, 14–15.
43 Alspach, "Does Your Boss Sound a Little Funny? It Might Be an Audio Deepfake."
44 "Fake Voices "Help Cyber-Crooks Steal Cash.""
45 Swerling, "Doctored Audio Evidence Used to Damn Father in Custody Battle."
46 Witten and Knudsen, "Why Seeing Is Believing," 489.
47 Li, Chang, and Lyu, "In Ictu Oculi."
48 Lee, "How Puny Humans Can Spot Devious Deepfakes."
49 deHaven-Smith, "State Crimes against Democracy in the War on Terror," 407.
50 "Algorithm of Harm: Facebook Amplified Myanmar Military Propaganda Following Coup."
51 Ayyub, "I Was the Victim of A Deepfake Porn Plot Intended to Silence Me."
52 Lee, "How Puny Humans Can Spot Devious Deepfakes."

53 Venema and Geradts, "Digital Forensics, Deepfakes, and the Legal Process," 16–17.
54 Chesney and Citron, "Deep Fakes," 1785–86.
55 Victor, "'Access Hollywood' Reminds Trump: 'The Tape Is Very Real.'"
56 Hao, "The Biggest Threat of Deepfakes Isn't the Deepfakes Themselves."
57 Chesney and Citron, "Deep Fakes," 1766–68.
58 Köbis, Doležalová, and Soraperra, "Fooled Twice," 6–7.
59 Anand and Bianco, "The 2021 Innovations Dialogue Conference Report: Deepfakes, Trust and International Security," 21–22.
60 Department of Homeland Security, "Phase 2: Increasing Threats of Deepfake Identities—Mitigation Measures," 26–27.
61 Huijstee et al., *Tackling Deepfakes in European Policy*, 25.
62 Kiri Eaglesfield, "It's Being Reported That Ukraine Attempted to Bomb the Russian Ships Transporting the Snake Island Soldiers to Maintain the Myth That They Died with Honour."
63 Avi, "Ppl Be Aware, #deepfake Vid of #RamzanKadyrov Is Floating. B4 Sharing or Retweet, Plz Ensure Its Not Fake. #Fake Vid Link Being Shared by Few #Ukranian, #US, & #Uk Handles—Https://Twitter.Com/Antiputler_news/Status/1497654759972655108?T=iqtyRbl-Ee_mN4Smi5cwiw&s=19. Original Vid Link Is Here - Https://Twitter.Com/JustGOOS/Status/1497669977092104198?T=HVvsfPc5cvUmKXCTLtoS1w&s=19. Sm Ppl Realised & Deleted."
64 Ling, "Russia Tries to Terrorize Ukraine with Images of Chechen Soldiers."
65 Philosopher Karl Popper was inspired by the teachings of Greek philosopher Plato, who argued that there is a *paradox of freedom*, where great restraint should be exercised in the absence of control since this absence would make "the bully free to enslave the meek." Popper, building on this, argued with this *paradox of tolerance* that "unlimited tolerance must lead to the disappearance of tolerance." Popper, Ryan, and Gombrich, *The Open Society and Its Enemies*, 581–82. Applying this philosophy to deepfakes, one could argue that in order to preserve the marketplace of ideas and the tolerance required for it to flourish, we need to be intolerant to those means that disrupt the marketplace, in this case deepfakes.
66 Samuels, "How Misinformation on WhatsApp Led to a Mob Killing in India."

Bibliography

Ajder, Henry, Giorgio Patrini, Francesco Cavalli, and Laurence Cullen. "The State of Deepfakes: Landscape, Threats, and Impact." Deeptrace, September2019.

Alspach, Kyle. "Does Your Boss Sound a Little Funny? It Might Be an Audio Deepfake." *Protocol*, August 18, 2022. https://www.protocol.com/enterprise/deepfake-voice-cyberattack-ai-audio.

Anand, Alisha, and Belén Bianco. "*The 2021 Innovations Dialogue Conference Report: Deepfakes, Trust and International Security.*" Geneva, Switzerland: United Nations Institute for Disarmament Research, 2021. https://unidir.org/sites/default/files/2021-12/UNIDIR_2021_Innovations_Dialogue.pdf.

Avi, @avi_bravim_0m. "Ppl Be Aware, #deepfake Vid of #RamzanKadyrov Is Floating." Tweet. *Twitter*, February 26, 2022. https://twitter.com/avi_bravim_0m/status/1497681400379875328?s=20&t=QJsdl3sButJ-4FRNJqGfbQ.

Ayyub, Rana. "I Was the Victim of A Deepfake Porn Plot Intended to Silence Me." *Huffington Post UK*, November 21, 2018. https://www.huffingtonpost.co.uk/entry/deepfake-porn_uk_5bf2c126e4b0f32bd58ba316.

Baker, Stephanie Alice. "From the Criminal Crowd to the "Mediated Crowd': The Impact of Social Media on the 2011 English Riots." Edited by Daniel Briggs. *Safer Communities* 11, no. 1 (January 13, 2012): 40–49. https://doi.org/10.1108/17578041211200100.

Bayer, Judit, Natalija Bitiukova, Petra Bárd, Judit Szakács, Alberto Alemanno, and Erik Uszkiewicz. "*Disinformation and Propaganda—Impact on the Functioning of the Rule of Law in the EU and Its Member States.*" Policy Department for Citizens" Rights and Constitutional Affairs, Directorate General for Internal Policies of the Union, February2019. http://www.europarl.europa.eu/supporting-analyses.

BBC. "Fake Voices "Help Cyber-Crooks Steal Cash,'" July 8, 2019, sec. Tech. https://www.bbc.com/news/technology-48908736.

Brewster, Thomas. "Fraudsters Cloned Company Director's Voice In $35 Million Bank Heist, Police Find." *Forbes*, October 14, 2021. https://www.forbes.com/sites/thomasbrewster/2021/10/14/huge-bank-fraud-uses-deep-fake-voice-tech-to-steal-millions/.

Burk, Scott William, and Gary Miner. *It's All Analytics! The Foundations of AI, Big Data, and Data Science Landscape for Professionals in Healthcare, Business, and Government.* 1st Edition. New York: Routledge, 2020.

Chesney, Robert, and Danielle Keats Citron. "Deep Fakes: A Looming Challenge for Privacy, Democracy, and National Security." *California Law Review* 107, no. 1 (2019): 1753–1820. https://doi.org/10.15779/Z38RVOD15J.

"Come Nasce Un Deepfake? Il Dietro Le Quinte Della Rubrica di Striscia," June 3, 2020. https://www.striscialanotizia.mediaset.it/news/come-nasce-un-deepfake-il-dietro-le-quinte-della-rubrica-di-striscia_9975.shtml.

CSL_BIBLIOGRAPHY Library of Congress. "Abraham Lincoln." Accessed February 5, 2022. http://www.loc.gov/pictures/item/2003654314/.

Damiani, Jesse. "A Voice Deepfake Was Used to Scam A CEO Out Of $243,000." *Forbes*, September 3, 2019. https://www.forbes.com/sites/jessedamiani/2019/09/03/a-voice-deepfake-was-used-to-scam-a-ceo-out-of-243000/#3937cfdd2241.

DeepMind. "*AlphaGo*." Accessed February 5, 2022. https://deepmind.com/research/case-studies/alphago-the-story-so-far.

DeepMind StarCraft II Demonstration, 2019. https://www.youtube.com/watch?v=cUTMhmVh1qs&ab_channel=DeepMind.

Department of Homeland Security, US. "Phase 2: Increasing Threats of Deepfake Identities - Mitigation Measures." Department of Homeland Security, August 10, 2022. https://www.dhs.gov/publication/2022-aep-deliverables.

Eaglesfield, Kiri. @zenxv. "It's Being Reported That Ukraine Attempted to Bomb the Russian Ships Transporting the Snake Island Soldiers to Maintain the Myth That They Died with Honour." Tweet. *Twitter*, February 27, 2022. https://twitter.com/zenxv/status/1497806347835830272?s=20&t=QJsdl3sButJ-4FRNJqGfbQ.

Forbes, Duncan. "Stefan Lorant." *History of Photography* 31, no. 1 (March2007): 90–91. https://doi.org/10.1080/03087298.2007.10443509.

France 24. "Deepfake Video of Former Italian PM Matteo Renzi Sparks Debate in Italy," August 10, 2019, sec. The Observers. https://observers.france24.com/en/20191008-deepfake-video-former-italian-pm-matteo-renzi-sparks-debate-italy.

Global Witness. "Algorithm of Harm: Facebook Amplified Myanmar Military Propaganda Following Coup," June 23, 2021. https://www.globalwitness.org/en/campaigns/digital-threats/algorithm-harm-facebook-amplified-myanmar-military-propaganda-following-coup/.

Glover, Claudia. "China Deploys AI News Anchor." *Tech Monitor*, May 21, 2020. https://techmonitor.ai/techonology/emerging-technology/ai-news-anchor-china#:~:text=China's%20national%20Xinhua%20news%20agency,Xinhua%20news%20agency%2C%20Zao%20Wanwei.

Hadhazy, Adam. "Is That Iranian Missile Photo a Fake?" *Scientific American*, July 10, 2008. https://www.scientificamerican.com/article/is-that-iranian-missile/.

Hao, Karen. "A Deepfake Bot Is Being Used to 'Undress' Underage Girls." *MIT Technology Review*, October 20, 2020. https://www.technologyreview.com/2020/10/20/1010789/ai-deepfake-bot-undresses-women-and-underage-girls/.

Hao, Karen. "The Biggest Threat of Deepfakes Isn't the Deepfakes Themselves," October 10, 2019. https://www.technologyreview.com/s/614526/the-biggest-threat-of-deepfakes-isnt-the-deepfakes-themselves/.

Huijstee, Mariëtte van, Pieter van Boheemen, Djurre Das, Linda Nierling, Jutta Jahnel, Murat Karaboga, and Martin Fatun. *Tackling Deepfakes in European Policy*. European Parliamentary Research Service, 2021. https://www.europarl.europa.eu/thinktank/en/document/EPRS_STU(2021)690039.

Humphries, Matthew. "Bruce Willis Deepfake to Star in Russian TV Ads." *PC Mag*, August 23, 2021. https://www.pcmag.com/news/bruce-willis-deepfake-to-star-in-russian-tv-ads.

Inhofe, James M. National Defense Authorization Act for Fiscal Year 2020, Pub. L. No. S.1790 (2019). https://www.congress.gov/bill/116th-congress/senate-bill/1790/text.

Johnson, Bobbie. "Deepfakes Are Solvable—but Don't Forget That 'Shallowfakes' Are Already Pervasive." *MIT Technology Review*, March 25, 2019. https://www.technologyreview.com/2019/03/25/136460/deepfakes-shallowfakes-human-rights/.

Köbis, Nils C., Barbora Doležalová, and Ivan Soraperra. "Fooled Twice: People Cannot Detect Deepfakes but Think They Can." *IScience* 24, no. 11 (November2021): 103364. https://doi.org/10.1016/j.isci.2021.103364.

Lee, Alex. "How Puny Humans Can Spot Devious Deepfakes." *Wired*, October 20, 2019. https://www.wired.co.uk/article/how-to-spot-deepfake-video.

Li, Yuezun, Ming-Ching Chang, and Siwei Lyu. "In Ictu Oculi: Exposing AI Created Fake Videos by Detecting Eye Blinking." In *2018 IEEE International Workshop on Information Forensics and Security (WIFS)*, 1–7. Hong Kong, Hong Kong: IEEE, 2018. https://doi.org/10.1109/WIFS.2018.8630787.

Ling, Justin. "Russia Tries to Terrorize Ukraine with Images of Chechen Soldiers." *Foreign Policy*, February 26, 2022. https://foreignpolicy.com/2022/02/26/russia-chechen-propaganda-ukraine/.

McCloskey, Joe, and David J. Mountain. "Guest Editor's Column." *The Next Wave*, 2019.

Mommers, Jelmer, Anoek Nuyens, and Rebekka De Wit. "Beste Mark Rutte, Zo Klink Je Als Je #klimaatleiderschap Toont." *De Correspondent*, October 28, 2021. https://decorrespondent.nl/12847/beste-mark-rutte-zo-klink-je-als-je-klimaatleiderschap-toont/1415855023-2bf2a907.

Oltermann, Philip. "European Politicians Duped into Deepfake Video Calls with Mayor of Kyiv." *The Guardian*, June 25, 2022, International edition. https://www.theguardian.com/world/2022/jun/25/european-leaders-deepfake-video-calls-mayor-of-kyiv-vitali-klitschko.

Parkin, Simon. "The Rise of the Deepfake and the Threat to Democracy." *The Guardian*, June 22, 2019, International edition. https://www.theguardian.com/technology/ng-interactive/2019/jun/22/the-rise-of-the-deepfake-and-the-threat-to-democracy.

Popper, Karl R., Alan Ryan, and E. H. Gombrich. *The Open Society and Its Enemies*. Princeton: Princeton University Press, 2013.

Roth, Andrew. "European MPs Targeted by Deepfake Video Calls Imitating Russian Opposition." *The Guardian*, April 22, 2021, International Edition. https://www.theguardian.com/world/2021/apr/22/european-mps-targeted-by-deepfake-video-calls-imitating-russian-opposition.

Russell, Stuart J., and Peter Norvig. *Artificial Intelligence: A Modern Approach*. Fourth edition. Pearson Series in Artificial Intelligence. Hoboken: Pearson, 2021.

Samuels, Elyse. "How Misinformation on WhatsApp Led to a Mob Killing in India." *Washington Post*, February 21, 2020. https://www.washingtonpost.com/politics/2020/02/21/how-misinformation-whatsapp-led-deathly-mob-lynching-india/.

Sedova, Katerina, Christine McNeill, Aurora Johnson, Aditi Joshi, and Ido Wulkan. "AI and the Future of Disinformation Campaigns: Part 1: The RICHDATA Framework." Center for Security and Emerging Technology, December2021. https://doi.org/10.51593/2021CA005.

Sedova, Katerina, Christine McNeill, Aurora Johnson, Aditi Joshi, and Ido Wulkan. "AI and the Future of Disinformation Campaigns: Part 2: A Threat Model." Center for Security and Emerging Technology, December2021. https://doi.org/10.51593/2021CA011.

Shearer, Elisa. "More than Eight-in-Ten Americans Get News from Digital Devices." Pew Research Center, January 12, 2021. https://www.pewresearch.org/fact-tank/2021/01/12/more-than-eight-in-ten-americans-get-news-from-digital-devices/.

Simonite, Tom. "Deepfakes Are Now Making Business Pitches." *Wired*, August 16, 2021. https://www.wired.com/story/deepfakes-making-business-pitches/.

Skansi, Sandro. *Introduction to Deep Learning: From Logical Calculus to Artificial Intelligence*. Undergraduate Topics in Computer Science. Cham, Switzerland: Springer, 2018.

Smith, LancedeHaven-. "State Crimes against Democracy in the War on Terror: Applying the Nuremberg Principles to the Bush–Cheney Administration." *Contemporary Politics* 16, no. 4 (December2010): 403–420. https://doi.org/10.1080/13569775.2010.523939.

Stokel-Walker, Chris. "How to Spot a Deepfake, According to Experts Who Clocked the Fake Persona behind the Hunter Biden Dossier." *Business Insider*, November 21, 2020. https://www.businessinsider.com/how-to-spot-a-deepfake-2020-11.

Swerling, Gabriella. "Doctored Audio Evidence Used to Damn Father in Custody Battle." *The Telegraph*, January 31, 2020. https://www.telegraph.co.uk/news/2020/01/31/deepfake-audio-used-custody-battle-lawyer-reveals-doctored-evidence/.

Synthesia. "How We Made David Beckham Speak 9 Languages," June 4, 2020. https://www.synthesia.io/post/david-beckham.

Venema, Agnes E. "Opinie: Je kunt dus doodleuk als deepfake met de Kamer vergaderen, en dat is géén grap." *De Volkskrant*, April 26, 2021. https://www.volkskrant.nl/columns-opinie/opinie-je-kunt-dus-doodleuk-als-deepfake-met-de-kamer-vergaderen-en-dat-is-geen-grap~b0662c22/.

Venema, Agnes E., and Zeno J. Geradts. "Digital Forensics, Deepfakes, and the Legal Process." *The SciTech Lawyer* 16, no. 4 (2020): 14–23.

Victor, Daniel. "'Access Hollywood' Reminds Trump: 'The Tape Is Very Real.'" *The New York Times*, November 28, 2017. https://www.nytimes.com/2017/11/28/us/politics/donald-trump-tape.html.

Wardle, Claire, and Hossein Derakhshan. "Information Disorder - Toward an Interdisciplinary Framework for Research and Policymaking." Council of Europe, September 27, 2017. https://rm.coe.int/information-disorder-report-version-august-2018/16808c9c77.

Widmer, Ted. "Lincoln Captured!" *The New York Times*, May 15, 2011. https://opinionator.blogs.nytimes.com/2011/05/15/lincoln-captured/.

Witten, Ilana B., and Eric I. Knudsen. "Why Seeing Is Believing: Merging Auditory and Visual Worlds." *Neuron* 48, no. 3 (November2005): 489–496. https://doi.org/10.1016/j.neuron.2005.10.020.

Yang, Kai-Cheng, Onur Varol, Clayton A. Davis, Emilio Ferrara, Alessandro Flammini, and Filippo Menczer. "Arming the Public with Artificial Intelligence to Counter Social Bots." *Human Behavior and Emerging Technologies* 1, no. 1 (January2019): 48–61. https://doi.org/10.1002/hbe2.115.

14

THE STRATEGIC LOGIC OF DIGITAL DISINFORMATION

OFFENCE, DEFENCE AND DETERRENCE IN INFORMATION WARFARE

H. Akin Unver and Arhan S. Ertan

Introduction

Governments, state agencies and foreign policy institutions are increasingly deploying organised disinformation to distract and confuse their adversaries. In their 2020 report, Oxford Computational Propaganda Project identified organised, state-sponsored disinformation campaigns in 81 countries with a rapidly increasing number of "cyber troops" (semi-officially employed individuals working on state-sponsored information operations) and campaign intensities.[1]

While disinformation has largely been constructed as a form of "attack" perpetrated by authoritarian countries against liberal democracies, more democratic countries too, have engaged in organised disinformation attempts abroad. For example, both France and Russia engaged in disinformation campaigns in Mali, Central African Republic and other Sahel region countries to build influence and discredit opponents.[2] US State Department had a long-running program of digital disinformation against Jihadi content online, inserting its analysts into extremist discussions via pseudonyms and sharing false information to misdirect the militant group's online efforts.[3] In Hungary, the government used disinformation against Romania in order to make the case internationally that the refugee crisis was Bucharest's fault. Similarly, both Belarus and Poland instrumentalized disinformation during the most recent Ukrainian refugee crisis.[4] A 2021 European Parliament report has indicated that disinformation between nations has become rampant in Western Balkans, disrupting the political stability of the region and generating significant discrediting momentum for the EU.[5] From Brazil, Argentina to South Africa, India and Australia, a broad range of countries and regime types have been involved in organised disinformation.[6]

Propaganda, manipulation and misdirection have been long-standing tactics of diplomacy and international competition. In the last decade, and especially around the 2016 US Presidential election, "fake news" and disinformation became buzzwords of sorts that led to a rediscovery of the role of information in political competition. The biggest difference between the traditional and more recent debates on the matter is the digitalisation of information warfare and the subsequent scale, volume and speed advantages brought by this

DOI: 10.4324/9781003190363-17

digitalization. The advent of Information and Communications Technologies (ICTs) brought about a faster information exchange medium where traditional gatekeepers like editors, censors or curators are of secondary importance and often irrelevant. While in more traditional media forms, broadcast is dependent on the approval of an intermediary individual or a group, with ICTs and social media, this approval is often hard to enforce with the sheer scale of information poured into such media venues. Although automated content moderation works in most cases, it can easily be circumvented.[7] With information gatekeepers out of the way, information becomes disintermediated (reduction in the use of intermediaries), with information suppliers (citizen journalists or anyone with access to social media) directly able to reach information consumers around the world, in real time.[8] The disintermediated nature of modern information exchange has rendered ICTs a conducive ground for misinformation (unintended spread of false information), disinformation (purposeful creation and dissemination of false information) and malinformation (deliberate use of accurate or inaccurate information with the purpose of harming an individual or people).[9]

To that end, the study of disinformation in the digital domain requires renewed attention as traditional studies of propaganda fail to address the speed, scale and the disintermediated nature of information sharing. Digital disinformation has been relatively-well studied in domestic political context, especially in the United States. However, opportunities for disinformation research within comparative politics and international relations (IR) fields are still very much untapped. Particularly, there is still no consensus in the field over how to conceptualise disinformation within the confines of IR: is it best understood as a "weapon", a "tactic", or is it simply a more robust form of propaganda?[10] How is digital disinformation different from disinformation in older media systems and how much does disintermediation affect the way people communicate and consume information, and as a result seek to alter or contribute to international affairs?[11] More importantly, why do countries choose to engage in disinformation against other countries and does disinformation as a foreign policy tool serve a different purpose than disinformation as a domestic political tool?

This chapter seeks to contribute to this emerging debate by exploring disinformation in international relations as a rational actor problem. It situates information warfare as a dyadic-dynamic interaction between the side that initiates disinformation (Attacker), the side that seeks to counter these efforts (Defender) and the international audience (IA) that affects the "winner" of this interaction. Ultimately, the chapter explores the payoff calculus of the Attacker and the Defender and aims to provide a path of exploration into the inner workings of deterrence in information warfare.

Why Do Countries Resort to Organised Manipulation? Unpacking the Demobilizing Logic of Disinformation

What accounts for the rapid explosion of digital disinformation in modern politics in the last decade? Although organised manipulation and propaganda have long been key tactics in statecraft and international politics, we do not yet have robust explanations for how mass digitalisation of communication changed its main parameters in inter-state political communication. Traditional works on propaganda treats political manipulation as a small part of a diverse array of communicative strategies, aimed to promote a political point of view, often through misleading and biased narratives.[12] However, such traditional works do not address the role of disinformation in political communication and how such disintermediation—coupled with vast size and lightning speed of ICTs—affects organised disinformation. A rapidly emerging, yet nascent body of work focuses overwhelmingly on the domestic implications of disinformation, particularly concentrating around election manipulation and foreign meddling in democratic processes of a small group of Western nations.[13] While disinformation research has so far been heavily focused on domestic political contexts of a small group of countries, there is still much to be done on fundamental conceptualization and case study research work on how digital disinformation alters existing communicative processes in international relations. This includes how to conceptualise disinformation in foreign affairs: is it a tactic, a weapon, a form of attack (similar to cyber-attacks), or a tool of diplomacy?[14]

Several countries have already contextualised disinformation as a national security threat. The US State Department defines disinformation as a *"quick and fairly cheap way to destabilise societies and set the stage for potential military action"*.[15] A joint US State Department and Joint Chiefs of Staff white paper later posited that: "*Information has been weaponized, and disinformation has become an incisive instrument of state policy*".[16] Russian General Valery Gerasimov wrote in a 2013 article portraying disinformation as crucial: "*The role of nonmilitary means of achieving political and strategic goals has grown, and, in many cases, they have exceeded the power of force of weapons in their effectiveness*".[17] Similarly United Nations Development Program discourse also constructs disinformation as a weapon—a trend, which began with the COVID-19 pandemic.[18] In the same vein, NATO's securitization of disinformation began with the first Ukraine war (2014) and remained as a crucial component of hybrid warfare in NATO doctrines.[19] The concept of a digital information war had already been included in Russian 2010 Military Doctrine, which was broadened in its 2014 update to include social media and ICTs.[20] Since then, both practices and allegations of disinformation have proliferated across other governments including China, UK, France, Italy, South Africa, Turkey and Kenya (among others), where disinformation is used domestically to discredit political opponents, a foreign policy tool to confuse international rivals and a key threat that has to be defended against—all at the same time.[21]

Indeed, disinformation is increasingly being used as a force factor during international crises. During the 2017 Gulf crisis between Saudi Arabia, Qatar and the UAE, a robust disinformation campaign, followed by cyber-attacks, brought them to the brink of limited conventional war and required significant diplomatic effort to disentangle the damage caused by disinformation.[22] In that case, disinformation had significantly aided in the escalation of the crisis, increasing the costs of backing down by raising audience costs. Following the shooting down of its SU24 in northern Syria, Russia had launched a major disinformation campaign against Turkey, drawing a wedge between Ankara and other NATO countries on the Syrian war, damaging intra-alliance cohesion and ultimately creating itself a relatively unchallenged information space as it entered the Syrian theatre militarily.[23] In Nigeria, the government employed a disinformation campaign in the last decade targeting international aid agencies and workers, significantly impairing the ability of those agencies to work on humanitarian relief in the country.[24] From click farms of Indonesia to the "black campaigning" in the Philippines, disinformation is being deployed as a distinct strategy to discredit and demobilize regional politics in Southeast Asia and more recently, to sow mistrust towards China's Sinovac-CoronaVac vaccine.

The most commonly agreed-upon purpose of disinformation in both domestic and international politics is to distract, confuse and demobilize adversaries.[25] The chain of thought is as follows: disinformation seeks to confuse an adversary to such a degree that even if it is debunked later on, the short-term distraction yields sufficient strategic payoff for the source of disinformation ("Attacker") in the form of demobilizing and dividing the other side ("Defender"). Since disinformation attempts often get fact-checked and debunked, their strategic utility is often short-term.[26] Therefore, disinformation works best during time sensitive events such as elections, diplomatic crises, emergencies and natural disasters.

In the case of an election, the "Attacker" attempts to boost the chances of the friendly candidate and weaken the hostile one in the "Defender" country, increasing the likelihood of the friendly candidate getting elected. If the hostile candidate's victory is inevitable, the purpose of disinformation becomes weakening the hostile candidate's winning margin so that their rule becomes more difficult and contested. This prevents the hostile candidate's ability to focus on the Attacker country after the election.

In the case of a diplomatic crisis or escalation, the logic works similar: by distracting and confusing an adversary, the Attacker seeks to gain short-term strategic payoff either by reducing the level of support for the Defender government or its policies, or slow down and demobilize its current course of action. In cases of direct armed conflict, disinformation is used to demoralize, demobilize, discredit and slow down the adversary's diplomatic and military efforts.

When an international dispute is pursued through non-military terms, disinformation can be used both to further escalate a crisis (especially when the Attacker is distinctly superior to the Defender), or to de-escalate one (when the Attacker is distinctly weaker than the Defender). Current empirical evidence and the relevant literature makes it hard to present a conclusive statement about whether disinformation is the "tool of the weak".[27] Stronger states have used disinformation against weaker adversaries as much as vice versa, and such disinformation has been deployed during active conflicts as much as non-militarized disputes. While there is a tendency to portray democracies as primary targets of disinformation, democracies have been sources of disinformation as well, preventing a clear regime type argument to materialize.[28]

Contrary to classical forms of communicative disruption—such as traditional media propaganda—digital disinformation has a shorter strategic utility span, but has a wider and faster reach.[29] Digital disinformation attempts can spread across a global audience within a matter of minutes and potentially alter short-term beliefs about key events, but is always subject to debunking and fact-checking after its immediate benefits. Moreover, in contrast to traditional propaganda, digital disinformation often focuses more on weakening the adversary's narrative and framing of events rather than strengthening their own side and it is this uniquely disruptive focus of digital disinformation that separates its strategic utility from that of classical propaganda efforts.[30]

Yet, the Attacker's advantage in disinformation is generally short-lived. Following its immediate effect on demobilising, demoralising and debilitating Defender's efforts, the Attacker gets debunked and the fact-checked version of its narrative spreads equally fast across ICTs and social media. This debunking generates a form of "reputational penalty" for the Attacker, who gets "named and shamed" in international platforms and suffers from an additional, secondary "suspicion penalty" on its successive communication efforts.[31] An Attacker that overuses disinformation suffers these reputational and suspicion penalties at an increasing rate, causing its factually correct public diplomacy and government communication efforts to be met with low interest and resistance, and being accused of crying wolf. In turn, each utilization of disinformation by an Attacker as a foreign policy tool, reduces the effectiveness of its successive communication efforts—disinformation and otherwise—lowering the net utility the Attacker gets from communicative manipulation at each successive turn. In simple terms, disinformation as a foreign policy tool has diminishing returns, whereby its overproduction results in increasing penalties in successive rounds of an iterative multi-stage game.

The core puzzle of disinformation is therefore that although it has short-term strategic advantages, it has more significant reputational and communicative costs to the Attacker in the medium-to-long term, yet Attackers nonetheless rely on disinformation knowing that they will suffer from these costs. To that end, the focus of this chapter is this *ex-post* inefficiency of disinformation: why do countries engage in disinformation even though they know that their *ex-ante* gains from it often dwarf in contrast to their *ex-post* reputational costs? In *Fearonian* fashion, an IR–Realist explanation of disinformation would fit into three broad categories: anarchy, preventive action and positive expected utility.[32] In terms of anarchy, the absence of a deterrence mechanism or a supranational authority creates an international disinformation environment that favours the Attacker. Similar to cybersecurity debates where the Attacker has a distinct advantage due to attribution problems and mediocre chance of reprisals,[33] disinformation too, is a domain where the Attacker has a distinct timing and first-mover's advantage against the Defender. Quite often, a successful disinformation campaign generates substantial short-term benefit around important time-sensitive events, even if such campaigns are debunked and a counter disinformation offensive begins. To that end, the world of global disinformation is a significantly anarchic domain.

In a system dominated by anarchy and first-mover advantages, preventative action becomes the norm. Attacker in an information war finds initiating a campaign more preferable if it perceives that information war unavoidable. Since the disinformation domain is anarchic and information war is perceived as inevitable, a government will find the payoff of initiating a campaign greater than the risk of not initiating, in order not to suffer from the penalties of becoming a Defender by moving in late. To that end pre-emption becomes a form of prevention: not of the information war, but of the costs of suffering from the initial salvo.

Finally, Attackers often play down the reputational costs of initiating a disinformation campaign and overestimate their short-term net utility compared to their medium- and long-term reputational costs originating from this action. Attackers may view short-term payoffs from initiating a disinformation campaign (by distracting and demobilising the Defender) more preferable compared to any obscure reputational and suspicion penalties later on—or believe that the Defender cannot debunk the claim. Indeed, the Defender may not always successfully debunk a disinformation claim, especially when the volume of disinformation is too high (such as a botnet campaign), or the disinformation is built in a sophisticated, hard-to-debunk fashion. If the Attacker believes that its disinformation campaign is sophisticated enough that it will be difficult to fact-check its claims, it will pursue disinformation believing that it will not suffer from reputational or suspicion penalties later on. Similarly, even if the Defender successfully debunks a claim, it may not be able to disseminate its fact-checked response widely or in time. Indeed, this was the core claim of Vosoughi, Roy and Aral's (2018) seminal work: in politically charged environments, false news spread faster and wider than accurate news across social media platforms.[34] Similarly, Saling et al. (2021) find that even users that regularly fact-check news online can still share disinformation inadvertently if the event is emotional and momentous enough.[35] To that end, in some cases, the Defender may debunk successfully, but it may fail to disseminate true claims sufficiently and as a result, may fail to generate international public reaction against the Attacker and fall short of "naming and shaming" the Attacker.[36]

Therefore, an Attacker may find initiating a disinformation campaign preferable, if:

a   it has a different calculus about reputational and suspicion costs of the interaction compared to the Defender and the international audience (IA),

b   it believes the Defender will not be able to "name and shame"—or at least on time, c it believes the Defender will not be able to successfully debunk and disseminate the claims,

d it will not suffer from significant reputational or suspicion costs beyond the short-term, or if both costs are not significant enough in comparison to its short-term payoff.

A Model of Disinformation in International Crises

In order to demonstrate the "offence-defence balance" in an information war, we offer a multi-stage "decision theoretic" model (as opposed to a "game theoretic" one), as in our model, there is one "agent" (player) who makes a strategic decision. This remains an iterative, multi-stage interaction where the decision-making agent considers the expected utility from all alternative outcomes, simultaneously considering the discounted value of utility from future interactions.

The following formal representation of the problem outlines the core interaction. There are three players, "Attacker" (A), Defender" (D) and the digital "international audience" (IA) which both A and D seek to convince. Attacker is the origin point of the information campaign and in our model, is the only strategic decision-maker in the model. We choose to model these dynamics as an infinitely repeated interaction among 3 parties:

1 The Attacker's repertoire of action consists of three moves:

- Action *TN*: Do not engage in disinformation—Share only **true news**.

    a This action is taken when the ground reality works in Attacker's favour and the Attacker doesn't feel a need to resort to disinformation. Or the Attacker believes that the short-term strategic gain from spreading fake news is too low compared to the reputation and suspicion penalties it will suffer later on. The latter case often happens when a militarily weaker country seeks the support of stronger, more democratic allies and has to keep its choices within an "acceptable" range in order not to attract the criticism of its more powerful allies.

- Action *FN*: Spread **fake news** and engage in an organised disinformation campaign.

    a This line of action happens when the ground reality works against the Attacker's interests and the Attacker feels a need to alter short-term perceptions of the Defender and the international audience either to gain a strategic edge, or to disrupt the level of mobilisation and consensus in the adversary camp. This choice materialises when the Attacker views short-term strategic utility from disinformation greater than the reputational and suspicion costs that will arise from the following rounds.

- Action *NN*: The Attacker does not engage in any information activity—accurate or inaccurate and remains silent.

    a Often, countries choose not to engage in any form of public diplomacy and choose to remain silent on emerging issues in order to calibrate their responses in later rounds of the game.

2 We assume that the dominant strategy of the defender is to **debunk** as many possible "attacks" of disinformation as its information capacity allows. While in some cases the Defender may also prefer to do nothing, or fire back with more disinformation rather than true information, these action types are rare, and will remain beyond the scope of this model.

- Action **DNS: Debunk** the fake news and "**name and shame**" the Attacker on the international stage.

  a   This debunking action happens when the Defender views the Attacker's disinformation campaign too damaging and resorts to countermeasures in the form of fact-checking and disseminating the true version of the events.

3   In this model, we also consider the international audience as an actor with agency, which also acts based on whether it believes the Attacker's version of events, or the Defender's. In our model, the international audience does not always believe in accurate news and may often believe and help sustain a disinformation campaign, especially if the disinformation campaign in question is a sophisticated one that is hard to debunk and triggers emotional sensitivities of the global audience.

- Action **B**: In this case, the international audience gravitates in favour of the fake news and the overall digital community converges upon **believing** the disinformation campaign. When this happens, the Attacker's short-term utility from spreading disinformation becomes sustained in the medium-term and yields the greatest payoff for the Attacker.
- Action **NB**: When successful debunking and fact-checking performance meets with a successful countermeasure campaign, the international digital community converges upon **rejecting** the disinformation campaign. When this happens, the Defender's "name and shame" strategy yields the greatest payoff, and the Attacker suffers from reputational and suspicion penalties early on.

Attacker initiates the first phase of the interaction. If conditions on the ground favour the Attacker's version of events, it engages in **TN** (share true news; or engage in public diplomacy) as it does not gain additional payoff by sharing fake news. For example, if the Attacker is meddling in the Defender's elections and the candidate that is friendly to the Attacker country has a distinct advantage in polls, the Attacker has no interest in destabilising the Defender and risk exposure. If the ground conditions work against the Attacker's favour, then the Attacker has an interest in disseminating disinformation and destabilising/demobilising the Defender. Let P_T be equal to how much value the Attacker assigns to the probability of ground conditions working in favour of its interests. This is a subjective value and is assigned by the Attacker on a case-by-case basis, so any value the Attacker assigns on ground reality not favouring its interests is $1—P_T$. In case the ground reality works in the Attacker's favour, the dominant strategy for the Attacker is to spread the accurate news and conduct public diplomacy communication (not disinformation) through that event. If the Attacker finds the ground truth as contrary to its interests, it has two options: a) do nothing and let the situation progress on its own, or b) engage in disinformation to alter short-term perceptions of the international and domestic audiences.

    In the next stage, the international audience (IA from now on) is either successfully misled by the disinformation attempt, or reject the information provided by the Attacker, based on their previous iterations of the information exchange. This choice of IA is contingent upon previous iterations of the information provided by the Attacker. While in our model, the interaction starts off with the Attacker's choice, it is important to keep in mind that there have been numerous interactions in the past, causing the IA to develop pre-existing beliefs about both the Attacker and the Defender. We can model the IA's probability of believing the disinformation attempt by the attacker and help it spread (P_B) in two alternative pathways:

a   IA's probability of believing the disinformation attempt by the Attacker and help it spread (P_B) either due to a successful disinformation attempt, or poor fact-checking practices, or IA's pre-existing beliefs in favour of the Attacker,

b   IA's probability of not believing in the disinformation attempt $(1-P_B)$ due to fast and high-volume fact-checking and IA's pre-existing beliefs in favour of the Defender.

These two alternative models are mathematically equivalent, so we continue with the first interpretation. Here we assume that P_B does not vary based on the news spread by the Attacker being actually true or false. This is likely at this stage, as the ground truth is not known by the international audience but only by the Attacker.

   In the final stage of the model, the Defender tries to debunk as many of the disinformation attempts as possible, as it is the apparent dominant strategy for them. In rare cases, the Defender may also choose to remain silent and not respond, or in other cases it may respond by initiating a disinformation campaign itself. But these rare attempts have a high risk of failure too, either because the Defender might not have resources/time to deal with the scale of disinformation, or even when it debunks successfully, it may not convince the international audience due to pre-existing beliefs, or weak dissemination performance and media capacity. In this case again, these two alternative scenarios are mathematically equivalent, and we choose to assign a probability (P_D) for a "successful debunk" of a disinformation attempt.

   It is important to note here that, both P_B *and* P_D are unknown *ex-ante* for all the parties except the Attacker, and the Attacker assigns a subjective probability for these options, as it does for P_T.

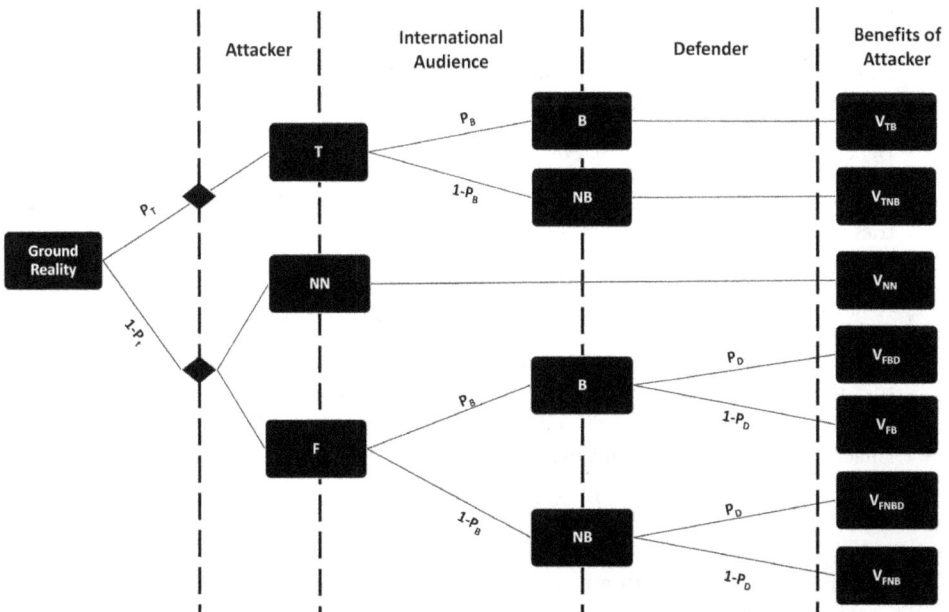

Figure 14.1 Decision Tree of the Information Interaction Model Between the Attacker, Defender and the International Audience

Evaluating Game Outcomes

As can be seen from the decision tree representation of our model in Figure-1, there are 7 potential outcomes of the interactions among the three "agents" of our model. Brief descriptions of the scenarios behind each outcome and corresponding value of each outcome for the Attacker are as follows:

- Outcome **TB**: Ground reality conforms to the Attacker's interests; hence it tries to spread it as part of public diplomacy, IA is convinced by this information policy.

  Benefit for Attacker: V_{TB}

- Outcome **TNB**: Ground reality conforms to the Attacker's interests; hence it tries to spread it as part of public diplomacy, but IA is not convinced by this information policy.

  Benefit for Attacker: V_{TNB}

- Outcome **NN**: Ground reality does not conform to the Attacker's interests, but it chooses to remain silent and not engage in any particular information action on the matter.

  Benefit for Attacker: V_{NN}

- Outcome **FB**: Ground reality does not conform to the Attacker's interests; hence it tries to engage in a disinformation campaign. IA is convinced by the disinformation campaign and the Defender cannot successfully debunk claims.

  Benefit for Attacker: V_{FB}

- Outcome **FBD**: Ground reality does not conform to the Attacker's interests; hence it tries to engage in a disinformation campaign. IA is convinced by the disinformation campaign, but then it is successfully debunked by the Defender. Defender successfully incurs "name & shame" punishment (**PUN**) against the attacker. IA's likelihood of believing the Attacker (**P_B**) decreases in all future interactions.

  Benefit for Attacker: V_{FB}_PUN

- Outcome **FNB**: Ground reality does not conform to the Attacker's interests; hence it tries to engage in a disinformation campaign. IA is not convinced by the disinformation campaign.

  Benefit for Attacker: V_{FNB}

- Outcome **FNBD**: Ground reality does not conform to the Attacker's interests; hence it tries to engage in a disinformation campaign. IA is not convinced by the disinformation campaign. Defender successfully incurs "name & shame" punishment (**PUN**) against the Attacker. IA's likelihood of believing the Attacker (**P_B**) decreases in all future interactions.

  Benefit for Attacker: V_{FNB}_PUN

It is important to underline that it is not realistic to expect the Attacker to make precise estimations about the values of these benefits; hence they are highly subjective estimations. As it will be apparent below, in our discussion of the dynamics of the model, the subjective nature of these benefits may cause the Attacker—and in some cases the Defender—to mis-calculate the pros and cons of each strategy.

Expected Utilities for the Attacker

The Attacker pursues disinformation only when the ground reality does not conform to its interests. Since we assume that the pay-off structure of the interactions among all three decision making agents remain static over time, the comparison of expected utilities does not change over time either, hence the Attacker continues with the same onset strategy in all future iterations of this interaction. When the ground reality does not conform to its interests, the Attacker has two strategies to choose from: **FN** or **NN**. Hence, a rational decision requires to compare the expected utilities resulting from each of those strategies.

So, the equations follow:

NN—doing nothing when the ground reality does not conform to the attacker's interests:

$$EU_{NN} = V_{NN} + \frac{1}{r}[((1 - P_T)xV_{NN}) + (P_T x EU_{TN})] \qquad (14.1)$$

FN—engaging in a disinformation campaign when ground reality does not conform to the attacker's interests:

$$EU_{FN}(P_B) = P_B x[(1 - P_D)xV_{FB} + P_D x(V_{FB} - PUN)] + (1 - P_B)$$

$$x[(1 - P_D)xV_{FNB} + P_D x(V_{FNB} - PUN)] + \frac{1}{r}EU\frac{LR}{FN} \qquad (14.2)$$

Where:
r = discount rate[37]

$$EU_{TN} = (P_B x V_{TB}) + ((1 - P_B)xV_{TNB})$$

$$EU\frac{LR}{FN} = (1 - P_T)x[P_D x EU_{FN}(P_{B_{low}}) + (1 - P_D)xEU_{FN}(P_B)]$$

$$+P_T x[P_D x EU_{TN}(P_{B_{low}}) + (1 - P_D)xEU_{TN}(P_B)]$$

$P_{B_{low}}$ = *value of IA's likelihood of believing the attacker after a successful debunk*

$$(14.3)$$

Since the Attacker decides between **FN** and **NN** by comparing the corresponding expected utilities of these two strategies, the direction of the effects of a number of important parameters on the values of outcomes for Attacker must be addressed. Variables that increase **EU_FN** and/or variables which cause a decrease in **EU_NN** makes engaging in a disinformation campaign more attractive for the Attacker. Therefore, as outlined in the preceding formulations, the expected utilities change in favour **EU_FN** (as opposed to **NN**)—causing Attacker to choose **FN** strategy when:

- **P_T** is lower, **P_B** is higher, **P_D** is lower
- **V_{NN}** is lower, **V_{FB}** is higher, **V_{FNB}** is lower, **V_{TB}** is higher, **V_{TNB}** is lower, **PUN** is lower.

It is crucial to note that both the subjective nature of probabilities as well as the benefits of the Attacker at alternative potential outcomes may cause the Attacker to choose a strategy which brings them a lower ex-post actual utility.

Discussion

Our model posits that deterrence in international information war can be attainable through two venues: first, if the Defender can successfully signal its debunking capacity to the Attacker, and second, if the Defender can demonstrate its ability to rally the international audience to its cause quickly enough. The first condition is attainable through building and maintaining a robust fact-checking ecosystem either within the government ranks, or the civil society, but ideally both. If a Defender regularly demonstrates rapid and successful debunking performance in past interactions, the Attacker's valuation of its short-term utility by deploying disinformation will likely be reduced. Since one of the core drivers of disinformation is the Attacker's high valuation of its short-term demobilising and distracting potential against its adversaries, those adversaries can in turn demonstrate that they can debunk and disseminate accurate news rapidly in order to reduce the Attacker's perceived payoffs.

The second condition is attainable through diplomatic alliances, media and cultural power, and fewer reliance on fake news as a government strategy in past information interactions. If the Defender has regularly used disinformation as a conscious strategy in past interactions—more so compared to the Attacker—it will be harder for the IA to rally behind the Defender's cause. To that end, not resorting to disinformation is a cumulative resource that countries "save" over time, and can "cash-in" during emergencies, by rendering the IA more receptive towards its cause.

This chapter has identified two important dynamics. First, the fact that disinformation may incur reputational and suspicion costs to the Attacker suggests that there must be greater short-term gains for the Attacker so that it prefers spreading fake news. This reasoning advances the prevalent wisdom that states engage in disinformation because it is a cheap way of demobilising an adversary and there are no repercussions against this form of action. Second, Defenders are not as completely vulnerable to disinformation campaigns as the mainstream debate suggests and the Attacker's advantage in information warfare is not absolute. The Attacker chooses to engage in organised disinformation because it believes that the Defender will not be able to debunk these claims on time and at scale. If the Defender demonstrates the opposite reliably, then the Attacker's decision to launch a disinformation campaign is not a given and automatic, and may be deterred from choosing this course of action. In addition, the Defender may deter the Attacker through its influence over the IA, by not regularly engaging in disinformation campaigns itself and optimising its cultural and media power, which accumulates over time, through "good practices".

We expect a number of criticisms on our model. First, we concede that the Defender may not be isolated into a single decision to automatically debunk all disinformation attempts. It can indeed choose not to take action, or spread disinformation itself to win the short-term information war. This is indeed an important point, but empirically, such cases have been so rare that modelling them within the same parameters of likelihood with the rest of our model can be misleading. Second, we anticipate that our decision to include the International Audience as a decision-making actor may incur too much agency on it, as in most cases IA's likelihood of believing disinformation or not is very deterministic: if the Attacker is skilful and can generate sufficient emotional triggers, the IA has a likelihood of believing the disinformation. Also, pre-existing beliefs about the Attacker and the Defender matters significantly in driving the momentum of the IA. However, we believe that the IA is the actual kingmaker in this interaction and its gravitational dynamics impact the outcome of the information war. To that end, we favour retaining the agency of the IA.

Ultimately, future research can be directed towards exploring the Defender's actions in more detail and building more extensive games with IA as an actor or as a bystander to the information war. Additionally, works that deploy our model in actual empirical cases from around the world (not just isolated to Russia-China versus the US and Europe) would enrich and modify the model significantly. We hope that future debates that explore the rationality and payoff structure of disinformation proliferate and generate increased attention.

Notes

1 Samantha Bradshaw, Hannah Bailey, and Philip N. Howard, "Industrialized Disinformation: 2020 Global Inventory of Organized Social Media Manipulation," Computational Propaganda Research Project (Oxford, UK: Oxford Internet Institute, 2021), https://demtech.oii.ox.ac.uk/research/posts/industrialized-disinformation.

2 "Rival Disinformation Campaigns Targeted African Users, Facebook Says," *The Guardian*, December 15, 2020, sec. Technology, https://www.theguardian.com/technology/2020/dec/15/central-african-republic-facebook-disinformation-france-russia.

3 Sohini Chatterjee and Peter Kreko, "State-Sponsored Disinformation in Western Democracies Is the Elephant in the Room | View," *Euronews*, July 6, 2020, sec. news_news, https://www.euronews.com/2020/07/06/state-sponsored-disinformation-in-western-democracies-is-the-elephant-in-the-room-view; Jacob Silverman, "The State Department's Twitter Jihad," *POLITICO Magazine*, July 22, 2014, https://www.politico.com/magazine/story/2014/07/the-state-departments-twitter-jihad-109234.

4 "Ukrainian Refugees and Disinformation: Situation in Poland, Hungary, Slovakia and Romania," *European Digital Media Observatory* (blog), April 5, 2022, https://edmo.eu/2022/04/05/ukrainian-refugees-and-disinformation-situation-in-poland-hungary-slovakia-and-romania/.

5 Samuel Greene et al., "Mapping Fake News and Disinformation in the Western Balkans and Identifying Ways to Effectively Counter Them" (Brussels: Policy Department for External Relations Directorate General for External Policies of the Union, February 2021).

6 Davey Alba and Adam Satariano, "At Least 70 Countries Have Had Disinformation Campaigns, Study Finds," *The New York Times*, September 26, 2019, sec. Technology, https://www.nytimes.com/2019/09/26/technology/government-disinformation-cyber-troops.html.

7 Tarleton Gillespie, "Content Moderation, AI, and the Question of Scale," *Big Data & Society* 7, no. 2 (July 1, 2020): 2053951720943234, https://doi.org/10.1177/2053951720943234.

8 Scott A. Eldridge, Lucía García-Carretero, and Marcel Broersma, "Disintermediation in Social Networks: Conceptualizing Political Actors' Construction of Publics on Twitter," *Media and Communication* 7, no. 1 (March 21, 2019): 271–85, https://doi.org/10.17645/mac.v7i1.1825.

9 Matteo Cinelli et al., "The Limited Reach of Fake News on Twitter during 2019 European Elections," *PLOS ONE* 15, no. 6 (June 18, 2020): e0234689, https://doi.org/10.1371/journal.pone.0234689.

10 Don Fallis, "The Varieties of Disinformation," in *The Philosophy of Information Quality*, ed. Luciano Floridi and Phyllis Illari, Synthese Library (Cham: Springer International Publishing, 2014), 135–61, https://doi.org/10.1007/978-3-319-07121-3_8.

11 Christina la Cour, "Theorising Digital Disinformation in International Relations," *International Politics* 57, no. 4 (August 1, 2020): 704–23, https://doi.org/10.1057/s41311-020-00215-x.

12 Ernst B. Haas, "The Balance of Power: Prescription, Concept, or Propaganda?," *World Politics* 5, no. 4 (July 1953): 442–77, https://doi.org/10.2307/2009179; Gary D. Rawnsley, "Radio Diplomacy and Propaganda," in *Radio Diplomacy and Propaganda: The BBC and VOA in International Politics, 1956–64*, ed. Gary D. Rawnsley, Studies in Diplomacy (London: Palgrave Macmillan UK, 1996), 6–17, https://doi.org/10.1007/978-1-349-24499-7_2; Ben D. Mor, "The Rhetoric of Public Diplomacy and Propaganda Wars: A View from Self-Presentation Theory," *European Journal of Political Research* 46, no. 5 (2007): 661–83, https://doi.org/10.1111/j.1475-6765.2007.00707.x.

13 Yochai Benkler, Robert Faris, and Hal Roberts, *Network Propaganda: Manipulation, Disinformation, and Radicalization in American Politics* (Oxford University Press, 2018); W Lance Bennett and Steven Livingston, "The Disinformation Order: Disruptive Communication and the Decline of Democratic

Institutions," *European Journal of Communication* 33, no. 2 (April 1, 2018): 122–39, https://doi.org/ 10.1177/0267323118760317; Chris Tenove, "Protecting Democracy from Disinformation: Normative Threats and Policy Responses," *The International Journal of Press/Politics* 25, no. 3 (July 1, 2020): 517–37, https://doi.org/10.1177/1940161220918740.

14 Alexander Lanoszka, "Disinformation in International Politics," *European Journal of International Security* 4, no. 2 (June 2019): 227–48, https://doi.org/10.1017/eis.2019.6.

15 "Disarming Disinformation: Our Shared Responsibility," United States Department of State, April 7, 2022, https://www.state.gov/disarming-disinformation/.

16 "Russian Strategic Intentions: A Strategic Multilayer Assessment (SMA) White Paper," SMA White Papers (Boston, MA: National Security Innovations, May 2019), https://nsiteam.com/sma-white-paper-russian-strategic-intentions/.

17 Molly K. Mckew, "The Gerasimov Doctrine," POLITICO Magazine, October 2017, https://politi.co/2KZQlKd.

18 "UNDP: Governments Must Lead Fight against Coronavirus Misinformation and Disinformation," United Nations Development Programme, June 10, 2020, https://www.undp.org/press-releases/undp-governments-must-lead-fight-against-coronavirus-misinformation-and.

19 Akin Unver and Ahmet Kurnaz, "Securitization of Disinformation in NATO Lexicon: A Computational Text Analysis," SSRN Scholarly Paper (Rochester, NY: Social Science Research Network, February 21, 2022), https://doi.org/10.2139/ssrn.4040148.

20 Michał Pietkiewicz, "The Military Doctrine of the Russian Federation," *Polish Political Science Yearbook* 47, no. 3 (2018): 505–20.

21 Daniel Funke and Daniela Flamini, "A Guide to Anti-Misinformation Actions around the World," Poynter Institute for Media Studies, August 2020, https://www.poynter.org/ifcn/anti-misinformation-actions/.

22 Marc Owen Jones, "The Gulf Information War| Propaganda, Fake News, and Fake Trends: The Weaponization of Twitter Bots in the Gulf Crisis," *International Journal of Communication* 13, no. 0 (March 15, 2019): 27; H. Akin Unver, "Can Fake News Lead to War? What the Gulf Crisis Tells Us," War on the Rocks, June 13, 2017, https://warontherocks.com/2017/06/can-fake-news-lead-to-war-what-the-gulf-crisis-tells-us/.

23 H. Akin Unver, "Russia Has Won the Information War in Turkey," *Foreign Policy* (blog), April 21, 2019, https://foreignpolicy.com/2019/04/21/russia-has-won-the-information-war-in-turkey-rt-sputnik-putin-erdogan-disinformation/; Akin Unver and Ahmet Kurnaz, "Russian Digital Influence Operations in Turkey 2015–2020," *Project on Middle East Political Science* 43 (August 5, 2020): 83–90.

24 Idayat Hassan and Jamie Hitchen, "Nigeria's Disinformation Landscape," *Social Science Research Council* (blog), October 6, 2020, https://items.ssrc.org/disinformation-democracy-and-conflict-prevention/nigerias-disinformation-landscape/.

25 Axel Gelfert, "Fake News: A Definition," *Informal Logic* 38, no. 1 (2018): 84–117, https://doi.org/10.22329/il.v38i1.5068; Michael Jensen, "Russian Trolls and Fake News: Information or Identity Logics?," *Journal of International Affairs* 71, no. 1.5 (2018): 115–24.

26 Peter J. Phillips and Gabriela Pohl, "The Hidden Logic of Disinformation and the Prioritization of Alternatives the Era of Dis-and-Misinformation," *Seton Hall Journal of Diplomacy and International Relations* 22, no. 1 (2021): 24–34.

27 Alexander Rozanov, Julia Kharlamova, and Vladislav Shirshikov, "The Role of Fake News in Conflict Escalation: A Theoretical Overview," SSRN Scholarly Paper (Rochester, NY: Social Science Research Network, May 10, 2021), https://doi.org/10.2139/ssrn.3857007.

28 Edda Humprecht, "Where 'Fake News' Flourishes: A Comparison across Four Western Democracies," *Information, Communication & Society* 22, no. 13 (November 10, 2019): 1973–88, https://doi.org/10.1080/1369118X.2018.1474241.

29 Manuel Castells, *The Rise of the Network Society*, vol. 1, The Information Age: Economy, Society and Culture (Hoboken, NJ: John Wiley & Sons, 2011).

30 Joshua A. Tucker et al., "Social Media, Political Polarization, and Political Disinformation: A Review of the Scientific Literature," SSRN Scholarly Paper (Rochester, NY: Social Science Research Network, March 19, 2018), https://doi.org/10.2139/ssrn.3144139.

31 Nathan Walter et al., "Fact-Checking: A Meta-Analysis of What Works and for Whom," *Political Communication* 37, no. 3 (May 3, 2020): 350–75, https://doi.org/10.1080/10584609.2019.1668894.

32 James D. Fearon, "Rationalist Explanations for War," *International Organization* 49, no. 3 (1995): 379–414.

33 Rebecca Slayton, "What Is the Cyber Offense-Defense Balance?: Conceptions, Causes, and Assessment," *International Security* 41, no. 3 (2016): 72–109.

34 Soroush Vosoughi, Deb Roy, and Sinan Aral, "The Spread of True and False News Online," *Science* 359, no. 6380 (March 9, 2018): 1146–51, https://doi.org/10.1126/science.aap9559.
35 Lauren L. Saling et al., "No One Is Immune to Misinformation: An Investigation of Misinformation Sharing by Subscribers to a Fact-Checking Newsletter," *PLOS ONE* 16, no. 8 (August 10, 2021): e0255702, https://doi.org/10.1371/journal.pone.0255702.
36 David M. J. Lazer et al., "The Science of Fake News," *Science* 359, no. 6380 (March 9, 2018): 1094–96, https://doi.org/10.1126/science.aao2998.
37 Note that, as it is an "infinite interaction" model, the equation to be used for calculating discounted present value of future values simplifies to the form in equation-1.

Bibliography

Alba, Davey, and Adam Satariano. "At Least 70 Countries Have Had Disinformation Campaigns, Study Finds." *The New York Times*, September 26, 2019, sec. Technology. https://www.nytimes.com/2019/09/26/technology/government-disinformation-cyber-troops.html.

Benkler, Yochai, Robert Faris, and Hal Roberts. *Network Propaganda: Manipulation, Disinformation, and Radicalization in American Politics*. Oxford University Press, 2018.

Bennett, W Lance, and Steven Livingston. "The Disinformation Order: Disruptive Communication and the Decline of Democratic Institutions." *European Journal of Communication* 33, no. 2 (April 1, 2018): 122–139. https://doi.org/10.1177/0267323118760317.

Bradshaw, Samantha, Hannah Bailey, and Philip N. Howard. "Industrialized Disinformation: 2020 Global Inventory of Organized Social Media Manipulation." Computational Propaganda Research Project. Oxford, UK: Oxford Internet Institute, 2021. https://demtech.oii.ox.ac.uk/research/posts/industrialized-disinformation.

Castells, Manuel. *The Rise of the Network Society*. Vol. 1. The Information Age: Economy, Society and Culture. Hoboken, NJ: John Wiley & Sons, 2011.

Chatterjee, Sohini, and Peter Kreko. "State-Sponsored Disinformation in Western Democracies Is the Elephant in the Room | View." *Euronews*, July 6, 2020, sec. news_news. https://www.euronews.com/2020/07/06/state-sponsored-disinformation-in-western-democracies-is-the-elephant-in-the-room-view.

Cinelli, Matteo, Stefano Cresci, Alessandro Galeazzi, Walter Quattrociocchi, and Maurizio Tesconi. "The Limited Reach of Fake News on Twitter during 2019 European Elections." *PLOS ONE* 15, no. 6 (June 18, 2020): e0234689. https://doi.org/10.1371/journal.pone.0234689.

Cour, Christina la. "Theorising Digital Disinformation in International Relations." *International Politics* 57, no. 4 (August 1, 2020): 704–723. https://doi.org/10.1057/s41311-020-00215-x.

Dang, Hoang Linh. "Social Media, Fake News, and the COVID-19 Pandemic: Sketching the Case of Southeast Asia." *Austrian Journal of South-East Asian Studies* 14, no. 1 (June 28, 2021): 37–58. https://doi.org/10.14764/10.ASEAS-0054.

United States Department of State. "Disarming Disinformation: Our Shared Responsibility," April 7, 2022. https://www.state.gov/disarming-disinformation/.

Eldridge, Scott A., Lucía García-Carretero, and Marcel Broersma. "Disintermediation in Social Networks: Conceptualizing Political Actors' Construction of Publics on Twitter." *Media and Communication* 7, no. 1 (March 21, 2019): 271–285. https://doi.org/10.17645/mac.v7i1.1825.

Fallis, Don. "The Varieties of Disinformation." In *The Philosophy of Information Quality*, edited by Luciano Floridi and Phyllis Illari, 135–161. Synthese Library. Cham: Springer International Publishing, 2014. https://doi.org/10.1007/978-3-319-07121-3_8.

Fearon, James D. "Rationalist Explanations for War." *International Organization* 49, no. 3 (1995): 379–414.

Funke, Daniel, and Daniela Flamini. "A Guide to Anti-Misinformation Actions around the World." Poynter Institute for Media Studies, August2020. https://www.poynter.org/ifcn/anti-misinformation-actions/.

Gelfert, Axel. "Fake News: A Definition." *Informal Logic* 38, no. 1 (2018): 84–117. https://doi.org/10.22329/il.v38i1.5068.

Gillespie, Tarleton. "Content Moderation, AI, and the Question of Scale." *Big Data & Society* 7, no. 2 (July 1, 2020): 2053951720943234. https://doi.org/10.1177/2053951720943234.

Greene, Samuel, Gregory Asmolov, Adam Fagan, Ofer Fridman, and Borjan Gjuzelov. "Mapping Fake News and Disinformation in the Western Balkans and Identifying Ways to Effectively Counter Them." Brussels: Policy Department for External Relations Directorate General for External Policies of the Union, February2021.

Haas, Ernst B. "The Balance of Power: Prescription, Concept, or Propaganda?" *World Politics* 5, no. 4 (July1953): 442–477. https://doi.org/10.2307/2009179.

Hassan, Idayat, and Jamie Hitchen. "Nigeria's Disinformation Landscape." *Social Science Research Council* (blog), October 6, 2020. https://items.ssrc.org/disinformation-democracy-and-conflict-prevention/nigerias-disinformation-landscape/.

Humprecht, Edda. "Where "Fake News" Flourishes: A Comparison across Four Western Democracies." *Information, Communication & Society* 22, no. 13 (November 10, 2019): 1973–1988. https://doi.org/10.1080/1369118X.2018.1474241.

Jensen, Michael. "Russian Trolls and Fake News: Information or Identity Logics?" *Journal of International Affairs* 71, no. 1.5 (2018): 115–124.

Jones, Marc Owen. "The Gulf Information War| Propaganda, Fake News, and Fake Trends: The Weaponization of Twitter Bots in the Gulf Crisis." *International Journal of Communication* 13, no. 0 (March 15, 2019): 27.

Lanoszka, Alexander. "Disinformation in International Politics." *European Journal of International Security* 4, no. 2 (June2019): 227–248. https://doi.org/10.1017/eis.2019.6.

Lazer, David M. J., Matthew A. Baum, Yochai Benkler, Adam J. Berinsky, Kelly M. Greenhill, Filippo Menczer, Miriam J. Metzger, et al. "The Science of Fake News." *Science* 359, no. 6380 (March 9, 2018): 1094–1096. https://doi.org/10.1126/science.aao2998.

Mckew, Molly K. "The Gerasimov Doctrine." *POLITICO Magazine*, October2017. https://politi.co/2KZQlKd.

Mor, Ben D. "The Rhetoric of Public Diplomacy and Propaganda Wars: A View from Self-Presentation Theory." *European Journal of Political Research* 46, no. 5 (2007): 661–683. https://doi.org/10.1111/j.1475-6765.2007.00707.x.

Phillips, Peter J., and Gabriela Pohl. "The Hidden Logic of Disinformation and the Prioritization of Alternatives the Era of Dis-and-Misinformation." *Seton Hall Journal of Diplomacy and International Relations* 22, no. 1 (2021): 24–34.

Pietkiewicz, Michał. "The Military Doctrine of the Russian Federation." *Polish Political Science Yearbook* 47, no. 3 (2018): 505–520.

Rawnsley, Gary D. "Radio Diplomacy and Propaganda." In *Radio Diplomacy and Propaganda: The BBC and VOA in International Politics, 1956–64*, edited by Gary D. Rawnsley, 6–17. Studies in Diplomacy. London: Palgrave Macmillan UK, 1996. https://doi.org/10.1007/978-1-349-24499-7_2.

The Guardian. "Rival Disinformation Campaigns Targeted African Users, Facebook Says," December 15, 2020, sec. Technology. https://www.theguardian.com/technology/2020/dec/15/central-african-republic-facebook-disinformation-france-russia.

Rozanov, Alexander, Julia Kharlamova, and Vladislav Shirshikov. "The Role of Fake News in Conflict Escalation: A Theoretical Overview." SSRN Scholarly Paper. Rochester, NY: Social Science Research Network, May 10, 2021. https://doi.org/10.2139/ssrn.3857007.

"Russian Strategic Intentions: A Strategic Multilayer Assessment (SMA) White Paper." SMA White Papers. Boston, MA: National Security Innovations, May 2019. https://nsiteam.com/sma-white-paper-russian-strategic-intentions/.

Saling, Lauren L., Devi Mallal, Falk Scholer, Russell Skelton, and Damiano Spina. "No One Is Immune to Misinformation: An Investigation of Misinformation Sharing by Subscribers to a Fact-Checking Newsletter." *PLOS ONE* 16, no. 8 (August 10, 2021): e0255702. https://doi.org/10.1371/journal.pone.0255702.

Silverman, Jacob. "The State Department's Twitter Jihad." *POLITICO Magazine*, July 22, 2014. https://www.politico.com/magazine/story/2014/07/the-state-departments-twitter-jihad-109234.

Slayton, Rebecca. "What Is the Cyber Offense-Defense Balance?: Conceptions, Causes, and Assessment." *International Security* 41, no. 3 (2016): 72–109.

Tenove, Chris. "Protecting Democracy from Disinformation: Normative Threats and Policy Responses." *The International Journal of Press/Politics* 25, no. 3 (July 1, 2020): 517–537. https://doi.org/10.1177/1940161220918740.

Tucker, Joshua A., Andrew Guess, Pablo Barbera, Cristian Vaccari, Alexandra Siegel, Sergey Sanovich, Denis Stukal, and Brendan Nyhan. "Social Media, Political Polarization, and Political Disinformation: A Review of the Scientific Literature." SSRN Scholarly Paper. Rochester, NY: Social Science Research Network, March 19, 2018. https://doi.org/10.2139/ssrn.3144139.

European Digital Media Observatory. "Ukrainian Refugees and Disinformation: Situation in Poland, Hungary, Slovakia and Romania," April 5, 2022. https://edmo.eu/2022/04/05/ukrainian-refugees-and-disinformation-situation-in-poland-hungary-slovakia-and-romania/.

United Nations Development Programme. "UNDP: Governments Must Lead Fight against Coronavirus Misinformation and Disinformation," June 10, 2020. https://www.undp.org/press-releases/undp-governments-must-lead-fight-against-coronavirus-misinformation-and.

Unver, H. Akin, and Ahmet Kurnaz. "Russian Digital Influence Operations in Turkey 2015–2020." *Project on Middle East Political Science* 43 (August 5, 2020): 83–90.

Unver, H. Akin, and Ahmet Kurnaz. "Securitization of Disinformation in NATO Lexicon: A Computational Text Analysis." SSRN Scholarly Paper. Rochester, NY: Social Science Research Network, February 21, 2022. https://doi.org/10.2139/ssrn.4040148.

Unver, H. Akin. "Can Fake News Lead to War? What the Gulf Crisis Tells Us." War on the Rocks, June 13, 2017. https://warontherocks.com/2017/06/can-fake-news-lead-to-war-what-the-gulf-crisis-tells-us/.

Unver, H. Akin. "Russia Has Won the Information War in Turkey." *Foreign Policy* (blog), April 21, 2019. https://foreignpolicy.com/2019/04/21/russia-has-won-the-information-war-in-turkey-rt-sputnik-putin-erdogan-disinformation/.

Vosoughi, Soroush, Deb Roy, and Sinan Aral. "The Spread of True and False News Online." *Science* 359, no. 6380 (March 9, 2018): 1146–1151. https://doi.org/10.1126/science.aap9559.

Walter, Nathan, Jonathan Cohen, R. Lance Holbert, and Yasmin Morag. "Fact-Checking: A Meta-Analysis of What Works and for Whom." *Political Communication* 37, no. 3 (May 3, 2020): 350–375. https://doi.org/10.1080/10584609.2019.1668894.

15

CONSPIRACY THEORIES AND UNFOUNDED RUMOURS IN CONTEMPORARY DEMOCRACIES

BEYOND TRUTH AND TRUST

Ruxandra Buluc

1 Setting the scene for the post-truth era

Societal systems in democracies such as healthcare, education, the military, the economy, government, etc. rely on specialists' expertise and citizens' trust for their well-functioning. If truth and knowledge, the bases of expertise, and trust are subverted, then democracies fail. Conspiracy theories and unfounded rumours compromise the foundation of truth and trust, while offering nothing to replace them. Democratic societies are in danger of falling apart at the seams as constructive dialogue and debate between experts and citizens become impossible due to the fact that conspiracy theories and rumours promote distrust of powerful elites, be they epistemic (i.e., scientists, researchers, experts in various fields, etc.) or institutional (i.e., public authorities, organisations, governments, etc.).

In order to understand how truth and trust have been subverted, one needs to examine the societal developments that have led to this. Questions about knowledge and truth as objective yardsticks that measure and promote human progress and societal development were prompted by postmodern tenets which stated that social discourse should not be seen as the expression of an objective reality, but as mere subjective discursive interpretations of plural realities that reflect the proponents' interests in their quest for power. Furthermore, since any public discourse is only an individual or organisational interpretation of reality and not reality itself, then who can say which and whose interpretation is correct or what is, in fact, true. Foucault[1] explained that knowledge and power could not be separated, knowledge was constructed through discourse, and as such it reflected the subjectivity of the proponent. Consequently, knowledge itself was seen as fragmented and became an instrument for those in power to choose what aspect or view served their interests and promote it, while suppressing dissenting aspects. Thus, the very notion of an attainable, knowable truth became debatable.

The person who transmuted the debate regarding discursivity and subjectivity into the scientific realm was Lyotard[2], who contested the subjugating grand narratives which he saw

DOI: 10.4324/9781003190363-18

as totalizing, autocratic, inflexible and noninclusive systems of reality representation that could not account for the fact that reality is actually fragmentary and individualistic, not grand and unifying. Continuing in Foucault's vein, he considered that knowledge and power (be it governmental, corporate, political) were inextricably linked, and the fact that the powerful used knowledge in their own interests meant that knowledge legitimacy was linked to power legitimacy and to the discourse that power used to legitimise itself. Knowledge became a commodity and a stake in the battle for power. Furthermore, Lyotard argued that science, although it tried to represent itself as objective and detached from power struggles, was, in fact, nothing more than another grand narrative which pretended to encapsulate the truth. In Lyotard's words "knowledge and power are simply two sides of the same question: who decides what knowledge is, and who knows what needs to be decided? In the computer age, the question of knowledge is now more than ever a question of government."[3] Lyotard connected knowledge to power and thus fuelled the debates regarding the objectivity of knowledge and the extent to which it could become an instrument for those in power to wield to serve their own interests.

Therefore, in the postmodern understanding, truth and power are two sides of the same coin as truth serves those in power to consolidate their positions and promote their agendas. Truth becomes ideological, infused with personal beliefs and intuitions, and more and more detached from scientific endeavours. Going to the extreme of embracing and promoting the idea that reality is constructed exclusively socially, discursively and there is nothing beyond people's perceptions, understandings and convictions gives rise to a radical form of vox populi which denies the existence of objective facts and claims that scientific theories are not grounded in reality, but rather subjective perceptions and interpretations. This postmodern questioning has laid the foundations for the post-truth era.

The *Oxford English Dictionary* defined post-truth in 2016 as "relating to or denoting circumstances in which objective facts are less influential in shaping public opinion than appeals to emotion and personal belief." What this definition implies is that in contemporary societies the truth has become irrelevant, has lost credence and valour in the face of personal opinions and emotions. As McIntyre explains, this is

> a corruption of the process by which facts are credibly gathered and reliably used to shape one's beliefs about reality. Indeed, the rejection of this undermines the idea that some things are true irrespective of how we feel about them, and that it is in our best interests (and those of our policy makers) to attempt to find them.[4]

Post-truth represents an era in which there is no objective truth, but truths constructed by individuals to serve their interests and reflect their beliefs, each of them equally valid and worthy of recognition. So the question arises, if there is no objective truth, but only individual narratives about subjective truths, and if powerful elites construct the truth to serve their interests and enhance their power, why should people trust official narratives, why should epistemic or institutional authorities' claims be taken at face value and accepted?

Where postmodernism shattered trust in science as a promoter of knowledge and truth, post-truth attacked the foundation of the last grand narrative: democracy. Science and democracy are both based on transparency, rationality and trust. If these are contested and undermined, democratic systems lose their footing. Nichols explains how democracy and science are connected and how their mutual progress is connected:

expertise and government rely upon each other, especially in a democracy. The technological and economic progress that ensures the well-being of a population requires the division of labour, which in turn leads to the creation of professions. Professionalism encourages experts to do their best in serving their clients, to respect their own boundaries, and to demand their boundaries be respected by others, as part of an overall service to the ultimate client: society itself.[5]

Along the same lines, Nicodemo[6] states that if the true and the false cannot be distinguished, distrust in institutions and experts is proliferated and the crisis targets the rational understanding of reality.

Moreover, democracy entails freedom of expression which is predicated not solely on one's right to express one's opinion but also on one's right to have that opinion matter and on one's responsibility for the consequences of that opinion. In principle, this freedom is the bedrock of democratic systems, and it has led to tremendous success in countering discrimination, promoting equal rights and building inclusive societies. However, if taken to the extreme and neglecting the responsibility side of it, it can also turn against itself and subvert the progressive endeavours in democratic societies. Conspiracy theories and unfounded rumours are the visible and vocal manifestations of public distrust in epistemic and institutional expertise, rejections of scientific truth, all promoted in the online environment and, especially, in social media. I would argue that the pervasiveness of conspiracy theories and unfounded rumours in contemporary societies is due to a perfect storm created by truth decay, trust decay, and changes in the informational environment. The coming sections analyse these factors to determine how they have created the fertile conditions for conspiracy theories and unfounded rumours to flourish and subvert democratic principles.

2 Truth decay

Researchers Kavanagh and Rich introduce the concept of "truth decay" and define it as

> a set of four related trends: 1. increasing disagreement about facts and analytical interpretations of facts and data; 2. a blurring of the line between opinion and fact; 3. the increasing relative volume, and resulting influence, of opinion and personal experience over fact; 4. declining trust in formerly respected sources of factual information.[7]

Truth decay erodes the constructive public debate because it annihilates the common ground of a shared understanding of what reality is, which the citizens need in order to debate on the means of generating progress and well-being in their respective societies.

Truth decay is predicated on the rejection of facts that do not suit personal beliefs and interests, as well as on the rejection of expertise meant to identify, catalogue, comprehend and transmit said facts. Several factors interact to produce truth decay: counterknowledge, death of expertise, and knowledge resistance.

Counterknowledge refers to "misinformation packaged to look like fact–packaged so effectively, indeed, that the twenty-first century is facing a pandemic of credulous thinking."[8] Thompson notices the ways in which fringe ideas, once considered absurd, have moved to mainstream communication. In his view, counterknowledge claims to be knowledge but fails empirical verification, as its ideas are disproved by evidence. It "misrepresents reality (deliberately or otherwise) by presenting non-facts as facts."[9] That is the true and the false are indistinguishable and can be exchanged one for the other.

Nichols goes even further in his analysis and focuses on the effects of knowledge rejection. He states that the death of expertise is more than the mere rejection of knowledge; it is "fundamentally a rejection of science and dispassionate rationality, which are the foundations of modern civilization."[10] At present, citizens believe they have information and knowledge on everything because information is so easily available and often confused with knowledge and expertise. However, Nichols points out that people cannot function in a complex and highly specialized society as contemporary developed ones without admitting that their knowledge has limits, even though this admission undermines their sense of autonomy and self-reliance. People are less and less inclined to accept the fact that there are experts in various fields who know and understand more than laypeople. This trend of reluctance to trust others' expertise is very detrimental at a societal level, as it disrupts the normal functioning of society. Nichols asserts that this is more than mere indifference to established knowledge, it is "the emergence of a positive hostility to such knowledge."[11] People feel angry and resentful that they should accept that specialists have more expertise than they do, and these disruptive emotions drown out the rationality of expertise and nullify its progressive endeavours.

In the same vein, Strömbäck et al speak of knowledge resistance which they define as "the tendency to resist available evidence, and more specifically empirical evidence"[12]. Empirical evidence refers to how things actually are. Therefore, rejecting it presupposes a form of irrationality[13] that denies the link between the empirical evidence and a claim or conclusion. The basis for this rejection is not diverging empirical evidence, but motivated reasoning that dismisses the existing empirical evidence which contradicts or alters something that the people believe. This leads to fact polarization[14] and to the corrosion of common knowledge and truth values in social debates.

What all these concepts capture is the decline of reliance on truth and expertise in contemporary democratic societies, a rejection of knowledge in favour of personal opinions and emotional beliefs. Truth decay subverts the very essence and promise of democratic systems which is to encourage and foster progress for all individuals.

3 Trust decay

Truth decay is correlated with what I will term "trust decay". If everything is shaped by powerful political, economic, media elites, then how could average people trust anyone? How could they know who is presenting facts and who is spinning conspiracy theories? And this brings the discussion to the issue of trust and its centrality in democratic systems. Trust legitimises democratic institutions because the officials are elected by the people in the hope and trust that they will act in the citizens' best interests to ensure societal progress. However, if this trust is shattered or even simply shaken, the very legitimacy of those institutions is questioned.

Hardin[15] defines trust on the basis of encapsulated-interest theory and of belief in the moral commitment of the trusted, which means that the trusted party proves trustworthy in order to maintain the relationship in the future and does so by aligning their interests with those of the trusting person. Furthermore, Larson et al[16] explain that trust is of vital importance when there is an "implicit imbalance of power due to a high level of information asymmetry, where trusting individuals accept a vulnerable position in relation to a trusted party" and the trusting party needs to make an active choice to believe that the trusted party has their best interests at heart and also the expertise to perform well to protect those interests as expected. Möllering[17] goes even further and explains that trust presupposes a mental leap on the part of the trusting party that takes them from interpretation to expectation irrespective of the unknowable. Therefore, trust presupposes an imbalance in power and/or knowledge and an alignment of interests between the parties involved. If the interests are perceived as misaligned, then the mental leap is suspended, and trust is forfeited.

From a societal point of view, trust is always needed because people do not have direct access to all the vast amount of knowledge that regulates the well-functioning of a society. People need to trust medical doctors to prevent them from getting sick, or cure them if they do; they need to trust economists to understand and propose regulations that will keep economies afloat; they need to trust politicians that they will adopt the best public policies to garner societal progress; they need to trust teachers that they will instruct their children in the best ways possible, etc.

Moore explains that democracies rely on a delicate and complex balance between trust and distrust, which can be termed the "paradox of democracy": "we need trust in order to enable effective democratic governance, but we need to implement institutions that suggest a deep distrust of what our legislators [and other officials] will do when offered an opportunity to control the levers of power."[18] These control mechanisms are constitutional (the institutional separation of powers); popular vigilance (citizens are encouraged in democracies to keep an eye on and signal instances in which institutions overstep their competence, act illegally or malfunction); partisan distrust (engenders productive competition among political parties so that the best policies win). If these regulatory mechanisms function appropriately, initial distrust can foster the development of trust. In democratic systems, citizens are sometimes forced to choose between trusting their representatives or completely withdrawing from public participation in democratic processes, which is why, in some cases, if trust appears to be completely misplaced, citizens resent the democratic institutions and may adopt conspiracy theories that demonise them.

However, when conspiracy theories affect/target democratic institutions, Moore explains that conspiracy politics emerge which

> is both premised on and generative of distrust. In contrast to the positive dynamic in which distrust motivates scrutiny and institutional protections that in turn promote trustworthy government, there is also the possibility of a negative dynamic, in which distrust promotes further distrust.[19]

This involution is caused by two main factors: a) partisan spirals of distrust because as parties try to appeal to their supporters, they find themselves in the situation of having to accept and even promote the conspiracy theories that those supporters believe in order to maintain said support. This can easily spiral out of control and promote a politics of fear and polarization that undermines the common ground for healthy debate among competing policies; b) a self-fulfilling prophecy of distrust involving a situation in which a misleading belief creates the conditions for its own fulfilment, where the belief becomes [...] the 'father [to] the reality'. If the authorities are constantly questioned, this enhances the citizens' feeling that distrust is required and the confirmation bias will cause them to nitpick until the smallest possible instances of apparently unaccounted for data are turned into a conspiracy theory.

I would argue that trust is multilevel in a democratic society in that it can be interpersonal, institutional and epistemic, and for every level people need to have trust that those operating in that environment will have their best interests at heart. Interpersonal trust refers to the trust that an individual places in the people that surround them and that they personally know and rely on for their well-being and personal development: family, friends, acquaintances, close community. Institutional trust refers to the trust the individual has in the various types of institutions that govern the functioning of any society, and, particularly, of democratic societies which are the focus of this chapter. Epistemic trust refers to trust in knowledge and implicitly in knowledge producers, scientists, that they have the citizens' best interests, well-being and societal development and improvement at heart.

What conspiracy theories and unfounded rumours do is undermine institutional trust (powerful secret actors with nefarious interests have corrupted said institutions) and epistemic trust (scientists are not objective, they do not engage in accurate research, but respond to financial stimulants). This creates a vicious circle in which conspiracy theories and unfounded rumours weaken trust, leading to a state of uncertainty, anxiety, perceived lack of control which enhance the need to find meaning and explanations that, in turn, are readily available and easily understandable in conspiracy theories and unfounded rumours.

It is important to notice that interpersonal trust remains intact. Its foundation is mainly emotional; therefore contestations of facts, scientific truth and explanations do not weaken it. As Kavanagh & Rich expound, "social relationships and networks play a large role in the formation of beliefs and attitudes"[20], however, they severely limit the diversity of information that one comes in contact with and reinforce echo chambers in which information is never externally verified and confirmation of even the most outrageous belief is readily available. In search of personalised content, people have personalised knowledge and facts, and while people are entitled to their own opinions, they are not entitled to their own facts. Unfounded rumours and conspiracy theories circulate freely in close(d) communities, in which people share them with their peers, who accept them on the basis of interpersonal trust.

This brings to the fore the discussion about the shift in communication that also facilitates the spread of conspiracy theories and unfounded rumours. Filter bubbles[21], selection algorithms[22], and echo chambers[23] have customised the content that users access based on their beliefs, and thus have personalised the information and knowledge they gain, as well as the interactions they engage in. This trend leads to accentuated truth decay since facts and opinions are not differentiated and the content is actually segregated to be in tune with the groups' pre-existing beliefs which are thus reinforced.[24] Consequently, ruptures and polarisation increase in societies, as people insulate themselves in communities with no contact with opposing or diverging views, where debate does not exist, only a spiral of confirmation and "tribal" belief reinforcement.[25] Nichols presents the results of a study performed at University College of London which revealed the fact that despite having more available sources of information than ever before, students limited their reading to the very first lines of an article and then moved to the next. He explains that this is not actually reading, but scrolling in search of confirmatory details for a pre-existing belief, and marks the unwillingness to engage in the attempt to follow and understand contradictory articles. Ultimately, this disengagement is detrimental to democracies because it "undermines the role of knowledge and expertise in a modern society and corrodes the basic ability of people to get along with each other in a democracy."[26] The factual common ground so necessary for informed democratic debates in the public sphere with respect to how and what societies should do is fractured, due to lack of adherence to common facts and consensual truth, and to lack of constructive debates. If one compounds truth decay and trust decay, one has the perfect fertile ground for conspiracy theories and unfounded rumours to disperse in society. The following section examines what unfounded rumours and conspiracy theories are and how they have evolved into democracy-corroding agents.

4 Unfounded rumours and conspiracy theories

Both rumours and conspiracy theories have been around for millennia. Many times, the two concepts have appeared in the same contexts, and they may seem to bear many resemblances. However, it is important to take note of the fundamental differences between them in order to better ascertain their impact on democratic societies and to understand how they can enhance each other's effects.

a. Rumours

One of the first definitions of rumours is offered by Allport and Postman as "propositions of faith on specific (or current) topics that pass from person to person, usually by word of mouth, without any evidence of their truth"[27]. They also explain that rumour transmission is fuelled by two conditions: importance and ambiguity, which relate in a multiplicative fashion. As far as motivation is concerned, Allport & Postman explain that rumours may serve to alleviate emotional urges as well as to offer intellectual rationalisations. That is they provide meaning, understanding and relief both to emotions and to cognitive dissonance.

Sunstein also proposes a definition of rumour which he considers to be

> claims of facts about people, groups, events, and institutions that have not been shown to be true, but that move from one person to another, and hence have credibility not because direct evidence is available to support them, but because other people seem to believe them.[28]

Rumours become widespread among people because they reinforce or are compatible with those persons' pre-existing beliefs and self-interests.

DiFonzo and Bordia define rumour in social and organizational contexts as "unverified and instrumentally relevant information statements in circulation that arise in contexts of ambiguity, danger or potential threat, and that function to help people make sense and manage risk"[29]. They focus on the ambiguity-alleviation, meaning-making and risk-management functions of rumours. Their assertion is that rumours appear when a group is trying to make sense of an ambiguous, uncertain, confusing and possibly dangerous situation, and constitute a collective effort on a group's part to come up with explanations, and thus to exert at least secondary control over a situation, in which they cannot exert primary control, that is act themselves to change it in a favourable manner to serve their interests.[30] Rumours have four characteristics: a) they are information statements, as an idea in conveyed in a meaningful statement; b) they are communicated information, shared among persons that are part of a group; c) they comprise of information that is "instrumentally relevant" to the persons involved in rumour transmission, that is the information relates to, affects or even threatens those persons in some way; d) they are important and novel information that is not verified.[31] Where Allport and Postman identified two drivers for rumour transmission, DiFonzo and Bordia name five: uncertainty, importance or outcome relevant involvement, lack of control, anxiety, and belief.[32] They also focus on the social motivations of rumour transmission: fact-finding, relationship enhancement, and self-enhancement. Ali[33] further explains that fact-finding refers to the people's need for epistemic understanding of their environment and the uncertain and potentially dangerous situations they find themselves in, while interpersonal and intrapersonal motivations focus on the need to maintain and strengthen relationships in the group and to protect and promote one's positive image as persons who hold some essential and precious information that others are not privy to.

Overall, a rumours' function is to help people make sense of an ambiguous situation, on the basis of little or not at all verified information, in order to allow them to adapt to potentially threatening or dangerous situations. Therefore, rumours are collective mean-ing-making activities, and they rely on the interaction among the group members in order to produce comprehension. However, several problems have been noticed in research regarding the efficiency and reliability of group decision-making. Sunstein and Hastie[34] categorise the problems with groupthink: a) they often amplify, rather than correct, individual errors in judgment; b) they fall victim to cascade effects, as members follow what others say or do; c) they become polarized, adopting more extreme positions than the ones they began with; d) they emphasize what everybody knows instead of focusing on critical information that only a few people know.

DiFonzo and Bordia, Ali and Giry[35] also explore the relationship between trust or distrust in institutions and rumour transmission. DiFonzo and Bordia[36] posit that trust is a key element in rumour transmission; if the organization is not trusted, then rumours spread more easily, and distrust can increase uncertainty, anxiety and, consequently, rumour transmission. Ali emphasises that rumours are created and shared when people feel powerless in face of dominating institu-tions and need to negotiate their own understanding of a certain phenomenon. Giry explains that "rumours reveal trust in the people while they express, as counter power, defiance against official institutions."[37]

If people do not listen to anybody outside their group who might have information disconfirming the rumour, if they do not trust anybody who does not believe the same things they do, then any rumour will turn into a belief. The battle for trust cannot be won if nobody listens to anybody other than their own in-group peers. And social media, with its filter algorithms, ensure that people are exposed to rumours that confirm their already held beliefs and censor information that may challenge those beliefs. Rumours have moved from person-to-person transmission towards online spread and the rumours that people believe tend to come from sources that they have something in common with: if they are persons, then they are from the same group (colleagues, family, friends); if they are sites, then they are sites they have already decided they trust.

Rumours flourish in crisis situations, when little is known for certain. If in these situations people do not trust institutional or epistemic authorities then they will reject any official explanations if they contradict what their close(d) group believes. In these cases, people will turn to their circle of acquaintances, where trust resides, to gain an understanding of what is happening. If rumours persist and are not assuaged, they can become endemic and turn into conspiracy theories.

b. Conspiracy theories

Conspiracy theories have always existed in societies, however, at present, they have gained momentum due to their easy spread and appeal in social media. Moreover, they have begun to corrupt people's understanding of the world and their willingness to listen to experts and authorities in times of crisis and not only, thus threatening not only the further development of societies but also the very health and security of the communities they live in.

One of the most challenging aspects in tackling and countering conspiracy theories is the fact that they are conspicuously hard to define. The reason for this is that conspiracies exist in the real world and have been proven time and again, from ancient times (the conspiracy to kill Caesar) to modern days (Watergate). As Uscinski explains, a conspiracy can be defined as: "a secret arrangement between two or more actors to usurp political or economic power, violate established rights, hoard vital secrets, or unlawfully alter government institutions to benefit themselves at the expense of the common good"[38] and it refers to events that proper authorities have determined that have taken place. The key words in this definition are the proper authorities, meaning those that have expertise and instruments at their disposal, that are tasked to investigate and determine how and why certain events have occurred. However, given the increasing distrust in expert knowledge as well as in authorities, the public is less and less willing to accept such official explanations and more and more prone to look for and entertain alternative explanations, that contradict the official expert reports and that promote conspiracy theories.

It is challenging to draw a line between what can be considered a real conspiracy (the activity of secretly planning with other people to do something bad or illegal that could be harmful to other people—definition according to Cambridge dictionary) and the imagined conspiracies that purport to have detected such secret nefarious plots, where the official explanations and narratives prove that none such infamous designs existed, or if they have existed, than the perpetrators have been exposed.

One of the most common definitions in the literature is the one proposed by Uscinski[39] who views conspiracy theory as referring "to an explanation of past, ongoing, or future events or circumstances that cites as a main causal factor a small group of powerful persons, the conspirators, acting in secret for their own benefit and against the common good." Keeley terms conspiracy theories as unwarranted and explains that they propose an "explanation of some historical event (or events) in terms of the significant causal agency of a relatively small group of persons—the conspirators—acting in secret."[40] Definitions along the same lines are proposed by van Prooijen & van Lange who state that "belief in conspiracy theories as explanatory beliefs about a number of actors who join together in secret agreement, and try to achieve a hidden goal which is perceived as unlawful or malevolent"[41]; and Douglas & Sutton who define conspiracy theories as "attempts to explain the ultimate causes of significant social and political events as secret plots by powerful and malicious groups."[42] These conspirators could be part of the government, NGOs, scientists, religious organizations, etc. and wield enough power to set events in motion. Their main goal is attaining and maintaining power by any means necessary, even if it entails sacrificing the lives of the people.

Conspiracy theories put forth an "accusatory perception which may or may not be true, and usually conflicts with the appropriate authorities."[43] Brotherton further explains that conspiracy theories "are easy ways of telling complicated stories", which ignore the complexity of world events and their plurality of causes, dismiss the chance factor, and thus "streamline reality" by transforming all events into clearly defined strings of causes and effects in which the evil perpetrator is clearly indicated, by dismissing randomness in an attempt to find a deeper meaning. "Conspiracy Theories render the inexplicable explicable, the complex comprehensible. They pave over messy, bewildering, ambiguous reality with a simple explanation: They did it."[44] The basic notion is that people in power are hiding the truth from the public with a view to attaining their own sinister goals to the detriment of the general public. Moreover, the appeal of conspiracy theories is given by the fact that they are "morality tales with all-knowing and all-powerful villains and naive victims who have no idea what is really going on until the truth is revealed by the Conspiracy Theorist."[45] They embody the classic underdog story in which random events are invested with a deeper significance, a story which reiterates and reinforces David's archetypal struggle against and defeat of Goliath.

One other important aspect of conspiracy theories is that they have ideological underpinnings[46] and this distinguishes them from rumours. While rumours are about understanding, conspiracy theories are about exerting power. They promote and advance political objectives via seductive and attractive explanations for major events which are meant to sway the public in a certain direction. They promote a political ideology and help achieve political objectives. Oliver & Wood[47] support the idea and further notice that people's pre-existing political beliefs influence and even dictate the conspiracy theories they support, and these theories in turn further influence their political allegiances.

Therefore, conspiracy theories can be defined as explanatory beliefs about significant events or crises, plotted by a group of powerful secret actors, whose goal is to achieve their own nefarious ends, with no regard for the interests of ordinary people who are exploited and kept in the dark. But what can be noticed in the current post-truth, post-trust environment is that conspiracy theories are (d)evolving.

Muirhead and Rosenblum[48] identify and define a new evolution in conspiracy theories: new conspiracism. In essence this trend refers to conspiracies without the theory, meaning without any attempts to prove that the assertions may be accurate and founded. The goal of new conspiracism is to delegitimise democracy. Classic conspiracism attempts to make sense of randomly occurring events by appointing agency to powerful and malevolent people and painstakingly collecting evidence to try and create patterns to explain what happened. New conspiracism no longer strives to uncover proof but rather replaces it with a different set of tools: accusations, innuendo, bare assertions, repetition. The catchphrases that synthesise the essence of new conspiracism are: a lot of people are saying it and it is true enough. New conspiracism is extremely dangerous because it is a virulent and direct attack on "shared modes of understanding and explaining things in the political world. It unsettles the ground on which we argue, negotiate, compromise, and even disagree. It makes democracy unworkable—and ultimately it makes democracy seem unworthy."[49] New conspiracism operates in a more decentralised manner since the point is not to produce an apparently coherent alternative explanation for official accounts, but delegitimise. It delegitimises healthy political opposition and knowledge-producing institutions (including the free press, the academia, the research communities) and thus corrodes the foundations of democracy.

In Muirhead and Rosenblum's view, new conspiracism does double damage: delegitimation and disorientation.[50] Delegitimation occurs by rejecting official explanations, knowledge, value and authority of democratic institutions, practices, processes, representatives, which makes them seem meaningless, worthless, and undesirable. These new conspiracism attacks may appear outrageous, absurd, confusing, unfounded, but they are agitating, relentless, constantly unexpected, requiring the attention and energy of institutional and epistemic authorities to debunk them. This continuous assault is cognitively and emotionally exhausting, not only for those who try to counter them, but also for those who witness these incessant attacks. This exhaustion leads to disorientation as people no longer know what and who to believe. However, they find themselves engaged in the spectacle and relinquish requirements for accuracy and factuality in exchange for excitement. As historian Timothy Snyder cautions, the danger is that "discussion shifts from the public and the known to the secret and the unknown. Rather than trying to make sense of what is around us, we hunger for the next revelation."[51] This emotionally driven need for the novel to assuage people's confusion with respect to what is going on makes them more prone to accept conspiracies. But one must bear in mind that conspiracy theories are about maintaining control[52] as well, as audiences become disengaged from democratic processes and debates if they are made to feel entirely and repeatedly powerless, faced with the never-ending plots of nefarious elites whose only goal is to cause them harm. If trust in authorities, be they institutional or epistemic, is already very low, and truth is discounted, then rumours or conspiracy theories will only go further to entrenching these beliefs and delegitimizing democracy.

Conclusion

We cannot underestimate the grave effects that unchecked conspiracy theories and unfounded rumours have on democratic systems. Their proponents clamour that they have the right to an opinion, and that their opinion should be heard and count for something at a societal level. Adding to this the speed of dissemination enhanced by the online environment and by social media in particular, and the result is a perfect storm in which conspiracy theories and unfounded rumours disrupt public debate, creating a false perception of an opposing side to the official one, drowning out pertinent and legitimate voices and subverting the public trust in established facts and even in the validity of an open, constructive debate. Without such dialogue, democratic societies, at least, are thwarted in their development by the polarization of the citizens who find themselves unable or unwilling to interact with others who have diverging opinions. If there is no common denominator of understanding and no common reference points, then debate becomes impossible, and arguments deteriorate into quarrels. All aspects of societal knowledge and function can be affected by unfounded rumours and conspiracy theories: science is altered when people believe that scientists are actually corrupted representatives of big corporations, the democratic processes suffer when people exercise their voting rights based on conspiracy theories and not facts and data, society is harmed when policies are enacted not based on knowledge but on conspirational beliefs, international relations suffer when disinformation outweighs facts and real events. Further research is required into the best approaches and practices to restore truth and rebuild trust in democratic societies. This research could focus along two main lines of development: communicative and educational. Firstly, trust cannot be built on a lack of understanding of what epistemic and institutional authorities are doing, of their responsibilities and areas of competence. Therefore, transparent, engaged, evidence-based communication could assist in building public comprehension as it could inform and educate the citizens simultaneously, and thus, in time, restore the public trust in these authorities. Furthermore, epistemic authorities (researchers, scientists, experts, etc.) could engage in more public debates, could communicate more clearly regarding the benefits of their work to the wider public so that their discoveries become widely known and accepted. Secondly, educational programs should focus on providing integrated, common knowledge with respect to scientific discoveries and progress so that young people would feel more in control, more capable of following the evolution of the world they live in, and less likely to accept the facile and erroneous explanations provided by unfounded rumours and conspiracy theories. Moreover, educational programs could also focus on the functioning of democratic systems, on the roles, competencies and limitations of the different institutions so that young people can understand how these systems work and how important citizens' engagement and participation are for the development of democracies, as well as how detrimental their disengagement could be.

Notes

1 Foucault, *The Order*.
2 Lyotard, *The Postmodern*.
3 Idem, 8–9.
4 McIntyre, *Post-Truth*, 20.
5 Nichols, *The Death*.
6 Nicodemo, *Disinformazia*.
7 Kavanagh & Rich, *Truth Decay*, x–xi.
8 Thompson, *Counterknowledge*, 8.
9 Idem, 9.

10 Nichols, *The Death*, xii.
11 Idem.
12 Strömbäck et al, "Introduction", 1
13 Glüer & Wikforss, "What is", 30.
14 Idem, 43.
15 Hardin, "Conceptions".
16 Larson et al, "Measuring".
17 Möllering, "The Nature".
18 Moore, "On the Democratic", 113.
19 Idem, 116.
20 Kavanagh & Rich, *Truth Decay*, 88.
21 Pariser, *The Filter*.
22 Oremus, "Who Controls"; Miller, "From policing".
23 Sunstein, *Echo*.
24 Kavanagh & Rich, *Truth Decay*; Nicodemo, *Disinformazia*.
25 Kavanagh & Rich, *Truth Decay*, Tobias Rose-Stockwell "How We", McIntyre, *Post-Truth*.
26 Nichols, *The Death*.
27 Allport & Postman, *The Psychology*; Allport & Postman, "An Analysis".
28 Sunstein, *On rumors*.
29 DiFonzo & Bordia, *Rumor Psychology*, 73.
30 Idem and Ali, "Impacts of"
31 DiFonzo and Bordia, *Rumor Psychology*, 16–17.
32 Idem, 69.
33 Ali, "Impacts of".
34 Sunstein & Hastie, *Wiser*.
35 DiFonzo and Bordia, *Rumor Psychology*; Ali, "Impacts of"; Giry, "A Specific".
36 Idem, 203.
37 Giry, "A Specific".
38 Uscinski, *Conspiracy*, 48.
39 Idem.
40 Keeley, "Of Conspiracy", 116.
41 van Prooijen & van Lange, *Power*.
42 Douglas & Sutton, "Why Conspiracy".
43 Uscinski, *Conspiracy*, 48; Levy "Radically socialized"; Keeley, "Of Conspiracy".
44 Brotherton, *Suspicious Minds*.
45 Cassam, *Conspiracy*.
46 Idem and Oliver & Wood, "Conspiracy".
47 Idem.
48 Muirhead & Rosenblum, *A Lot of*,
49 Idem, 6–7.
50 Idem.
51 Snyder, *The Road*, 37.
52 Pomerantsev, *This Is Not*, 68.

Bibliography

Ali, Inayat. "Impacts of rumors and conspiracy theories surrounding COVID-19 on preparedness programs". *Disaster Medicine and Public Health Preparedness*, 2022, 16.1: 310–315.

Allport, Gordon W., and Leo Postman. *The Psychology of Rumor*. New York: Henry Holt, 1947.

Allport, Gordon W., and Leo Postman. "An analysis of rumor". *Public Opinion Quarterly*, 1946, 10(4), 501–517.

Brotherton, Rob. *Suspicious minds: Why we believe conspiracy theories*. Bloomsbury Publishing, 2015.

Cassam, Quassim. *Conspiracy Theories*. Polity Press, 2021.

DiFonzo, Nicholas, and Prashant Bordia. Rumor psychology: Social and organizational approaches. *American Psychological Association*, 2007.

Douglas, Karen M.; Sutton, Robbie M. "Why conspiracy theories matter: A social psychological analysis." *European Review of Social Psychology*, 2018, 29.1: 256–298.

Foucault, Michel. *The Order of Things. An Archaeology of the Human Sciences.* New York: Vintage House, division of Random House Inc., 1994.

Giry, Julien. "A specific social function of rumors and conspiracy theories: strengthening community's ties in trouble times. A multilevel analysis." in *Ethnology in the 3rd Millennium: Topics, Methods, Challenges,* 2016.

Glüer, Kathrin & Åsa Wikforss. "What is knowledge resistance" (29–48) *Knowledge Resistance in High-Choice Information Environments.* Strömbäck, J., Wikforss, Å., Glüer, K., Lindholm, T., & Oscarsson, H. (eds.). New York & London: Routledge, 2022.

Hardin, Russell. "Conceptions and Explanations of Trust" (3–39) in Cook, Karen, ed. *Trust in society.* New York: Russell Sage Foundation, 2001.

Kavanagh, Jennifer & Rich, Michael D. *Truth Decay. RAND report.* Santa Monica, California: the RAND Corporation, 2018.

Keeley, Brian. L. "Of Conspiracy Theories", *The Journal of Philosophy,* 1999, Vol. 96, No. 3, pp. 109–126.

Larson, H. J., Clarke, R. M., Jarrett, C., Eckersberger, E., Levine, Z., Schulz, W. S., & Paterson, "Measuring trust in vaccination: A systematic review". *Human Vaccines & Immunotherapeutics,* 2018, 14.7: 1599–1609.

Levy, Neil. "Radically socialized knowledge and conspiracy theories". *Episteme,* 2007, 4.2: 181–192.

Lyotard, Francois. *The Postmodern Condition: A Report on Knowledge.* Minneapolis: University of Minnesota Press, 1979.

McIntyre, Lee. *Post-truth.* Cambridge MA: MIT Press, 2018.

Miller, Carl. From policing to news, how algorithms are changing our lives, 2016, available at https://www.thenational.ae/arts-culture/from-policing-to-news-how-algorithms-are-changing-our-lives-1.177049.

Moore, Alfred. "On the democratic problem of conspiracy theories" (111–134) in *Conspiracy Theories and the People Who Believe Them,* ed. Joseph E. Uscinski. Oxford University Press, 2018.

Möllering, Guido. The nature of trust: From Georg Simmel to a theory of expectation, interpretation and suspension. *Sociology,* 2001, 35. 2: 403–420.

Muirhead, Russell & Nancy L. Rosenblum. *A Lot of People Are Saying: The New Conspiracism and the Assault on Democracy.* Princeton and Oxford: Princeton University Press, 2019.

Nichols, Tom. *The death of expertise: The campaign against established knowledge and why it matters.* Oxford University Press, 2017.

Nicodemo, Francesco. *Disinformazia. La comunicazione al tempo dei social media.* Venezia: Marsilio Editori, 2017.

Pariser, Eli. *The filter bubble: How the new personalized web is changing what we read and how we think.* Penguin, 2011.

Oliver, J. Eric; Wood, Thomas J. "Conspiracy theories and the paranoid style (s) of mass opinion." *American Journal of Political Science,* 2014, 58. 4: 952–966.

Oremus, Will. Who Controls Your Facebook Feed. A small team of engineers in Menlo Park. A panel of anonymous power users around the world. And, increasingly, you. 2016, available at http://www.slate.com/articles/technology/cover_story/2016/01/how_facebook_s_news_feed_algorithm_works.html?via=gdpr-consent.

Pomerantsev, Peter. *This is not propaganda: Adventures in the war against reality.* PublicAffairs, 2019.

Rose-Stockwell, Tobias. How We Broke Democracy. Our technology has changed this election, and is now undermining our ability to empathize with each other, 2016, available at https://medium.com/@tobiasrose/empathy-to-democracy-b7f04ab57eee

Snyder, Timothy. *The Road to Unfreedom: Russia, Europe, America.* Tim Duggan Books, 2018.

Strömbäck, Jesper, Åsa Wikforss, Kathrin Glüer, Torun Lindholm and Henrik Oscarsson. "Introduction. Toward understanding Knowledge Resistance in High-Choice Information Environments" (1–28) in *Knowledge Resistance in High-Choice Information Environments.* Strömbäck, J., Wikforss, Å., Glüer, K., Lindholm, T., & Oscarsson, H. (eds.). New York & London: Routledge, 2022.

Sunstein, Cass. "The daily we: Is the internet really a blessing for democracy". *Boston Review,* 2001, 26.3: 4–9.

Sunstein, Cass R. *On rumors: How falsehoods spread, why we believe them, and what can be done.* Princeton University Press, 2014.

Sunstein, Cass R., and Reid Hastie. *Wiser: Getting beyond groupthink to make groups smarter.* Harvard Business Press, 2015.

Thompson, Damian. *Counterknowledge: How we surrendered to conspiracy theories, quack medicine, bogus science and fake history.* New York, London: W. W. Norton & Company, 2008.

Uscinski, Joseph E. *Conspiracy theories and the people who believe them.* Oxford University Press, USA, 2018.

Van Prooijen, Jan-Willem, and Paul AM Van Lange. *Power, politics, and paranoia: Why people are suspicious of their leaders.* Cambridge University Press, 2014.

16

COGNITIVE WARFARE

UNDERSTANDING THE THREAT

Alina Bârgăoanu and Flavia Durach

1 Introduction: no good historical precedent for a period such as ours

Even before the COVID-19 pandemic and before the war in Ukraine, there was no good historical precedent to account for the rapid developments taking place in the information and communication ecosystem and going beyond that information ecosystem itself, which are social, cultural, demographic, technological, economic, political, geopolitical at the same time. Among such changes, the most prominent are: challenges to multilateralism and to the institutions put in place after World War II; great power competition (including its corresponding phenomena, loss of monopoly over the information space from Western actors, competitive story-telling and narratives' arms race); East-West divide in the trans-Atlantic world and EU internal competition especially in the Brexit aftermath; a generalized crisis of (political) representation and a generalized feeling of discontent with "the establishment"; the decline of trust in traditional institutions (families, parties, mainstream mass media) and in expert systems (including science and academia) across the Western world.

Adding to these, both the COVID-19 pandemic and the war in Ukraine have exposed trends in the information and communication ecosystem that entitles us to premise our work on the idea that we are dealing with new phenomena, for which historical precedents provide little guidance. These changes, which are intrinsically technological in nature, have blurred the distinctions between external and internal boundaries, foreign policy and internal politics, war and peace, between private and public, reality and fiction, online and offline environments, between persuasion, propaganda, information and entertainment. Such blurring distinctions, and many others, have brought about new, challenging terms, such as "unpeace" (Leonard, 2021), "intermestic operations" (both domestic and international) (Vériter, 2021), "computerized reality" (Bjola, 2017), or ampliganda (DiResta, 2021).

Out of these trends, the main one is the domination of information and communication ecosystem by global digital platforms. It is true that, as EDMO (2022a) report on Disinformation and the war in Ukraine acknowledges that right now, disinformation is "not only a platform problem", but spreads via social media, mainstream media or word of mouth via political, cultural and entertainment content and via the speech of high-profile politicians and influencers. At the same time, according to our understanding, it is the structural features of the trans-national

DOI: 10.4324/9781003190363-19

platform-driven ecosystem that has created the condition for the cognitive warfare waged against facts and reality both in the context of the COVID-19 pandemic and of the War in Ukraine.

Digital platform technologies allow for a global system of production, distribution and marketing of unreliable information, involving broad geographical spaces and a transnational network of sources, audiences, and communication vectors, even when "the targets for misinformation are citizens of one country" in particular (Howard, 2020, xiii). Besides, that new platform-, data-, algorithm- driven ecosystem is optimized for engagement, attention grabbing, polarization and migration of fringe topics to the mainstream. It creates transnationally fragmented, polarized, yet homogenous publics (that is, polarized publics inside one country, which are homogenous at a trans-national level). The emergence of these two strikingly homogeneous blocs—"us vs. them", two "mainstreams" that are tightly polarized along facts, identities and worldviews in prominent liberal democracies represents a new threat to the integrity of public discourse and public spaces in these societies and the breeding ground for waging cognitive warfare, for attacking a society's ability to operate with a shared epistemology.

2 "Technology is the message"

There is a strong intellectual tradition acknowledging the role of prevalent technologies and means of communication in shaping societies and their public sphere (Harold Innis, Marshall McLuhan, Neil Postman, Niall Ferguson). While the founder of this tradition, Harold Innis, puts forth a rather normatively neutral reading of the succession of civilizations and empires based on the characteristics of their prevalent media (1950/2007; 1951/2008), Neil Postman takes a harsh normative approach towards the way in which society and culture surrender to technology (Postman, 1993). For McLuhan (1962/2011; 1964/1994; 1967/2001), technologies are extensions of the human body and the evolution of communication and, through it, human history, is marked by two major technological revolutions: the invention of the printing press, and of electricity. Building on this, Niall Ferguson places the emergence of social media/ digital technologies next to the invention of the printing press and considers that our era is that of the "Second Networked Age", with the computer in the role of the printing press (Ferguson, 2018).

For the purposes of this chapter, what we retain from the extensive work done by the authors belonging to this line of thought is the importance of technology and of the prevalent communication infrastructure in shaping public discourse and public life, what P. N. Howard aptly synthetizes in his recent book as

> Any framework for understanding politics that is simply about elections, political parties, and government, and or that assigns technology a minor role in explaining current affairs, won't come with compelling explanations for political outcomes … Any sensible definition of democracy—or authoritarianism—must include elements of its information infrastructure (Howard, 2020, x).

By placing the discussion on the contemporary communication and information phenomena in this intellectual tradition that gives technology central stage in shaping public discourse and public life, we argue against the idea that what we are dealing with today is a mere upgrade of old propaganda or disinformation by technological means, a mere coupling between some old techniques and new means of communication. Instead, we argue that contemporary disinformation phenomena (disorder, chaos, warfare, cognitive and emotional polarization) are technology-driven, are co-substantial with social media/ digital technologies and take place in a new information, communication and persuasion ecosystem that is platform, algorithm- and big-data driven.

Beside this, by embracing this intellectual tradition, we go beyond the idea that connectivity and digital, Internet-based technologies are double-edged swords, whose good or bad effects are "in the way that you use it"; rather, they are an inherently destabilizing for the current world order, giving birth to "connectivity conflicts", which are harder to recognize, to describe and to withstand. Consequently, our understanding of the spirit of the times needs to change: "We are having to get used to an unstable, crisis-prone world of perpetual competition and endless attacks between competing powers. Welcome to the age of unpeace" (Leonard, 2021, 8).

The COVID-19 pandemic and the war in Ukraine have shown that everything can be and has been weaponized, from tangible things such as masks, medical gear and treatment, vaccine shots, refugee flows, food and energy prices to more intangible ones such as scientific inquiry, socio-economic identities, critical thinking, critical inquiry and doubt. But, above all, what is weaponized is the very connectivity between countries, people and public spaces: "In geopolitics, it is trade, finance, the movement of people, pandemics, climate change and above all the Internet that are being weaponized [...] It is connectivity itself that gives people the opportunity to fight, the reasons to compete, and the arsenal to deploy" (Leonard, 2021, 3).

3 Cognitive warfare—hijacking public discourse altogether

Cognitive warfare, encompassing various forms of false, misleading or manipulative content, lies, hoaxes, conspiracy theories, has been around for some time, accompanying war efforts, social disruptions, political turmoil or economic upheaval. However, contemporary efforts towards turning the human mind into an international battlefield are structurally different from similar efforts deployed only 50 or 70 years ago, for example.

We conceptualize cognitive warfare as a new, technology-driven phenomenon, based on at least three lines of reasoning, two having to do with its inherent features and one with its effect. First, it has the objectives to disrupt public conversations, confuse public opinion, cultivate doubt and suspicion rather than to convince or to persuade. Second, it interferes both with content and with its circulation. And finally, its effect is to create polarization of a specific kind, which is not political, policy-or subject-matter driven, but it is cognitive and emotional, all-or-nothing, irrespective of specific topics.

Let us go deeper into each of these three.

Differently from previous efforts targeting human cognition, attitude and opinion-building, which took place under circumstances of information scarcity, cognitive warfare takes places under circumstances of information over-abundance (Bawden & Robinson, 2020). The explosion of digital technologies allows for the massive instantaneous dissemination of multiple contradictory narratives, with little to no consideration for possible contradictions, in order to create information overload/ information fatigue (Hoyle et al, 2022; Lanoszka, 2019, 17; de Liedekerke & Zinkanell, 2020; Daniel & Eberle, 2021).

According to EDMO's report (EDMO, 2022a),

> Russia's war in Ukraine has exposed a variety of shifting and evolving disinformation tactics, while at the same time indicating that the stable goals of disinformation, irrespective of these tactics and irrespective of the subject matter, are to sow division,

create confusion, alter the terms of public conversations in liberal democracies and ultimately hijack them altogether.

Its major goal is to attack a society's capacity to act on a shared epistemology, on a shared reality, to create an emotional environment where everybody can feel entitled to their own facts:

> We lose our ability to distinguish the real from the unreal, the actual from the imagined, or the threat from the conspiracy. The powers working to disrupt democracies through memetic [cognitive] warfare understand this well. Contrary to popular accounts, they invest in propaganda from all sides of the political spectrum. The particular narratives they propagate through social media are less important than the immune reactions they hope to provoke.
>
> *(Rushkoff, 2021, 38).*

Contemporary cognitive warfare is dwelling in provoking reactions and in the viralisation of reactions, no matter what the nature of these reactions are (rejection, approval, outrage, excitement, confusion, panic): "Social media provide a megaphone to both/ all sides of the debate" (Corbu et al., 2021); they dilute, amplify and move them to the extremes up to the point where mainstreaming is no longer possible, and the center of public conversation can no longer hold.

It is thus important to realize that the main goal of contemporary disinformation, propaganda, influence and hostile information endeavors is not necessarily to persuade the targeted publics of something, but to strategically shape public discourse, create confusion, sow distrust and create an emotional environment where people no longer care about facts, they disregard them, what J. Kavanagh and M. D. Rich (2018) call "truth decay" (Kavanagh & Rich, 2018). In practical terms, when trying to document the effects of cognitive warfare, looking for persuasive effects (e.g. how many people endorse Kremlin narratives) can be misplaced in comparison to looking for signals of information fatigue, low trust levels, generalized suspicion and confusion and prevalence of inflammatory, fringe or extremist content.

Second, cognitive warfare relies both on new tools for interfering with content (increased possibilities for content creation and for text, photo and video manipulation and integration) and on tools for interfering with content circulation, targeting and amplification (increased possibilities to amplify content by means of automated accounts, trolls, bots, fake profiles, fake crowds, click factories and by means of micro-targeting). Understanding the new phenomenon of interfering both with content (along the true—false continuum) and with its circulation, targeting and amplification (interfering with the "social" component of social media, hacking the algorithms and recommendation engines to go make the content go viral) is essential for documenting it and designing efforts to tackle it (Bârgăoanu & Radu, 2018). Thus, cognitive warfare no longer operates along a true/ false binary classification, but along a wide spectrum of content types: non-factual, but plausible; non-factual, but realistic; non-factual, but political/ cultural/ identity-based; factual, but misleading (facts that are framed in a misleading way, or taken out of context, or cherry-picked); factual, but hyper-partisan, sensationalized (clickbait). At the same time, by taking advantage of the technological affordances of digital platforms, can be waged with content whose factual character is quasi-irrelevant: that is, content (factual or not) that is amplified on an industrial scale with the help of various tools and practices: bots and networks of bots, sock puppet accounts, fabricated and automated accounts, like factories, troll farms, click farms, fake followers/ fake crowds (both for validation and defamation), SEO for fabricated search results; or content (factual or not) that is hyper-personalized/ micro-targeted and algorithmically-amplified, increasingly driven by machine learning and AI (deepfakes, generative language models, conversational AI chatbots, data-driven and data-enabled audience segmentation).

So, the new, technology-enabled cognitive warfare weaponizes both content, creating an information- ultra rich ecosystem where for each fact there is a plausible counter-fact, for each narrative, a plausible counter-narrative, thus amplifying confusion and the implosion of the mainstream (see above) and our digital behavior/ fingerprint (using our personal data and our previous digital behavior in order to feed us only with hyper-personalized, bias-confirming content) (Ghosh & Scott, 2018; Singer & Brooking, 2020). Based on the digital fingerprint of each and every user, recommendation algorithms prioritize the more familiar post to platform users, and push content based rather on popularity, or likelihood of engagement than accuracy, factual character or public interest.

Therefore, we are looking at a complex phenomenon which, while exploiting our old cognitive biases, our mental shortcuts in dealing with information, our shorter and shorter attention span, our predisposition to give in to group pressure, is enabled, on a massive, industrial scale, by technology (tools, technologies and dominant practices for big data-driven amplification, algorithmic engagement, precision segmentation, micro-targeting, psychographics profiling, computational persuasion) (Ghosh & Scott, 2018).

It is essential to understand that contemporary cognitive warfare is a fraud with the content as such (blurring the distinctions between facts and fiction, between reality and pseudo-reality) and a fraud with the users' engagement, which can be manufactured/ amplified by resorting to big data, algorithms, recommendation engines, and other automated means (de Cock Buning, 2018; Bontridder & Poullet, 2021; Duan et al. 2022; Hoffmann et al., 2019; Singer & Brooking, 2020).

Finally, the conjoint effect of these two features is to create, perpetuate and reinforce polarization, that is, a specific type of emotional, exclusionary polarization not around certain topics that could not be bridged, but around any issue, along identity lines irrespective of specific topics, not accepting any compromise/middle ground. Strategic polarizing by high profile political or public figures notwithstanding, we consider that what we are dealing with a phenomenon which is enabled by such structural features of social media platforms as online syndication, algorithmic amplification of inflammatory content and failure to prioritize content moderation and other enforcement mechanisms especially outside English-speaking communities (Bessi et al., 2015; Narayanan et al, 2018; Spohr, 2017; Sunstein, 2017; Lauer, 2021).

Thus, the power of cognitive warfare (waged via disinformation, propaganda, influence operations or hostile information campaigns) lies in its capacity to disorient and, especially, to disconnect people. All these types of endeavors become a formidable "technology for the facilitation of social polarization and the destruction of social ties" (Asmolov, 2018). Their major interference is not necessarily with the notion of truth, but, as the author of this article emphasizes, "the impact of disinformation should be examined in the context of the social relationship between people who read, respond to, and share news in a situation of conflict" (Asmolov, 2018, 70).

4 "The weaponization of everything": a case study from the war in Ukraine

In this section, we discuss and illustrate the most important features of cognitive warfare by means of a case study in Romania. The traditional approach to cognitive warfare and its means (be them disinformation, propaganda, influence operations or hostile information campaigns) is to look at it as if it was an issue-specific, persuasive phenomenon. In our experience, a case-by-case approach to identify and then tackle cognitive warfare is meant to have only a limited success. Cognitive warfare relies on "the weaponization of everything" (Galeotti, 2022), that is, by the systematic, long-term exploitation of societal vulnerabilities, even when there is no immediate strategic communication objective to be met. Consistently pushing over time distorted world views upon targeted societies has an indirect effect on how future issues, even those that cannot be anticipated, will be perceived by the population. Consequently, the rapid success of misleading, hostile narratives on specific issues may be rooted in the strategic narratives deployed in the past, indirectly working in favor of the disinformation actor.

In order to illustrate this point, we look at a case study on the war in Ukraine. We focus on Romania, one of the countries bordering the conflict, and one of the important points of refuge for Ukrainians fleeing the war. Just like the entire Central and Eastern Europe (CEE) region, Romania is one of the targets of Kremlin information-related operations. We will show that, in terms of vulnerability to issue-specific disinformation narratives that explicitly refer to Russia, Ukraine, NATO/US in the context of the war, there is no reason for immediate concern. However, if we look at narratives meant to weaken the relation of trust between Romania and NATO/US, the EU, and the West in general, a very different picture emerges. We will discuss the implications of the findings by showing that the latter narratives weaponize historical dissatisfactions related to the perceived East-West divide, a topic that is prevalent and polarizing in the Romanian society (Bârgăoanu et al., 2020; Bârgăoanu & Durach, 2020; Bârgăoanu et al., 2019; Volintiru, et al., 2021). Although explicit pro-Russia attitudes were not significant in Romania at the beginning of the war, the high levels of agreement with disinformation narratives on the unfair treatment of Romania by the West indirectly support Russia's strategic communication objectives.

4.1 *Method*

In order to unveil the main disinformation narratives and attitudes towards the actors directly or indirectly involved in the Ukraine war, we conducted a national survey using an online panel (N=1000), representative of the population of Romania that has access to the Internet and is aged 18 or higher, using quotas for gender, age, and education.[1] The mean age in the sample is 43.19 years (SD=13.16). The sample comprises 52.3% women and 47.7% men. In terms of education, the sample comprises 55.2% low educated people and 44.8% people with high education. People living in urban areas represent 86.7% of the sample. The national survey was conducted by Dynata; data were collected during March 1–9, 2022.

First, we looked at opinions about Russia, Ukraine, NATO/US and the participants' own country (Romania), by measuring the level of agreement with a number of misleading narratives circulating in the Romanian digital ecosystem. These narratives strategically target the afore-mentioned actors and were selected based on a number of reports from fact checkers, specialized analyses, and public opinion surveys of relevance (Durach & Volintiru, 2022; EDMO, 2022b; EUvsDisinfo, 2022; INSCOP, 2022; Kapm, 2016; Bârgăoanu, 2022), as well as the authors' own fact-checking experience.

In the next step of the analysis, we measured overall attitudes based on levels of agreement with the disinformation narratives on Russia, Ukraine, NATO/ the US and Romania. For this purpose, we grouped the aforementioned statements in unidimensional scales, using factorial analysis (measured for internal consistency using Cronbach's alpha). We labelled the factors mentioned in this analysis "the attitude towards …", although the dominant narratives circulating in the context of the Ukraine war have many nuances that do not strictly reflect attitudes. For the ease of interpretation, we labelled them as attitudes (or global trends):

- Attitudes towards Russia—factor loadings from.697 to.865 (α=.900, M=2.59, SD=1.67) [scale from 1 to 7, 1 = anti-Russia; 7 = pro-Russia]
- Attitudes towards Ukraine—factor loadings from.888 to.914 (α=.879, M=2.81, SD=1.88) [scale from 1 to 7, 1 = pro-Ukraine (anti-Russia); 7 = anti-Ukraine (pro-Russia)]
- Attitudes towards NATO/US—factor loadings from.857 to.916 (α=.870, M=3.15, SD=1.90) [scale from 1 to 7, 1 = pro-NATO/US (anti-Russia); 7 = anti-NATO/US (pro-Russia)]
- Attitudes towards Romania—factor loadings from.694 to.886 (α=.926, M=4.14, SD=1.77) [scale from 1 to 7, 1 = positive attitude towards Romania in the context of the current conflict; 7 = negative attitude towards Romania in the context of the current conflict]

4.2 Results

The dominant attitude is of disagreement with the main misleading narratives favoring Russia (Figure 16.1). The narratives best supported by the respondents are those that justify Russia's aggression and project an image of power: "Russia is a superpower that needs to be respected." (23,7% agreement), "Due to NATO aggressive expansion, Russia is now 'surrounded by enemies' and needs to defend itself." (22,1%), "Current tensions between Ukraine and Russia are the result of the aggressive behavior of Ukraine and its Western allies." (15,9%), "Russia is only defending its legitimate interests and bears no responsibility for the current conflict." (15,8%), The least popular narratives are that "Vladimir Putin is a true patriot, defending his country's interests." (15%), and "the US interfered in Kosovo, so Russia should be allowed to interfere in Ukraine; Kosovo and Crimea are similar situations." (14,3%).

Public opinion on Russia

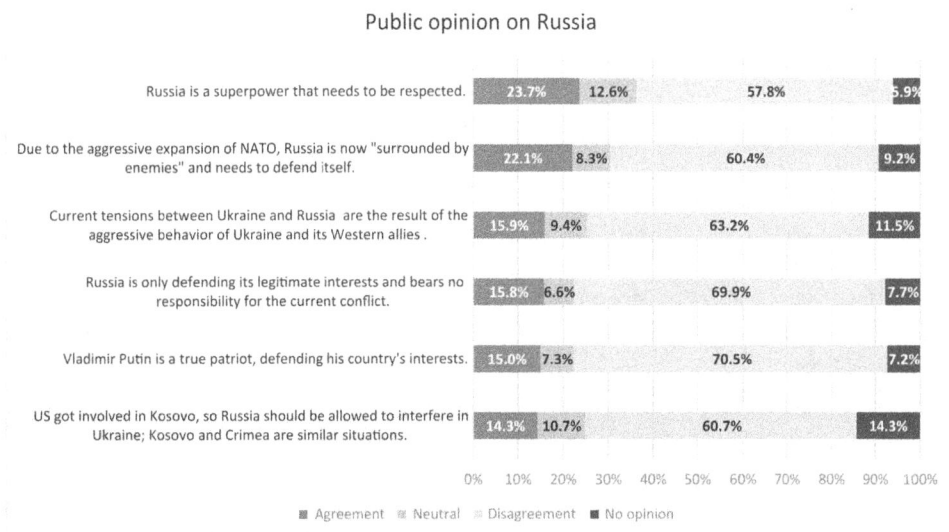

Figure 16.1 Public opinion on Russia. Authors' calculations.

Global attitudes towards Russia are negative for almost three quarters of the respondents (74.6% anti-Russia—Figure 16.2). Pro-Russia attitudes are expressed by 15,3% of the sample, and neutral attitudes, by 10,1%.

Respondents express relative disagreement with the misleading narratives on Ukraine (Figure 16.3). The claim that "*Ukraine is a puppet of NATO and the US.*" was the most supported (21,7%), followed by a narrative meant to exploit an older historical territorial dispute between Ukraine and Romania: "Ukraine took territories from Romania; therefore Romania should not get involved in the conflict with Russia." (21%). Lastly, 17% of respondents agreed that "*Ukraine is an artificial, failed state, making no progress in its 30 years of independence.*"

Attitudes towards Russia - globally

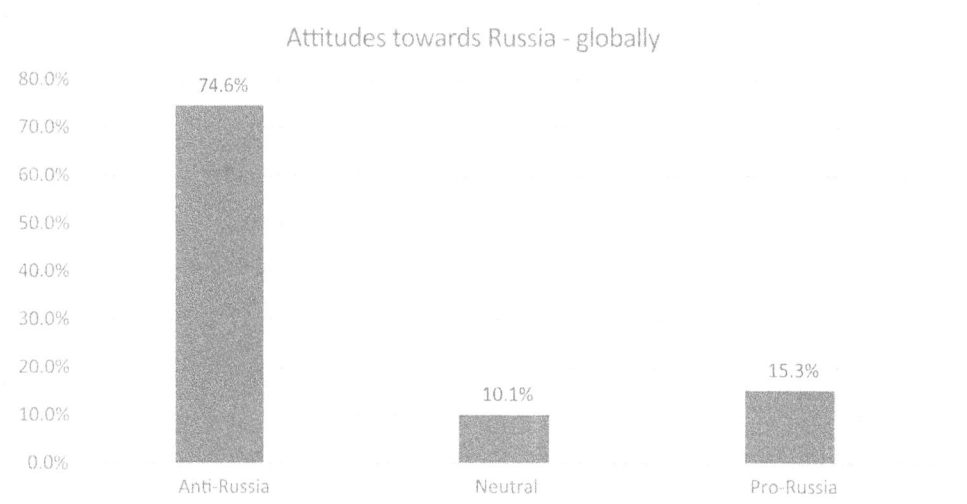

Figure 16.2 Attitudes towards Russia—globally. Authors' calculations.

Public opinion on Ukraine

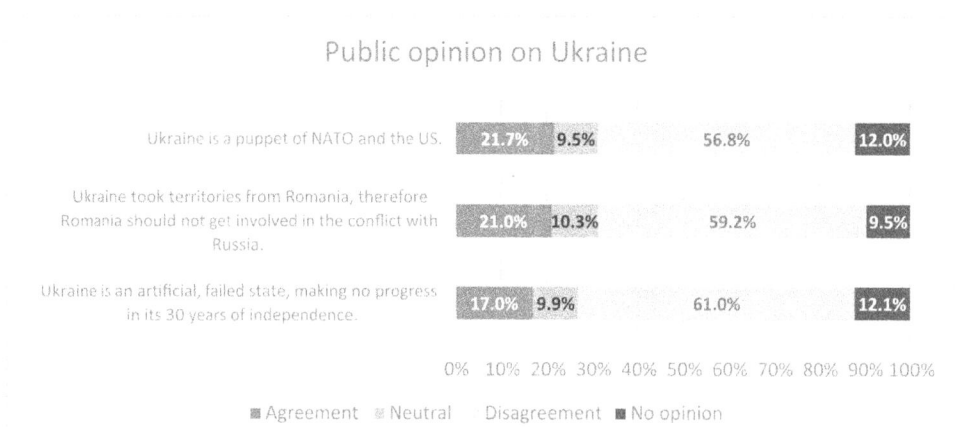

Figure 16.3 Public opinion on Ukraine. Authors' calculations

The dominant global attitudes towards Ukraine are positive (68,1% pro-Ukraine and anti-Russia). Additionally, 21,4% of respondents expressed anti-Ukraine and pro-Russia attitudes, while 10,5% are neutral (Figure 16.4).

When looking at misleading narratives targeting NATO and the US, we note that opinions are more nuanced than in the previous items, in the sense that disagreement with the given statements is less prominent compared neutral attitudes or agreement (Figure 16.5). For instance, 27,5% of respondents agree that "NATO/ the US got too close to Russia's borders.", 21,4% that "The EU, the US and the West are in decline.", and 20% that "NATO/ the US have been plotting against Russia ever since the end of the Cold War."

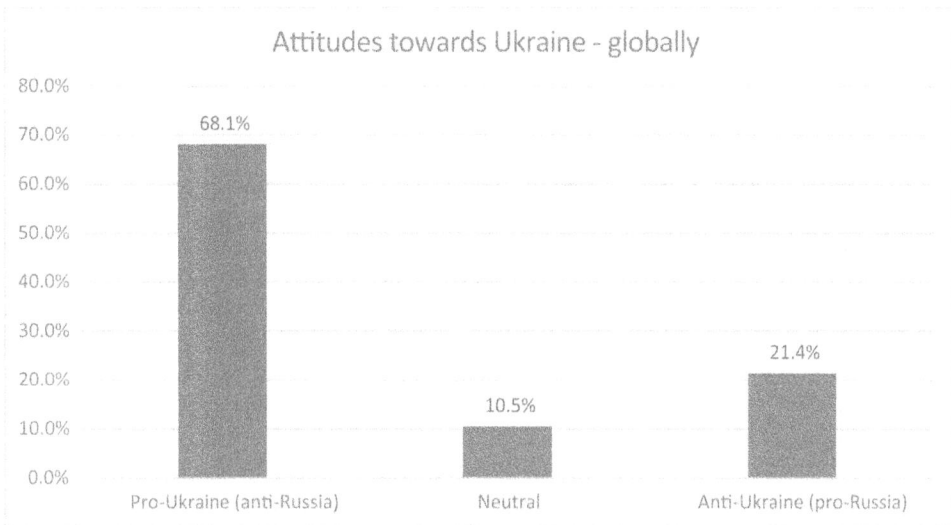

Figure 16.4 Attitudes towards Ukraine—globally. Authors' calculations.

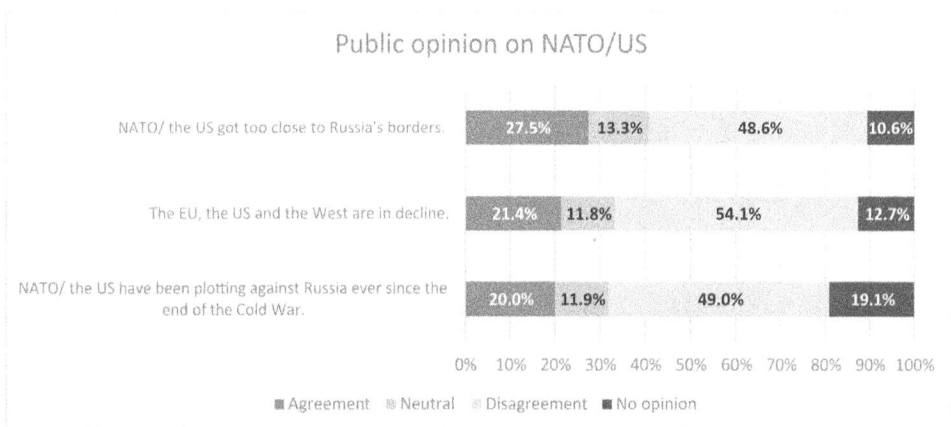

Figure 16.5 Public opinion on NATO/US. Authors' calculations.

The dominant global attitudes towards NATO-US are positive (58,8% pro NATO/US and anti-Russia) (Figure 16.6). Anti-NATO/US attitudes were observed for 24,4% of responses, and neutral attitudes, for 16,9%.

The data presented up to this point are not exceedingly worrisome. Based only on these elements, we are to assume that support for Ukraine and anti-Russian attitudes are high. Nevertheless, if we look at another set of disinformation narratives on Romania's alleged peripheral status compared to Western counterparts, we notice completely different orientations of the public opinion.

In what Romania is concerned, for six out of the nine misleading/ hostile narratives we tested, the percentage of people who agree with the respective statement is larger than the percentage of people who disagree with it (Figure 16.7). Furthermore, the most popular claim, that "In the last 30 years, politicians contributed to the sabotage of the national economy. They sacrificed Romania's interests, they are traitors." is met with agreement by 65,7% of respondents, and with disagreement by only18,1%. Other popular claims are related to Romanian's perceived status of inferiority, subordination even, compared to Western counterparts: "Romanians no longer decide for themselves, foreigners rule over us and exploit our resources. Romania is a colony of the West." (48,7% agreement), "In matters of foreign policy, Romania has no say. Romania's foreign policy is irrelevant." (44,7%), "Romania is sold to the West." (42,8%), "Romania gave up its sovereignty after EU accession." (36,3%). Another set of narratives express doubts on the security guarantees offered through Romania's NATO membership: "Romania will be sacrificed in the context of the Russia-Ukraine war." (3,5%), "NATO/ the US will not defend Romania in the case of a Russian attack." (27,1%). Lastly, narratives stroking fears and calling for Romania's neutrality are also relatively successful: "Romania's army is useless, Russia can occupy Romania in no time." (43,3%), "Romania should not get involved/ should remain neutral in the conflict between Ukraine and Russia, and between Russia and the US/ Western powers." (42%).

When we look at global attitudes towards Romania in the context of the war in Ukraine, we note that the dominant attitude is negative (43,7%), and only 36,75 positive, with 19, 6% of the sample being neutral (Figure 16.8).

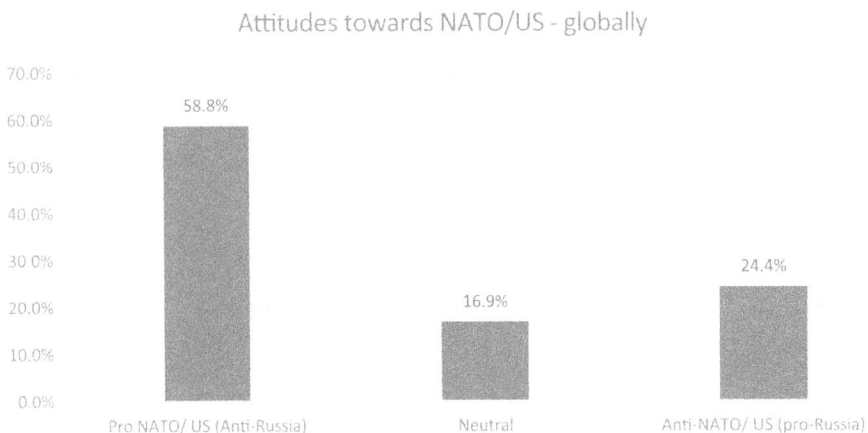

Figure 16.6 Attitudes towards NATO/US—globally. Authors' calculations.

Public opinion on Romania

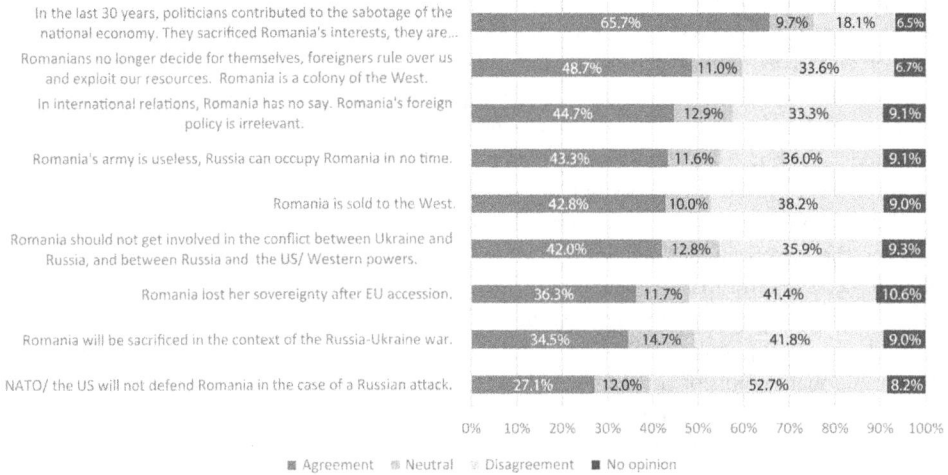

Figure 16.7 Public opinion on Romania. Author's calculations.

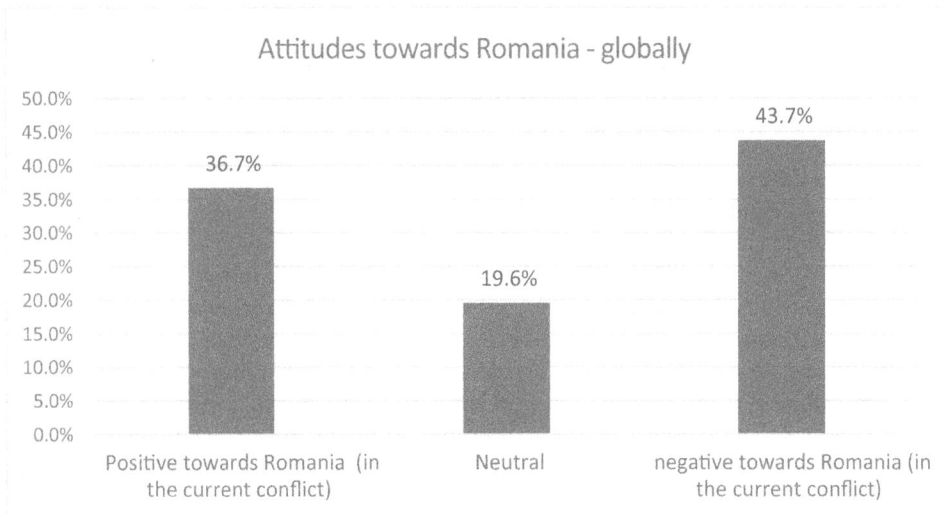

Figure 16.8 Attitudes towards Romania—globally. Authors' calculations.

4.3 Discussion

These results show that, given historical grievances and Romania's proximity to the conflict, both geographically and geopolitically, open support for Russia runs relatively low (although pro-Russia attitudes in 15% of the sample is not a completely negligible value). However, the vulnerability to other types of narratives, which indirectly work towards eroding consensus on the benefits of EU membership, positive attitudes about own country, and undermining unity within NATO/the EU, is strikingly higher.

As part of cognitive warfare, the latter narratives weaponize long-standing drivers of disinformation in the Romanian society, especially misconceptions, sometimes genuine grievances about the East-West divide in the EU. As shown elsewhere (Bârgăoanu, 2022), this type of narratives exploits the CEE ambivalence towards the West, by this ambivalence meaning that "considerable segments of the population of these countries do not feel as they are part of the Western Block entirely—and this <<in-between>> identity can be easily exploited" (Krekó, 2020, 2). Recent polling data (Durach & Volintiru, 2021) show that the top three most pervasive misleading narratives in Romania at the end of 2021 were related to perceptions of the country's undeserved position of inferiority and the incompatibility between the European and national identities (i.e., "Romania was accepted in the EU for its labor force, cheap primary resources and the market access", "Romania is a colony of the EU", "Romanians' national identity is threatened").

While the most popular misleading or hostile claims in this study are not openly pro-Russia, they fit the main points of anti-EU, anti-NATO narratives. Instead of promoting obvious pro-Kremlin stances, these narratives amplify anti-Western views using certain rhetorical devices, questioning the advantages of being a member of the EU or NATO, and spreading generalized suspicion. As statistical data suggest (Volintiru et al., 2021), the East-West divide is rooted in socio-economic factors, but it is frequently narrated in political/civilizational terms, as a clash over values (Bârgăoanu and Durach, 2020). Taking advantage of this, disinformation narratives, as part of cognitive warfare, weaponize doubt and uncertainty, emphasize existential security dilemmas, unsolved issues, and historical anxieties.

5 Implications for understanding, documenting and tackling cognitive warfare

Using the permeability to different types of disinformation, misleading or hostile narratives in Romania as a case study, we can draw some valuable insights for understanding the nature of contemporary cognitive warfare. Firstly, it becomes apparent that propaganda and disinformation campaigns do not hesitate to weaponize any domestic or cross-national issue of relevance for the target audience. While efforts to counter disinformation tend to be issue-based (e.g., the COVID-19 pandemic; the EP elections; the war in Ukraine), disinformation, misleading or hostile narratives will exploit any convenient subjects than lend themselves to complexity and controversy. The specific issues to be discussed (in misleading or inaccurate ways) are secondary; what matters is their potential to fit the greater narrative and to trigger strong reactions by exploiting the structural features of the information and communication ecosystem (amplifying reactions, grabbing attention, hacking algorithms with inflammatory content). The goal is not to convince/ influence on particular subjects, but to continuously feed a state of social distrust, confusion, and growing all-out polarization.

Furthermore, the pervasiveness of disinformation, misleading or hostile narratives, like the ones mentioned here, is an indicator of our weaknesses, as societies, governments and research communities. Despite our complacency and our ongoing efforts starting from 2015–2016 in the EU and in the US alike, we keep failing in our action plans for building resilience to this new type of warfare, for which the human mind and emotions are an international battlefield. Romania, and by extension the CEE region, is particularly vulnerable to cognitive warfare; here, weak immunity against disinformation is exploited by feeding feelings of insecurity and inferiority and amplifying anti-establishment, anti-Western views.

Lastly, the persistence of problematic content in countries that are rather overlooked by efforts to curb disinformation, such as Romania, allows us to assume that measures so far remain insufficient (and/or inefficient) to prevent social, political and geopolitical harm. Continuing to address the issue on a case-by-case basis, instead of focusing on the overarching strategic narratives discussed in this chapter leads to little progress in building resilience to the technologies, strategies and tactics of contemporary cognitive warfare.

Several types of responses can prove fruitful in this context. One set of measures are communication-related, having to do with understanding and acting upon the emerging or already established features of the communication and information ecosystem, while another, inter-related, one goes beyond this prerequisite of proper conceptualization, understanding and communicative action.

Since the nature of contemporary cognitive warfare is inseparable from the abuses of the new communication and information ecosystem, decisive responses involve changes in the regulation of the digital media ecosystem, investments in digital and information literacy, better situational awareness, including awareness of the features of the new information and communication ecosystem, improvements in cyber defense. What is also needed is a strong focus on strategic communication in order to counter misleading or hostile narratives, build positive narratives, discourage the appeal to binary frames and world views, practice inclusive leadership and avoid strategic polarizing for political or electoral gains.

On the other hand, there is a need for "hard" responses to social discontent and to key societal vulnerabilities and measures to reduce the disengagement between citizens and the national and supranational leadership. Investments in resilience should address internal vulnerabilities: tackle inequality and sources of resentment; invest in robust media and information systems; invest in good governance and the delivery of public goods, including education, social justice and the perception of social justice.

6 Conclusions

The chapter explored a relatively new phenomenon, known in the community of research and practice by many names, from online disinformation to information warfare, or cognitive warfare. The novelty of contemporary cognitive warfare—one that turns the human mind into an international/ transnational battlefield—is that it takes advantage of a completely changed communication and information ecosystem. The fact that the global digital platforms evolved to dominate the communication and information eco-system means, in practice, that every major issue in a society can be weaponized thanks to the very connectivity the large-scale use of the digital platforms entails, and to the availability of numerous technological means to artificially amplify the reach of the desired content.

We found three core arguments to support our claim that cognitive warfare should be understood as a new, technology-driven phenomenon: that it has the objective to disrupt public conversations, and to cultivate doubt, rather than to persuade in a certain direction; that it interferes both with content and with its technologically-amplified circulation; and lastly, that its effect is to create an all-out polarization, which is both cognitive and emotional in nature, irrespective of the issue(s) at hand.

The pervasiveness of strategic misleading or hostile narratives in the last decade, culminating with those massively deployed in the context of the war in Ukraine in 2022, is an indicator of the relative failure of countermeasures implemented across the Western world so far. We seek to contribute to the topic by advocating for a better understanding of what the technologies, strategies, tactics and effects of contemporary cognitive warfare are. We look at the phenomenon through the lenses of its capacity to interfere both with content and with its circulation, thus going beyond the traditional disinformation studies that pay attention to possibilities for manufacturing content in the first place. Instead of taking a case-by-case approach, we look at the phenomenon through the lenses of its capacity not to hijack a candidate, a political party, a specific subject, but rather to weaponize every major issue of concern in a society ("hot button issues") and hijack public conversation altogether. Finally, we advocate for conceptualizing contemporary cognitive warfare as a technological problem, whose solution is not solely technological, but technological, social, psychological and political, and communicative at the same time.

Note

1 According to the National Institute of Statistics, data for 2020 (from a national face-to-face survey on 10188 households) show that the Internet penetration rate in Romania is estimated at 91.3%. 78.2% of Romanian households have access to the Internet from home. The socio-demographics of the general population of Romania with access to the Internet from home are as follows: 87.0% men vs. 84.9% women, 60.9% people from urban areas, 98.2% people with higher education. More details are available at: https://insse.ro/cms/sites/default/files/field/publicatii/accesul_populatiei_la_tehnologia_informatiei_si_comunicatiilor_romania_2020.pdf.

Bibliography

Asmolov, G. (2018). The disconnective power of disinformation campaigns. *Journal of International Affairs*, 71(1.5), 69–76.

Bârgăoanu, A. (2022, March 2). The Fight Against Russia's Undermining Narratives. *Visegrad Insight*. https://visegradinsight.eu/the-fight-against-russias-undermining-narratives/.

Bârgăoanu, A., Buturoiu. R. & Durach, F. (2019). The East-West Divide in the European Union. A Development Divide Reframed as a Political One. In Dobrescu, P. (ed.), *Development in Turbulent Times. The many faces of inequality within Europe*. Berlin: Springer, ISBN 978-973-030-11360-5, pp. 105–118.

Bârgăoanu, A., Durach, F. (2020). Clashing Views between the Old and New Members of the European Union. A Socio-economic Divide Narrated as a Political One. In Abraham, F. (ed.). *1989 Annus Mirabilis. Three decades after: desires, achievements, future*. București: comunicare.ro., pp. 303–332.

Bârgăoanu, A., Durach, F., Buturoiu, R. (2020). Reshaping the European Public Sphere: Preliminary Insights into the European Backlash in Central and Eastern Europe. In Winiarska-Brodowska, M. (2020). *In Search of a European Public Sphere: Challenges, Opportunities and Prospects*, Cambridge Scholars Publishing, 90–115. ISBN (10): 1-5275-4705-4701.

Bawden, D. & Robinson, L. (2020). *Information Overload: An Overview. In: Oxford Encyclopedia of Political Decision Making*. Oxford: Oxford University Press. doi:10.1093/acrefore/9780190228637.013.1360.

Bessi, A., Zollo, F., Del Vicario, M., Scala, A., Caldarelli, G., & Quattrociocchi, W. (2015). Trend of narratives in the age of misinformation. *PloS one*, 10(8), e0134641.

Bjola, C. (2017, November 21). Digital Diplomacy 2.0: From Social to Computerised Reality. *Becoming Human: Artificial Intelligence Magazine*. https://becominghuman.ai/digital-diplomacy-2-0-from-social-to-computerised-reality-388d8a83b809.

Bontridder, N., & Poullet, Y. (2021). The role of artificial intelligence in disinformation. *Data & Policy*, 3. doi:10.1017/dap.2021.20.

Corbu, N., Bârgăoanu, A., Durach, F., & Udrea, G. (2021). Fake News Going Viral: The Mediating Effect of Negative Emotions. *Media Literacy and Academic Research*, 4(2).

Daniel, J., & Eberle, J. (2021). Speaking of hybrid warfare: Multiple narratives and differing expertise in the 'hybrid warfare'debate in Czechia. *Cooperation and Conflict*, 56(4), 432–453. https://doi.org/10.1177/00108367211000799.

de Cock Buning, M. (2018). *A multi-dimensional approach to disinformation: Report of the independent high level group on fake news and online disinformation*. Publications Office of the European Union.

de Liedekerke, A., & Zinkanell, M. (2020). *DECEIVE AND DISRUPT: DISINFORMATION AS AN EMERGING CYBERSECURITY CHALLENGE*. AIES STUDIES Nr. 13 June 2020 https://www.aies.at/download/2020/AIES-Studies-2020-13.pdf.

DiResta, R. (2021, October 11). It's Not Misinformation. It's Amplified Propaganda. *The Atlantic*. https://www.theatlantic.com/ideas/archive/2021/10/disinformation-propaganda-amplification-ampliganda/620334/.

Duan, Z., Li, J., Lukito, J., Yang, K. C., Chen, F., Shah, D. V., & Yang, S. (2022). Algorithmic Agents in the Hybrid Media System: Social Bots, Selective Amplification, and Partisan News about COVID-19. *Human Communication Research*. Volume 48, Issue 3, July 2022, Pages 516–542, https://doi.org/10.1093/hcr/hqac012.

Durach, F. & Volintiru, C. (2022). *Disinformation, societal resilience and COVID-19*. Report of Aspen Institute Romania, in partnership with Eurocomunicare Association. https://www.eurocomunicare.ro/wp-content/uploads/2022/05/DISINFORMATION-SOCIETAL-RESILIENCE-AND-COVID19_Report-FINAL-1.pdf.

EDMO. (2022a, June 29). *10 Recommendations by the Taskforce on Disinformation and the War in Ukraine*. https://edmo.eu/2022/06/29/10-recommendations-by-the-taskforce-on-disinformation-and-the-war-in-ukraine/.

EDMO. (2022b, April 19). *A Ukraine-related disinformation tsunami hit Europe in March*. Monthly brief no. 10—EDMO fact-checking network. https://edmo.eu/wp-content/uploads/2022/02/Tenth_Fact-Checking_Report_Apr_2022_H.pdf.

EUvsDisinfo. (2022, January 24). *Disinformation about the current Russia-Ukraine conflict—seven myths debunked*. https://euvsdisinfo.eu/disinformation-about-the-current-russia-ukraine-conflict-seven-myths-debunked/.

Ferguson, N. (2018). *The Square and the Tower: Networks and Power, from the Freemasons to Facebook*. Penguin Press.

Galeotti, M. (2022). *The Weaponisation of Everything: A Field Guide to the New Way of War*. Yale University Press.

Ghosh, D., & Scott, B. (2018). *The technologies behind precision propaganda on the Internet*. https://scholar.harvard.edu/files/dipayan/files/digital-deceit-final-v3.pdf.

Hoffmann, S., Taylor, E., & Bradshaw, S. (2019). The market of disinformation. *OXTEC Oxford Techonology & Elections Commission, Oxford Internet Institute*, 49. https://demtech.oii.ox.ac.uk/research/posts/oxtec-the-market-of-disinformation/.

Howard, P. N. (2020). *Lie Machines: How to Save Democracy from Troll Armies, Deceitful Robots, Junk News Operations, and Political Operatives*. Yale University Press.

Hoyle, A., Powell, T., Cadet, B., & van de Kuijt, J. (2022). Web of Lies: Mapping the Narratives, Effects, and Amplifiers of Russian Covid-19 Disinformation. In *COVID-19 Disinformation: A Multi-National, Whole of Society Perspective* (pp. 113–141). Springer, Cham.

Innis, H. (1950/2007). *Empire and Communications*. Dundurn Press.

Innis, H. (1951/2008). *The Bias of Communication*. University of Toronto Press.

INSCOP. (2022, January 26). *IANUARIE 2022: Neîncrederea publică: Vest vs. Est, ascensiunea curentului naţionalist în era dezinformării şi fenomenului ştirilor false—Ediţia a IV-a. Capitolul 1: Securitate militară. Capitolul 2: Vest vs. Est, Încrederea în ţări şi lideri internaţionali*. https://www.inscop.ro/ianuarie-2022-neincrederea-publica-vest-vs-est-ascensiunea-curentului-nationalist-in-era-dezinformarii-si-fenomenului-stirilor-false-editia-a-iv-a-capitolul-1-securitate-militara-capi/.

Kapm, K.H. (2016). *Russia's Myths about NATO: Moscow's propaganda ahead of the NATO Summit*. Security Policy Working Paper, No. 15/2016, Federal Academy for Security Policy< Germany. https://www.baks.bund.de/sites/baks010/files/working_paper_15_2016.pdf |

Kavanagh, J. & Rich, M. D. (2018). *Truth Decay: An initial exploration of the diminishing role of facts and analysis in American public life*. Rand Corporation.

Krekó, P. (2020). *Policy Brief: Drivers of Disinformation in Central and Eastern Europe and Their Utilization During the Pandemic*. Centre for Democracy & Resilience, GLOBSEC. https://www.globsec.org/wp-content/uploads/2020/06/Drivers-of-disinformation-in-Central-and-Eastern-Europe.pdf.

Lanoszka, A. (2019). Disinformation in international politics. *European journal of international security*, 4(2), 227–248. doi:10.1017/eis.2019.6.

Lauer, D. (2021). Facebook's ethical failures are not accidental; they are part of the business model. *AI and Ethics*, 1(4), 395–403. https://doi.org/10.1007/s43681-021-00068-x.

Leonard, M. (2021). The Age of Unpeace: How Connectivity Causes Conflict. Bantam Press.

McLuhan, M. (1962/2011). *The Gutenberg Galaxy. The Making of the typographic man*. University of Toronto Press.

McLuhan, M. (1964/1994). *Understanding Media: The Extensions of Man*. The MIT Press.

McLuhan, M. (1967/2001). *The Medium Is the Massage: An Inventory of Effects*. Gingko Press.

Narayanan, V., Barash, V., Kelly, J., Kollanyi, B., Neudert, L. M., & Howard, P. N. (2018). *Polarization, partisanship and junk news consumption over social media in the US*. arXiv preprint arXiv:1803.01845.

Postman, N. (1993). *Technopoly: The Surrender of Culture to Technology*. Vintage.

Rushkoff, D. (2021). *Team Human*. W. W. Norton & Company.

Singer, P. W. & Brooking, E. T. (2020). *LikeWar: The Weaponization of Social Media*. Eamon Dolan/ Houghton Mifflin Harcourt.

Spohr, D. (2017). Fake news and ideological polarization: Filter bubbles and selective exposure on social media. *Business information review*, 34(3), 150–160. https://doi.org/10.1177/0266382117722446.

Sunstein, C. (2017). *#Republic: Divided Democracy in the Age of Social Media*. Princeton University Press

Vériter, S. (2021, December 14). *European Democracy and Counter-Disinformation: Toward a New Paradigm?* Carnegie Endowment for International Peace. https://carnegieendowment.org/2021/12/14/europea n-democracy-and-counter-disinformation-toward-new-paradigm-pub-85931.

Volintiru, C., Bârgăoanu, A., Ştefan, G., Durach, F. (2021). East-West Divide in the European Union: Legacy or Developmental Failure? *Romanian Journal of European Affairs*, Vol. 21, No. 1, June 2021.

PART IV

Countering Disinformation and Building Resilience

17

JOURNALISTIC APPROACHES TO INFORMATION SOURCES, FACT-CHECKING, VERIFICATION, AND DETACHED REPORTING

Pablo Hernández-Escayola

The online credibility problem

The Internet and social networks have completely changed the relationship between the public and the sources of information. Before these communication technologies appeared, citizens knew where the messages they received came from. Each person chose the newspaper, radio station or television channel where they wanted to get their information and had a rough idea of the political stance of each medium and how reliable their news was. The audience knew who was speaking to them. Social networks have totally changed this scheme. Now people move in a disordered, chaotic, fragmented and non-hierarchical environment in which they are not aware of who is the author of much of the content they consume. These new rules of the game in the world of communication have undoubted positive aspects. Due to new communication technologies, citizens have many more possibilities to express themselves and have their voices heard. In addition, they can easily satisfy any demand for information with a few simple searches and create virtual communities with people who share their interests. But at the same time, the access of millions of people to active public debates makes it much more difficult to distinguish between valuable information and attempts of manipulation by those who use new technologies to disinform. Furthermore, social networks facilitate anonymity, which makes it even more difficult to assess the reliability of messages.

The media are recognizable sources and traditionally acted as gatekeepers of information. They had the job of selecting among all the information available the pieces that are worth becoming news and shape them to serve them to the public. But in the new communicative ecosystem they are losing that role. They have gone from being referral sources to having difficulty building trust (Pickard, 2020). The polls show the erosion of credibility they are suffering. According to the Ipsos Global Trustworthiness Monitor 2022 only 19% believe the media is trustworthy[1]. And the 2022 Edelman Trust Barometer shows that only 46% of those

DOI: 10.4324/9781003190363-21

surveyed trust the journalists compared to 75% who trust scientists[2]. There are several reasons why both the media and journalists have trouble getting the public to trust them.

Bad journalistic practices

The process of adapting traditional media to the high choice environment created by the Internet and social networks is not being easy. The struggle to gain an audience leads to resorting to practices such as sensationalism and clickbait that can produce short-term results but also negatively affect the prestige of the media outlet that uses them (Molyneux, 2019). Another way to get public attention is to adopt a clear political agenda. Audiences on social networks are highly segmented and bringing the editorial line closer to a specific political ideology can serve to reach specific niches. But the radicalization of the message makes the public perceive the media as part of the process of polarization of society. A Reuters Institute investigation (Toff et al, 2021) detected that perceptions of bias and hidden agendas in news coverage were prevalent in the audiences and that prominent media personalities were often perceived as polarizing figures.

Confusion with other sources

Clickbait and sensationalism are often an attempt to gain attention in an environment where the amount of content offered is so high that it is difficult to get an audience. They are solutions that end up being counterproductive in the long term, but they are a reaction to a real problem that the media faces: journalistic brands are diluted in the online world. The public has difficulty knowing who is the author of the messages that reach them. The audience has lost the direct reference that they got, for example, with a newspaper in their hands and now they move in scenarios in which a tweet from that newspaper looks exactly the same as that of an ordinary citizen or an anonymous account. Faced with this situation, users of social networks use a series of mental shortcuts to find out if they can trust a specific informative content or not. Ross Arguedas et al (2022), has studied them and has detected six:

- Pre-existing ideas they held about news in general or the specific news brands where the information was coming from.
- Social cues from family and friends who shared or engaged with the news.
- The tone and wording of headlines
- The use of images or videos.
- The presence of advertising and whether or not information appeared to be sponsored.
- Platform-specific cues such as number of likes or labels.

The journalistic brand is only one of the six drivers used to measure the credibility of content. In addition, these mental shortcuts vary depending on the platform and the journalistic brand can lose importance. For example, according to this research, on Facebook and WhatsApp social clues such as who shares the information are the most important driver of credibility, and on Google the order of the search results is the main cue. According to this study many people have problems identifying where the information was coming from on all three platforms. The difficulties for brand attribution in online environments were also probed by Kalogeropoulos and Newman (2017). Through a survey, they measured the effects of indirect exposure to content via algorithmically driven feeds or recommendations and discovered that less than half of the participants could remember the name of the news brand for a particular story when coming from search or social media. That means that the public remembers the information and where they found it, but struggles to remember the media outlet that published that information.

Fake media

The characteristics of the new technological and communicative scenario also facilitate the appearance of pseudo media that act with a very marked political agenda or that function as mere propaganda agents. Their main interest is not to inform but to spread very specific narratives and to do so they resort to all kinds of disinformation and manipulation. The impact of these pseudo media is harmful on several levels. On the one hand, they contribute to the spread of hoaxes on social networks and increase social polarization. On the other hand, by formally presenting themselves as media outlets, they affect the reputation of the media outlets that are doing informative work. This impact is very strong because part of the public does not know how to distinguish between an informative media outlet and a propagandistic web. This lack is especially serious among young people, who are less accustomed to consuming information and struggle to identify journalistic sources. A study by the Universities of Loyola and Seville (Herrero-Diz et al., 2020) discovered that young people gave the same credibility to an article from La Vanguardia, a well-known Spanish media outlet, as to one from a fringe outlet as Caso Aislado. The hypothesis in this study is that the young people did not know either of the two outlets so they made their trust choices based only on the format of the contents.

Some of these pseudo media go beyond the dissemination of self-serving disinformation. They also directly attack the traditional media, accusing them of being at the service of power. That allows them to present themselves as the alternative to reveal the truth that the powerful do not want to be known. This idea that the information provided by the traditional media is biased and controlled by the elites ends up influencing some parts of the society that start to distrust all kinds of official sources. This feeling of mistrust in the authorities and the legacy media is linked to the use of alternative outlets that reflect anti-establishment worldviews (Hameleers et al, 2022).

This type of pseudo media can present themselves as information media because there is no clear, evidence-based definition of what a media outlet is. A set of guidelines to differentiate between journalistic initiatives from those that try to imitate the format of a media outlet but are only used to promote a determined agenda without adhering to any ethical standards. This is a pending exercise for the journalistic profession, and it would have to set some essential minimum requirements to be a media outlet on the basis that a media outlet does not lie systematically and consciously.

One way to address the trust in media problem is the creation of projects such as the Journalism Trust Iniciative[3] or NewsGuard[4]. They try to give a seal of quality to the media that meet certain conditions. Its intention is to create spaces of trust around the media that comply with its guidelines and differentiate outlets committed to journalism from pseudo media.

But the problem of the credibility of sources in the online environment goes beyond the media and its crisis of confidence. On the internet and social networks, information sources of which the public hardly have references and, therefore, do not know if they are reliable, multiply. In fact, social networks allow the users themselves to act as sources. Before the social media, citizens were simply the recipients of information content, now they have an active role in the communication process. Not only can they create content, but their reactions are also essential in the distribution chain. Each user decides whether or not to share content. The impact that content achieves depends on a cascade of such decisions. Users of social networks thus gain the power to increase the dissemination of certain content and even increase their credibility when they show their support through likes (Luo et al., 2020). The problem is that this action of sharing content is moved by emotional impulses (Lu et al., 2022) that the creators of disinformation know perfectly well. Disinformation is designed so that the receiver acts without thinking, to provoke a reaction guided by emotions and without questioning whether the content that is helping to spread or gain popularity is true or false. Pennycook et al. (2020) have found that people share false claims about COVID-19 partly because they fail to think sufficiently about whether or not the content is accurate when deciding what to share.

In this communicative scenario in which gate keepers are experiencing a crisis of confidence, there is a lot of information available but few references to its reliability, and malicious actors appear in order to take advantage of this situation to spread disinformation, it is where the work of the fact-checkers is essential.

The evolution of fact-checking

If the concept of "fact-checking" is interpreted as the mere verification of what is going to be published in a communication medium, it could be seen as a basic journalistic practice. But fact-checking has taken on a different and much broader dimension in recent years, with the emergence of the internet and social networks and the proliferation of disinformation in online environments. Lucas Graves in his book Deciding what's true (2017) explains how this evolution has come about. During the 20th century, the first fact-checking departments appeared, mainly in large American media outlets such as The New Yorker or The New York Times. These were independent departments within the newsroom that had the mission of making sure that everything that was published was verified. The objective was to protect the credibility of the medium.

Around the year 2000, the mission of fact-checkers began to evolve. Journalists rethought their relationship with politicians and their speech. They considered that journalistic work goes beyond transmitting what politicians say and that it was necessary to establish a control on whether their messages were real. Fact-checkers responded to this need. Some media created specific teams to carry out this type of verification. But the first independent agencies dedicated to fact-checking public discourse also appeared. Therefore, in this second stage, the main fact-checker's job is no longer to carry out an internal control of the quality of the information of a medium but to qualify the reliability of external sources: politicians and their speeches.

The third stage of fact-checking responds to the communicative revolution brought about by the popularization of the internet and social networks. Fact-checkers continue to monitor politicians' speech, but their work goes much further. The great challenge now is to stand up to a tsunami of disinformation that uses the communication channels that new technologies have generated. False messages that try to manipulate public opinion no longer come only from the mouths of politicians or public figures, they now circulate on social networks or instant messaging applications. Pursuing them is especially complicated because they go from one social network to another, they change formats and, on many occasions, it is not possible to know for sure who is behind them. For this reason, the focus of fact-checking initiatives increasingly turns to the most viral or most dangerous online disinformation content. The field of action passes from the conventional political debate to the entire space occupied by social networks. In these environments, disinformation goes beyond the limits of politics and, as has been seen during the COVID-19 pandemic, it has a serious impact in other areas such as health. But debunking is not enough. As we have seen, the public does not know who to trust when moving through online environments. For this reason, fact-checkers have to make citizens perceive them as reliable sources so that their task makes sense. A verification that the public does not believe in is completely useless.

How fact-checkers work to be reliable sources

Transparency is the essential requirement to gain public trust. Citizens must know exactly who fact-checkers are, how they are financed, what they verify, why they do it and how they do it. Transparency is not only a fundamental ethical exercise; it is also an essential resource due to the type of work that fact-checkers face. Much of their work consists of unmasking the disinformation of those who want to manipulate society through lies. The response of these bad actors when one of their hoaxes is debunked is usually a direct attack on the credibility of fact-checkers. They are often accused of a lack of independence or of acting motivated by hidden interests dictated by someone powerful. To prevent this type of attack from casting doubt on the fact-checkers, it is essential that the financing methods and work methodology are transparent and available to the public. The international organizations that are in charge of coordinating the efforts of fact-checkers at an international level take this very much into account. The main one is the International Fact-Checking Network (IFCN). In 2015, it drew up a Code of Principles that defines what characteristics an organization dedicated to fact-checking must have, and which has become the main reference on this subject[5]. The Code consists of five major commitments:

- A commitment to non-partisanship and fairness. Signatories will follow the same process for every fact-check. They do not take policy positions on the issues they are fact-checking and will not concentrate their verifications on any one side.
- A commitment to standards and transparency of sources. Signatories provide all sources in enough detail that readers can replicate their work, except in cases where a source's personal security could be compromised.
- A commitment to transparency of funding and organization. Signatories are transparent about their funding sources.
- A commitment to standards and transparency of methodology. Signatories explain the methodology they use to select, research, write, edit, publish and correct their fact-checks. They encourage readers to send claims to fact-checks and are transparent on why and how they fact-check.
- A commitment to open and honest corrections policy. Signatories publish their corrections policy. They correct clearly and transparently, seeking as far as possible to ensure that readers see the corrected version.

The path marked out by the IFCN Code of Principles is followed by other international associations or organizations that try to generate joint initiatives between fact-checkers. For example, the European Digital Media Observatory (EDMO), an organization created to coordinate the action of European fact-checkers and promote collaborative research between them, directly admits the signatories of the Code of Principles of the IFCN. For those who are not part of the IFCN and, therefore, do not have to abide by the Code, it does require them to at least make an effort of transparency[6]. Specifically, they are obliged to disclose and avoid any potential conflict of interest, must make public their organizational and proprietary structure, and be free to influence or control over them by political parties or movements.

How codes are applied

The Codes of good practice of these international organizations, mainly the IFCN, establish a common ground that marks the essential minimum for a fact-checking initiative. But each fact-checker develops its own internal codes that delve into the guidelines set at the international level. These internal codes are the explicit commitment that each organization acquires with the public. They give details of the specific mechanisms that each organization uses to carry out its work and, in compliance with the principle of transparency, they are published on their web pages.

An example of this adaptation process is Maldita.es, a Spanish non-profit foundation dedicated to fact-checking and a signatory of the IFCN. In its internal code, it broadens the bases established by the International Organization's Code of Principles and introduces new mechanisms to ensure the transparency, independence and quality of its work. Maldita.es develops specific methodologies for three different areas of work: the verification of hoaxes[7], disinformation on scientific issues[8] and the monitoring of messages from politicians[9]. All three methodologies are published on their website.

The main element of this internal code is what determines how hoaxes are verified. The first thing it defines is the criteria to choose which disinformation is debunked. Maldita.es explains that the selection is made based on virality and dangerousness. If a hoax is widespread or if it is considered to be harmful to health or attacks a vulnerable group, work begins to debunk it. If it is not viral or dangerous, it is not debunked to prevent the verification itself from inadvertently amplifying the spread of the hoax.

During the investigation of the alleged hoaxes, Maldita.es takes special care with how it handles the sources. They are all identified and cited, whether natural persons, entities or official bodies. If a source does not want to be cited, it is not used in the debunk, it can only serve to give context on the matter of disinformation. Once the investigation has been carried out, the reporter presents it to the editing team, who evaluate how the verification process has been handled. According to the methodology of Maldita.es, at least four editors have to give their approval of an article before it is published. In addition, any of the eleven members of the editorial team can veto a fact-check if he or she believes that it does not meet the necessary requirements.

This methodology is the basis of the work of Maldita.es, but there are specific sections that have their own methodology that adapts to their needs. The science section makes it clear that as science progresses, scientific knowledge may change. The debunks of Maldita.es are based on the best scientific evidence of the moment, but it asks that the publication dates of the information be taken into account because the scientific process may mean that some of the article's claims have been outdated. In the case of the methodology for monitoring political statements, it is explained that opinions are not verified unless they include references to specific facts, documents, statistics or data. The sources they use in the verifications are linked so that the audience can check what the debunk is based on.

The elaboration and publication of these methodologies is an exercise of transparency in order to put in the hands of the public a knowledge that allows them to better evaluate the work of Maldita.es, but also to detect possible errors and inconsistencies. In the event that an error is detected in its articles, Maldita.es also has a commitment to correct it on its website and on its social networks.

Each fact-checking organization introduces its own peculiarities in its internal codes to act transparently and clarify to its followers the commitments they make when carrying out their work. For example, Politifact focuses its activity on the statements of politicians, which is why in its code it places special emphasis on the need for independence and political neutrality of its journalists and the processes they follow to conduct their verifications[10]. Africa Check operates in an environment where the quality of the data sources is not too high. So they commit to using the best available data for their verifications but clarify that they rate the accuracy of statements, but they do not set out to identify minor inconsistencies[11].

Fact-checkers and sources

The main mission of fact-checkers is to reduce the pernicious effects of information disorders. As described by Wardle and Derakhshan (2017), these disorders can be divided into three categories:

- Disinformation. Information that is false and deliberately created to harm a person, social group, organization or country.
- Misinformation. Information that is false, but not created with the intention of causing harm.
- Mal information. Information that is based on reality, used to inflict harm on a person, organization or country.

As we can see, the intention of the source that shares the false content is relevant when qualifying false content in one category or another. In addition, in the environment of social networks and messaging applications, it can be difficult to distinguish whether the source that sends us a content is the one who has created it or if it is simply limited to sharing it. Both processes, evaluating the intentions of sources and discovering who is the creator of a particular content, are difficult and problematic for fact-checkers. Keep in mind that the methodologies of fact-checkers require them to be very sure of all their statements and have evidence to make them. So fact-checkers are very cautious about attributing a particular disinformation to a specific source. However, the verification work does reveal some details of the behavior of the agents that usually appear linked to disinformation. There are three main types of motivations for spreading hoaxes:

- Economic interest. They try to get an economic return from the clicks. They use flashy hoaxes or misleading headlines to attract visits to their web pages or increase the views of advertising content.
- Ideological interest: They try to modify the public's perception of reality to present distorted facts that favor their predisposition to support a certain ideology. For example, disinformation about migration.
- Interest in creating chaos or trolling: There are hoaxes for which it is difficult to find an explanation beyond the desire to cause chaos or get a massive deception of the population for fun. However, behind them there may be a deeper intention because this disinformation that conveys the feeling that we cannot trust anything we find on social networks can also make us distrust everything that comes to us from the media and official sources.

This general mistrust of any authority source reaches the extreme in those who follow conspiracy theories. These types of beliefs were traditionally related to disorders close to paranoia. However, in recent years, conspiracy theories have gained followers and spread around the world. Researchers have found that at the root of the problem is something called epistemic mistrust (Pierre, 2020). It is no longer about the ideas themselves, but about a deep and visceral distrust towards people or institutions. When something happens, instead of looking for the different causes, it is directly attributed to hidden and malicious forces. Official sources are not only suspicious, but are at the service of evil.

The rejection of official sources is also common among denialist movements. These science deniers accept evidence only if it confirms their prior beliefs, so they try to attack scientific consensus (Schmid and Betsch, 2019). To attack these consensuses, they try to generate an "alternative truth" that they reinforce with the support of false experts (Diethelm & Mckee, 2009). The deniers present such false experts as the true reliable sources in matters in which citizens have a hard time knowing who to trust. The action of denialism goes beyond generating false sources, they also attack experts who support the scientific consensus with evidence and data. This was confirmed by a Nature investigation that compiled the attacks suffered by scientists during the COVID-19 pandemic[12]. More than two-thirds of scientists who had commented on the coronavirus reported negative experiences doing so, and nearly a quarter said they had received death threats. These attacks on the source's image of authority are a relevant social clue when it comes to shaping public opinion. The clues that the elites send have relevant effects on how society perceives problems such as climate change (Brulle et al., 2012) or on the measures taken to combat COVID-19 (Dada et al., 2021).

Disinformers also try to erode the credibility of fact-checkers by creating confusion around their work. We have seen it during Russia's invasion of Ukraine. An investigation by Propublica and Clemson University revealed that false fact-checks were being used[13]. They found a series of alleged verifications circulating in pro-Russian circles who claimed Ukrainian media were using manipulated images to accuse the Russian military of committing atrocities. However, using the metadata of the videos, the researchers discovered that the manipulation of the images and the fact-check had been put together with the same computers and concluded that the same people created the hoax and debunked it. These practices of impersonating fact-checkers and falsifying verifications are especially dangerous because they add even more confusion and can call into question the work of fact-checkers who are acting according to the best ethical standards.

Alert citizens

As we have seen, citizens are no longer the passive receivers of information and in the new scenario they are a fundamental part of the communication process. Therefore, the solutions to the problem of sources also go through helping them make better decisions when they move in online environments. Hoaxes spread faster than true news (Vosoughi et al., 2018), and one reason for this is that people are overconfident in their ability to spot disinformation (Lyons et al., 2021). Media literacy actions are designed for citizens to better understand the environment of social networks, discover the techniques used by disinformation spreaders and improve their ability to detect reliable sources.

The basic analysis that must be done to detect disinformation focuses on three fundamental aspects:

- The language. Hoaxes usually use sensationalist language to attract attention and to directly appeal to emotions
- The evidence. Most hoaxes never provide any evidence such as links or concrete data.
- The sources. Disinforming content does not indicate its sources either because there are none or nobody knows about them.

Regarding the specific sources, it is convenient to compare the content that we want to check with other alternative sources on the same topic or news. Information should be distrusted in the following cases (Erviti et al., 2022):

- If the source is anonymous or is from an unknown medium, unreliable or with a history of falsehoods.
- If the information lacks external references or does not come from expert and authorized sources on that specific topic.
- If it includes expressions such as "experts say", without specifically and expressly identifying them.

In any case, the basic advice for citizens is not to act impulsively and to reflect before sharing. Simply stopping to think about who wrote the article influenced the extent to which they believed it has been shown to increase skepticism and reduce vulnerability to disinformation (Kim and Dennis, 2018).

The fact-checker and his community

Another way to make citizens more resistant to disinformation is to involve them directly in the fight against it. Media literacy gives resources to detect lies and manipulations. But there are citizens concerned about disinformation who are willing to go further. One way to do this is to use tiplines that multiple fact-checkers use. Through these tools, people can alert of suspicious content found on the internet, social networks or instant messaging applications (Palomo and Sedano, 2018). Once fact-checkers have confirmed that it is a hoax, since the user of the tipline knows where they have found the disinformation, they can also share the verification there. This practice of getting the debunk to where the disinformation was found is critical to increasing its effectiveness.

There are fact-checking initiatives that take citizen participation further. This is the case of Maldita.es, which offers them the possibility of becoming a source of their debunks. Many of the followers of this fact-checking organization are experts in something, be it a scientific discipline, an academic discipline, and they simply know how to speak an uncommon language. The knowledge of these people can be very useful for certain debunks. Furthermore, as we have already seen, the public has a low level of trust in the media and journalists, but it does trust experts. For example, they are the most trusted sources on covid for 86% of Spaniards compared to the 46% who trust news organizations, according to the Reuters Institute (Nielsen et al., 2021). In the United Kingdom, 84% believed in scientists, doctors and experts compared to 51% who believed in the media. For this reason, Maldita.es proposes to its followers that they donate what they call their "superpowers". That is the knowledge they have and can be used to make a debunk. Those followers who donate their "superpower" become a source for Maldita.es and begin to act directly on the disinformation that affects their area of knowledge and that they know why it is a lie and how to prove it. In this way, they contribute their expertise and the prestige that the experts have and Maldita.es gives them publicity and ensures that the debunks that are prepared with their help go further than they would if they tried to unmask the lies by themselves.

Notes

1 Ipsos Global Trustworthiness Monitor 2022, 17 January 2022, https://www.ipsos.com/en/global-trustworthiness-monitor-2021

2 2022 Edelman Trust Barometer, https://www.edelman.com/trust/2022-trust-barometer
3 Journalism Trust Initiative, https://www.journalismtrustinitiative.org/es/about
4 NewsGuard https://www.newsguardtech.com/ratings/rating-process-criteria/
5 International Fact Checking Network Code of Principles https://www.ifcncodeofprinciples.poynter.org/
6 European Digital Media Observatory Admission Criteria https://edmo.eu/admission-criteria/
7 https://maldita.es/metodologia-de-maldito-bulo/
8 https://maldita.es/metodologia-de-maldita-ciencia/
9 https://maldita.es/metodologia-de-maldito-dato/
10 The Principles of the Truth-O-Meter: PolitiFact's methodology for independent fact-checking https://www.politifact.com/article/2018/feb/12/principles-truth-o-meter-politifacts-methodology-i/
11 How we rate claims in AfricaCheck https://africacheck.org/how-we-fact-check/how-we-rate-claims
12 Bianca Nogrady, "'I hope you die': how the COVID pandemic unleashed attacks on scientists". Nature, 13 October 2021 https://www.nature.com/articles/d41586-021-02741-x
13 Craig Silverman and Jeff Kao. In the Ukraine Conflict, Fake Fact-Checks Are Being Used to Spread Disinformation. ProPublica, March 8 2022 https://www.propublica.org/article/in-the-ukraine-conflict-fake-fact-checks-are-being-used-to-spread-disinformation

Bibliography

Arguedas, A. R., Badrinathan, S., Mont'Alverne, C., Toff, B., Fletcher, R., & Nielsen, R. K. *Snap Judgements: How Audiences Who Lack Trust in News Navigate Information on Digital Platforms.* (2022) https://reutersinstitute.politics.ox.ac.uk/snap-judgements-how-audiences-who-lack-trust-news-navigate-information-digital-platforms.

Brulle, Robert J., Jason Carmichael, and J. Craig Jenkins. "Shifting public opinion on climate change: an empirical assessment of factors influencing concern over climate change in the US, 2002–2010." *Climatic Change* 114.2 (2012): 169–188. https://doi.org/10.1007/s10584-012-0403-y.

Dada, S., Ashworth, HC., Bewa, MJ. and Dhatt, R. "Words matter: political and gender analysis of speeches made by heads of government during the COVID-19 pandemic." *BMJ global health* 6.1 (2021): e003910. http://dx.doi.org/10.1136/bmjgh-2020-003910.

Diethelm, Pascal, and Martin McKee. "Denialism: what is it and how should scientists respond?." *The European Journal of Public Health* 19.1 (2009): 2–4. https://doi.org/10.1093/eurpub/ckn139.

Erviti, M.C., Salaverría-Aliaga, R., León, B., et al. *"Mentiras contagiosas. Guía para esquivar la desinformación en salud."* (2022). https://doi.org/10.15581/978-84-8081-720-2.

Graves, Lucas. *Deciding what's true*. New York. Columbia University Press. 2017.

Hameleers, Michael, Anna Brosius, and Claes H. de Vreese. "Whom to trust? Media exposure patterns of citizens with perceptions of misinformation and disinformation related to the news media." *European Journal of Communication* (2022):.https://doi.org/10.1177/02673231211072667.

Herrero-Diz, P., Conde-Jiménez, J. and Reyes de Cózar, S.. "Teens' motivations to spread fake news on Whatsapp." *Social Media+ Society* 6.3 (2020): https://doi.org/10.1177/2056305120942879.

Kalogeropoulos, A., and Newman, N. *I saw the news on Facebook': Brand attribution when accessing news from distributed environments* (Digital News Project 2017). https://www.digitalnewsreport.org/publications/2017/i-saw-news-facebook-brand-attribution-distributed-environments/.

Kim, Antino, and Alan R.Dennis. "Says who? The effects of presentation format and source rating on fake news in social media." *MIS quarterly* 43.3 (2019): 1025–1039. http://dx.doi.org/10.2139/ssrn.2987866.

Lu, X., Vijaykumar, S., Jin, Y., & Rogerson, D.Think before you Share: Beliefs and emotions that shaped COVID-19 (Mis) information vetting and sharing intentions among WhatsApp users in the United Kingdom. *Telematics and Informatics*, 67.(2022) https://doi.org/10.1016/j.tele.2021.101750.

Luo, M., Hancock, J. T., & Markowitz, D. M.Credibility perceptions and detection accuracy of fake news headlines on social media: Effects of truth-bias and endorsement cues. *Communication Research*, 49(2), (2022) 171–195. https://doi.org/10.1177/0093650220921321.

Lyons, Benjamin A., Montgomery, JM., Guess, AM., Nyhan, B. and Reifler, J. "Overconfidence in news judgments is associated with false news susceptibility." *Proceedings of the National Academy of Sciences* 118.23 (2021). https://doi.org/10.1073/pnas.2019527118.

Molyneux, Logan, and Mark Coddington. "Aggregation, clickbait and their effect on perceptions of journalistic credibility and quality." *Journalism Practice* 14.4 (2020): 429–446. https://doi.org/10.1080/17512786.2019.1628658.

Nielsen, Rasmus Kleis, Anne Schulz, and Richard Fletcher. "An ongoing infodemic: how people in eight countries access news and information about coronavirus a year into the pandemic." *Reuters Institute for the Study of Journalism* (2021). https://reutersinstitute.politics.ox.ac.uk/ongoing-infodemic-how-people-eight-countries-access-news-and-information-about-coronavirus-year#footnote-003.

Palomo, Bella, and Jon Ander Sedano Amundarain. "WhatsApp como herramienta de verificación de fake news. El caso de B de Bulo." *Revista latina de comunicación social* 73 (2018): 1384–1397. https://doi.org/10.4185/RLCS-2018-1312.

Pennycook, G., McPhetres, J., Zhang, Y., Lu, J. G., and Rand, D. G. "Fighting COVID-19 misinformation on social media: Experimental evidence for a scalable accuracy-nudge intervention." *Psychological science* 31.7 (2020): 770–780. https://doi.org/10.1177/0956797620939054.

Pickard, Victor. "Restructuring Democratic Infrastructures: A Policy Approach to the Journalism Crisis" *Digital Journalism*, 8:6,(2020) 704–719, DOI:10.1080/21670811.2020.1733433.

Pierre, Joseph M. "Mistrust and misinformation: A two-component, socio-epistemic model of belief in conspiracy theories." *Journal of Social and Political Psychology* 8.2 (2020): 617–641. https://doi.org/10.5964/jspp.v8i2.1362.

Schmid, Philipp, and Cornelia Betsch. "Effective strategies for rebutting science denialism in public discussions." *Nature Human Behaviour* 3.9 (2019): 931–939. https://doi.org/10.1038/s41562-019-0632-4.

Toff, B., Badrinathan, S., Mont'Alverne, C., Ross Arguedas, A., Fletcher, R., & Nielsen, R.Listening to what trust in news means to users: qualitative evidence from four countries. (2021) https://reutersinstitute.politics.ox.ac.uk/listening-what-trust-news-means-users-qualitative-evidence-four-countries.

Vosoughi, Soroush, Deb Roy, and Sinan Aral. "The spread of true and false news online." *Science* 359.6380 (2018): 1146–1151.10.1126/science.aap9559.

Wardle, C., & Derakhshan, H. *Information disorder: Toward an interdisciplinary framework for research and policymaking.* (2017) Council of Europe report 27

18

A NATIONAL SECURITY PERSPECTIVE ON INFORMATION LEAKS

Joseph A. Cannataci and Aitana Radu

Today, we believe more than ever that we have a "right to know", when it comes to what is happening in the world outside. Wars, as seen most recently with the 2022 resumption of the Russian invasion of Ukraine,[1] are fought as much on television channels and social media as on the field. Knowing and sharing information about an event is proxy to taking part in it and, in today's information-rich world, citizens feel this is a given right.

What happens, however, when the action in question belongs to the traditionally closed field of national security? Clearly, cases such as those of Chelsea Manning[2] and Edward Snowden[3] seem to indicate that there are individuals, who believe that the citizen's right to know extends even there, overruling national security considerations in a democratic society. This raises some important questions such as: Is leaking sensitive national security information ever justified, and if yes, how could we test this? Does it matter who leaks the information and for what purpose? Does it matter if the leak is related to national or foreign intelligence operations? How important is the impact of the leak and how could we balance the presumed negative impact on intelligence assets with the potential benefits on preserving the democratic fabric of our societies? This chapter will explore each of these questions in more detail with the purpose of further developing the discussion about building the outlines of a universal framework for assessing information leaks.

The natural place to start is by trying to understand how can we judge whether an information leak is in the public's interest? Presumably a leak that puts the strategic intelligence assets of a country at serious risk can cause more short and long-term damage to its citizens' security and welfare that benefits for the preservation of democratic processes. Such a leak may also potentially cause harm to a number of operatives in the field, whose names and locations would be revealed and their mission thwarted, thus ultimately possibly producing a result, which is not beneficial for upholding the democratic values overall promoted by a given democracy.

Recent events have thus compelled researchers to try and develop more systematic frameworks of analysis when it comes to information leaks. One that stands out among these is Sagar's 2016 framework of five criteria for assessing whether the disclosure of classified information is justifiable or not: (1) the disclosure must reveal wrongdoings by a government official or agency; (2) unauthorized disclosure must be based on clear and

DOI: 10.4324/9781003190363-22

convincing evidence; (3) unauthorized disclosures should not impose a disproportionate burden on national security; (4) the leaker should utilize the least drastic measures of disclosure and (5) the leaker must be willing to disclose his/her identity. While Sagar's framework might appear easy to use at a first glance, the reality is that it only goes further to show the complexity of the issue.[4]

Take the first criteria for example, namely that the disclosure must reveal wrongdoings by a government official or agency. The first and obvious question that comes to mind and one that Sagar himself pointed out is who gets to decide what constitutes a wrongdoing? This challenge is further compounded in the case of what Kwoka labels as "deluge leaks", such as Wikileaks or the Snowden revelations, where massive quantities of government information were leaked, making it difficult to assess whether each and every piece of information revealed referred to a wrongdoing or not.

Beyond the difficulty posed by the quantity of data there is also the question on who would get to decide on whether something constitutes a wrongdoing or not and on the basis of what. Presumably, this would be left to an independent body such as the judiciary who would not only have to look at the legality of the action referred to in the leaks, as per national legal framework and international conventions, but it would also be advisable to assess whether the action passed the test of proportionality and legitimacy. For example, in the absence of an internationally agreed convention on government surveillance, spying on your allies may be permissible by national legislation, but whether each particular instance of spying is necessary and proportional would be another discussion, which in all circumstances would require a careful analysis.

The issue of "wrongdoing" is rendered even more complex by the distinctions between "right" and "wrong" and "legal" and "illegal". Some actions may of themselves be "enabled by law" and thus "legal" but could still produce an effect which is harmful to society. For the "wrongdoing" criteria to work better, it would need to be interpreted broadly and under-stood as a leak illustrating a situation which is actually or potentially harmful to society. Thus, for example, some of the activities revealed by Snowden may actually have been legal (while others have since been declared by some courts to be illegal) but, although possibly justifiable on one legal basis or another, taken individually or together with others, they may be shown to be harmful to the democratic fabric of society, infringing on a number of fundamental freedoms including privacy, freedom of expression and freedom of association. The Snowden revelations helped drive fundamental reforms of laws governing the oversight of surveillance in France, Germany, The Netherlands and the UK and thus it may be argued that the leaks achieved an important beneficial outcome for society, but only if "wrongdoing" is inter-preted as having a meaning broader than "illegal activity". Indeed, it may be further argued that the significant reforms carried out post-Snowden and the creation of the post of UN Special Rapporteur on Privacy are the most concrete form of proof that Snowden was cor-rect in his judgement that the activities he revealed constituted wrongdoing that could harm the fabric of a democratic society which is bound by the rule of law.

The importance of being able to clearly identify whether the leaks point to a wrongdoing is also closely connected to the extent to which a leaker should face punishment. There are those who argue that if the disclosure provides evidence of illegal (or at least harmful) action by the government this should positively influence the treatment of the source as their actions would be justifiable as serving the public good. However, the judges in the Manning case have disagreed with this approach, declaring that the leaker's subjective intent is irrele-vant.[5] The morass of sometimes contradictory judgements in the US and elsewhere point to a clear need for clarification of the statute's intent and a more holistic approach to legislation.

Since most whistleblower legislation around the world is primarily designed to protect the employment of the employee and related confidentiality conditions, in some countries such as the USA the whistleblower legislation is out of synch with other laws such as the Espionage Act. If the latter and other laws are amended they would need to be able to do so e.g. placing a clear burden of proof on the Government where revelation is carried out as part of a whistle blower action. As will be seen in the next part of the chapter, the public good argument is also not a clear cut one, raising some important questions of whether the source can in any circumstances be absolved from guilt.

The third criteria,[6] referring to the burden that the leaks have on national security is yet another controversial one as more often than not intelligence operations are shrouded in a culture of secrecy that makes it difficult if not impossible to assess the true impact of an information leak on national security. A case in point are the Snowden leaks. In their immediate aftermath, the retiring NSA Director General Keith B. Alexander had called them "the greatest damage to our combined nations' intelligence systems that we have every suffered".[7] This assessment appears to have been at least implicitly rebutted by Alexander's successor, Admiral Michael S. Rogers, who argued that, while the leaks were clearly a negative occurrence, there was no indication that "the sky is falling".[8] In the time since then, several experts and former intelligence practitioners have provided various examples of how Wikileaks and Snowden have negatively affected US national security and interests abroad. These examples fall into three main categories: (a) losing access to human sources in different parts of the globe; (b) disclosing information on information collection techniques and equipment to adversaries and (c) foreign policy set-backs. However, even according to former members of the CIA the magnitude of the impact is difficult to quantify due their long-term implications and the fact that the information coming from the leaks often adds to pre-existing knowledge that adversaries may already possess.[9] Notwithstanding the examples given, there have been journalists, such as some writing in the New York Times, who did point out that no hard evidence of the actual impact was actually put forth.[100]

Is the culture of secrecy a problem in these cases? Certainly yes, especially when also linked to the problem of "overclassification". Evidence has shown that,[11] instead of protecting national security, "overclassification" has the opposite effect, by discouraging intelligence sharing and thus making it more difficult to establish connections and prevent threats. The lack of incentives to avoid overclassification compared to the significant negative consequences for improper disclosures is an incentive in itself to err on the side of caution or, in some circumstances, to play the system in order to hide government failures or misconduct. However, in spite of the many recommendations aimed at reducing overclassification[12] the problem is still rife today, as shown by a March 2020 Pentagon memo implementing a new information restriction category "Controlled Unclassified Information", which, while avoiding the term classified, still aims to restrict public release of information.[13]

So, what are the options in such a case, as it is clear that calling for total transparency when it comes to intelligence is both unfeasible and undesirable? In addition to ongoing reforms aimed at reducing overclassification, which have proved more or less successful, one other available solution is to be found in the way intelligence oversight bodies are set-up. A strong and independent intelligence oversight body, which is given unbridled access to intelligence documents, as is the case with IPCO in the UK would go a long way in ensuring that secrecy cannot be employed as a cover for wrong-doing. While its privileged access would still not be sufficient to fully be able to estimate impact, its independent opinion could be considered as at least partial evidence of the impact derived from the leak.

The fourth criteria put forth by Sagar is for the leaker to utilize the least drastic measure of disclosure, which raises some interesting reflections on how recent technology developments change the way information leaks are carried out and whether the current legal framework for whistleblower protection is adequate. Though information leaks have existed for a long time we cannot separate recent technological evolutions from how this phenomenon has evolved if we look at the sheer volume of information that had been leaked by both Manning and Snowden by comparison to their predecessors. It is clear to everyone that digitization of data makes it easier for individuals to gain access to, leak and widely distribute national security information. This makes Kwoka (2015) raise an additional question on whether the leaker, in the case of what she labels as "deluge leaks",[14] has actually read and/or understood all the information being leaked, which is in fact an essential pre-requisite for them to meet Sagar's limitation criteria.

It also raises some important questions on the objectives of the leakers and the extent to which we can distinguish between mass information leaks and targeted ones. Much as it happens in the debate on whether mass surveillance should be permissible to the same degree as targeted surveillance, when it comes to information leaks there is the question on whether mass leaks can ever be legitimate. By comparison to "targeted leaks", such as Ellsberg's disclosure of information on U.S. policy in Vietnam, mass leaks include large quantities of information on various topics from foreign policy to covert intelligence operations. If weighing the legitimacy of a targeted leak is a complex affair, trying to do the same for a mass leak, which is aimed at various governmental actions can appear to be impossible, when trying to use Sagar's framework. Of course, one could argue that it was that very volume of information that made it possible for the public to fully understand the complexity of the reality revealed (such as the scope and depth of the US surveillance programme) and provided enough compelling evidence.

On the other hand, one cannot discuss choice of means without briefly examining whether the existing legal framework on whistleblowers is fit for purpose. One of the most used arguments against Snowden, including by President Obama,[15] has been his choice of leaking the information directly to the public and not following the official channels available for whistle-blowers. This is a hugely problematic argument in practise since it may also be claimed that the experience of most whistle blowers is not a positive one. On the contrary, blowing the whistle by following prescribed official channels appears to often lead to the whistleblower being at minimum discouraged then consequently marked out, ostracised, prosecuted, demoted and shifted sideways or even losing their employment.[16] There is very little to suggest that in most recent cases whistle-blowers are positively appreciated within the intelligence communities. On the contrary there is plenty of evidence of the intelligence establishment and the Governments closing ranks and doing everything to discourage most forms of disclosure including whistleblowing.[17]

All the evidence from the past 15 years suggests a growing climate within Government circles which is a disincentive towards information leaks. In recent history, the Obama administration, in turn followed by the Trump administration, have systematically increased the number of classified leaks subjected to criminal investigation. In 2017, then-Attorney General Jeff Sessions condemned what he called "a staggering number of leaks" from within the Trump administration, with referral going up by 400% from the previous year. According to the Justice Department this was meant to act as a deterrent against further leaks, though some argued that this action against those leaking information also had a political dimension.[18] The fallacies of the public good argument

Though Sagar's criteria are useful, at least on a theoretical level, to create a better framework for understanding information leaks, it would seem that, without further refinement, they cannot be operationalized sufficiently in order to create a workable test. The reason being the difficulty in addressing the old argument on how to assess whether a leak serves the public good more that it harms it, which is something that criteria 1, 3 and 4 seems to circle back to.

If we look at the arguments invoked in the Manning, Assange or Snowden cases, we see that all three leakers perceived themselves as defenders of public interest, who wanted to inform the public of what the government was doing in secret and which they thought to be wrong. From their point of view the leaks served their purpose of revealing multiple governmental wrongdoings, something which, they believed, could not be achieved by other means.

However, if we are to judge by legal outcomes, research has shown that, at least in the US, by 2017, no individual had ever been acquitted based on a finding that the public interest in the released information was so great that it justified an otherwise unlawful disclosure. This in spite of the fact that there have been cases when the government was unable to secure convictions or has dropped or significantly reduced the criminal charges against the alleged leakers.[19] While this US example refers to a single country and its legal system and cannot necessarily be extrapolated to universal applicability, this does not make it less relevant in trying to understand how the argument of public interest works or doesn't in a court of law.

One of the inherent limitations is that the public interest considered by the courts in the USA has been limited to the citizens of its own country, while it can be argued that revelations such as Snowden's had significant impact beyond US borders. The positive outcome has been that countries such as the UK, Germany and France, among others, have been forced to reconsider government secrecy and the manner in which they carry out intelligence operations. Likewise, the consideration of the public interest may yield different results in regional courts as opposed to strictly national courts. Thus, the European Court of Human Rights and the European Court of Justice did take the public interest into account when considering issues of necessity and proportionality in landmark cases which depended at least partially on undisputed outcomes of the Snowden and other revelations. The *Schrems* decisions in the ECJ and *Big Brother Watch et al. vs UK* in the ECtHR are some examples of wrongdoing being decided upon in a manner favouring society outside the UK and the US. On the negative impact side, the revelations did not only affect US intelligence capabilities, but also does those of their allies, in addition to sowing international mistrust and harming cooperation relationships among Western countries.

The other inherent limitation can be found in the difficulty in assessing the potential benefit to the public interest, especially as it can always be argued that the same benefit, however it may have been quantified, could have been otherwise achieved through legal means.

Preserving democratic values

The problem of information leaks is further compounded by the fact there is no one universally accepted definition of what may be in the public interest and relative metrics, especially so in the case of intelligence community leaks. Dahl (2015) argues that for an association to be fully democratic it needs to meet five criteria, one of which being enlightened understanding. According to him "citizens must enjoy ample and equal opportunities for discovering and affirming what choice in a matter before them would best serve their interests".[20] Did the ordinary American citizen have the same understanding of the US' intelligence capabilities and operations before and after the Snowden revelations? Probably not.[21] An interesting question to ask is whether the citizens in question actually felt that their interest was defended.

A Pew Center survey from 2014 had shown that Americans were almost equally divided over whether the leaks had served or harmed the public interest, and most believed that Snowden should be prosecuted.[22] However, another survey taken in 2015 by the same Center showed that nearly 1/3 of American adults had taken steps to protect their information from government surveillance programs. The survey also showed that while 57% of those interviewed considered it unacceptable for the government to monitor the communications of U.S. citizens, roughly the same percentage (54%) considered it acceptable to monitor the communications of people from other countries. This latter finding provides some insight as to the ambivalent receipt of the revelations. Were one to use this as a benchmark for measuring public interest we would have to introduce a "proximity criteria", as in "the leaks serve public interest if referring to actions that directly impact the citizens of that country" versus "they don't serve public interest if they refer to actions outside national borders". Certainly, such an argument, which is unfortunately used even when drafting national legislation on surveillance, does not hold well against the universality of human rights, though it serves to show that citizens' themselves cannot be considered good judges for whether an action serves the public interest or not.

Going back to Dahl's criteria, is the knowledge acquired sufficient to argue that the benefit to democracy outweighed the costs? Again, probably not. A key condition of a functioning social contract is for citizens to surrender a part of their rights in exchange for prosperity and security. If, however, information leaks hinder the state's capacity to provide security, including by damaging citizens' trust in government, then it makes it difficult for one to argue that democracy has been served.

Information leaks as tools of public policy

What happens when the leaks are not meant for a national public alone? Since the most recent outbreak of hostilities in Ukraine started in February 2022 we are witnessing a controlled release of information leaks by the US as a policy-making tool.[23] Its aim is presumably to act both as a deterrent to Russia, which is losing its strategic advantage by seeing its classified military information being disclosed in the media and to raise awareness among global citizens of the conflict and thus again put indirect pressure on the Russian government.

The controlled leaking of information as a policy-making tool is by no means new, having seen examples of it dating back to the Cold War, but it does raise a question of whether the executive is, or perhaps even shouldn't, be put under the same scrutiny in terms of the leaks serving the public good, as individual leakers.

Using the law for business-process re-engineering when dealing with information leaks

Given the above, the test, if we can call it such, should not only be composed of criteria, such as the ones outlined by Sagar, however further refined e.g. by a sufficiently broad understanding of wrongdoing. Having criteria alone, however sound as a start, is not enough to ensure that a country has all the right conditions in place in order to properly decide whether a national security relevant information leak was sufficiently in the public interest. Instead, it is perfectly conceivable to create an eco-system characterised by the existence of institutions that would help operationalize these refined criteria. This may be achieved by a detailed legal framework which would include a package of measures from which one would be unable to cherry-pick. The key term here would be "package" since the strength of the system would lie in a number of measures which, when taken together, are complementary and supportive of each other.

Democracies should therefore be encouraged to research and undertake an in-depth reform of all applicable regulation such that:

a. strong legislation legislation is created or reinforced, thus protecting the rights of whistle-blowers by ensuring that individuals who witness wrongdoings feel that they have a realistic legal path they can pursue, rather than one which would lead to their being ostracised. This necessarily also entails synchronisation between whistle-blower law and all other laws, especially those concerning espionage and the leaking of official secrets. Most importantly would be the introduction into the relevant national laws of provisions on the lines of

> It shall be a defence to claim that any unauthorised disclosure of information about wrongdoing by the state to bona fide media was done in the public interest. Where such a defence is raised, this matter shall be decided by the court of law specially designated in this Act and the onus shall lie upon the prosecuting state to prove that the public interest was insufficiently advanced by a leak of information relevant to information security. Likewise, the burden of proof that national security interests were actually or potentially harmed lies with the prosecuting state authority. Where the court is satisfied that the leaking of information was indeed carried out in the public interest, it may take and order all measures intended to protect the employment and welfare of the person concerned, including compensation to be paid by the state in recognition of the public service thus rendered by an accused leaker of information relevant to national security.

It may be argued that this type of legal reform would, in practise, further deter wrongdoing by the state given that the hitherto absolute blanket of national security may no longer be abused since the risk of wrongdoing being revealed grows in direct inverse proportion to the reduction of risk of legal sanction of the whistleblower. Almost needless to say, such legislative reform alone, however extensive, is not enough. It needs to be accompanied and supported by regional and national executive policies strongly establishing and reinforcing an institutional culture promoting transparency and accountability. In the same way as a culture promoting legality, necessity and proportionality needs to be "baked into" every level of a national security entity such as an intelligence service, the corresponding culture of transparency and accountability would help deter possible abuses of the immense power that such a national security entity enjoys;

b. an independent independent court with judges enjoying full security clearance is established and/or assigned by statute to examine the evidence brought before it, including evidence of concrete negative impact on national security;

c. a truly independent oversight body, entrenched at law and combining legal, technological and operation expertise, which could authoritatively consider the legality, necessity and proportionality of intelligence operations such that it would be in a position to bring before the assigned court all the evidence required to assess the public interest of the leak.

It may be seen that the above proposals would use the law to operationalise criteria for determining whether a leak of information relevant to national security can be justifiable. Legal mechanisms and principles are thereby used to re-engineer the safeguards afforded to *bona fide* whistleblowers, shift the burden of proof on to the state, and create/reinforce those institutions which would have a role in determining whether a leak was justifiable in revealing wrongdoing or avoiding further harm. By doing so, strong democracies can actually improve the esteem and trust that their national security agencies should aspire to.

Notes

1 Continuing in an even more intense manner a conflict initiated by the Russian invasion of the Crimean Peninsula in 2014 and its erstwhile annexation by Russia.
2 See United States v. Private First-Class Chelsea Manning; and articles on Chelsea Manning in *The Conversation*.
3 Greenwald, "Edward Snowden: the whistleblower".
4 Sagar, *Secrets and Leaks*,132.
5 Brennan Center for Justice, "National Security Whistleblowing".
6 Criteria two and five of Sagar's framework have not been addressed in this chapter as there are perhaps less controversial by comparison to the others.
7 MacAskill, "New NSA chief".
8 Ibid.
9 Myre, "WikiLeaks U.S. National Security?".
10 Editorial Board New York Times, "Edward Snowden, Whistle-Blower".
11 The 9/11 Commission found that the "[c]urrent security requirements nurture overclassification and excessive compartmentation of information among agencies".
12 Goiten et al., "Reducing Overclassification through Accountability".
13 Capaccio, "Military Officer Bemoans Pentagon's".
14 Deluge leak is according to Kwoka a new category of leak encompassing vast quantities of records that the leaker likely knows nothing about, the common threat being simply that the leaker has access to them. See Kwoka, "Leaking and Legitimacy".
15 White House Press Release, "Remarks by the President".
16 See Giglio, "The problem with Whistle-Blower"; Howley. "Violence, Intimidation and Incarceration"; Kiriakou "Fate of America's Whistleblowers".
17 See Connon, "Are Intelligence-Community Leakers". For more historical perspective including examples in law enforcement and elsewhere in analogous cases even outside the intelligence community see Glazer, "Whistleblowers How They Fared"; and Martin, "Illusions of Whistleblower Protection".
18 Klippenstein, "Trump Administration Criminal Investigation".
19 See Benkler, "A Public Accountability Defense", 281, for an analysis of high-profile leak incidents that includes individuals who were not prosecuted.
20 Dahl, *On Democracy*, 45.
21 The state of privacy in post-Snowden America Pew Research Center survey from 2016 showed that 87% of those interviewed had heard about the surveillance programs. See Pew Research Center, "The state of privacy".
22 Pew Research Center, "Obama's NSA Speech".
23 Toosi, "Spy world wary".

Bibliography

Benkler, Yochai. "A Public Accountability Defense for National Security Leakers and Whistleblowers", *Harvard Law & Public Review*, 281 (2014): 311–320.
Brennan Center for Justice, "*National Security Whistleblowing: A Gap in the Law*".
Capaccio, Anthony. "No. 2 Military Officer Bemoans Pentagon's Excess Classification", *Bloomberg*, October 28, 2021. https://www.bloomberg.com/news/articles/2021-10-28/no-2-military-officer-bemoans-penta-gon-s-excess-classification..
Connon, Eric. "Are Intelligence-Community Leakers Internationally Protected or Simply "Whistling in the Dark"? Assessing the protection afforded to Whistleblowers under International Law", *Case Western Law Review* 67 (2017):3 https://scholarlycommons.law.case.edu/caselrev/vol67/iss3/14/..
Dahl, Robert A. *On Democracy 2ⁿᵈ Edition*. Yale: Yale University Press, 2015.
Editorial Board New York Times, "Edward Snowden, Whistle-Blower", *New York Times*, January 1, 2014. https://www.nytimes.com/2014/01/02/opinion/edward-snowden-whistle-blower.html.
Giglio, Mike. "The problem with the Whistle-Blower System", *The Atlantic*, September 21, 2019. https://www.theatlantic.com/politics/archive/2019/09/problem-whistleblower-system/598555/.
Glazer, Myron. "Ten Whistleblowers and How They Fared", *The Hastings Center Report*, vol. 13, no. 6 (1983):33–41. https://www.jstor.org/stable/3560742..

Goiten, Elizabeth and David Shapiro. "Reducing Overclassification through Accountability", *Brennan Center*, October 5, 2011. https://www.brennancenter.org/our-work/research-reports/reducing-over-classification-through-accountability.

Howley, Kevin. "Violence, Intimidation and Incarceration: America's War on Whistleblowers" *Journal for Discourse Studies*, 7.3 (2019): 265–284. https://scholarship.depauw.edu/commtheatre_facpubs/3/.

Kiriakou, John. "The sad fate of America's Whistleblowers", *Otherworlds*, October 14, 2015. https://otherwords.org/the-sad-fate-of-americas-whistleblowers/.

Klippenstein, Ken. "Trump Administration Referred a Record Number of Leaks for Criminal Investigation", *The Intercept*, March 3, 2021. https://theintercept.com/2021/03/02/trump-leaks-criminal-investigation/.

Kwoka, Margaret B. "Leaking and Legitimacy", University of Denver Sturm College of Law Legal Research Paper Series, Working Paper No. 14–49, vol 48:1387, April 24, 2015. https://ssrn.com/abstract=2494375.

MacAskill, Ewen. "New NSA chief says "sky not falling down' after Snowden revelations", *The Guardian*, June 30, 2014. https://www.theguardian.com/world/2014/jun/30/nsa-chief-michael-rogers-snowden-intelligence-leaks.

Martin, Brian. "Illusions of Whistleblower Protection", *UTS Law Review*, 5 (2003):119–130. https://documents.uow.edu.au/~bmartin/pubs/03utslr.html.

Myre, Greg. "How Much Did WikiLeaks Damage U.S. National Security?", *NPR*, April 12, 2019. https://www.npr.org/2019/04/12/712659290/how-much-did-wikileaks-damage-u-s-national-security.

Pew Research Center. "Obama's NSA Speech Has Little Impact on Skeptical Public". January, 20, 2014. https://www.pewresearch.org/fact-tank/2016/09/21/the-state-of-privacy-in-america/.

Pew Research Center. "The state of privacy in post-Snowden America". September 21, 2016. https://www.pewresearch.org/fact-tank/2016/09/21/the-state-of-privacy-in-america/.

Sagar, Rahul. *Secrets and Leaks: The Dilemma of State Secrecy*. Princeton and Oxford: Princeton University Press, 2016.

The 9/11 Commission Report. *Final Report of the National Commission on Terrorist Attacks Upon the United States*. July 22, 2004. https://www.9-11commission.gov/report/911Report_Exec.pdf.

Toosi, Nahal. "Spy world wary as Biden team keeps leaking Russia intel", *Politico*. February 8, 2022. https://www.politico.com/news/2022/02/08/spy-world-biden-leaking-russia-intel-00006956.

White Press Release. "Remarks by the President in a Press Conference", August 9, 2013. https://obamawhitehouse.archives.gov/the-press-office/2013/08/09/remarks-president-press-conference.

19

AN ETHICAL UNDERSTANDING OF MILITARY STRATEGIC COMMUNICATION, PUBLIC RELATIONS, AND PERSUASION

Irena Chiru and Ruxandra Buluc

> We must remember that in time of war what is said on the enemy's side of the front is always propaganda, and what is said on our side of the front is truth and righteousness, the cause of humanity and a crusade for peace
>
> *(Walter Lippmann)[1]*

Conceptual proximities: strategic communication, public relations, and persuasion

The conceptual nexus between various communication practices (i.e., strategic communication, public relations, media relations, crisis management, marketing, publicity, advertising, manipulation, lobby, etc.) has been extensively approached by the literature in the field of communication. The underlying process for all these communicative practices reunited under the umbrella of strategic communication is persuasion. We will analyse the conceptual differences between two of the most relevant practices for the military domain: public relations and strategic communication, as well as the role persuasion plays in their operation.

Public relations (PR) is "a strategic communication process that builds mutually beneficial relationships between organisations and their publics." Edward L. Bernays, considered the father of modern public relations, assigned three important meanings to the concept: a) information the public receives; b) persuasion aimed at the public to alter attitudes and actions; c) efforts to align an organisation's attitudes and actions with the public and vice versa.[2] These three facets still hold true at present, and signal the struggle for public support that is at the core of public relations. At the same time they raise ethical questions regarding the interests at play in the endeavour of shaping public consent. Bernays explains that public relations function as a two–way instrument of adjustment and integration between the organisational goals and the public's understanding and support.[3] He also focused on what characterises the public, namely the demand for information and the expectation that they be consulted in matters that concern society, lest they try to find their information elsewhere and choose undesirable courses of action for the organisation.[4]

DOI: 10.4324/9781003190363-23

The literature in the field has been very productive, advancing several definitions for public relations and their scope. From the common points in definitions it becomes clear that public relations is a dynamic two-way strategic communication process, part of the management function of an organisation, whose goal is to foster mutually beneficial relationships between the organisation and its publics, making use of influence and thus allowing the organisation to smoothly conduct its business; it includes research, planning, implementation and evaluation[5]. Therefore, we could argue that PR focuses mainly on communication aimed at maintaining a positive relationship between organisations and the community, and, to this end, makes use of influence and persuasion.

Strategic communication encompasses public relations, as it has a broader scope, referring not only to building positive relationships with publics, but also to understanding and managing the target audiences, the informational environment, the ways in which narratives expounding the organisation's values, interests and objectives can be created, transmitted, mediated in our currently digitally connected and controlled environment, and how the publics could interact and interfere with these narratives. A simple definition of strategic communication is "the practice of deliberate and purposive communication that a communication agent enacts in the public sphere on behalf of a communicative entity to reach set goals."[6]

However, behind these simple terms, one can infer the complex scope of this type of communication, of the processes involved, of the interests, values and objectives promoted, of the long-term goals it sets forth. Strategic communication has the ability to shape, modify, change, instil or extinguish behaviours, beliefs, convictions, attitudes in the audiences it addresses, and consequently, if used effectively, it could alter not only social groups or communities but the very fabric of societies. This is the reason why strategic communication has become an integral part of governmental bodies' communication enterprises, and the military is no exception.

What was less obvious in definitions of public relations, but becomes overtly expressed in strategic communication, is the role that persuasion plays in bringing about the desired effects. Persuasion has been defined in various fashions over the years, starting from Aristotle's definition that identified its three main elements: logos (the reasoning), ethos (the responsibility) and pathos (the emotions). Gass & Seiter investigated numerous definitions of persuasion and identified the following common elements: it is a continuous and interactive process in which a communicator attempts to form, modify, strengthen, affect, extinguish the audience's beliefs, attitudes, behaviours through messages that are transmitted via various channels.[7]

Charles U. Larson[8] provides a definition of persuasion which implies that persuasion requires intellectual and emotional participation between both persuader and persuadee that leads to shared meaning and co-created identification. He explains that persuasion can raise ethical dilemmas as people use it to alter other people's beliefs, attitudes, values, actions; it requires the persuader to make "conscious choices" with respect to the goals and means employed; it involves judgement on the part of the receiver, the persuader or other independent persons[9]. In this light, he approaches the issue of "responsible persuasion" and sets three prerequisites for it to occur: (1) both parties involved in persuasion have "approximately equal access to the media of communication available"[10], neither of them is censored or gagged; (2) persuaders reveal their agendas to the audience beforehand so the latter are aware they are subject to persuasion; (3) receivers of persuasion are responsible for testing the persuader's assertions, for seeking alternative sources, for not judging before careful consideration of the issue; hence the call for the public to become "critical receivers". The issue of responsible persuasion lies at the core of debates regarding the ethical nature of strategic communication, in the military field and beyond, as we will explore further.

Projecting and Perceiving the Military: An Understanding of the Military Public Relations and Strategic Communication

Military organisations or armed forces are organisational entities shaped for a wide range of purposes which are generally centred around providing national security, understood as the "pursuit of freedom from threat and the ability of States and societies to maintain their independent identity and their functional integrity against forces of change, which they see as hostile"[11]. Using the framework for understanding provided by E. Forster, armed forces play different functions vis à vis society[12], which have varying implications for their relationship with society and their bases for legitimacy. As critical national security defence entities, the armed forces' main prerogative is the controlled use of violence, but this is continuously (re) shaped in relation to functional requirements, as well as to social and cultural variables. Hence, since before recorded history itself, the social function of armed forces has been decisive for their own identity and functionality.

One of the means of expression for this social function of the military is through public relations or public affairs as part of the civil-military relationship and as an expression of the calls for military "image making". Military activities at the end of the 20th and substantially the beginning of the 21st century have been divided between the opportunities and the pressures of preserving credibility and legitimacy by projecting solid public images and parlaying an organisation's values and actions with a compelling story. Mirroring these macro-societal trends, the relationship between the military and the society has been driven by the requirements and public expectations of the "perception is reality" principle.

The main catalysing factor in "military image making"[13] has been given by the professional interests of the military organisations together with their increased involvement in government policies and different societal sectors. Hence, the development and growth of military public relations can be understood in the larger framework of the performance and efficiency evaluated not merely by the results in times of war or peace, but by the extent to which these results are publicly explained and apprehended, and public support for them is garnered. In other words, public opinion management has become a goal of military public relations.

Historically, military public relations have been pioneered in the US and relied on the pragmatic needs dictated by the Revolutionary War. The strategies and tactics used to swell the ranks of patriots dedicated to the Revolutionary cause are examples of early public relations. PR remained in its infancy until World War I when the Committee on Public Information (CPI) was established in order to mobilise public opinion behind the war effort[14]. The committee used every available means of mass communication—from newspapers and vivid images in posters to speeches in churches and service clubs to instil fear through war propaganda. The CPI often blurred Wilson's political goals with the national interest. Following the end of the war in 1918, the committee started to be criticised for mixing political goals with national interests and for overselling the war. At the end of the conflict, a new public information unit was organised in the Army's Military Intelligence Division, which, in 1929, became the Public Relations Branch. Later on, distancing himself from these practices and considering them an example of mistakes to be avoided, President Franklin D. Roosevelt created the Office of War Information to promote World War II. The office documented American life and culture by showing aircraft factories, members of the armed forces, and women in the workforce and largely used photographs and captions with emotional content to inspire patriotic fervour in the American public[15].

Also in the UK, the development of PR practices is related to the First and the Second World Wars and later on to the "The Projection of England"[16] and the reflection of the emergent concept of cultural diplomacy embodied by the British Council. The system of military-media relations incorporating the national media and reporters into the war effort was institutionalised in 1937 as the War Office's Director of Public Relations. From this period, the PR techniques have been progressively applied, including formal training in giving media interviews during operations into the Royal Military Academy (1970), training soldiers rather than civilians as media escorts on operations, creating the Territorial Army Pool of Information Officers[17], and the explicit acknowledgement of constructive media relationships as an essential variable impacting the morale of the UK deployed forces[18].

Two important observations must be made here which both refer to and reflect the "opinion management" line of action directly connected to public relations. Firstly, such a retrospective look directly connects the beginning of the PR practices to the auspices, culture and necessity of conflict and war propaganda, and to influencing and tailoring public opinion of the conflict. Hence, the dual nature embedded in the concept and, in particular, in the practical applications of the principles and techniques of PR—PR employ communication, while being driven by an extended portfolio of latent political and ideological goals. Secondly, the embryos of both PR and the strategic shaping of public opinion are to be found in or close to military and intelligence organisations and leadership which could raise questions about their ethical foundations.

What was dictated by circumstantial war propaganda requirements has in time been reshaped and become a widely embraced practice[19], either as public relations or as public affairs. More or less publicly overt and transparent, all military organisations produce some form of persuasive communication "to try to motivate men, women and occasionally even children to serve and fight"[20]. In other words, communication has been central to the ways in which military organisations and their actions have been projected and perceived, to how public opinion has been managed, particularly in the twentieth and twenty-first centuries, along with the unprecedented development of the mass- and multimedia technologies.

Most armed forces in most democratic countries have set up public communication policies as part of their own engagement to keep the public opinion informed and facilitate the conditions that lead to public trust in the readiness and righteousness of the military to conduct operations in peacetime, conflict and war. They are meant to provide information on the role of the military in general or during specific operations, maintain awareness, facilitate recruitment and retention, and counter social perceptions and fears which are frequently associated with military interventions and their humanitarian impact. A brief look at the corpus of armed forces' strategies in the field allow us to conclude that the rule of law, transparency, accountability, human rights, and genuine openness towards communication are the key-words of the military public engagement with the media and public opinion.

For example, in the field of defence, the US Department of Defence expressed for the first time its explicit will to reach out to its key audiences under the umbrella of strategic communication defined as

> focused United States Government efforts to understand and engage key audiences to create, strengthen, or preserve conditions favourable for the advancement of United States Government interests, policies, and objectives through the use of coordinated programs, plans, themes, messages, and products synchronised with the actions of all instruments of national power[21].

This definition encompasses the attempts to coordinate and give strategic coverage to the various departments involved in publicly transmitting government messages regarding US strategic values and interests to external as well as internal audiences.

Soon after, NATO also took a keen interest in employing strategic communication to further its mission and objectives both at a strategic level to societies in general, be they from NATO member countries or third parties, and at operational and tactical levels in theatres of operations.[22] This focus on strategic communication reflects the need NATO felt to adapt its influence strategies to the requirements of the contemporary operational environment; it was particularly driven by NATO forces' inability to gain the support of the Afghan populations to the extent hoped and within the timetable set[23].

Designed to reinforce the influence exercised by NATO communications over its audiences, the current NATO StratCom aims to (1) contribute to the successful implementation of NATO operations, by incorporating strategic communications at all levels, (2) build public awareness, understanding, and support for specific NATO policies, operations, and other activities in all relevant audiences, internal and external (3) contribute to general public awareness and understanding of NATO as part of a broader and on-going public diplomacy effort[24]. The NATO Strategic Communication concept is designed to ensure that audiences receive fair, clear and opportune information regarding its actions, and that the interpretation of the Alliance's messages is not left alone to NATO's adversaries or other audiences[25]. NATO military strategic communication is developed to reflect the challenges of the informational environment in which the campaigns are undertaken. As Hotzhausen and Zerfass explain, "what sets the current public sphere apart from that of the 20th century is that it is more participative rather than representative."[26] This has led to a change in the way in which strategic communication campaigns are designed and implemented; the focus is now equally split between understanding audiences, creating and sending the appropriate messages, managing the informational environment, monitoring public reactions to them and adapting them accordingly.

Starting with the NATO Military Concept for Strategic Communication[27], continuing with the Strategic Communications Directive[28], and up to the most recent version, the NATO Military Policy on Strategic Communications[29], strategic communication from NATO perspective groups public diplomacy, military public affairs, psychological operations and information operations under the umbrella of strategic communication. All the types of communication mentioned use persuasion to various degrees, as their ultimate goal is to influence perceptions, attitudes and behaviour, affecting the achievement of political and military objectives which can be the subject of a discussion in ethical terms. James P. Farwell[30] explains that the basic difference between public affairs, public diplomacy and psychological operations lies in the approach they take to the facts they present.

More precisely, public affairs use influence indirectly, as their main goal is to inform the public. They focus on accuracy and reporting the facts as comprehensively as possible. But as, Christopher Paul also explained, there is "no magically, neutrally informing"[31]. Once informing the public begins, a certain stance is adopted from which the facts are presented. Farwell also points out that public affairs fully respect the truth "but use it actively to drive narratives favourable to our actions that influence the media and their audiences."[32] This is what constitutes smart public affairs which "*always* seeks to influence, if for no other reason than to bolster credibility."[33] As we will explain further, credibility is the foundation of trust, and maintaining the trust of the audiences is key to ethical strategic communication.

Psychological operations have a different approach to presenting facts, meaning that their aim is not, first and foremost, to inform, but "to further the interests of our military personnel and their endeavours, not those of the target audience."[34] In the case of PsyOps, there is a different target audience than in the case of public affairs. PsyOps are not undertaken on domestic soil, which means that the public is foreign and possibly not friendly. In fact, PsyOps are the most controversial in terms of ethics as Farwell explains[35], because they may employ truth selectively, may omit facts and even mislead the audience. The goal of PsyOps is not to present accurate facts in the best interests of the target audience, but to support the objective of military operations. For this reason, PsyOps are often conducted in secret, unlike public affairs and public diplomacy.

When it comes to public diplomacy, there is again a different approach. Public diplomacy influences foreign publics in order to advance national goals and interests, and in this respect it is similar to PsyOps. However, it takes place in the open, it addresses more general issues, trends, attitudes, beliefs that it attempts to mould by using persuasion openly. It strives to create "long-term relationships and addresses an entire country or region, and entails open and public conduct. Public diplomacy fosters dialogue, with the process of sending and receiving to promote understanding."[36] Public diplomacy combines communicative approaches from public affairs and diplomacy as such in order to obtain a measurable impact on the attitudes of a larger foreign audience.

In its 2019, the British MoD proposed an updated definition of strategic communication as "advancing national interests by using Defence as a means of communication to influence the attitudes, beliefs and behaviours of audiences."[37] As such, it is based on a) Planning Defence activities focusing on how and what they communicate and on the interpretation of those activities by those audiences whose behaviour is to be modified in some way; b) consistency of actions, images and words so as not to create the communicative gaps through which disinformation could spread. The document also places great emphasis on target audience analysis and on an information environment analysis which are considered essential for designing the most appropriate activities, to reach the desired effects and results[38]. If these analyses are absent or mistaken, the effects could be damaging to national interests, which could be affected by disinformation campaigns.

Our analysis of the evolution of military public relations and strategic communication, based on open, publicly available documents,[39] has revealed the fact that the US, NATO and the UK have dealt more extensively and openly with the issue, in an attempt to make its persuasive role and mechanisms more transparent and easily understandable by the public. This is, in itself, an ethical stance with respect to the ways in which strategic communication (public relations included) could play an active role in shaping public opinion on military issues and operations, and a testament to the overarching tenet that in order for persuasion to reach its goals, the public must be aware, informed and understand the need for certain attitudes, behaviours or actions to be changed, accepted or extinguished.

The Contentious Discussion on Military Strategic Communication

Achieving information dominance while preventing the competition from taking advantage of the information environment has always been a goal in promoting national security, both during peace and war times. If in the past this informational competition was mainly fought by military organisations in order to achieve a military qualitative or quantitative advantage, more recently the weaponisation of information and disinformation has transcended the frontiers and has been actively applied in politics, economy and culture, thus initiating what we would term an era of "mess communication". Hence, information and disinformation are presently one of the most used tactics of power expression, exertion and control either by the military or by other entities in security and defence.

In this context and looking at the military public communication practices from an ethical perspective, several observations can be made. More than ever, the distinction between the role of military organisations in using information, on one hand, and fighting disinformation, on the other hand, is indistinct. As previously mentioned, the explicit undertaking of PsyOps and information operations as part of the strategic communication doctrine reinforces this frail demarcation, especially in cases in which they do not operate fully transparently due to the classified nature of the operations.

To that, one can add the complexity brought forth by the involvement of private companies or strategic communications firms offering computational propaganda as a service or by the use of computational propaganda to manipulate public opinion as part of several governments' policies. For example, 25 countries were working in 2019 with private companies for disinformation purposes and 70 countries were using computational propaganda to manipulate public opinion[40]. Hence, although the principles of war prove to be enduring and still apply, the circumstances of their application have changed.

In addition, military operations are nowadays under media and social scrutiny, while many countries are confronted with the challenge of gaining consensus and proving their compatibility with public interest. Against this unprecedented multiplication of entities that are using (dis)information offensively or defensively, the role and impact of military organisations as actors in facing and countering disinformation must be understood and possibly reconsidered, through the lens of strategic communication meant to provide accurate accounts of military operations and shape public understanding in favour of the military organisations.

There is a pervasive dimension in the "opinion management" actions conducted by military organisations or by the government bodies through the armed forces. Its most explicit expression is geared by the most visible military action which is war. Looking at military actions at least for the past 100 years, publicly justifying the causes of conflict and seeking to gain support for the burdens of casualties and its costs and to maintain public acceptance seem to have been the rule and not the exception[41]. Therefore, as part of foreign policy, war "is not as not so much about facts, but about perception and public opinion"[42].

Recent examples provided by the Gulf War, the Iraq War or the weapons of mass destruction case demonstrated that waging a sustained war cannot be done or even conceived without the support of its people. It takes all "persuasive powers to achieve this"[43], from the use of manifest to latent various instruments such as news outlets, sensational narratives or revelations. The reference to these recent and largely debated examples is not intended as value or ethical judgements concerning who is right and who is wrong, but rather to illustrate the challenges in investigating the mechanisms involved in the political call for military intervention or potentially controversial security policy.

To conclude, strategic communication has been central to the ways in which wars and military service have been projected and imagined, particularly in the twentieth and twenty-first centuries. However, simply acknowledging the power of strategic communication and its persuasive undertones as a means for the military to maintain its public support is not enough. Both the credibility and legitimacy of the military significantly depend on truthful accounts of activities that are carried out in the name of the people.

Reconsidering Ethics in the Era of "Mess Communication"

From an ethical perspective, what we need to analyse is a complicated landscape in which the military organisations have learnt and been taught by the historical circumstances to promote themselves and use persuasive techniques in selling the cause of national security and the need for action. In order to determine what constitutes ethical and unethical military strategic communication approaches used as a strategy in the information or counter-disinformation warfare we propose both a deontological—intentions-based and a teleological—consequences-based perspective.

Two ethical considerations with respect to military strategic communication and their use of persuasion can be advanced here. Firstly, the enduring issue of trust. Does the communicator have the public's best interest at heart? And to what extent does the public accept the fact that the communicator has their best interest at heart? This is the ethical linchpin that can either bring the whole strategic communication process together or undermine it entirely. Trust and credibility underlie the answer to both questions. The trust the public places in the authority that communicates and the credibility that said authority consequently has are two sides of the same coin. If that trust is frayed (possibly due to previous miscommunication or misinformation or even blatant disinformation activities), then the ethical nature of strategic communication is questionable and, therefore, it cannot reach its persuasive goal. Transparent and responsible persuasion requires the public to be open and receptive to the transmitted messages (as well as to the values and interests contained therein). In order to be open, the public needs to trust the source; otherwise they will be impervious to any persuasion attempts.

Trust relies on perceived legitimacy, a clean track record, consistency between messages and deeds, which, in turn, require effective communication and public relations. In an increasingly polarised world, where information and disinformation cohabitate and address their own separate and even distinct audiences, where trust is a very rare commodity for authorities, the ethics of strategic communication comes under scrutiny more and more, as people question the very values and interests the communicators base their endeavours on.

By default, military strategic communication has an advantage with respect to other types of public communication undertaken by various public institutions. The military's underlying interest and goal is to provide security and foster peace, which most people, irrespective of their political views, would agree are in their best interest. Even so, the extent to which people would allow themselves to become the subjects of persuasion depends on the credibility and track record of the entities perceived as engaging in military strategic communication, on the side of the conflict they find themselves on, and on what other values and interests come associated with security and peace (such as democracy, gender equality, freedom of religion and speech, etc.) which may not be desirable for foreign audiences.

Secondly, armed forces cannot be as transparent as other government agencies and therefore they adopt what might be called "controlled transparency"[44]. This means that although military organisations assume a more active role in public space, certain topics and areas of debate must remain completely secret or will be addressed only in terms of national interest that entails secrecy. From an ethical perspective, the question is: is there any relevant information withheld from the public due to need-to-know and secrecy constraints? And if that information would come to be revealed later, would it prove nefarious to the public? Due to secrecy requirements, military operations cannot be fully overt, and not all information can be released to the general public. However, at present, people are used to having unlimited access to information any time and any place and thus the public feels entitled to be presented the whole picture, even with respect to covert military operations, and are less likely to trust that their representatives know best. If and when the public perceives there is a gap in knowledge, it opens the door to disinformation which can weave a weaponized narrative that further undermines legitimacy and trust.

An Ethical by Design Military Strategic Communication Framework

Shannon A. Bowen explains that ethical communication relies on every side involved having access to "full, truthful, and accurate information as well as voluntary participation and free choice."[45] Communication ethics requires that moral autonomy and freedom to choose are guaranteed in order to foster good, pro-active dialogue, facilitate change, uncover truth, manage risks, create understanding. In other words, for any goal of strategic communication, ethics should play a role in maintaining an open, transparent and autonomous relationship among those involved in the process. Moreover, "ethical strategic communication fosters relationships and trust,"[46] which in turn allows a more facile attainment of set goals.

Traditionally, public relations ethics are based on five pillars which provide an analytical instrument to help frame and solve controversies: Veracity (to tell the truth), Non-maleficence (to do no harm), Beneficence (to do good), Confidentiality (to respect privacy), Fairness and social responsibility (to be fair and socially responsible).[47] However, these pillars need to be re-examined in light of controlled transparency previously mentioned. Military strategic communication has certain limitations that cannot be transcended, and which, if not well explained to the public, might undermine its persuasive function. In this era of open and pervasive communication and vast amounts of freely available data, it is vital for military strategic communication practitioners to become an active part of the discussion that explains both the benefits as well as the constraints of engaging in public dialogue on issues of national security.

Irrespective of how accurate and desirable the above-mentioned traits of ethical strategic communication are, they are quite difficult to operationalize in a given context. Therefore, we advance a framework that allows for concrete statements, which applied to particular situations, could provide the necessary guidance to achieve ethical military strategic communication. Focusing both on practitioners, the deontological perspective, and on the public (citizens), the teleological perspective, the framework will embrace a two-fold, integrative approach to determining the holistic ethical underpinnings of strategic communication, while asserting the following principles:

- **Veracity and accountability**—Accountability in front of the public refers to the extent to which those approving military operations them feel they could defend their decisions before the public if their actions became public. This test could be very well incorporated in strategic communication. Strategic communicators should attempt to "tell it like it is" more often than not and to the extent that that information can be made public or explain why some information cannot be released. In this sense, D. Omand[48] provides as a test of ethics the following question: to what extend "those approving (...) feel they could defend their decisions before the public if their actions became public?"
- **Non-maleficence and beneficence**—The desired advantages as a result of strategic communication should be justifiable in terms of the public good, meaning the people's best interests are protected/considered. The possible harmful side-effects are taken into account, as well as their extent.
- **Confidentiality**—The persuasive leverage to be used in communication must be proportionate to the seriousness of the situation to be publicly discussed within the framework of human rights. People's rights should be protected as much as possible from limitations or invasion, and only in dire, and well explained cases, could the deployment of the security apparatus of the state, with its attendant moral hazards, be justified in relation to the scale of potential harm to national interests that is to be prevented.

- **Fairness and social responsibility**—Costs-benefits analyses are needed, but not just in terms of "own costs" but also societal, meaning the acceptable risks of unintended consequences, such as collateral damage during and after military operations.

Furthermore, in order to help cultivate "critical receivers" of persuasion, we reinforce the need for the audience to be aware of and signal possible mishandlings of information by strategic communication practitioners. To this end, we consider that existing points of concern in the literature[49] which were initially designed to signal when journalism mishandles descriptions of violence, could become very useful starting points for guidelines to be used by critical receivers of persuasion with respect to the military. Consequently, when being exposed to military strategic communication, citizens should actively try to answer the following questions:

- **Are military operations decontextualised in the process of communication**? Are only the enemy's irrational intentions presented without looking at the greater underlying reasons for conflict and dispute? Is the enemy's demonisation the only narrative promoted and are one's own nation's interest and gains shaded?
- **Does the messaging reduce the number of parties in a conflict to two**, when often more are involved? Does the story just ignore the role played by other outside or external forces such as foreign governments and transnational companies and thus oversimplify the narrative?
- **Does the narrative present military operations and interventions as inevitable** by omitting alternatives?
- **Does it fail to explore peace proposals** and to offer images of peaceful outcomes and focuses solely on conflict as a means to resolve the situation? To what extent does it ignore reconciliation attempts which could heal the society in question? When information about attempts to resolve conflicts are absent, fatalism is reinforced, fatalism being a mitigating factor in justifying future conflicts.

Employing these guidelines consistently could help ensure that the public is also actively engaged and aware of the persuasive enterprises, and feels empowered to control the process. Thus the public may be more willing to accept that, when strategic communication messages do not raise questions with respect to any of these issues, they can be considered trustworthy and in their best interests.

Final Remarks

Ethical communication does not depend solely on transparency and speaking the truth, but its foundation lies at the very core of the undertaken actions themselves. Those actions must be ethical in and of themselves so as to be presented as such to the public. In the case of the military, actions speak as loudly as words, and if discrepancies appear, they will have deep effects on the state's ability to ensure citizens' security. Even if some actions may at first glance appear unethical, it is the role of military strategic communication practitioners to come to the fore of the public discussion and to explain what the undesirable consequences of not taking those actions would have been. There are many "catch 22" situations in military operations; however, if openly acknowledged and discussed, they could become opportunities to foster public trust and ensure effective future communication.

In addition to the existing string of academic theorisation, the military public communication and the media function as an agent of power require further consideration, which should allow a systematic re-evaluation of the values and ethics of both messages but also military actions themselves. One of the issues that deserves to be problematized in the same ethical context is the responsibility involved in the process of attracting new recruits and the degree to which employing techniques derived from advertising and commercial marketing makes military communication more persuasive while ethical. Also, an evaluation of those representations of the military in cultural artefacts aimed at raising the attractiveness of the service in armed forces would be needed. Last but not least, the way in which ardent topics such as redressing injustice, loneliness or global crises are employed and the shift from more traditional representations of the military service to new ones are probably to be discussed in the future.

Notes

1 Lippmann, Walter. Public Opinion. Free Press Paperbacks, 1997.
2 Bernays, *Public Relations*, 12.
3 Idem, 16.
4 Bernays, *Crystallizing*.
5 For an extended list of definitions see Lehmann, *Public Relations* and Smith, *Strategic Planning*, 57–58.
6 Holtzhausen & Zerfass, "Strategic Communication", 283, 74.
7 Seiter & Gass, *Perspectives on Persuasion*,
8 Larson, *Persuasion*, 22–24.
9 Idem, 43.
10 Idem, 23.
11 Buzan, "New Patterns", 432.
12 Forster, *Introduction*.
13 Stanhope, The Soldiers, 294.
14 Neumann, "Committee".
15 Koppes & Black, "What to Show",
16 Sir Stephen Tallents' pamphlet *The Projection of England* (1933) argued for the creation of a school for "national projection"—a propaganda body that would defend the British "sense of fair play" from the challenges of America, fascist Europe and the Soviet Union. Tallents' notion of cultural diplomacy led to the creation of the British Council (1934).
17 Badsey, "The Media", 122.
18 Pyper, "The media".
19 A representative example can be found in the FM 3–61.1 Public Affairs Tactics, Techniques and Procedures, HEADQUARTERS, DEPARTMENT OF THE ARMY, available at https://irp.fas.org/doddir/army/fm3-61-1.pdf.
20 Maartens & Bivins, *Propaganda*.
21 Strategic Communication Joint Integrating Concept.
22 NATO Military Concept for Strategic Communications.
23 Reding, Weed & Ghez, *NATO's Strategic Communications*.
24 Stratcomcoe, About Strategic Communications.
25 Reding, Weed & Ghez, *NATO's Strategic Communications*.
26 Holtzhausen & Zerfass, "Strategic Communication", 5.
27 NATO Military Concept for Strategic Communication.
28 ACO Strategic Communications.
29 NATO Military Policy on Strategic Communications.
30 Farwell, *Persuasion and Power*.
31 Paul, *Strategic Communication*, 14.
32 Farwell, *Persuasion and Power*, 40.
33 Idem, 40.
34 Idem, 5.
35 Idem, 5.

36 Idem, 52.
37 Joint Doctrine, 4.
38 Idem, 8.
39 There may be other documents pertaining to strategic communication in the military, but they are classified and, therefore, inaccessible.
40 Howard & Bradshaw, *The Global Disinformation*.
41 Welch & Fox, *Justifying*.
42 Simons, "Policy", 2.
43 Knightley, "The Role of the Media", 372.
44 Dumitru, "Building".
45 Bowen, "Strategic communication".
46 Idem.
47 Parsons, *Ethics*, 26.
48 Omand, *Securing*, 287.
49 Galtung & Ruge, "Structure".

Bibliography

About Strategic Communications, available at https://stratcomcoe.org/about_us/about-strategic-comm unications/1.

ACO Strategic Communications, May 2012.

Badsey, Stephen. *The Media and International Security*. Matthew Midlane Psychology Press, 2000.

Bernays, Edward L. *Public Relations*. University of Oklahoma Press, 1952.

Bernays, Edward L. *Crystallizing Public Opinion*. New York: Boni And Liveright, 1934.

Bowen, Shannon A. "Strategic Communication, Ethics of" in *The International Encyclopedia of Strategic Communication*. Robert L. Heath and Winni Johansen (Editors-in-Chief), Jesper Falkheimer, Kirk Hallahan, Juliana J. C. Raupp, and Benita Steyn (Associate Editors). JohnWiley & Sons, Inc.: 2018.

Bradshaw, Samantha & Philip N. Howard. *The Global Disinformation Disorder: 2019 Global Inventory of Organised Social Media Manipulation*. Working Paper 2019.2. Oxford, UK: Project on Computational Propaganda.

Buzan, Barry. "New Patterns of Global Security in the Twenty-First Century." *International Affairs (Royal Institute of International Affairs 1944-)* vol. 67, no. 3, 1991, pp. 431–451, https://doi.org/10.2307/2621945. Accessed 20 Jul. 2022.

Dumitru, Irena. "Building an Intelligence Culture From Within: The SRI and Romanian Society", *International Journal of Intelligence and CounterIntelligence* 27:3 (2014): 569–589, doi:10.1080/08850607.2014.900298.

Farwell, James P. *Persuasion and Power. The Art of Strategic communication*. Washington: Georgetown University Press, 2012.

Forster, Anthony. "Introduction: Armed Forces and Society in Europe" in *Armed Forces and Society in Europe*. Palgrave Macmillan, London, 2006, 1–18. https://doi.org/10.1057/9780230502406_1.

Galtung, Johan and Mari Holmboe Ruge. "The Structure of Foreign News", *Journal of Peace Research* 2 No. 1 (1965): 64–91.

Ghilani, Jessica L. "The Army Just Sees Green. Utopian Meritocracy, Diversity, and United States Army Recruitment in the 1970s" in *Propaganda and Public Relations in Military Recruitment Promoting Military Service in the Twentieth and Twenty-First Centuries*. Brendan Maartens and Thomas Bivins (eds). London &New York: Routledge, 2021, 145–157.

Holtzhausen D. & A. Zerfass. "Strategic Communication, Opportunities and Challenges of the Research Area" in *The Routledge Handbook of Strategic Communication*, Derina Holtzhausen and Ansgar Zerfass (eds.). Routledge Taylor and Francis, 2015.

Holtzhausen, D. R., & Zerfass, A. (2013). "Strategic communication—Pillars and perspectives on an alternate Paradigm." In K. Sriramesh, A. Zerfass, & J.-N. Kim (Eds.), *Current Trends and Emerging Topics in Public*. Relations and Communication Management. New York, NY: Routledge, 2013, 283–302.

Joint Doctrine Note 2/19 Defence Strategic Communication: *An Approach to Formulating and Executing Strategy*, UK Ministry of Defence (MOD)

Knightley, Phillip and Welch, David (2012). *Justifying War: Propaganda, Politics and the Modern Age*, Palgrave Macmillan.

Koppes, Clayton R., and Gregory D. Black. "What to Show the World: The Office of War Information and Hollywood, 1942–1945." *The Journal of American History* 64, no. 1 (1977): 87–105. https://doi.org/10.2307/1888275.

Larson, Charles U. *Persuasion. Reception and Responsibility.* Wadsworth: Cengage Learning, 2010.

Lehmann, Whitney. *Public Relations Writer's Handbook.* New York & London: Routledge, 2020.

Maartens, Brendan. "Narratives of service, sacrifice and security. Reflecting on the legacy of military recruitment" in *Propaganda and Public Relations in Military Recruitment Promoting Military Service in the Twentieth and Twenty-First Centuries.* Brendan Maartens and Thomas Bivins (eds). London &New York: Routledge, 2021.

NATO Military Policy on Strategic Communications, July2017.

NATO Military Concept for Strategic Communication, July2010.

Neumann, Caryn E. *Committee on Public Information,* available at https://www.mtsu.edu/first-amendment/a rticle/1179/committee-on-public-information.

Omand, David. *Securing the state.* Oxford: Oxford University Press, 2010.

Parsons, Patricia J. *Ethics in Public Relations. A Guide to Best Practice.* London: Kogan Page Limited, 2008.

Paul, Christopher. *Strategic Communication. Origins, Concepts, and Current Debates.* Santa Barbara: Praeger, 2011.

Pyper, H.H. *The media in Modern Warfare—Friend of Foe.* In Hawk, Bracknell RAF Staff College.

Reding, Anais, Kristin Weed, and Jeremy Ghez. *NATO's Strategic Communications concept and its relevance for France.* Santa Monica, CA: RAND Corporation, 2010 available at https://www.rand.org/pubs/technical_ reports/TR855z2.html.

Seiter, John S. & Robert H. Gass. *Perspectives on Persuasion, Social Influence, and Compliance Gaining.* Pearson, 2003.

Simons, Greg. "Policy and Political Marketing: Promoting Conflict as Policy", *Journal of Political Marketing*, 0:1–19, 2020, doi:10.1080/15377857.2020.1724426..

Smith, Ronald D. *Strategic Planning for Public Relations*, Sixth edition. Abingdon: Routledge, 2021.

Stanhope, H. *The Soldiers.* London: Hamish Hamilton, 1979.

Strategic Communication Joint Integrating Concept (2009) available at https://www.jcs.mil/Portals/36/ Documents/Doctrine/concepts/jic_strategiccommunications.pdf?ver=2017-12-28-162005-353.

Welch, David & Jo Fox (eds.). *Justifying War: Propaganda, Politics and the Modern Age.* London: Palgrave Macmillan, 2012.

20

EMOTIONAL DIPLOMACY IN TIMES OF UNCERTAINTY AND DISINFORMATION

Tereza Capelos and Grete Krisciunaite

> Sometimes people ask me—with tongue in cheek—so who is the enemy? And I respond with conviction—confusion, instability, unpredictability.
>
> *President George W. Bush (National Security Archive, 1990: 3)*

Introduction

This chapter engages with uncertainty in international politics in the context of disinformation and beyond. We offer an in-depth analysis of types of uncertainty and their psychological properties, and discuss how understanding uncertainty can be valuable for the practice of emotional diplomacy in the context of disinformation. Disinformation, as a tool of political warfare, originates in the practice of "dezinformatsiya" by intelligence bodies in the Soviet Union. It is a deliberate, high frequency campaign built on narratives, constructed with the aim to deceive (Starbird, 2020). It targets wide and specialised audiences (Paul and Matthews, 2016; Mareš and Mlejnkova, 2021) and compromises societal, environmental, economic, military, and political security (Buzan et al., 1998). At the national level, disinformation portrays political actors as liars who betray or misgovern their country (Mareš and Mlejnkova, 2021). At the international level, disinformation focuses on the calculated spread of false information by non-state actors or foreign countries seeking to confuse or spark disagreement among other states (Gerrits, 2018).

Despite its high-risk implications, the role of disinformation in international politics is significantly under-researched, with extant studies primarily focusing on deception and its consequences (Lanoszka, 2019). Deception studies examine domestic or bureaucratic affairs (Ripsman, 2004), war making processes, and mobilisation of public support towards controversial policy legislations (Schuessler, 2015), while prominent IR scholars suggest that countries periodically bluff but rarely deceive each other (Mearsheimer, 2011). The few studies on disinformation focus mainly on Russia and the policy challenges it initiates (Darczewska, 2014). Scholars also show that disinformation vulnerability depends on the domestic circumstances of a country, with polarised societies (e.g., with significant minority diasporas) and winner-takes-all electoral systems being more susceptible to disinformation manipulation compared to homogeneous societies and multi-party systems (Gerrits, 2018).

DOI: 10.4324/9781003190363-24

We argue that to understand disinformation and its impact, it is important to unpack assumptions about the uncertainty that accompanies it. While disinformation at first instance can stimulate *false certainty* by presenting information that is false as if it were true, once deception is suspected or demonstrated, disinformation stimulates *uncertainty* (Richards, 2021). Disinformation can spark uncertainty even when it is not revealed, by overcomplicating an issue, debate or problem, by altering perceptions of the probability of an event or an outcome, or by introducing ambiguity. Because uncertainty arises in the context of complexity, probability, and ambiguity, it is tightly connected to disinformation.

Uncertainty is a prevalent experience in the international realm which can also exist independently of disinformation. Characteristically, what is known as *unresolvable uncertainty* (Wheeler, 2018) is seen as a ubiquitous attribute of the political world stemming from the actor's subjective cognitive judgments of the objective political reality (Alvarez and Franklin, 1994) and is considered threatening and unpleasant (Booth and Wheeler, 2008; Bas et al., 2017; Kydd, 2005; Smith and Stam, 2004; Mearsheimer, 2001; Jervis, 1976).

Uncertainty is approached as *fear of asymmetric information* (Waltz, 1979; Grieco, 1993; Mearsheimer, 1994), *ignorance* (Keohane and Nye, 1977; Lake and Powell, 1999), *ambiguity* (Tetlock, 1998; Goldgeier and Tetlock, 2001), and *indeterminacy* of identities and actions (Finnemore, 1996; Wendt, 2000). Examinations of nuclear deterrence, terrorist violence, nuclear brinkmanship, ceasefires in civil conflicts, and power shifts (Katzenstein and Seybert, 2018: 46; Hammerstad and Boas, 2015; Petersen, 2011) see risk in uncertainty, denoting an intrinsic desire for control (Eidinow, 2011).

Understanding uncertainty, either as part of, or aside from disinformation, is relevant for understanding diplomacy. The need for diplomacy comes in uncertain times (Holmes, 2018; Wheeler, 2018), and diplomacy as a process and practice contains high levels of uncertainty, spanning from the intentions and capabilities of counterparts to the probability of conflict or peace (Katzenstein and Seybert, 2018; Bas and Schub, 2017; Bas, 2012; Rathbun, 2007). Indeed, diplomacy scholars discuss uncertainty in the context of culture (Seong-Hun, 2008), military conflict (Bas et al., 2017), world system (Gunaratne, 2005), institutional order (Piskarska, 2016), hybrid warfare (Gaufman, 2017), digital transition (Manor, 2019), cross-country political trends (Surowiec and Manor, 2021) and communication strategies (Miskimmon et al., 2017).

To recap, we are interested in understanding uncertainty, because it is central to understanding how disinformation affects perceptions, because it can precede disinformation and therefore contextualise it, and because it can exist independently from disinformation in the anarchic international politics realm. We elaborate on extant conceptualizations of uncertainty in international politics, borrowing insights from international relations (IR), psychology, sociology, neuroscience and organisational studies. Next, we examine the sources and effects of uncertainty in the context of international diplomacy. Finally, we offer a novel interpretation of uncertainty which could be useful for diplomats and other political actors as they navigate institutional, national and international crises of disinformation.

Uncertainty in international politics

The *complexity* of the international realm creates ripe conditions for uncertainty (Bas et al., 2017). The multiple rival countries with diverse intentions and signals, the scattering of government bodies, the absence of overarching international authority, result in a "noisy" environment with high levels of uncertainty (Waltz, 1979; Mearsheimer, 2019). Different uncertainty source paths inevitably blend, so that individuals (and countries) operating in complex environments encounter ambiguous information about probabilistic events, and therefore become uncertain (Hogg and Blaylock, 2011).

Probability is another trigger of uncertainty. Uncertain persons are rarely ignorant about events and hence derive their assumptions from subjective calculations of probability (Baratgin et al., 2018). "Uncertainty as probability" corresponds to what scholars of IR call "uncertainty as risk", reflected in the efforts of individuals or countries seeking to calculate the probability of uncertain events (Bas et al., 2017). While Schelling (1966: 93) highlights that "violence, especially war, is a confused and uncertain activity, highly unpredictable", others argue that war is a sequence of probabilistic fights (Smith, 1998; Wagner, 2000), containing costs of cooperation (Yared, 2010) and the undetermined benefits of peacemaking (Chassang and Padro i Miguel, 2009). In uncertain contexts, actors who exhibit risk-prone behaviour are more likely to engage in disputes (Bas and Schub, 2017).

Ambiguity over an adversary's intentions and military capabilities is another salient source of uncertainty (Smith and Stam, 2004; Johnson, 2004; Keohane, 1984). The ambiguity of information inevitably requires actors to make subjective decisions prior to interpretation. Furthermore, even when relying on concrete rules or agreements, ambiguity is still a factor since rules and arguments can be subject to specific interpretations (Best, 2012). Ambiguity triggers uncertainty because of the presence of competing interpretations (Zong and Demill, 2015).

The above indicates that the international relations environment is complex, probabilistic, ambiguous, and rife with uncertainty. Disinformation stimulates and accentuates complexity, probability, and ambiguity, heightening uncertainty and promoting false certainty. While recognising the value of uncertainty, different theoretical approaches within IR adopt conceptualizations informed by specific assumptions about the state of the world, which ultimately provide a fragmented account of uncertainty. We review these conceptualizations and their implications for understanding uncertainty below.

The realist school of IR sees uncertainty as the *fear of missing important information* (Waltz, 1979; Grieco, 1993). Here the biggest challenge is the shortage of information towards the intentions and behaviour of adversaries. Mearsheimer (1994: 10) notes that "a state may be reliably benign, but it is impossible to be certain of that judgement because intentions are impossible to divine with 100 percent certainty". Paradoxically, even though realists perceive uncertainty as lack of information, only the information that verifies an opponent's malign intentions is accepted as reliable (Ibid.). Information demonstrating peaceful or conciliatory signals of others is recognised as suspicious. Walt (1987) identifies such strategy as an issue of selectivity. Realists argue it is the amount and type of information that matters for political actors to escape the fear of not knowing one's enemy and the uncertainty around it. Although not explicitly acknowledged, such selectivity of information corresponds to what psychologists label "confirmation bias", a situation where individuals emphasise the information corresponding to their expectations while ignoring contradictory information (Wason, 1960).

Rationalists perceive uncertainty as complete lack of knowledge which they equate to *ignorance* (Fearon, 1995). The former US Secretary of Defense Donald Rumsfeld in 2002 famously said, "there are things we do not know; but there are also unknown unknowns - the ones we don't know we don't know… It is the latter category that tends to be the difficult one" (cited in Zak, 2021). Importantly, the term "ignorance" does not allude to the limitations of the abilities of decision-makers, but rather to the general state of not knowing (Rathbun, 2007). Rationalists respond to uncertainty as ignorance by collecting *all* possible information rather than focusing on expectation-consistent information (Keohane, 1993).

Similarly, scholars exploring the "bargaining model of war" (Morrow, 1989; Fearon, 1995) see uncertainty as the *dearth of information* over the enemy's capabilities. Realists and rationalists postulate that without parsing evidence to evaluate adversaries' intent, the shadow of uncertainty cannot be escaped, and this is the rationale of IR scholarship on the security dilemma or spiral models (Booth and Wheeler, 2008; Waltz, 1979; Jervis, 1976, 1978; Mearsheimer, 1994). Paradoxically though, additional information amplifies, rather than reduces, the likelihood of conflict (Fey, 2015; Arena and Wolford, 2012). This is an important and interesting implication questioning the expectation stemming from psychological studies that information acquisition limits uncertainty (Mussweiler and Posten, 2012). In psychology, uncertainty and ignorance are ubiquitous accompaniments of incomplete information (Smithson, 1989), yet scholars caution uncertainty is "the conscious awareness or subjective experience of ignorance" (Anderson et al., 2019: 2). Uncertainty is seen as "a higher-order metacognition", a perceptual acceptance that one lacks knowledge (Ibid.; Smithson, 1989, 2012).

Cognitivists and constructivists in IR favour the conceptualization of uncertainty as *ambiguity*, highlighting the inability of political representatives to correctly interpret signals from the political environment (Tetlock, 1998). This discrepancy between the message of the signal and the receiver's interpretation occurs due to cognitive errors, noises and biases—limitations inherent to individuals' mental abilities (Iida, 1993; Goldgeier and Tetlock, 2001). As Jervis (1997) noted, while individuals might expect the linearity and simplicity of events, reality can be the opposite. Political events occur in an exponential order and with uncertainty. Ambiguity intensifies due to the amount of information one must process, the variety of duties of the policymakers, and the interconnection of international affairs (Haas, 1980).

Studies in psychology agree individuals can feel lost when faced with an insurmountable amount of information (Goldgeier and Tetlock, 2001). Here it is important to distinguish between uncertainty, ambiguity, and risk. *Ambiguity* emerges in response to irregular, doubtful, and insufficient information (Ellsberg, 1961). The famous Ellsberg paradox (Ibid.) suggests that people react more aversely to ambiguous rather than risky or probabilistic events (Bammer and Smithson, 2012). Individuals opt for risky steps with familiar probabilities instead of unknown steps, even when the utility of unknown probabilities is higher. Furthermore, actors are more pessimistic about their choices when operating in ambiguous environments (Binmore et al., 2012). While uncertainty manifests uneasiness about a future-oriented event (Kahneman et al., 2021; Grenier et al., 2005), ambiguity unfolds in the situation occurring at the present time (Furnham and Marks, 2013; Greco and Roger, 2001). Another interesting distinction sees ambiguity as the attribute of the impetus, and uncertainty as the affective state triggered by the latter (Ellsberg, 1961; Buhr and Dugas, 2002). Other studies like Greco and Roger (2001) suggest that ambiguity and uncertainty have the same origins, such as "unfamiliarity", "incomprehensibility", and "instability", with the framing of the uncertainty depending on how the subject acts in the circumstances of uncertainty (Hillen et al., 2017: 65). Returning to IR, less frequent accounts of uncertainty see it as *indeterminacy* (Wendt, 2000). Constructivist scholars argue every phenomenon is a socially constructed outcome (Finnemore, 1996). The world is a dynamic place where political players doubt their identity and goals and cannot be sure what behavioural strategy (norms) to employ towards constantly changing events in the international stage (Keck and Sikkink, 1998). Indeterminacy can surround many attributes, such as the identity of states, norms of society, duties of institutions, and the behaviour of those working within them (Reus-Smit, 1997). According to this approach, many subjects, objects, and processes within the international realm are unsettled, and consequently an individual faces uncertainty through the subjective redefinition of various interpretations (Levy, 1994).

The experience of uncertainty as risk and its alternatives

Scholarship in IR "has mobilised uncertainty as riskness" (Mills, 2021: 283), disregarding Knight's (1921) distinction between risk as a measurable probabilistic entity and uncertainty as an unmeasurable unknown. According to Knight, an uncertain individual cannot make any calculations since every event is different, and therefore uncertainty cannot be mixed with risk (Ibid.). Interestingly, the conflation of risk and uncertainty in IR rests on the ambiguity over the intentions of others (Ramsay, 2017). This demonstrates how complex and multi-dimensional the experience of uncertainty can be, involving elements of risk as well as ambiguity. Seldom do IR scholars base their works on Beck's (2000) framework of *risk society*, the idea that the modern society itself generates the risks, for instance, through techno-economic decision-making. Scholarship on weapons of mass destruction (Riqiang, 2013), nuclear deterrence strategies (Kroenig, 2013), terrorist attacks (Kydd and Walter, 2006), and military disputes (Fortna, 2003), is inattentive to the difference between uncertainty and risk, using the two terms interchangeably, even though the concept of uncertainty is much broader (Marchau et al., 2019).

Moving beyond IR, we come across social science conceptualizations of uncertainty which approach it as an aversive, paralysing, challenging, puzzling, slippery, indefinite, suspicious, and disturbing experience (Anderson et al., 2019; Frost, 2019), but also as an affectively neutral, or even affectively positive experience. We discuss these alternative conceptualizations below.

Affect-neutral conceptualizations see uncertainty as "lack of sure knowledge" (Downs, 1957: 77), which is not perceived as positive or negative. It can be "a dimension of attitude strength" (McGraw et al., 2003: 426), "an acknowledgment of what one does not know, but also that one does not know" (Anderson et al., 2019: 2), "a notion that negative events may occur and there is no definitive way of predicting such events" (Carleton, 2012: 106), "a state or an objective feature of the world" (Wakeham, 2015: 719), or the time frame of prediction before an aversive event (Greco and Roger, 2001). A related concept is *uncertainty tolerance*, defined by scholars as a state of "negative and positive psychological responses—cognitive, emotional, and behavioural—provoked by the conscious awareness of ignorance about particular aspects of the world" (Hillen et al., 2017: 70; Herman et al., 2010), a dispositional characteristic of specific individuals (Anderson et al., 2019) or both (Wright, 2010).

Uncertainty is also conceptualised as *delay* (Rachlin, 1989). Experiences or events that occur instantly are observed as certainties, while those delayed become uncertainties (Luhmann et al., 2008). Kahneman and Tversky (1979) find that delay and uncertainty have similar impact: individuals avoid risk-taking when offered delayed gains, but are more risk-oriented when encountering delayed misfortunes. However, behavioral psychologists demonstrate that although reaction to delay is like the one triggered by uncertainty, individuals spotlight the uncertainty aspect only when considering their manoeuvres in the presence of an uncertain and delayed event, and the relationship between the two is tenuous (Blackburn and El-Deredy, 2013).

Uncertainty is also frequently approached as *ambivalence* (McGraw et al., 2003; Gross et al., 1995). Ambivalence triggers the human body and involves cognitive and affective attributes since it is an experience of "having both positive and negative [...] feelings at the same time" (Schneider et al., 2021: 570; Schneider and Schwarz, 2017). According to Alvarez and Brehm

(2002), uncertainty can be mitigated through laborious computation, but ambivalence cannot. An individual who lacks information to make an informed decision feels uncertain, whereas an ambivalent person holds the information he or she needs, yet feels unable to select one of the available options. Henceforth, ambivalence puts uncertain events on the horizon (Rothman et al., 2017) and helps individuals make unbiased judgments while considering multifaceted aspects of an ambivalent issue (Schneider et al., 2021).

Uncertainty can also be an affectively positive experience (Scoones, 2019; Smithson, 2012). Uncertainty *per se* is an abstraction and therefore, to make it more manageable, individuals "tunnel" it, by focusing on what it clearly is, while becoming oblivious to what it could have been (Taleb, 2010: 203). Taleb argues that certainty is nothing more than a mere recalibration of uncertainty (Ibid.). This recalibration can build an interesting momentum of relishing uncertainty as *opportunity* (Griffin and Grote, 2020). Bammer and Smithson (2012) write that although a familiar scenario offers a psychologically more comfortable environment, the phantom of uncertainty shows us the open windows where others see closed doors. Busch (2020) employs the concept of *serendipity* to refer to the "unexpected good luck resulting from unplanned moments in which proactive decisions lead to positive outcomes" (Ibid.,7). In this sense, accepting uncertainty is "an essential condition for a free, productive and happy human life" (Shorokhov, 2020: 1).

Scholars of functionalist theories portray uncertainty as *knowledge*. The lack of cognitive and perceptual competencies leads to uncertainty which stimulates information seeking and knowledge gain (Lewandowsky et al., 2015), either from revisiting/remembering existing experiences, or through the observation of new signals in the environment (Busch, 2020). Uncertain individuals become proactive while seeking to find the missing part of the puzzle, and thus uncertainty transforms into knowledge (Mackey, 2012).

Penrod's (2007: 661) phenomenological research on experiences of uncertainty offers a counter-intuitive conclusion: uncertainty can stimulate confidence and a sense of control, evoking an experience of *personal growth*. Individuals who find ways to progress forward in uncertain environments can feel powerful in future similarly unpredictable situations. When recognizing uncertainty, individuals overstep their perceived limitations and thrive in the unknown (Morrison, 2016), adding to their mental toolkit (Gordon, 2001). Echoing Fromm (1947: 45), "uncertainty is the very condition to impel man to unfold his powers."

Morrison (2016) refers to the "new normal" to illustrate that uncertainty can become a normative state of human life. Uncertainty as *inevitability* and "a way of life" is appropriate for individuals frequently faced with uncertainty and is not considered "a problem to be overcome" (Scoones, 2019: 12; Bauman, 2013). Batteaux and his colleagues (2020) show that individuals who realise that the accurate predictability of events is impossible, change their trust in certainty and accept uncertainty as a fact. Capelos et al. (2021) link the conceptualizations of trust and uncertainty, arguing that accepting uncertainty is the fundamental first step to establishing trust relations.

Lessons from uncertainty in the context of disinformation

Uncertainty is a key outcome of disinformation, but can also precede it, and exist independently of it. We have provided a comprehensive account of the conceptualizations and experiences of uncertainty in IR and beyond, noting uncertainty as *fear* (Waltz, 1979), *ambiguity* (Hillen et al., 2017; Rosen et al., 2014), *probability* (Smithson, 2012; Schelling, 1963), *ignorance* (Smithson, 1989; Fearon, 1995), *indeterminacy* (Wendt, 2000), *ambivalence* (McGraw et al., 2003), *delay* (Rachlin, 1989) *opportunity* (Griffin and Grote, 2020), *knowledge*

(Lewandowsky et al., 2015), *growth* (Morrison, 2016) and *inevitability* (Morrison, 2016; Scoones, 2019: 12; Bauman, 2013). This rich understanding of uncertainty challenges the dominant view of uncertainty as aversive and risky, and invites the consideration of its neutral, positive or affectively ambivalent content (Capelos et al., 2021), as well as its potential to stimulate knowledge and growth.

Uncertainty is ubiquitous in diplomacy environments and the international stage (Holmes, 2018; Wheeler, 2018). While diplomats and policy makers are not immune to uncertainty, recognizing the different types of uncertainty can promote more effective negotiations and more fruitful dispute resolutions. Diplomats and policy makers who can distinguish different types of uncertainty, can also engage with uncertainty more effectively, make its nuances known to others, and calibrate strategies to contain or maintain uncertainty. Studies show emotional awareness in mediators leads to the adoption of strategies that open effective avenues of cooperation (Capelos and Smilovitz, 2008). We see *uncertainty awareness* as an essential skill-set worth honing. To reach a consensus, diplomats must consider their opponents' uncertain preferences, and this is difficult to do unless they clearly understand and identify their own relationship with uncertainty. As Gorbachev astutely noted, "politics is sometimes a search for possible in the sphere of unfamiliar" (National Security Archive, 1990: 10) and this search succeeds when one attempts to understand the nature of the unfamiliar.

Bibliography

Alvarez, R. M. and J. Brehm. *Hard Choices, Easy Answers*. Princeton, NJ: Princeton University Press, 2002.

Alvarez, R. M. and C.H. Franklin. "Uncertainty and Political Perceptions." *The Journal of Politics* 56, no. 3 (1994): 671–688.

Anderson, E. C., Carleton, R. N., Diefenbach, M. and P. K. Han. "The Relationship Between Uncertainty and Affect." *Frontiers in Psychology* 10, (2019): 2504.

Arena, P. and S. Wolford. "Arms, Intelligence, and War." *International Studies Quarterly* 56, no. 2 (2012): 351–365.

Bammer, G. and M. Smithson. *Uncertainty and Risk: Multidisciplinary Perspectives*. Routledge, 2012.

Baratgin, J., Politzer, G., Over, D. E. and T. Takahashi. "The Psychology of Uncertainty and Three-valued Truth Tables." *Frontiers in Psychology* 9, (2018): 1479.

Bas, M., McLean, E. and T. Whang. "Uncertainty in International Crises." *The Korean Journal of International Studies* 15, no. 2 (2017): 165–189.

Bas, M. A. "Measuring Uncertainty in International Relations: Heteroskedastic strategic models." *Conflict Management and Peace Science* 29, no. 5 (2012): 490–520.

Bas, M. A. and R. Schub. "The Theoretical and Empirical Approaches to Uncertainty and Conflict in International Relations." In *Oxford Research Encyclopedia of Politics*, edited by W.R. Thompson, 177–194. Oxford: Oxford University Press, 2017.

Bauman, Z. *Liquid Times: Living in an Age of Uncertainty*. Chichester: John Wiley and Sons, 2013.

Beck, U. "Risk Society Revisited: Theory, Politics and Research Programmes". In *The Risk Society and Beyond. Critical Issues for Social Theory*, edited by B. Adam, U. Beck, and J. Van Loon, 211–229. London: Sage, 2000.

Best, J. "Ambiguity and Uncertainty in International Organizations: A History of Debating IMF Conditionality". *International Studies Quarterly* 56, no. 4 (2012): 674–688.

Binmore, K., Stewart, L. and A. Voorhoeve. "An Experiment on the Ellsberg Paradox." *Journal of Risk and Uncertainty* 45, (2012): 215–238.

Blackburn, M. and W. El-Deredy. "The Future is Risky: Discounting of Delayed and Uncertain Outcomes." *Behavioural Processes* 94, (2013): 9–18.

Booth, K. and N. J. Wheeler. *The Security Dilemma: Fear, Cooperation and Trust in World Politics*. Basingstoke, UK: Palgrave Macmillan, 2008.

Brothers, D. *Toward a Psychology of Uncertainty: Trauma-centered Psychoanalysis*. New York: Analytic Press, 2008.

Buhr, K. and M. J. Dugas. "The Intolerance of Uncertainty Scale: Psychometric Properties of the English Version." *Behaviour Research and Therapy* 40, (2002): 931–945.

Bull, P. *Posture and Gesture*. Oxford: Pergamon, 1987.

Busch, C. *The Serendipity Mindset: The Art and Science of Creating Good Luck*. Penguin UK, 2020.

Buzan, B., Wæver, O., Wæver, O. and J. De Wilde. *Security: A New Framework for Analysis*. Lynne Rienner Publishers, 1998.

Capelos, T., Wheeler, N. J., Cervacio, C. and S. Chrona. "Trust and Uncertainty in International conflict." Paper presented in International Society of Political Psychology Annual Conference. Montreal, 2021.

Capelos, T. and J. Smilovitz. "As a Matter of Feeling: Emotions and the Choice of Mediator Tactics in International Mediation." *The Hague Journal of Diplomacy* 3, (2008): 63–85.

Carleton, R. N. "The Intolerance of Uncertainty Construct in the Context of Anxiety Disorders: Theoretical and Practical Perspectives." *Expert. Rev. Neurother* 12, (2012): 937–947.

Chassang, S. and G. Padro i Miquel. "Economic Shocks and Civil War." *Quarterly Journal of Political Science* 4, no. 3 (2009): 211–228.

Darczewska, J. "The Anatomy of Russian Information Warfare: The Crimean Operation, A Case Study." *OSW Point of View* 42, 2014.

Downs, A. *An Economic Theory of Political Action in A Democracy*. New York, 1957.

Eidinow, E. *Luck, Fate, and Fortune: Antiquity and Its Legacy*. New York: Oxford University Press, 2011.

Ekman, P. "What Scientists Who Study Emotion Agree About." *Perspectives on Psychological Science* 11, no. 1 (2016): 31–34.

Ellsberg, D. "Risk, Ambiguity, and the Savage Axioms." *Quarterly Journal of Economics* 75, (1961): 643–669.

Fearon, J. D. "Rationalist Explanations of War." *International Organization* 49, no. 3, (1995): 379–414.

Fey, M. "*Learning to Fight: Information, Uncertainty, and the Risk of War*." Working paper, 2015.

Finnemore, M. *National Interests in International Society*. Ithaca: Cornell University Press, 1996.

Fortna, V. P. "Inside and Out: Peacekeeping and the Duration of Peace after Civil and Interstate Wars." *International Studies Review* 5, no. 4, (2003): 97–114.

Fromm, E. *Man for Himself*. New York: Henry Hold, 1947.

Frost, J. "Certainty, Uncertainty, or Indifference? Examining Variation in the Identity Narratives of Nonreligious Americans." *American Sociological Review* 84, no. 5 (2019): 828–850.

Furnham, A. and J. Marks. "Tolerance of Ambiguity: A Review of the Recent Literature." *Psychology* 4, no. 9 (2013): 717–728.

Gaufman, E. *Security Perceptions and Public Perception: Digital Russia and the Ukraine crisis*. 1st Edition. Springer International Publishing: Imprint: Palgrave Macmillan, 2017.

Gerrits, A. W. "Disinformation in International Relations: How Important Is It?" *Security and Human Rights* 29, no. 1–4 (2018): 3–23.

Goldgeier, J. M. and P.E. Tetlock. "Psychology and International Relations Theory." *Annual Review of Political Sciences* 4, no. 1, (2001): 67–92.

Greco, V. and D. Roger. "Coping with Uncertainty: The Construction and Validation of a New Measure." *Personality and individual differences* 31, no. 4 (2001): 519–534.

Grenier, S., Barrette, A. M. and R. Ladouceur. "Intolerance of Uncertainty and Intolerance of Ambiguity: Similarities and Differences." *Personality and Individual Differences* 39, (2005): 593–600.

Grieco, J. M. "Understanding the Problem of International Cooperation: The Limits of Neoliberal Institutionalism and the Future of Realist Theory." In *Neorealism and Neoliberalism: The Contemporary Debate*, edited by D. Baldwin, 301–338. New York: Columbia University Press, 1993.

Griffin, M. A. and G. Grote. "When is More Uncertainty Better? A Model of Uncertainty Regulation and Effectiveness." *Academy of Management Review* 45, no. 4 (2020): 745–765.

Gross, S. R., Holtz, R. and N. Miller. "Attitude Certainty." In *Attitude Strength: Antecedents and Consequences*, edited by R.E. Petty and J.A. Krosnick, 215–246. Mahwah, NJ: Erlbaum, 1995.

Gunaratne, S. A. "Public Diplomacy, Global Communication and World Order: An Analysis based on Theory of Living Systems." *Current Sociology* 53, no. 5 (2005): 749–772.

Haas, E. B. "Why Collaborate?: Issue-Linkage and International Regimes." *World Politics* 32, no. 3, (1980): 357–405.

Hammerstad, A. and I. Boas. "National Security Risks? Uncertainty, Austerity and Other Logics of Risk in the UK Government's National Security Strategy." *Cooperation and Conflict* 50, no. 4 (2015): 475–491.

Herman, J. L., Stevens, M. J., Bird, A., Mendenhall, M. and G. Oddou. "The Tolerance for Ambiguity Scale: Towards a More Refined Measure for International Management Research." *International Journal of Intercultural Relations* 34, (2010): 58–65.

Hillen, M. A., Gutheil, C. M., Strout, T. D., Smets, E. M. A., and P.K.J. Han. "Tolerance of Uncertainty: Conceptual Analysis, Integrative Model, and Implications for Healthcare." *Social Science Medicine* 180, (2017): 62–75.

Hogg, M. A. and D.L. Blaylock. *Extremism and the Psychology of Uncertainty*. John Wiley and Sons, 2011.

Holmes, M. *Face-to-face Diplomacy: Social Neuroscience and International Relations*. Cambridge University Press, 2018.

Iida, K. "Analytic uncertainty and international cooperation: Theory and Application to International Economic Policy Coordination." *International Studies Quarterly* 37, no. 4 (1993): 431–457.

Jervis, R. *Perception and Misperception in International Politics*. Princeton: Princeton University Press, 1976.

Jervis, R. "Cooperation Under the Security Dilemma." *World Politics* 30, no. 2 (1978): 167–214.

Jervis, R. *System Effects: Complexity in Political and Social Life*. Princeton: Princeton University Press, 1997.

Johnson, D. D. *Overconfidence and War*. Harvard University Press, 2004.

Kahneman, D. and A. Tversky. "Prospect Theory: An Analysis of Decision Under Risk." *The Econometric Society* 47, no. 2, (1979): 263–291.

Kahneman, D., Sibony, O. and C.R. Sunstein. *Noise: A Flaw in Human Judgment*. Little Brown, 2021.

Katzenstein, P. and L. Seybert. "Uncertainty, Risk, Power and the Limits of International Relations Theory." In *Protean Power: Exploring the Uncertain and Unexpected in World Politics*, edited by P. Katzenstein and L. Seybert, 27–56. Cambridge: Cambridge University Press, 2018.

Keck, M. and K. Sikkink. *Activists Beyond Borders: Advocacy Networks in International Politics*. Ithaca: Cornell University, 1998.

Keohane, R. O. *After Hegemony: Cooperation and Discord in the World Political Economy*. Princeton, NJ: Princeton University Press, 1984.

Keohane, R. O. "Institutionalist Theory and the Realist Challenge After the Cold War." In *Neorealism and Neoliberalism: The Contemporary Debate*, edited by D. Baldwin, 269–301. New York: Columbia University Press, 1993.

Keohane, R. O. and J. Nye. *Power and Interdependence*. New York: Harper Collins, 1977.

Kydd, A. H. *Trust and Mistrust in International Relations*. Princeton, NJ: Princeton University Press, 2005.

Kydd, A. H. and B. F. Walter, "The Strategies of Terrorism." *International Security* 31, no. 1 (2006): 49–80.

Knight, F. H. *Risk, Uncertainty, and Profit*. New York: Houghton Mifflin, 1921.

Kroenig, M. "Nuclear Superiority and The Balance of Resolve: Explaining Nuclear Crisis Outcomes." *International Organization* 67, no. 1 (2013): 141–171.

Lake, D. and R. Powell. *Strategic Choice and International Relations*. Princeton: Princeton University Press, 1999.

Lanoszka, A. "Disinformation in International Politics." *European Journal of International Security* 4, no. 2 (2019): 227–248.

Levy, J. S. "Learning and Foreign Policy: Sweeping a Conceptual Minefield." *International Organization* 48, no. 2 (1994): 279–312.

Lewandowsky, S., Ballard, T. and R.D. Pancost. "Uncertainty as Knowledge. Philosophical Transactions of the Royal Society: A. Mathematical, Physical and Engineering Sciences." *Article* 373, (2015): 20140462.

Luhmann, C. C., Chun, M. M., Yi, D. J., Lee, D. and X.J. Wang. "Neural Dissociation of Delay and Uncertainty in Intertemporal Choice." *Journal of Neuroscience* 28, no. 53 (2008): 14459–14466.

Mackey, J. "Musical Improvisation, Creativity and Uncertainty." In *Uncertainty and Risk*, 123–132. Routledge, 2012.

Manor, I. *The Digitalization of Public Diplomacy*. New York: Springer International Publishing, 2019.

Marchau, V. A., Walker, W. E., Bloemen, P. J. and S.W. Popper. *Decision Making under Deep Uncertainty: From Theory to Practice*. Springer Nature, 2019.

Mareš, M. and P. Mlejnková. "Propaganda and Disinformation as a Security Threat." In *Challenging Online Propaganda and Disinformation in the 21st Century*, (2021): 75–103.

McGraw, K. M., Hasecke, E. and K. Conger. "Ambivalence, Uncertainty, and Processes of Candidate Evaluation." *Political Psychology* 24, no. 3 (2003): 421–448.

Mearsheimer, J. J. *The Tragedy of Great Power Politics*. New York: W. W. Norton, 2001.

Mearsheimer, J. J. "Bound to Fail: The Rise and Fall of the Liberal International Order." *International Security* 43, no. 4 (2019): 7–50.

Mearsheimer, J. J. "The False Promise of International Institutions." *International Security* 19, no. 3 (1994): 5–49.

Mearsheimer, J. J. *Why Leaders Lie: The Truth about Lying in International Politics*. Oxford: Oxford University Press, 2011.

Mills, L. "Managing Uncertainty: The Everyday Global Politics of Post-9/11 US Public Diplomacy." In *Public Diplomacy and the Politics of Uncertainty*, edited by P. Surowiec and I. Manor, 277–303. Palgrave Macmillan, Cham, 2021.

Miskimmon, A., O'Loughlin, B. and L. Roselle. *Forging the World: Strategic Narratives and International Relations*. University of Michigan Press, 2017.

Morrison, D. T. "Being with Uncertainty: A Reflective Account of a Personal Relationship with an Asylum eeker/refugee." *Counselling Psychology Review* 31, no. 2, (2016): 10–21.

Morrow, J. D. "Capabilities, Uncertainty, and Resolve: A Limited Information Model of Crisis Bargaining." *American Journal of Political Science* 33, no. 4 (1989): 941–972.

Mussweiler, T. and A.C. Posten. "Relatively certain! Comparative Thinking reduces Uncertainty." *Cognition* 122, no. 2 (2012): 236–240.

National Security Archive. *Excerpt from the Second Conversation between M. S. Gorbachev and G. Bush.* Washington, White House, May 31, 1990, https://nsarchive.gwu.edu/document/16135-docum ent-21-record-conversation-between.

Paul, C. and M. Matthews. "The Russian "Firehose of Falsehood" Propaganda Model." *Rand Corporation* 2, no. 7 (2016): 1–10.

Penrod, J. "Living with Uncertainty: Concept Advancement." *Journal of Advanced Nursing* 57, no. 6 (2007): 658–667.

Petersen, K. L. "Risk Analysis: A Field within Security Studies?" *European Journal of International Relations* 18, no. 4 (2011): 693–717.

Piskarska, K. *The Domestic Dimension of Public Diplomacy: Evaluating Success through Civic Engagement.* Palgrave, 2016.

Rachlin, H. *Judgement, Decision, and Choice: A Cognitive/Behavioural Synthesis.* WH Freeman/Times Books/ Henry Holt and Co, 1989.

Ramsay, K. W. "Information, Uncertainty, and War." *Annual Review of Political Science* 20, (2017): 505–527.

Rathbun, B. C. "Uncertain about Uncertainty: Understanding the Multiple Meanings of A Crucial Concept in International Relations Theory." *International Studies Quarterly* 51, no. 3 (2007): 533–557.

Reus-Smit, C. "The Constitutional Structure of International Society and the Nature of Fundamental Institutions." *International Organization* 51, no. 4 (1997): 555–589.

Richards, J. "Fake News, Disinformation and the Democratic State: A Case Study of the UK Government's Narrative." *Icono14* 19, no. 1 (2021): 95–122.

Ripsman, M. N. "The Politics of Deception: Forging Peace Treaties in the Face of Domestic Opposition." *International Journal* 60, no. 1 (2004): 189–216.

Riqiang, W. "Certainty of Uncertainty: Nuclear Strategy with Chinese Characteristics." *Journal of Strategic Studies* 36, no. 4 (2013): 579–614.

Rosen, N. O., Ivanova, E. and B. Knauper. "Differentiating Intolerance of Uncertainty from Three Related but Distinct Constructs." *Anxiety Stress Coping* 27, (2014): 55–73.

Rothman, N. B., Pratt, M. G., Rees, L. and T.J. Vogus. "Understanding the Dual Nature of Ambivalence: Why and when Ambivalence Leads to Good and Bad Outcomes." *Academy of Management Annals* 11, no. 1 (2017): 33–72.

Schelling, T. C. *The Strategy of Conflict.* Oxford University Press, 1963.

Schelling, T. C. *Arms and influence.* Yale University Press, 1966.

Schneider, I. K. and N. Schwarz. "Mixed Feelings: The case of Ambivalence." *Current Opinion in Behavioural Sciences* 15, (2017): 39–45.

Schneider, I. K., Novin, S., van Harreveld, F. and O. Genschow. "Benefits of Being Ambivalent: The Relationship between Trait Ambivalence and Attribution Biases." *British Journal of Social Psychology* 60, no. 2 (2021): 570–586.

Schuessler, J. M. *Deceit on the Road to War: Presidents, Politics, and American Democracy.* Ithaca, NY: Cornell University Press, 2015.

Scoones, I. "What is Uncertainty and Why Does it Matter?" STEPS Working Paper 105. Brighton: STEPS Centre, 2019.

Seong-Hun, Y. "Cultural Consequences on Excellence in Public Diplomacy." *Journal of Public Relations Research* 20, no. 2 (2008): 207–230.

Shorokhov, A. "Uncertainty as an Important Determinant in Psychological Science and Practice." *E3S Web of Conferences* 210, (2020): 20008.

Smith, A. "Fighting Battles, Winning Wars." *Journal of Conflict Resolution* 42, no. 3, (1998): 301–320.

Smith, A. and A.C. Stam. "Bargaining and the Nature of War." *Journal of Conflict Resolution* 48, no. 6 (2004): 783–813.

Smithson, M. *Ignorance and Uncertainty: Emerging Paradigms.* Springer-Verlag, New York, 1989.

Smithson, M. "The Many Faces and Masks of Uncertainty." In *Uncertainty and Risk: Multidisciplinary Perspectives*, edited by G. Bammer and M. Smithson, 13–25. London, England: Earthscan, 2012.

Starbird, K. "Disinformation Campaigns are Murky Blens of Truth, Lies and Sincere Beliefs - Lessons from the Pandemic." *The Conversation.* Jul 23, 2020, https://theconversation.com/disinformation-campaigns-are-murky-blends-of-truth-lies-and-sincere-beliefs-lessons-from-the-pandemic-140677.

Surowiec, P. and I. Manor. *Public Diplomacy and the Politics of Uncertainty.* Palgrave Macmillan, 2021.

Taleb, N. N. *The Black Swan: The Impact of the Highly Improbable.* Random House, 2010.

Tetlock, P. E. "Social Psychology and World Politics." In *Handbook of Social Psychology*, edited by D.T. Gilbert, S.T. Fiske and G. Lindsey, 866–912. New York: McGraw, 1998.

Wagner, R. H. "Bargaining and War." *American Journal of Political Science* 44, no. 3, (2000): 469–484.

Wakeham, J. "Uncertainty: History of the Concept." In *International Encyclopedia of the Social and Behavioral Sciences*, edited by J.D. Wright, 716–721. Amsterdam: Elsevier, 2015.

Walt, S. M. *The Origins of Alliances.* Ithaca: Cornell University, 1987.

Waltz, K. *Theory of International Politics.* New York: Random, 1979.

Wason, P. C. "On the Failure to eliminate Hypotheses in a Conceptual Task." *Quarterly Journal of Experimental Psychology* 12, no. 3 (1960): 129–140.

Wendt, A. *Social Theory of International Politics.* Cambridge: Cambridge University Press, 2000.

Wheeler, N. J. *Trusting Enemies: Interpersonal Relationships in International Conflict.* Oxford: Oxford University Press, 2018.

Wright, R. "Self-Uncertainty and Its Cousins." In *Handbook of the Uncertain Self*, edited by R. Arkin, K. Oleson and P. Carroll, 421–443. New York: Psychology Press, 2010.

Yared, P. "A Dynamic Theory of War and Peace." *Journal of Economic Theory* 145, no. 5 (2010): 1921–1950.

Zak, D. "'Nothing ever ends': Sorting through Rumsfeld's Knowns and Unknowns." *The Washington Post*, Jul 1, 2021. Accessed Aug 1, 2021, https://www.washingtonpost.com/lifestyle/style/rumsfeld-dead-words-known-unknowns/2021/07/01/831175c2-d9df-11eb-bb9e-70fda8c37057_story.html.

Zong, Z. and B. Demil. "From Uncertainty to Ambiguity: A Discursive Approach of Emerging Field." *Management international/International Management/Gestiòn Internacional* 20, no.1, (2015): 63–175.

21

OPEN-SOURCE INFORMATION FOR INTELLIGENCE PURPOSES: THE CHALLENGE OF DISINFORMATION

Veli-Pekka Kivimäki

Open source information has historically played an important role in the production of secret intelligence. In the present day, the types of available open information sources have diversified due to digitalisation, and the subsequent explosion in availability of both user generated content and new commercial information services. As a result, Williams & Blum have observed intelligence produced solely from open sources, or open source intelligence (OSINT) is also undergoing a generational shift.[1]

Open source information is generally defined as publicly available information which can lawfully be acquired either through a collection activity or purchase. Unlike most other sources of intelligence, open sources produce "second hand information" already made public, which means the acquired information needs to be carefully vetted and verified.[2] Indeed, open source information used for intelligence purposes has always been subject to close scrutiny, as the basic assumption in intelligence analysis has been that open source information may be falsified with the intent to deceive.[3]

Disinformation is false information that is knowingly distributed. The word's roots lie in the 1920s activities of the new Soviet security apparatus (*dezinformatsiya*), and, in its early forms, it involved the creation of false documentary materials and press articles.[4] In KGB's later parlance, disinformation was considered part of "active measures", a collection of activities intended to undermine Soviet adversaries.[5] Disinformation could be disseminated overtly, e.g. through state ("white") propaganda outlets, or covertly, e.g. by planting fabricated news stories in foreign media outlets.[6]

In its Soviet roots, disinformation was closely linked to "black" propaganda—the spread of false information with concealed or misattributed sourcing.[7] In the United States, "white" propaganda and "black" psychological warfare were in the 1940s conceptualised as part of the political warfare toolkit.[8] A feature of both disinformation and black propaganda is that the false information is created with a purpose, with the intent to deceive.

Beyond propaganda use, in Soviet military thinking disinformation was considered an element of *maskirovka*, or military deception.[9] The use of disinformation in military deception

DOI: 10.4324/9781003190363-25

is by no means a Soviet invention, and its use can be traced back to pre-biblical times, illustrating the persistent power of weaponised lies in war.

Thomas Rid has described four waves for disinformation spanning the past century: the 1920s era of initial operations, a second wave of professionalising activities in the post-WW2 era, late 1970s era of well-resourced and institutionalised operations, and a fourth wave in 2010s enabled by the technological developments of the modern era.[10] The technological innovations that bring us new forms of open source information also enable new forms of disinformation, as well as new channels for its dissemination.

In our present day information environment, institutionally controlled distribution channels and institutionally generated open source information is complemented and even challenged by user generated content.[11] This also means the forms of disinformation are evolving. Disinformation is no longer an activity of only governments or political movements, but it is also manifests itself in the context of other societal issues, as evidenced by the COVID-19 pandemic. Propaganda is also no longer confined to official outlets or public spaces, but it is entering our private lives through digital devices and services. Asmolov & LeJeune have observed that propaganda has even become a participatory activity, as conflicts are increasingly socialized through new information technologies.[12] The Internet and social media have marked the beginning of a new era of digital disinformation.

With this background in mind, we can examine disinformation from two perspectives: as a means to influence public opinion through presentation of false information, and as an activity which is intended to conceal facts or true intentions in order to mislead the public, decision-makers and analysts about the true nature of events. Common to both, in intelligence analysis there is first a need to detect the disinformation and secondly, seek to understand the motive behind it—why the false information was created in the first place. Detection of disinformation leads to the question of the actor behind the activity, which may involve disentangling networks of disinformation purveyors, as to understand the roots of the disinformation activity. Long term propaganda analysis can lead to an understanding about the narratives being used and how they evolve over time,[13] meaning identified and known sources of false information may also be of interest for intelligence analysts.

Open source information

Historically, open source information has consisted largely of written publications, print media, and broadcast media. With the advent of the Internet, existing information sources became increasingly available online, but the Internet also enabled new types of data services not previously available to the public. Starting with home pages and blogs, self-publishing to the whole world became a possibility to anyone with Internet access. A new era of user generated content truly started when social media combined with ubiquitous mobile Internet access, meaning events in the world can be immediately captured, and shared with the world.

Social media has also entered the realm of international relations. "Twiplomacy", or the conduct of foreign relations using social media, has entered the diplomatic lexicon[14], meaning that foreign intelligence organisations cannot avoid what is happening in the social media space, when world leaders give statements and engage other nations via this medium. Digital diplomacy gives an additional layer to analysis of world events, as officials may bypass traditional foreign policy apparati and engage with the world directly via social media. This also means countries engaging in overt propaganda activities can do so directly to international audiences.

In a world inundated with user generated content, traditional media has sought ways to utilise this material, and new working practices have been established to find newsworthy social media material.[15] New forms of digital open source investigation entities not affiliated with existing institutions have also emerged. As an example, collaborative investigation group Bellingcat emerged in this environment, undertaking investigations which utilised digital eyewitness material, open satellite imagery, and other open digital data sources to analyse the conflicts in Ukraine and Syria in the mid-2010s.[16]

A peculiar type of new data source are leaks from hacked organisations and online services. The contents of these leaks range from user data and personally identifying information to confidential internal documents of organisations. Whether such materials are usable for intelligence purposes poses both ethical and legal questions. Data leaks have been used by researchers to identify intelligence operatives[17], but have also resulted in serious breaches of sensitive personal health data.[18]

In the 2010s, space-based capabilities have become increasingly available as commercial services, bringing them into the realm of open source. This includes electro-optical imagery, meaning the satellite images we can see on services like Google Maps and Apple Maps, but also infrared imaging, synthetic aperture radar (SAR) imaging, and radio frequency emission geolocation. These capabilities give us unprecedented visibility into our physical world. The interpretation of this data however also requires proficiency, as it's possible to misinterpret what the sensors are telling, if the related phenomenology is not understood. As an example, SAR image acquisition systems have variable parameters which affect the output.[19]

Russia's 2022 offensive into Ukraine gives an example of where these new open sources came together to bring an unprecedented level of visibility to unfolding world events. Leading up to the invasion in February 2022, movements of Russia military equipment were tracked through eyewitness sightings on social media, at times giving an almost real time view of events on the ground. Commercial satellite imagery providers published imagery of force concentrations near Ukraine's border before the war. Where weather or coverage didn't permit for traditional imagery to be acquired, open SAR imagery was used to pinpoint force concentrations.[20] Commercial satellite images challenged Russia's narrative that no attack was being planned.

New types and forms of digital open source information are continuously becoming available. From the perspective of analysis, it is critical to understand the threat of how such information might be tainted.

Digital disinformation

For the current generation of disinformation, UNESCO handbook on *Journalism, Fake News and Disinformation* proposes a simple working definition: "information that is false, and the person who is disseminating it knows it is false".[21] It is the purposeful misleading which separates disinformation from *misinformation*. Misinformation may be a result of e.g. inter-correct interpretation of data, or misunderstanding. While the information may be false, there is no ulterior motive to deceive.

A core mechanism of deception is misdirection, which applies to both deception in the physical realm, as well as deception with disinformation. Misdirection is conducted with the intent to shape perceptions or beliefs of the target audience, and it can be achieved through disguise or distraction.[22] In the digital world, disguise can take the form of fake users created solely for the purposes of influence campaigns,[23] or front organisations controlled by intelligence organisations to support disinformation activities.[24] Distraction in turn seeks to shape perceptions of events, such as providing contradictory falsified testimony or evidence to ongoing events.

According to a Swedish research study, target audiences for disinformation operations can be broadly categorised on three levels: mass audience, sociodemographic groups, or individual physchographic level. Individual level targeting means the tailoring of messages as enabled by present day social media platforms, similar to the way that advertising profiles are built based on an individual user's behavior. While demographic targeting might be based on age or education level, psychographic targeting could be based on political views inferred by the social media platforms, based on user behavior and preferences.[25]

Digital disinformation operations have also been conceptualised as part of so-called hybrid threats. In this context, they are part information influence measures, which may for example be intended to coerce and undermine a target society's sense of security.[26] In a hypothetical example, psychographic (i.e. microlevel) targeting could utilise the data provided by social media platforms, combined with data obtained via cyber breaches or espionage operations, allowing disinformation to be selectively targeted at susceptible audiences during crises.[27]

To illustrate the real world functioning of disinformation campaigns, an example from the 1980s detailed in CIA's journal *Studies in Intelligence* bears relevance still today. The Soviet KGB organised a disinformation campaign which illustrates the core features of such efforts. Dubbed operation "INFEKTION" by East German foreign intelligence which supported the KGB, the campaign sought to spin the emerging AIDS virus as resulting from US government biological warfare experiments. The campaign started in July 1983 when a niche Indian newspaper published an article, which was assessed to be the handiwork of the KGB due to the specific claims made linking to earlier Soviet disinformation narratives on bioweapons. The initial thrust produced little effect, but a renewed effort started in 1985 by Soviet media. Repetition was seen to be a key to success in the disinformation campaign, and it was achieved by utilising various methods, including the use of forgeries.[28]

East German contributions to the AIDS disinformation campaign involved the enrollment of a scientist to produce a report, which mixed actual research with the disinformation claims on the virus' origins from bioweapons programs. The report was broadly disseminated at a conference of the Non-Alignment Movement, attended by delegates from over 100 countries. Related disinformation was also supplied by East German intelligence to a prominent author, who incorporated elements of it to a best-selling book. The "scientific" evidence went on to spread in the media in over 80 countries by the end of 1987.[29]

In 2020, researchers investigating a large-scale Russian disinformation activity running since 2014 dubbed the operation "Secondary Infektion", due to parallels with its predecessor. The operation did not have a single theme, but rather it focused on several themes which could be seen as supporting Russian foreign policy objectives. The themes included presenting Ukraine as a failed state, NATO as aggressive, Europe as weak, and the Russian government as victim of Western plots.[30] The content produced by the Secondary Infektion campaign was spread over 300 different platforms. The campaign utilised fake personas, forged documents and content, and in addition to mainstream platforms, content was also spread through a disinformation outlet with links to Russia's state security service, the FSB.[31]

Digital disinformation campaigns may also involve the use of forgeries, much like their predecessors a century ago. Mika Aaltola has described five stages of election meddling, where disinformation plays a role in the first stage to amplify suspicions and division in the target area. The second stage is the stealing of sensitive data, through cyber espionage operations. The third state is the leaking of data through intermediaries.[32] It is between stages two and three where forgeries might be employed, to manipulate the data being leaked. In the case of the Macron email leaks of 2017, metadata of some files in the leak was found to contain the digital fingerprint of a user, who could later be linked to a cyber unit of Russia's military intelligence.[33]

Attempts at election meddling have involved the use of psychographic targeting through the use of targeted advertising.[34] Visibility for disinformation content and activities has been sought by buying advertising for the content or accounts which act as disseminators of the false information. Such advertising can be tailored using the profiling opportunities which social media platforms offer for customising advertising. Taken further, advertising can also tie into a user's other internet activity, such as web browsing habits, or information collected by a user's device, giving targeting further granularity.[35]

Disinformation and deception are activities not only reserved for state actors. Insurgent and terrorist organisations such as have also employed deception, and in one study ISIL was found to engage in three types of deception: substantive deception focused on false or misleading content, source deception where the information's origin is obscured, and spread deception where the support of the activities are artificially amplified, to give a false impression of an ideology's organic spread.[36] This typology closely links to the definitions of disinformation and black propaganda discussed previously. The spread of deception in particular is where the digital environment and social media create new opportunities, as false supporters for an ideology can be created, and messages artificially amplified by networks of these false supporters.

A common pattern we can find from these examples is that in addition to the motive to deceive, in organised disinformation activities attempts are made to artificially amplify the message through various mechanisms. In the pre-Internet era, this was achieved for example through niche and propaganda outlets, use of authoritative figures to lend disinformation claims credence, and attempts to introduce disinformation elements into mainstream discourse by use of forgeries.

These above mentioned elements have equal relevance in the Internet era, in a digital form. In the present day, amplification may also be easier to achieve due to social media, where content can spread crossing social network community boundaries, becoming viral.[37] Virality of social media content resonates with the principle of "social proof", introduced by Robert Cialdini. According to this principle, our view of the acceptability of behavior is shaped by what others perceive as correct.[38] In social media, this translates to likes and shares.

When a large number of people share certain messages or viewpoints, we may be tricked to assess it based on its perceived popularity. Disinformation campaigns may also seek to utilise this effect, by exploiting false amplification of the disinformation content, for example by using networks of social media "bots", artificial users pretending to be real ones.

Visual disinformation

Social media services have become increasingly visual, providing a constant flow of still imagery and video. One form of distraction is the use of active misdirection by evoking emotional stimulus.[39] Images and video in turn can be very effective in triggering emotional responses, which disinformation campaigns attempt to turn to their advantage. Emotionally charged social media messages have been found to be shared more often and more quickly than messages with neutral content.[40]

The simplest form of disinformation use of images and video is miscontextualisation. Images or video taken during some event is repurposed, and reshared with a misleading description during a different event.[41] This effect can be seen for example during conflicts, where footage from prior conflicts may be passed around to make claims about military losses or attacks.[42]

On social media platforms, content aggregators have emerged. They act as proxies for receiving and sharing information, and may also scour the networks for information themselves. These aggregators may operate as personal accounts, anonymously, or with an organisational affiliation. Especially with content related to conflicts, such content aggregators may exacerbate the disinformation problem, if they do not conduct proper verification prior to sharing. Equally, malicious content aggregators could attempt to seed this type of disinformation, while pretending to be impartial observers.

Old imagery has also been used to paint one side of a conflict or societal issue as brutal. During the Catalan independence referendum of 2017, old imagery from 2012 riots was used to claim instances of police brutality.[43] In a situation like this, the originator knows they are passing along false information, with the intent to frame one party in a harmful way. At the same time, people sharing such material may not be knowingly participating in dissemination of disinformation, but reacting to the emotional trigger of perceived wrongdoing.

In another example, in 2014 a video of Syrian boy rescuing a girl under gunfire, before being gunned down himself, caused an uproar and went viral on YouTube. While there is ample evidence of atrocities having been committed in Syria[44], this video was a fake, created by a Norwegian director who wished to generate discussion about children in war zones. Blurring of fiction and reality was intended.[45] Such fabrications may have an unintended side effect: they can enable the actual perpetrators of atrocities to argue that real evidence of their crimes is also faked.

Beyond staged fakes, there's imagery manipulation; common, and already a longer term known issue for intelligence analysts. In 1969, veteran CIA imagery analyst Dino Brugioni noted that faking photographs had been done since photography was invented, and given their value as an important intelligence source, analysts need to be aware of the possibility photographs may be forged.[46] Today, the digital era has brought with it more flexible image editing possibilities, for example using applications such as PhotoShop. Some nations, such as North Korea, have taken advantage of digital editing in their own state propaganda.[47] Whether analogue or digital fakery, the detectability of such edits was and is largely dependent on the skill of the person conducting the edits, as editing can leave behind telltale forensic markers.

The 2010s brought with them the technical enablement of machine learning solutions on consumer level hardware. Deep neural networks are a subclass of machine learning (ML) systems, and they have been utilised for generating images, based on data an ML system has been taught with. Deep neural networks were popularised when they were used to create new forms of image fakes, dubbed "deep fakes". Using deep fake generation software, with a modest amount of skills, it was now possible for virtually anyone with a modern day personal computer to start generating false imagery. One of the first use cases was replacing a person's face with a different person's face. Since this technology could be used to generate new video clips, a concern was raised that deep fakes could be used for disinformation campaigns to, for example, fake videos of politicians making statements they never actually made.[48]

The technology behind deep fakes have extended to other use cases, as well. In addition to training systems to generate human faces, they have also been taught to change the style of images, for example by changing the visible season from summer to winter.[49] In yet another application potentially challenging open source analysts, an experimental system was created for generating completely artificial satellite images.[50]

The proliferation of tools and techniques for digital fakery presents new challenges for intelligence analysts. Literal information and imagery derived from open sources need to be carefully validated, especially if the claims presented by such material are extraordinary. As described above, such material may be miscontextualised, manipulated, or even generated for the purpose for deceiving.

Implications for the intelligence process

During the COVID-19 pandemic, where the spread of wrong information about the disease and its spread was rampant, renowned medical journal The Lancet Infectious Diseases observed in an editorial that "misinformation dilutes the pool of legitimate information".[51] The UNESCO handbook on disinformation notes that journalists risk being drowned in cacophony in the flood of mis- and disinformation,[52] and the same equally applies to intelligence analysts needing to process open source information. Indeed, the volume of data in today's information environment makes filtering out what's likely true and what isn't the first challenge for intelligence analysis.

In a major study on deception, Barton Whaley observed that deception varies greatly across time, cultures, disciplines, and practitioners.[53] These parameters are especially relevant when dealing with social media. Disinformation campaigns, their composition and themes can change over time. Culture has an impact on how and what kind of information is presented and can be expected to be ingested. Different online actors have their own modus operandi, and the individual style and competence of practitioners varies. Another important observation by Whaley was that historically highly sophisticated deception is rare. This is an important caveat also in the assessment of disinformation activities, as lower grade, even crude disinformation activities greatly outnumber the sophisticated ones also in the digital environment.

Establishing motive is the starting point for separating out misinformation from disinformation, and the disinformers. Pherson & Pherson note that deception is most likely to happen when the source has previous history of using it, when the timing of the disinformation is such that it can have significant impact, when the information cannot be verified from other sources, and when the information upends previously held beliefs.[54]

At the level of the source, Clark & Mitchell call for evaluating the source competence, access, and what vested interests or biases they may hold.[55] In the world of social media, this may involve assessing the history of an account, community or website which is distributing the material. The biases may be obvious, as some disinformation outlets focus only on posting information which is supportive of their chosen narratives.

In some cases, e.g. with rumours, tracing back to the original source of disinformation can be hardest part, and may require supplementary investigative activity.[56] Network analysis methods may assist in understanding the scope and spread of disinformation, and identify key nodes of disinformation spread.[57] The tracing of origin is made more difficult by the fact that content may jump across platforms, and language barriers. Thus, the originator of disinformation may be in a different language area and social media platform, than where the collector or analyst has made the initial observation about the existence of such disinformation content. If this is the case, the source and content need to be assessed in the correct local context, as otherwise the motive may be misinterpreted.

An example from history on how disinformation may jump from one context to another: during World War 2 forged German leaflets intended to demoralise soldiers were created by the Office of Special Services (OSS) and spread behind enemy lines. The forged documents were later found on German prisoners of war, thought by Army intelligence to be authentic, and the information given to be published by American newspapers. The intended audience of the forgeries were German soldiers, but story had appeal to mainstream audiences in the U.S.[58] In today's social media world, a similar scenario might take place, where disinformation is planted by country A into country B's social media spaces, where it might be picked up and reshared by country A's own social media users.

For analysing disinformation narratives, context-dependent analysis models can be identified, based on an actors modus operandi. In the case of Russian disinfomation, Ben Nimmo has described a 4D model of how Russia builds its disinformation narratives and reacts to claims against itself: dismiss the critics, distort the facts, distract from the issue, and dismay the audience.[59] Across these, it can be argued that misdirection, earlier mentioned as a core part of deception, is a mechanism present in all four. The target audience of these 4Ds might however be different. While foreign audience are receiving messages of dismissal, domestic audiences could be distracted from an issue to minimise it. Analytical frameworks of disinformation activities should address the channels, means, and target audiences of the disinformation. Overarching disinformation narratives may be supported by operations utilising different channels and means for various different target audiences.

A problem that needs to be addressed in the intelligence process is the impermanence of digital evidence.[60] Content may be removed from platforms by the users themselves, or by the social media platforms hosting the content. In order to determine whether disinformation activity constitutes a coordinated campaign, it may be necessary to collect and preserve related evidence over a longer period of time. While it is healthy that social media platforms purge their platforms of disinformation, it does also have implications to analysis, as the content of interest may disappear without warning.

For content distributed in file format, metadata may reveal details about when, where and how a piece of content was produced. Image files typically are embedded with metadata by the device or application where the image is created, but many social media services strip out such metadata when content is uploaded. But all do not. In February 2022, before Russia's overt major land attack into Ukraine, a video was distributed online purporting to show body camera footage of "Polish speaking mercenaries" attempting to destroy chlorine containers, supposedly in a makeshift chemical attack. The video file was uploaded to the messaging service Telegram, where metadata in the file was retained, and it was possible to determine that the video had bore markings of a commercial video editing software, and was created significantly before it was claimed to be filmed.[61]

Beyond metadata, there are forensic means to examine whether visual material being assessed is real or not. In the era of generated fakes, this is especially important as generated material may lead analysts onto dead end tracks if it is not detected, wasting time. From the technical point of view, there's a need to separate the old generation of edited digital fakes from the new generation of generated digital fakes. The underlying technology is different, which means the forensic solutions required are also different. Latest image editing software tools also utilise generative machine learning techniques, which aim to make the end result look as natural as possible.

Forensic examination processes and tool chains need to be developed with the different types of fakes in mind, in order to avoid false negatives in verification. If visual content is verified with wrong forensic solutions that do not match the underlying technology used to create the fakes, the verification does not yield a usable result. Imagery forensics is a specialised skill, which requires the ability to choose the right tool for the purpose, and interpret the results of the tools, with their caveats.

In interpreting and analysing visual evidence, it should be noted that material which evokes strong reactions, and which is felt strongest may have an outsized impact on the analyst, compared to its information value. This effect has been dubbed vividness weighting,[62] and it closely relates to the earlier discussed effect of visual material being more likely to be shared if it is emotionally triggering. Analysts of disinformation should pay special focus to material evoking particularly strong emotional responses, as it also may be an indicator of crafted disinformation for a purpose.

Conclusions

Forms of disinformation have changed over time, and they will continue to evolve with new technological developments. The scale of disinformation activities have grown in the social media age, and they can today be conducted by both state and non-state actors with a relatively low barrier of entry.

For intelligence practitioners, a key issue is separating out the real from the false in the new information environment, rich with different forms of open data sources. With false information intended to misdirect, motive and malicious intent separate misinformation from disinformation. Traditional principles of source evaluation still hold true, but digital investigations require specialised approaches in order to effectively dissect organised disinformation activities.

In assessing disinformation activities, from an intelligence point of view, the interesting question may not be whether a piece of information is true or not, but why the disinformation exists in the first place. Was it created with the intent to harm a specific group or individual? Was it timed to have effect in a specific situation? Is it impossible to independently verify? Is the same core message being repeated across multiple platforms? Does it link to a broader narrative, or fit a longer term pattern? Questions like these help us better understand the *raison d' être* of a disinformation activity.

An understanding about the nature of a disinformation leads us towards attribution—who would benefit from running of such a damaging campaign. Here, intelligence organisations can employ the other resources at their disposal, such as cyber and signals intelligence capabilities, to conduct attribution of disinformation operations with higher levels of confidence.

Notes

1 Williams and Blum, *Defining Second Generation Open Source Intelligence for the Defense Enterprise,* 40.
2 Jardines, "Open Source Intelligence," 6.
3 Grabo, *Warning Intelligence,* 287.
4 Krzak, "Operational Disinformation of Soviet Counterintelligence during the Cold War," 2–3.
5 Andrew and Mitrokhin, *The Mitrokhin Archive,* 292.
6 Shultz and Godson, *Dezinformatsia,* 2.
7 Jowett and O'Donnell, *Propaganda and Persuasion,* 21.
8 National Security Council, *269. Policy Planning Staff Memorandum.*
9 Beaumont, *Maskirovka: Soviet Camoflage, Concealment and Deception,* 3.
10 Rid, Active Measures, 6–7.
11 Williams and Blum, *Defining Second Generation Open Source Intelligence for the Defense Enterprise,* 11.
12 Asmolov and LeJeune, "The Effects of Participatory Propaganda", 13.
13 Jowett and O'Donnell, Propaganda and Persuasion, 314.
14 Green, "The Rise of Twiplomacy and the Making of Customary International Law on Social Media," 1.
15 Hänska-Ahy, Wardle and Browne, "Social media & journalism: reporting the world through user generated content."
16 Higgins, *We Are Bellingcat,* 6–7.
17 Ibid., 163.
18 Ralston, "They Told Their Therapists Everything. Hackers Leaked It All."
19 Simms, *SAR Image Interpretation for Various Land Covers,* 5–11.
20 The Economist, "A new era of transparent warfare beckons."
21 Ireton and Posetti, *Journalism, "Fake News" & Disinformation,* 45–46.
22 Malin et al., *Deception in the Digital Age,* 3–4.
23 Nimmo et al., *Sekondary Infektion,* 66.
24 U.S. Department of Treasury, "Treasury Escalates Sanctions Against the Russian Government's Attempts to Influence U.S. Elections."
25 Pamment et al., *Countering Information Influence Activities,* 25.
26 Giannopoulos, Smith and Theocharidou, *The Landscape of Hybrid Threats,* 32–33.

27 Stockton, *Defeating Coercive Information Operations in Future Crises*, 79.
28 Boghardt, *Soviet Bloc Intelligence and Its AIDS Disinformation Campaign*, 4–7.
29 Ibid., 7–14.
30 Nimmo et al., *Secondary Infektion*, 4.
31 NewsFront, identified in Graphika's report as a vehicle for the Sekondary Inflection campaign's disinformation materials, was attributed by the U.S. Government as being FSB-controlled in April 2021, and sanctioned for acting on behalf of the FSB.
32 Aaltola, *Democracy's Eleventh Hour: Safeguarding Democratic Elections Against Cyber-Enabled Autocratic Meddling,* 3–4.
33 Higgins, *We Are Bellingcat*, 157.
34 Permanent Select Committee on Intelligence, "Exposing Russia's Effort to Sow Discord Online."
35 Estrada-Jiménez et al., "Online Advertising: Analysis of Privacy Threats and Protection Approaches," 36.
36 Milton, "Truth and Lies in the Caliphate: The Use of Deception in Islamic State Propaganda," 224–231.
37 Weng, Menczer and Ahn, "Virality Prediction and Community Structure in Social Networks," 1.
38 Cialdini, *Influence: The Psychology of Persuasion*, 115–116.
39 Malin et al., Deception in the Digital Age, 5.
40 Stieglitz et al., "Emotions and Information Diffusion in Social Media—Sentiment of Microblogs and Sharing Behavior," 241.
41 Barot, "Verifying Images," 35.
42 Toler, "How to Verify and Authenticate User-generated Content," 195–198.
43 El País, "Fake Images from the Catalan referendum Shared on Social Media."
44 See e.g. reporting by the United Nations Human Rights Council's Independent International Commission of Inquiry on the Syrian Arab Republic.
45 Tomchak and McDonald, "Syrian "hero boy" video faked by Norwegian director."
46 Brugioni, "Spotting Photo Fakery," 57–58.
47 Taylor, "Is This North Korean Hovercraft-Landing Photo Faked?"
48 Waldemarsson, *Disinformation, Deepfakes & Democracy*, 8–11.
49 Liu, Breuel and Kautz, "Unsupervised Image-to-Image Translation Networks."
50 Zhao et al., "Deep fake geography? When geospatial data encounter Artificial Intelligence."
51 The Lancet Infectious Diseases, "The COVID-19 infodemic," 875.
52 Ireton and Posetti, Journalism, "Fake News" & Disinformation, 8.
53 Whaley, *The Prevalence of Guile*, 76.
54 Pherson and Pherson, *Critical Thinking for Strategic Intelligence*, 177–178.
55 Clark and Mitchell, *Deception, Counterdeception and Counterintelligence*, 176.
56 United Nations, *Berkeley Protocol on Digital Open Source Investigations*, 62–3.
57 Clark, *Intelligence Analysis: A Target Centric Approach*, 363.
58 *War Report, Office of Strategic Services, Vol. 2: Operations in the Field*, 97.
59 Nimmo, *Anatomy of an Info-War: How Russia's Propaganda Machine Works, and How to Counter It.*
60 Fiorella, Godart and Waters, "Digital Integrity: Exploring Digital Evidence Vulnerabilities and Mitigation Strategies for Open Source Researchers," 149–151.
61 Harding, Roth and Walker, "'Dumb and lazy': the flawed films of Ukrainian "attacks" made by Russia's "fake factory'."
62 Clark and Mitchell, *Deception, Counterdeception and Counterintelligence*, 186.

Bibliography

Aaltola, Mika. "*Democracy's Eleventh Hour: Safeguarding Democratic Elections Against Cyber-Enabled Autocratic Meddling*". Briefing Paper 226, Finnish Institute of International Affairs, 2017.
Andrew, Christopher and Vasili Mitrokhin. *The Mitrokhin Archive: The KGB in Europe and the West.* London, UK: Allen Lane, 1999.
Asmolov, Gregory and Lorrie LeJeune. "The Effects of Participatory Propaganda: From Socialization to Internalization of Conflicts." *Journal of Design and Science* 6 (2019), doi:10.21428/7808da6b.833c9940..
Barot, Trushar. "Verifying Images". Chap. 4 in *Verification Handbook: An Ultimate Guideline on Digital Age Sourcing for Emergency Coverage*, edited by Craig Silverman, 35–46. European Journalism Centre, 2014.
Beaumont, Roger. *Maskirovka: Soviet Camouflage, Concealment and Deception.* Center for Strategic Technology, Texas Engineering Experiment Station, Texas A&M University System, 1982.

Boghardt, Thomas. "Operation INFEKTION: The Soviet Bloc AIDS Disinformation Campaign". *Studies in Intelligence* 53, 4 (2009): 1–24. Center for the Study of Intelligence, Central Intelligence Agency.

Brugioni, Dino. "Spotting Photo Fakery". *Studies in Intelligence* 13, 1 (1969): 57–67. Center for the Study of Intelligence, Central Intelligence Agency.

Cialdini, Robert. *Influence: The Psychology of Persuasion*. New York, NY: HarperCollins, 2007.

Clark, Robert M. *Intelligence Analysis: A Target-Centric Approach*, 6th edition. Thousand Oaks, CA: SAGE, 2020.

Clark, Robert M. and William L. Mitchell. *Deception, Counterdeception and Counterintelligence*. Thousand Oaks, CA: SAGE, 2019.

Estrada-Jiménez, José, Javier Parra-Arnau, Ana Rodríguez-Hoyos and Jordi Forné. "Online Advertising: Analysis of Privacy Threats and Protection Approaches". *Computer Communications* 100 (2017): 32–51. doi:10.1016/j.comcom.2016.12.016..

Fiorella, Giancarlo, Charlotte Godart and Nick Waters. "Digital Integrity: Exploring Digital Evidence Vulnerabilities and Mitigation Strategies for Open Source Researchers". *Journal of International Criminal Justice* 19 (2021): 147–161. doi:10.1093/jicj/mqab022..

Giannopoulos, Georgios, Hanna Smith and Marianthi Theocharidou, eds. *The Landscape of Hybrid Threats: A Conceptual Model*. European Commission, Ispra, PUBSY No. 123305, 2020.

Grabo, Cynthia and Jan Golman. *Handbook of Warning Intelligence*. London, UK: Rowman & Littlefield, 2015.

Green, James A. "The Rise of Twiplomacy and the Making of Customary International Law on Social Media." *Chinese Journal of International Law* 21, 1 (2022): 1–53. doi:10.1093/chinesejil/jmac007..

Harding, Luke, Andrew Roth and Shaun Walker. "'Dumb and lazy': the flawed films of Ukrainian "attacks" made by Russia's "fake factory'". *The Guardian*, 21 February2022. https://www.theguardian.com/world/2022/feb/21/dumb-and-lazy-the-flawed-films-of-ukrainian-attacks-made-by-russias-fake-factory.

Higgins, Eliot. *We Are Bellingcat: An Intelligence Agency for the People*. London, UK: Bloomsbury, 2021.

Hänska-Ahy, Maximillian, Claire Wardle and Malachy Browne. "Social media & journalism: reporting the world through user generated content". In Milojevic, Ana and Lucia Vesnic-Alujevic, eds. *Audience interactivity and participation: interview/essays with/on journalists and politicians*. Transforming audiences, transforming societies working group 2, 2013. http://eprints.lse.ac.uk/49186/.

Ireton, Cherilyn and Julie Posetti, eds. *Journalism, "Fake News" & Disinformation: Handbook for Journalism Education and Training*. Paris, France: UNESCO, 2018.

Jardines, Bradley. "Open Source Intelligence". In *The Five Disciplines of Intelligence Collection*, edited by Mark Lowenthal and Robert Clark, 5–44. Thousand Oaks, CA: SAGE, 2016.

Jowett, Garth S. and Victoria O'Donnell. *Propaganda and Persuasion*, 6th edition. Thousand Oaks, CA: SAGE, 2015.

Krzak, Andrzej. "Operational Disinformation of Soviet Counterintelligence during the Cold War." *International Journal of Intelligence and Counterintelligence* 35, 2 (2022): 265–278. doi:10.1080/08850607.2021.2014280..

Liu, Ming-Yu, Thomas Breuel and Jan Kautz. "Unsupervised Image-to-Image Translation Networks". 31st Conference on Neural Information Processing Systems (NIPS 2017), Long Beach, CA, USA. arXiv:1703.00848v6, 2018.

Nimmo, Ben, Camille Francois, C. Shawn Eib, Lea Ronzaud, Rodrigo Ferreira, Chris Hernon, and Tim Kostelancik. *Secondary Infektion*. Graphika, 2020. https://secondaryinfektion.org/.

Malin, Cameron H., Terry Gudaitis, Thomas J. Holt and Max Kilger. *Deception in the Digital Age: Exploiting and Defending Human Targets Through Computer-Mediated Communications*. London, UK: Elsevier, 2017.

Milton, Daniel. "Truth and Lies in the Caliphate: The Use of Deception in Islamic State Propaganda". *Media, War & Conflict* 15, 2 (2020): 221–237. doi:10.1177/1750635220945734..

Nimmo, Ben. "Anatomy of an Info-War: How Russia's Propaganda Machine Works, and How to Counter It". StopFake, 19 May, 2015. https://www.stopfake.org/en/anatomy-of-an-info-war-how-russia-s-propaganda-machine-works-and-how-to-counter-it/.

Pamment, James, Howard Nothhaft, Henrik Agardh-Twetman and Alicia Fjällhed. *Countering Information Influence Activities: The State of the Art*. Research report, Lund University. Swedish Civil Contingencies Agency, 2018.

Pherson, Katherine Hibbs and Randolph H. Pherson. *Critical Thinking for Strategic Intelligence*, 2nd edition. Thousand Oaks, CA: SAGE, 2017.

Ralston, William. "They Told Their Therapists Everything. Hackers Leaked It All". *Wired*, May 4, 2021. https://www.wired.com/story/vastaamo-psychotherapy-patients-hack-data-breach/.

Rid, Thomas. *Active Measures: The Secret History of Disinformation and Political Warfare*. New York, NY: Farrar, Straus and Giroux, 2020.

Simms, Elizabeth. *SAR Image Interpretation for Various Land Covers: A Practical Guide*. Boca Raton, FL: Taylor & Francis, 2020.

Shultz, Richard H. and Roy Godson. *Dezinformatsia: Active Measures in Soviet Strategy*. New York, NY: Berkeley Publishing Group, 1986.

Stieglitz, Stefan & Linh Dang-Xuan. "Emotions and Information Diffusion in Social Media—Sentiment of Microblogs and Sharing Behavior". *Journal of Management Information Systems* 29, 4 (2013): 217–248. doi:10.2753/MIS0742-1222290408..

Stockton, Paul. *Defeating Coercive Information Operations in Future Crises*. National Security Perspective, Johns Hopkins University Applied Physics Laboratory, 2021.

Taylor, Adam. "Is This North Korean Hovercraft-Landing Photo Faked?" *The Atlantic*, March 26, 2013. https://www.theatlantic.com/photo/2013/03/is-this-north-korean-hovercraft-landing-photo-faked/100480/.

Toler, Aric. "How to Verify and Authenticate User-generated Content". Chap. 9 in *Digital Witness*, edited by Dubberley, Sam, Alexa Koenig and Daragh Murray, 185–227. Oxford, UK: Oxford University Press, 2020.

Tomchak, Anne-Marie and Charlotte McDonald. "Syrian "hero boy" video faked by Norwegian director". *BBC News*, 14 November, 2014. https://www.bbc.com/news/blogs-trending-30057401.

Waldemarsson, Christoffer. "*Disinformation, Deepfakes & Democracy: The European response to election interference in the digital age*". Alliance of Democracies Foundation, 2020.

Weng, Lilian, Filippo Menczer & Yong-Yeol Ahn. "Virality Prediction and Community Structure in Social Networks". *Scientific Reports* 3, 2522 (2013). doi:10.1038/srep02522..

Whaley, Barton. *The Prevalence of Guile: Deception through Time and across Cultures and Disciplines*. Foreign Denial & Deception Committee, National Intelligence Council, Office of the Director of National Intelligence, Washington, DC, 2007.

Williams, Heather J. and Ilana Blum. *Defining Second Generation Open Source Intelligence (OSINT) for the Defense Enterprise*. Santa Monica, CA: RAND Corporation, 2018.

Zhao, Bo, Shaozeng Zhang, Chunxue Xu, Yifan Sun and Chengbin Deng. "Deep fake geography? When geospatial data encounter Artificial Intelligence". *Cartography and Geographic Information Science* 48, 4 (2021): 338–352. doi:10.1080/15230406.2021.1910075, 2021.

"269. Policy Planning Staff Memorandum," National Security Council of the United States, May 4, 1948. https://history.state.gov/historicaldocuments/frus1945-50Intel/d269.

"A new era of transparent warfare beckons". *The Economist*, February 18, 2022. https://www.economist.com/briefing/2022/02/18/a-new-era-of-transparent-warfare-beckons.

Berkeley Protocol on Digital Open Source Investigations. Human Rights Center, UC Berkeley School of Law and Office of the High Commissioner for Human Rights, United Nations, 2022.

"Exposing Russia's Effort to Sow Discord Online: The Internet Research Agency and Advertisements". Permanent Select Committee on Intelligence, U.S. House of Representatives, 2018. https://intelligence.house.gov/social-media-content/.

"Fake Images from the Catalan referendum Shared on Social Media". *El País*, October 6, 2017. https://english.elpais.com/elpais/2017/10/06/inenglish/1507278297_702753.html.

"The COVID-19 Infodemic". *Lancet Infectious Diseases* 20, 8 (2020): 875. doi:10.1016/S1473-3099(20)30565-X..

"Treasury Escalates Sanctions Against the Russian Government's Attempts to Influence U.S. Elections". Press release, U.S. Department of Treasury, April 15, 2021.

War Report, Office of Strategic Services, Vol. 2: Operations in the Field. History Project, Strategic Services Unit, U.S. War Department, 1949. National Archives and Records Administration, Identifier: 24461636.

22

PROTECTIVE FACTORS AGAINST DISINFORMATION

Cristina Ivan

Culture as protective factor against disinformation

Information manipulation in general and disinformation in particular have been approached as an object of study from the perspective of security studies, strategic communication, psychology, the sociological study of narratives and last but not least ethical studies. Generally, disinformation has been understood as either partially true or false "content produced to generate profits, pursue political goals, or maliciously mislead" (Humprecht, Esser and Aelst, "Resilience to Online Disinformation", 2). However, in defining disinformation, we need to also understand that disinformation is part of a constellation of information pathologies that include multiple dysfunctional or malevolent approaches to information and that generate multiple dysfunctions of the democratic system—information manipulation, click-bait, fake news, propaganda, disinformation, misinformation, malinformation etc. Essentially, in establishing the nature of disinformation, however, one must first and primordially delineate it from misinformation, which also implies generation and dissemination of false content, but is unintentional in nature, and malinformation, "where genuine information is shared to cause harm, for example, by disclosing private information to the public"(idem, 3).

In security studies, explanations have been advanced as to the risks information manipulation and disinformation in particular incur for the democratic system and its well-functioning (McKay and Tenove, Disinformation as a Threat) (Aro, The cyber-space war) (Martin, Disinformation as instrumentality, 2010) (Edward and Pomeranzev, Winning the Information). Evidence based studies highlighted the fact that propaganda and disinformation instrumentalized as political weapons by enemy countries and/or entities have impacted with major consequences e.g. public perceptions and decision making in electoral processes (Weisburd, Watts and Berger, Trolling for Trump) (Ivan, Chiru and Arcos, A whole of), the public opinion during the Brexit campaign or the more recent public perception of the Covid pandemic (Disinformation: how to) (Candaele, Coronavirus is a political). (EEAS SPECIAL REPORT).

Experts in communication framed disinformation as an adversary of strategic communication and a dishonest mechanism of hijacking opinions and circumventing the truth to frame alternative realities to gain power capital. (Edward and Chomsky, A Propaganda Model) (Chekinov and Bogdanov, The Nature and Content) (Bradshaw and Howard, Why does Junk).

DOI: 10.4324/9781003190363-26

Psychologists in their turn approached the disinformation phenomenon from the perspective of e.g. cognitive errors that have to be exposed and corrected, such as the confirmation bias or motivated reasoning, psychological traits that predispose to adopting conspiracy theories and evidenced the link between misinformation and misperceptions. (Helmus, Russian Social Medi) (Hinchchliffe, Exposing echo chambers) (Humprecht, Esser, et al., The sharing of disinformation) Some researchers even went further to make a plea for the development of technological solutions that incorporate psychological principles to create new approaches to countering disinformation through "technocognition". (Lewandowsky, Ecker and Cook, Beyond Misinformation: Understanding).

Subsequent studies have proven that "the misinformed were more likely to be confident in their beliefs than the correctly informed" (Guess and Lyons, "Misinformation, Disinformation, and", 11), which opened up an entire new field of investigation about how the development of critical thinking and sound judgment could actually counter the effects of post truth, less rational, emotionally loaded partisan discourses that manipulate perceptions into false beliefs. (Foresman, Fosl and Watson, "The Critical Thinking") (Paul and Elder, "Critical Thinking: Tools").

And finally, critical security studies experts and cultural anthropologists looked at (manipulated) narratives and how they act as frames of knowledge and facilitators of sorts in the dynamic of disinformation. It has been argued "that(...)discourses facilitate confrontational rhetoric by creating a hostile 'Other'" (Baumann, "Propaganda Fights and", 1), while artificial intelligence and digital tools help foster virtual echo-chambers where beliefs are reinforced by circular exposure and argumentation of peer residents that build similar mental frameworks and maps of meaning. (TaskForce, "Trends of the Week") (University of Oxford, The Computational Propaganda) (Ivan, Chiru and Arcos, A whole of society) As a result, today we benefit from an entire array of strategic efforts and practical tools aimed at containing and countering the effects of disinformation.

Furthermore, in the past decade, with the increasing emergence of digital disinformation in the social media, the intervention of AI in turning disinformation viral and the advent of eco-chambers where citizens dive deep into circular reasoning and legitimation of own views at the expense of critical thinking, many studies have also been dedicated to how society can be prepared to become resilient to disinformation.

Resilience to disinformation—an ambivalent option

Resilience is a widely used term in physics where it describes the ability of an object to bounce back to its original state after being put to pressure. Once embraced by psychology, this term has come to describe

> the capacity of the individual or group to recover or come up whole and functional, perhaps more spiritually, mentally and emotionally enriched, after having been exposed to chronic or acute stress or trauma. Creating a so called invulnerability or high resistance to later trauma is also a goal shared by all resilience thinkers designing models of intervention.
>
> *(Ivan, "Resilience—The X")*

Hence, when referring to the ability to resist disinformation, the term resilience imposed itself quite rapidly, numerous studies looking at the individual and community characteristics that might strengthen resilience in the face of an adverse, disinformation dominated, ecosystem such as those emerged in the fringes of social networks. Yet, framing the citizens' ability to resist the allure of disinformation discourses as resilience has several advantages and shortcomings alike.

A number of cross national studies have shown that "resilience factors are country specific and are highly dependent on the respective political and information environments" (Humprecht, Esser, et al., "The sharing of disinformation"). The cornerstone in all these approaches however is placed on the formation of "an educated or knowledgeable public", able to demystify alternative realities through media literacy and digital skills. (Hicks-Goldston and Ritchart, "The New Digital"). Such models insist on educating the public in propaganda analysis and disinformation analysis, on how financing the media can affect lens of production, how cognitive biases can be identified and exposed, and how technologies can be used to create traction for fake messages and accounts on social media networks etc.

Other studies insist on larger socio-economic and political factors that may limit resilience, such as polarization of society, emergence of populist discourse, low trust in news, niche and partisan fringe media development, fragmented and atomized audiences, high use of social media; all these aspects have been quoted to affect and multiply effects of disinformation. (Humprecht, Esser and Aelst, "Resilience to Online"). As a result, the ability to remain immune to attempts to manipulate perceptions and cognitions implies that one needs to take into account the fundamental dimension of rationality vs emotion, but also digital literacy, political education, ability to understand the value and communicational force of symbols and embedded norms that may be employed to deceive our ability to read through disinformation strategies etc.

Hence, we may conclude that so far, studies that produced models of prevention and countering of propaganda and disinformation have focused on the ability to detect cognitive biases, to use critical thinking in detecting the fake logic behind arguments, the ability to fact check information and validate sources. The abilities requested from the model of an informed citizen also include a sharp mind, knowledgeable detective skills, and availability to always check and compare information at hand. A resilient citizen is always rational, able and willing to take time and use his digital skills to detect the work of trolls, bots, clone sites and fake social media accounts. He/she is also responsible and reflexive, takes formation of his/her own beliefs and convictions seriously, is aware of the process and ready to flexibly change whenever there is proof that he/she might have been mistaken.

And here comes the shortcomings of this type of approach, which starts from the premise that the norm in media, social and political discourse implies an inherent objectivity, that distortions can and have to be corrected by the appeal to rationality. The main strategy in this type of approach is to expose the emotionally loaded discourse, that manipulates facts only to provide a simplified and easily believable reality, and in which cognitive dissonance and complexity have been erased to comfort the anxious citizen. Building awareness, exposing false facts, building a strong citizen engagement in exposing disinformation are undoubtedly useful and should be strengthened. The question is whether these lines of action are enough and whether we may not be overlooking other actionable dimensions of the collective response. How many of the scholars that produce research on disinformation can actually claim to match the ideal prototype of the informed and responsible citizen?!

If we are to accept that correcting disinformation is rather an ideal to be worth fighting for and that we shall never attain a definite truth centered communication arena, then we might be able to frame the permanent fight for accuracy and relevance on a broader landscape! As some researchers rightfully remark, "the post-truth problem is not a blemish on the mirror. The problem is that the mirror is a window into an alternative reality. (Lewandowsky, Ecker and Cook, "Beyond Misinformation: Understanding", 3). In the post-truth regime, fake news become alternative facts often fed into large conspiracy theories. What scholars and practitioners alike must approach is indeed an alternative reality in which correction of mis- and disinformation are rarely fully effective: that is, despite being corrected, and despite acknowledging the correction, people by and large continue to rely at least partially on information they know to be false. This phenomenon is known as the continued-influence effect, and it has been observed across a broad range of materials and modes of testing, persisting even when participants are warned at the outset that they may be misinformed (Lewandowsky, Ecker and Cook, "Beyond Misinformation: Understanding", 3).

Hence, the underlying problems that seem to resist remedy through rationality, critical thinking, digital and media literacy converge on a path whose milestones are the declining trust in governments' ability to address social grievances, declining trust in science and authority, low values of social integration, low open-mindedness, intolerance of uncertainty coupled with a lack of understanding of the impact AI and cutting edge technologies have on shaping public discourse via social media channels. (Lewandowsky, Ecker and Cook "Beyond Misinformation: Understanding") (Stoica and Umbres, "Suspicious minds in times") (Maftei and Holman, "Beliefs in conspiracy theories")

At the opposite end, studies concerned with identifying those traits that make an individual more susceptible to disinformation seem to pathologize a whole category of audience and oversimplify the problem by stripping it of its dynamic and overlapping social, political, cultural and economic context (Culloty and Suiter, "Disinformation and Manipulation", 55).

In between, most recent models of intervention aimed to prevent and counter effects of propaganda balance citizen education and empowerment with systemic regulations and industry interventions aimed to insure a more ecological model of information dissemination and consumption. The 2022 Code of Practice on Disinformation that follows the European Commission guidance (2021) is the second document in a series that was initiated in 2018 and that brings together online platforms, the advertising industry, fact-checkers, academia, researchers and civil society, in an attempt to design a multi-dimensional model of commitments, joint actions and specific measures aimed to prevent and counter the effect disinformation. The areas covered by the code include "measures to reduce manipulative behavior used to spread disinformation (e.g. fake accounts, bot-driven amplification, impersonation, malicious deep fakes)", demonetization of disinformation purveyors, stronger transparency measures and efficient, labeling, media literacy initiatives, recognition and labeling of propaganda, empowering of researchers and the fact-checking community etc. ("The 2022 Code"). Hence, the citizen empowerment through media literacy and critical thinking is complemented by an entire array of tech savvy measures, it embraces legal recommendations, industry regulation and pragmatic measures aimed to demonetize, expose and neutralize. A promising line of action targets support from signatories for "robust access to platform data by the research community and adequate support for their activities as part of an effective strategy for tackling Disinformation" (idem, 26), as this recommendation responds to a rather urgent need. As Culloty and Suiter well remark,

> research on countermeasures is in its infancy and tends to be concentrated on the Global North. More fundamentally, the platforms have largely declined to share relevant data with independent researchers, which greatly impedes efforts to assess the scale and nature of the problem and to evaluate the effectiveness of interventions.
>
> *(Culloty and Suiter, "Disinformation and Manipulation", 66)*

Hence, in the near future, with the advent of innovative technological solutions that rely on artificial intelligence, machine learning, text and image analysis, as well as algorithms aimed to detect anomalies and outliers, technological approaches and research initiatives might add an optimal complement to the audience focused response detailed above. Should we then consider that where audience/citizen upskilling cannot go further in detecting disinformation, artificial intelligence will do the work? Can we safely rely on technology and education interventions, both top down and grassroots oriented to provide a sufficiently efficient formula? As evidenced by research mentioned above, individual and collective factors that make the audience susceptible to disinformation are highly complex and embedded both in the singular circumstances of individual biography (education, emotional intelligence, tolerance to uncertainty) but also on the built-in faith society has in media, science, authorities, drivers which respond to social and cultural factors that favor social polarization and populist discourses and that seem to grow, just like conspiracy theories, in the shadows of encapsulated communities with alternative reality views at the fringe of cyberspace. Last but not least, the lived cultural experience has shown us that such groups sometimes create social movements which, in their turn, sometimes evolve below radar, at the margins of society, until they are ready and willing to negotiate power capital. For winning a capital of trust is often just an antechamber to power.

A potential path of exploration—protective factors against disinformation.

So far we have discussed a series of factors that, when rightfully instrumentalized, enhance resilience to disinformation. We have concluded that research in the field is still in its incipient phase. Some factors have emerged as valid in comparative analysis of different national contexts, others seem to have been deeply embedded in the local and national context. Methodologies at hand addressed these resilience enhancing factors by looking at how disinformation operates, shedding light on their fabricated truths, building awareness on both contents and methods employed by disinformation promoters and engineering solutions in reverse. The key words are exposure of disinformation agents' and tactics and enhancing general public's awareness. Should disinformation use emotionally loaded accounts, oversimplify reality and offer a convenient scapegoat, then it is the task of the digital literate, rational, tech savvy user to debunk fake arguments, spot the lies and reinstate complexity into his/her version of reality, aided as he is, more and more, by technological solutions and fact checking entities. An arena is hence established where the good and bad actors' transaction credibility, trust and power by outsmarting each other and permanently innovating content and viral dissemination methods. Reverse engineering often works, and serious games designed to expose information manipulation tactics have been some of the most successful tools in creating awareness and upskilling competences on the part of the audience. (Jeon, et al., "ChamberBreaker: Mitigating the"); (Gertrudis-Casado, et al., "Los serious games como")

Another source of resilience building that might also offer promising prospects for research, again inspired by psychology, is that of the protective factors. These factors are defined as a combination of variables that are able to limit risky behavior.

> In recent years, it has been found that protective factors act both by promoting personal abilities useful in overcoming the various developmental tasks and by promoting greater well-being through the reduction, balancing, neutralization, or compensation of risk factors. Therefore, it is clear that there is a dynamic interaction between risk factors and protective factors.
>
> *(Bonino, Cattelino and Ciairano, "Adolescents and Risk. Behaviour", 83)*

While we do not hold at the moment evidence based research to suggest a concrete list of protective factors that might prove efficient in the case of disinformation exposure, psychological research in the field related to antisocial, criminal behaviour and radicalization seem to suggest two large fields of intervention that are unanimously agreed upon as areas which have a great potential to reduce or eliminate dysfunctional behaviors—(1) empowering individual agency and (2) building collective communities of practice with positive cultural norms and strong identities. (idem, 82) (Pressman and Ivan, "Internet Use and Violent") Such an approach would imply we no longer aim to clear the spots on the mirror of reality, which is inherently distorted by subjectivity and, as is the case with disinformation, by malevolent ends. By contrary, it would mean complementing the existing strategy focused on addressing risk factors, with another type of approach, one in which we willingly give up the idealist attempt to build up an objective mirror, only to focus on those protective factors that might strengthen our immunity system to resist to more and more complex forms of disinformation.

Hence, in the next part of this chapter, we would like to propose a case study approach that may offer a glimpse into future research of protective factors to disinformation. We shall address the Ukrainian / global grassroots cultural response to Russian propaganda and disinformation.

Cultural studies perspective—a methodology to investigate protective factors in the making

One of the fundamental tenets of cultural studies is that "all understanding of the world is mediated by cultural images and discourses" (Gray, "Research practice for cultural", 12) and that cultural studies are primarily "interested in the meaning making".(ibid. 17) In this view, the relationship created between meaning making, identity, subjectivity and the lived experience of our culture is fundamental to understanding how resilience to disinformation can emerge within the complex dynamic of the social fabric. In the history of cultural studies, the framing realized by the anthropologist that observed lived culture, only to write it down and fix it into a coherent and supposedly objective meaning, has become a rich tradition shared throughout the 19th and the 20th century. Yet, as Ann Gray observes in her seminal study of research practices for cultural studies, this writing down in history "has been seen as an operation of power with the ethnographer fixing his or her gaze on different cultures and rendering them visible, through published work, for the gaze of his or her community of readers." (idem, 18) Hence, we understand that framing lived culture and assigning it meaning in the larger social context in which it emerged is in itself a performative act that is indelibly linked to power, an act that can be either weaponized by disinformation agents, or used with protective ends by the disinformation target audience—citizens and communities. Therefore, when dealing with instrumentalized narratives used in disinformation and covert information operations, one must also take into account the elements of cultural cohesion / destruction that such narratives operate with and the meaning making process that they either create or intend to disintegrate at the level of the targeted audience. For it is a proven fact that propaganda and disinformation can destroy social cohesion, enhance polarization, radicalize in groups and transform outgroups into demonized others to be discriminated against. (Bradsma n.d.)

Another aspect that must be noted in this endeavor is the problematic embeddedness of the researcher that, according to the lens and methodology of the new ethnographer/cultural studies researcher, observes lived experience while positioning himself/herself as simultaneously in and out of the object of study. Admitting that "texts and practices are both products of and constitutive of the social world" (idem, 16), including here the media, social networks, school, state institutions etc., the aim of the cultural studies researcher is to focus his/her gaze and perform a deep dive analysis of these performative texts and how they create meaning, how they relate to identity, both individual and collective, and especially how they transgress norms to generate change in the social world. Unlike sociology that focuses on observing dominant trends fixed with large numbers and unlike the cultural anthropologist that focuses on a large diachronic depicting of norms, values and behaviors of a community, the cultural studies researcher will focus on identifying small numbers and emerging mutations as well as the discursive game changers that account for in-progress meaning making and validation. This research method "requires periods of intense investigation into meaning production, rather than extended periods of observation." (idem, 17). It also requires a significant effort to explore simultaneous interactions between meaning makers in the social world, multimodal productions and their intertextual dialogue, all of these features being likely to shed light on identity and power relations. With this lens in mind, we shall then proceed to an in-depth analysis of several multimodal cultural productions that seem to enter in dialogue and establish an intertextual relationship and that concur to the creation of a master narrative with promising potential to become an instantiation of the above mentioned protective factors.

Case study

In selecting our relevant case study material, we have opted for cultural productions that emerge out of the participatory digital culture, where creation is often anonymized, while co-production and non-attribution are widely shared behaviors. At the same time transgressive and empowering, these cultural productions derive their force from the amount of user interaction generated and the real time meaning making process they foster and encourage with digital users. Most rely on a media account of real life events only to then transgress into the symbolic regime and start to generate meaning(s) by the engagement of the audience. The reason behind this choice is that within the larger framework of cultural productions, these are the ones that record changes while in the making and offer a great opportunity to the researcher to observe grass root shifts in the social world. Such environments offer most prolific X-rays of every day communication and interaction within communities, also linking up individuals at a global scale.

In the image—text analysis we shall focus our attention on the language, the interpretative frames that build up on the primary event and its manifold constructed significances, and the cultural repertoire used to translate experience into a viable, forceful meaning. Finally, by making appeal to archetypal analysis, we shall attempt to prove that collectively produced stories engage the full force of a protective factor and create mechanisms of resilience at community level.

The topic of choice is, as mentioned above, the Ukrainian / global grassroots cultural response to Russian propaganda and disinformation as depicted by the "blue tractor" narrative.

The Russian invasion of Ukraine started on February 24, 2022. By March 1st, the Ukrainians have already organized resistance and war battles were entering full force. In this context, a short video released on social media channels like FB, Twitter and TikTok, depicting a blue tractor towing away a Russian tank with the apparent Russian driver running behind, became viral and has been viewed millions of times.

Apparently, the first 2 versions of the video were posted on twitter as early as February 27, by two users, Arslon_Xudosi and yeg0rpetrov, and together summed up around 150.000 views and more than 14.000 likes. One of the first Twitter accounts to retweet them was that of a Ukrainian diplomat, Olexander Sherba, with the text: "If true, it's probably the first tank ever stolen by a farmer… Ukrainians are tough cookies indeed." The accompanying hashtags included #StandWith Ukraine and #russiagohome. Later on, multiple recordings of tractors doing similar towings or even plowing in the mined fields to keep life going in the villages of Ukraine emerged on the wide web. The video is a low definition phone camera recording of an anonymous man that comments and laughs on the background.

Five months onwards, by end of July 2022, a simple search on google related to the terms " blue tractor" and "Ukraine" has delivered videos, images and 104 text results. Similar queries like "Ukrainian tractor cartoon", "a short history of tractors in Ukrainian", "blue tractor video", "Ukrainian tractor memes", "blue tractor in the field" were suggested as similar popular queries. In just a few months, the tractor has been made subject to videos, movies, posters, memes mockumentaries, cartoons, graphic art, games (Ukrainian Farmy), a postal stamp emission, T-shirts, embroidery patches and… a lot of debates! The blue tractor turned it into a symbol of Ukrainian resistance. Its narrative highlighted the regular citizen's answer of common sense bravery against an abusive invader and outsized aggressor.

Subsequent cultural productions based on the viral video depict distinct features that add up to the narrative.

Cartoons

On Youtube, there has been shared a 30 second animated cartoon with more than 2000 views as of July 2022. The clip profiles the blue tractor as a happy, friendly, humanized character resembling a Cars movie hero that jumps along while towing a Russian Z marked tank. The character sings a childrens' song in Ukrainian that translates as: "In the fields, in the fields, The blue tractor on big wheels, You can guess and win, Come on, let's begin, Guess who, guess who, guess who, guess who Sings this happy song." The artist, self-entitled Синий Трактёр (Blue Tractor) seems to have borrowed the character form a Kids TV show and gave it a new cultural meaning accommodating war realities for children and enhancing the popularity of the vehicle turned weapon of resistance, turned hero.

Video games

Another interesting production is that of a gamified representation of the same event. A group of self-described game developers form Ukraine has released a video game entitled "Ukrainian Farmy", depicting a shiny armoured blue tractor with an Ukrainian flag attached, that freely rides the country roads hunting for tanks and towing them away from action. The game script reads as follows: "In UKRAINIAN FARMY, you can experience being one of the scariest combatants on the Ukrainian battlefield—the Ukrainian tractor driver. Challenge yourself against time and "high-precision" artillery bombardments, and capture as many Russian tanks, trucks, and APCs as you can." ("Ukrainian Farmy")

This half humorous, half gamified framing of the "blue tank motif" is telling for a strategy that addresses the gaming community to create traction, gain a capital of trust and solidarity with the Ukrainian resistance. The simple and unproblematized narrative puts in the foreground another depiction of the every day hero, the farmer, and invites the gamer to side with its courage and defiance. Normalizing bravery as an ability to outsmart the enemy, producing heroes out of "humble farmers", praising the ingenious transformation of regular objects into weapons at hand are all narrative lines that add up to an already complex narrative and make the archetypal symbol take flight. In its limits, this narrative contributes to the juxtaposition of the Ukrainian hero figure on the archetype of *the apparently lesser hero, whom nobody expects to win the confrontation, and who manages to turn the odds in his favor through ingenuity and courage*—a solar archetype valorized in the classical stories of David and Goliath, Hercules, Gilgamesh etc.

Memes

This symbolic image has also attracted attention to the meme community, whose field literature have proven to be in today's world another form of capital based on subcultural knowledge, unstable equilibriums and discursive weapons. (Nissembaum and Shifman, "Internet memes as contested", 2). The term meme, refers to groups of digital items (such as images or videos) that share common characteristics, are created with awareness of each other, and are distributed online by multiple participants (Dawkins apud Nissembaum and Shifman).

The *meme* community exploited the narrative and codified it according to its own culture, adding up new layers of significance and attaching the symbol to a larger narrative putting in contrast the **Ukrainian Farmers vs. Russian Army. The Knowyourmeme** community, for instance, provides a track record of the posts[1] and the channels that were used to popularize the blue tractor meme:

> memes about Ukrainian farmers stealing tanks surfaced on multiple social media platforms like TikTok and Instagram, going into early March 2022, referencing other memes like Devious Licks, Ben Affleck Smoking and But It's Honest Work, among others. However, most Ukrainian farmer memes were shared on Reddit.

Same portal statistics show a sharp increase of popularity for the narrative in March 2022 and an equally sudden decline by July 2022.

Among the most interesting message carrier memes in the Farmers versus Russian Army series is one creating an anthesis between the Russian call for surrender on the battlefield, denied by the highly equipped and masked figure of a Ukrainian soldier, and the so called "Russian free tank" that has been abandoned to the Ukrainian farmer. The meme adds to the narrative the ironic approach of the Russian propaganda framing the invasion as a special operation "to instate freedom in Ukraine". The "free tank" abandoned on the fields by a supposedly unwilling to fight soldier is hence placed in ironic contrast to the Ukrainian farmer, rake in hand, calmly watching the spectacle.

The second meme (see: https://i.kym-cdn.com/photos/images/original/002/321/653/300) represents an indirect reference to the blue tractor story, providing another humorous follow up to the narrative line. It depicts a screenshot from the global commercial site E bay, where the supposedly captured tank was put on sale as a "used Russian T—72 Tank- fully functional, for a price of 400,000.00 USD". Another version of the same meme mentions the only disadvantage of bad smell in the cabin. The meme adds a capital of sympathy to the avatar of the nice, brave, good trading farmer that makes use of all resources to support his fight against the invaders.

A third meme (see: https://knowyourmeme.com/photos/2321601-ukrainian-farmers-vs-russian-army) provides a similar line of interpretation depicting the Ukrainian farmer as a cool, back street businessman, sunglasses down, contemplating opportunities for more profit. The two narrative strands concur against the background of the original video clip, pointing to a larger, more problematic opposition between state imposed violence and citizen response ingenuity on the one hand, and the highlight of a cathartic type of humor that exorcizes evil by diminishing its force against the otherwise powerless citizen with no soldier skills and weapons.

Among several Ukrainian tractor meme compilations posted on YouTube, one in particular multiplies the effect of the blue tractor to epic proportions (https://www.you

tube.com/watch?v=hheLODstezM n.d.). The tractor is seen towing the badly hit Russian admiral ship Moskwa from the deep sea, a giant submarine or a set of rockets; another meme shows a black and white drawing of a tractor with a colored Ukrainian flag atop towing a tank while the script reads the famous Putin declaration "everything is going according to plan"; in another photo of the ship Moskwa, the author has added a small green tractor towing the ship though the sea, while an apparent Lego edition features the same truck but in the traditional red color of the brand overshadowing what seems to be another towed tank.

A game meme shows a golden tractor "unlocked" in the game after successfully stealing 10,000 Russian tanks; a most humorous meme pictures a tractor towing a white long table in a fake Ikea add (see: https://www.heraldscotland.com/opinion/19976619.meme-day-herald-diary/) reminiscing the exaggeratedly long table used by Putin in televised discussions with peer European diplomats or his own staff. And the list could go on.

This meme series shifts the significance of the narrative towards the creation of a superhero human / machine character that produces miracle deeds with its apparent superpowers. The comic undertone maintains the cathartic function, adding a lighter tone of hope in the miracles that this already global narrative plot could announce. Exorcising the more serious undertones of the war, violence and crime, the meme series dwells on the creation of therapeutic contrast with the oppressor figure and kindles the ray of hope.

Postage stamps

The popularity of the topic has made the Ukrainian state tap in and as of July 2022, a new philatelic edition was issued featuring the famous tractor. The design resulted out of a competition caried out months before and is designed in the blue and yellow colors of the Ukrainian flag.

A recent MFA of Ukraine tweet announces the new postage stamps and confirms the iconic value assigned to it. The blue drawing on a sunset landscape echoing the sky lit by bombardment fires is another powerful image of survival and resilience, which are the current values most endeared by the Ukrainian people.

Mockumentary

Last but not least, we should mention the 45 seconds mockumentary entitled "The 'Natural' Death Of A Russian Tank" released on Twitter in May 2022, which then went viral on Youtube. The author is Tetyana Denford, a writer that claimed to have collaborated "with a very talented producer and writer". The mockumentary shows a badly beaten, fuming tank, filmed from all angles against a green field. The voiceover imitating the legendary Sir David Attenborough narrates the following story: "This is a Russian tank. Like the Pacific salmon, Russian tanks migrate long distances from the abyss of Russia to end their lives in Ukraine's beautiful fields. Driven by an evolutionary desire to end their lives in a better place, all species of Russian tank die."[2] In the clip, 2022 is called "a great year for predators such as the special operation forces TB2 Bayraktar Drones and local farmers on tractors".

Conclusions

The case study has shown how a modest telephone camera recording of an apparently humorous moment on the streets of Ukraine has become the generator of a cultural movement tapped in by common people, artists and state authorities, to create a preferred mediated reality, where courage, resistance and hope are the undercurrent. The high performativity of the image symbol, its innuendoes and multiplied forms of expression, the way the narrative is shaped and reshaped collectively, with its cathartic and humorous tone, also demonstrates a rich intertextual dialogue between the different cultural productions that seem to mimic, imitate, add up to each other, only to reflect the creation of a myth powerfully linked to the collective memory and identity.

What does this mean for our study of protective factors against disinformation? Let us multiply the effect of this narrative with the power of tens of other series of stories similarly written upon the social life of Ukraine and the West. We can only assume that in the life of a community, the writing of history's darkest times in such resilient tones is likely to produce a kind of resistance unmatched by that encouraged with logic, facts and figures. If propaganda and disinformation are aimed to induce false beliefs and destroy social cohesion of a community, collective cultural productions, generated in multiple modes and tones, by artists and common people alike, spanning across different spheres of the digital space, from meme generators to Youtube, social networks and even NFT's, is likely to prove an indomitable rival to the destructive power of propaganda and disinformation.

While investigation of our research premise is in its early stages and would require many more research endeavors for validation, yet, we believe the case study has shown we cannot overrule its potential to build protective factors to disinformation.

Notes

1 https://knowyourmeme.com/memes/ukrainian-farmers-vs-russian-army
2 https://www.ladbible.com/news/latest-fake-david-attenborough-voice-narrates-russian-tank-death-20220511

Bibliography

Aro, Jessika. 2016. "The cyber-space war: propaganda and trolling as warfare tools." *European view* 121–132. doi:doi:10.1007/s12290-016-0395-5..

Baumann, Mario. 2020. "Propaganda Fights' and 'Disinformation Campaigns': the discourse on information warfare in Russia-West relations." *Contemporary Politics* (Routledge) 1–20. doi: https://doi.org/10.1080/13569775.2020.1728612.

Best, Shivali. 2017. "The spread of fake news on Facebook and Twitter is made worse by social network algorithms." *Mail Online*, 20 iunie. http://www.dailymail.co.uk/sciencetech/article-4621094/Are-Facebook-Twitter-ENCOURAGING-fake-news.html.

Bonino, Silvia, Elena Cattelino, și Silvia Ciairano. 2003. *Adolescents and Risk. Behaviour, Functions and Protective Factors*. Torino: Springer.

Bradshaw, Samatha, și Philip P. Howard. 2018. "Why does Junk News Spread so Quickly across Social Media? Algorithms, Advertising, and Exposure in Public Life." http://comprop.oii.ox.ac.uk/research/working-papers/why-does-junk-news-spread-so-quickly-across-social-media/.

Bradsma, Bart. fără an. *Inside Polarisation*. https://insidepolarisation.nl/en/.

Candaele, Kelly. 2020. "Coronavirus is a political problem, not just a health problem. Remember that when you vote." *The Guardian*, March. https://www.theguardian.com/commentisfree/2020/mar/19/coronavirus-political-problem-health-voting-elections.

Chekinov, S.G., și S.A. Bogdanov. fără an. "The Nature and Content of a New-Generation War." *Military Thought*. http://www.eastviewpress.com/Files/MT_FROM%20THE%20CURRENT%20ISSUE_No.4_2013.pdf.

Culloty, Eileen, și Jane Suiter. 2021. *Disinformation and Manipulation in Digital Media*. Routledge.

"Disinformation: how to recognise and tackle Covid-19 myths." *News, European Parliament*. 30 March 2020. https://www.europarl.europa.eu/news/en/headlines/society/20200326STO75917/disinformation-how-to-recognise-and-tackle-covid-19-myths.

Edelman's Trust Barometer, Trust Inequality. fără an. "Edelman." http://edelman.edelman1.netdna-cdn.com/assets/uploads/2016/01/2016-Edelman-Trust-Barometer-Global-_-Mounting-Trust-Inequality.pdf.

Edward, Herman, și Noam Chomsky. fără an. "A Propaganda Model." În *Manufacturing Consent*, de Herman Edward și Noam Chomsky.

Edward, Lucas, și Pter Pomeranzev. 2016. *Winning the Information War*. Center for European Policy Analysis.

2020. "EEAS SPECIAL REPORT UPDATE: Short Assessment of Narratives and Disinformation Around the COVID-19 Pandemic." *EU vs Dinsinfo*. 01 April. https://euvsdisinfo.eu/eeas-special-report-update-short-assessment-of-narratives-and-disinformation-around-the-covid-19-pandemic/.

EU vs. Disinfo. 2016. "Estonia is building a concentration camp for its Russian-speaking citizens." *EU vs Disinfo*. https://euvsdisinfo.eu/report/estonia-is-building-a-concentration-camp-for-its-russian-speaking-citizens/.

Farmy, Ukrainian. fără an. https://ukrainian.itch.io/ukrainian-farmy.

Foresman, Galen A., Peter S. Fosl, și Jamie Carlin Watson. 2017. *The Critical Thinking Toolkit*. Wiley Blackwell.

Gertrudis-Casado, María-del-Carmen, María-del-Carmen Gálvez-de-la-Cuesta, Juan Romero-Luis, și Manuel Gértrudix Barrio. 2022. "Los serious games como estrategia eficiente para la comunicación científica en la pandemia de la Covid-19." *Revista Latina de Comunicacion Social*. doi:https://doi.org/10.4185/RLCS-2022-1788.

Gray, Ann. 2003. *Research practice for cultural studies. Ethnographic methods and lived cultures*. London: Sage Publications.

Grejdeanu, Tamra. 2017. "Propaganda rusă în Moldova. Cum funcționează?" *Radio Europa Liberă*. 28 aprilie. Accesat iulie 30, 2018. https://www.europalibera.org/a/propaganda-rusa-in-moldova/28457231.html.

Guess, A. M., și B. A. Lyons. fără an. Guess, A. M., & Lyons, B. A. (2020). Misinformation, Disinformation, and Online Propaganda. *Social Media and Democracy*, 10–33. Accesat 07 21, 2022. doi:doi:10.1017/9781108890960.003..

Helmus, Baron, Radin, Magnuson, Mendelsohn, Marcellino, Bega, Winkelman. 2018. *Russian Social Media Influence. Understanding Russian Porpaganda in Eastern Europe*. Rand Corporation.

Hicks-Goldston, Christina, și Amy Ritchart. fără an. "The New Digital Divide: Disinformation and Media Literacy in the US." 1–60.

Hinchchliffe, Tim. 2020. "Exposing echo chambers to eradicate the plague of propaganda." *The Sociable*. https://sociable.co/social-media/exposing-echo-chambers-to-eradicate-the-plague-of-propaganda/.

fără an. *https://www.youtube.com/watch?v=hheLODstezM*.

Humprecht, Edda, Frank Esser, Peter Van Aelst, Anna Staender, și Sophie Morosoli. 2021. "The sharing of disinformation in cross-national comparison: analyzing patterns of resilience." *Information, Communication & Society*. doi:https://doi.org/10.1080/1369118X.2021.2006744.

Humprecht, Edda, Frank Esser, și Peter Van Aelst. 2020. "Resilience to Online Disinformation: A Framework for Cross-National Comparative Research." *The International Journal of Press/Politics* 1–24.

Ivan, Cristina. 2013. "Resilience—The X Factor of the Organisational Endurance." In *Intelligence in the Knowledge Society, Proceedings of the XVIIIth International Conference*, de Irena Chiru and Teodoru Stefan, 161–172. ANIMV Publishing House.

Ivan, Cristina, Irena Chiru, și Rubén Arcos. 2021. "A whole of society intelligence approach: critical reassessment of the tools and means used to counter information warfare in the digital age." Intelligence and National Security495–511. doi:doi:10.1080/02684527.2021.1893072..

JamNews. 2017. Fake news in Moldova: fires, droughts, terror attacks and discredited politicians. 2017 septembrie. https://jam-news.net/?p=59912.

Jeon, Youngseung, Bogoan Kim, Aiping Xiong, Dongwon Lee, și Kyungsik Han. 2021. "ChamberBreaker: Mitigating the Echo Chamber Effect and Supporting Information Hygiene through a Gamified Inoculation System." *Proceedings of the ACIM on Human Computer INteraction*. 1–26. doi:https://doi.org/10.1145/3479859.

Lewandowsky, Stephan, Ullrich K.H. Ecker, și John Cook. 2017. "Beyond Misinformation: Understanding and Coping with the "Post-Truth" Era." Editor Elsevier. *Journal of Applied Research in Memory and Cognition*.

Lewandowsky, Stephan, Ullrich K.H. Ecker, și John Cook. 2017. "Beyond Misinformation: Understanding and Coping with the "Post-Truth" Era." *Journal of Applied Research in Memory and Cognition*. Journal of Applied Research in Memory and Cognition.

Maftei, Alexandra, și Andrei Corneliu Holman. 2022. "Beliefs in conspiracy theories, intolerance of uncertainty, and moral disengagement during the coronavirus crisis." *Ethics and Behaviour* 1–11. doi:https://www.tandfonline.com/action/showCitFormats?doi=10.1080/10508422.2020.1843171.

Martin, L. John. 2010. "Disinformation: an instrumentality in the propaganda arsenal." *Political Communication* 47–64.

McKay, Spencer, și Chris Tenove. 2020. "Disinformation as a Threat to Deliberative Democracy." *Political Research Quarterly*. doi:10.1177/1065912920938143..

Mediacritica, primul portal de educație mediatică. 2018. Moldova—teren fertil pentru fake news. 11 iulie. Accesat iulie 30, 2018. http://mediacritica.md/ro/moldova-teren-fertil-pentru-fake-news/#prettyPhoto.

fără an. "Multiculturalism." *Oxford Dictionaries*. Accesat 08 5, 2014. http://www.oxforddictionaries.com/definition/english/multicultural.

Nissembaum, Assaf, și Limor Shifman. 2015. "Internet memes as contested cultural capital: The case of 4chan's /b/ board." *SagePub Journals*1–19. doi:doi:10.1177/1461444815609313..

Paul, Richard, și Linda Elder. 2014. *Critical Thinking: Tools for Taking Charge of Your Professional and Personal Life*. New Jersey: Pearson Education.

Polygraph.info. 2018. "Polygraph." 26 April. Accesat August 30, 2018. https://www.polygraph.info/a/fake-news-in-hungary/29194591.html.

Pressman, D. Elaine, și Cristina Ivan. 2019. *Internet Use and Violent Extremism: A Cyber-VERA Risk Assessment Protocol*. IGI Global.

Stoica, Cătălin Augustin, și Radu Umbres. 2020. "*Suspicious minds in times of crisis: determinants of Romanians' beliefs in COVID-19 conspiracy theories*." *European Societies*S246–S261. doi:https://doi.org/10.1080/14616696.2020.1823450.

TaskForce, EU East StratCom. 2020. *Trends of the Week. Throwing Coronavirus disinfo at the wall to see what sticks*. EU StratCom Task Force.

2022. "The 2022 Code of Practice on Disinformation." *European Commission*. 2 July. file:///C:/Users/User/Downloads/2022_Strengthened_Code_of_Practice_Disinformation_TeAETn7bUPXR57PU2FsTqU8rMA_87585.pdf.

United Nations General Assembly. 2015. "*Plan of Action to Prevent Violent Extremism, Report of the Secretary-General*." A/70/674. Accesat September 20, 2020. https://www.un.org/en/ga/search/view_doc.asp?symbol=A/70/674.

University of Oxford. 2018. *The Computational Propaganda Project. Algorithms, Automation and Digital Politics*. http://comprop.oii.ox.ac.uk/.

Weisburd, Andrew, Clint Watts, și JM Berger. 2016. "Trolling for Trump: How Russia Is Trying to Destroy Our Democracy." *War on the Rocks*. https://warontherocks.com/2016/11/trolling-for-trump-how-russia-is-trying-to-destroy-our-democracy/.

PART V

General Trends and Regional Specificities in Countering Disinformation

23

THE EU APPROACH TO COMBATING DISINFORMATION: BETWEEN CENSORSHIP AND THE "MARKET FOR INFORMATION"

Valentin Stoian

1 Introduction

The interest of the European Commission in the phenomenon of disinformation was caused by several major events in the mid-2010s: the Russian annexation of Crimea in 2014 and the accompanying disinformation campaign, as well as the the effect that information manipulation had on the two major electoral events of 2016 (the US presidential elections and the Brexit referendum), later revealed through the Cambridge Analytica scandal. The Commission took note of the deleterious effects of disinformation on electoral processes in democracies and commissioned a report elaborated by a group of experts in the field of communication studies. The Report of the independent High Level Group on fake news and online disinformation, entitled "A multi-dimensional approach to disinformation"[1] discussed how disinformation is spread online, as well as the possible solutions to address the phenomenon. On the basis of the conclusions of the report, several strategic documents were elaborated, including several *Communications* and follow-up reports produced by both the Commission or the Commission and the High Representative for Foreign Affairs and Security Policy. The emergence of the COVID-19 pandemic and of specific forms of disinformation in this context led to a new *Communication* being issued by the Commission and the HRVP on combating disinformation adapted to the crises generated by the pandemic.

This chapter will comprise an analysis of these documents and will delineate the main European-level policies adopted to combat disinformation. Also, within the chapter, academic literature will be employed in order to offer the relevant framework on the basis of which policy documents were analyzed. The literature identified focuses both on synthesizing and conceptualizing the existing policies as well as on offering suggestions for improvement and adaptation.

The chapter argues that the EU Commission adopts a strategy of regulating the information space as a market, in which high quality information is too "expensive" to produce and

DOI: 10.4324/9781003190363-28

consume, while disinformation is "cheap" on both dimensions. Thus, the Commission looks to reduce the costs of consuming high quality information and increase the costs of disinformation. However, the Commission does not make value judgments on what constitutes "good" and what is "bad" information but delegates that decision to independent gatekeepers (fact-checkers and recognized journalists). Conversely, whenever the High Representative is involved in the elaboration of policy documents, the approach shifts to "message coherence" and strategic communication to counter disinformation narratives. Thus, a distinction between the "EU-as-regulator" versus the "EU-as-actor" can be discerned in the approach of the two institutions.

The chapter proposes to employ process-tracing[2] as a methodology to analyze the relevant moments in which the European Commission decided to adopt measures to combat disinformation, as well as to assess the way that the various policies were implemented. The documents issued by the Commission will thus be used creatively to organize a timeline of the relevant policy responses.[3] These will also be grouped according to the different categories defined by the academic literature.

2 Actors and policies in the struggle against disinformation

Literature on European policies combating disinformation approaches several aspects: the actors involved in adopting the policies, the content of these policies and the relationship between combating disinformation and the freedom of speech. Some authors discuss all issues in their work while others only focus on only one of them. When discussing the relationship between actors involved in policy adoption, researchers employ concepts such as multi-level governance, co-regulation and epistemic communities to explain this process. Conversely, when focusing on what the policies contain, authors argue that the European Commission focuses on the increasing the "costs" of disinformation. Finally, when analyzing the relationship between disinformation and freedom of speech, authors argue that strong approaches that involve forbidding disinformation might not be acceptable under current human rights regimes.

The topic of actors' interplay in the process of policy adoption is discussed by Saurwein and Spencer-Smith[4], who analyze European and national-level policies against disinformation by using two theoretical frameworks. On the one hand, the rise disinformation is described as a socio-technical assemblage composed of producers and sharers of disinformation, who are either politically or financially motivated, social media users who engage and re-share disinformation, social media companies whose algorithms keep people in the same "content bubble" automatically and the technology which allows people to re-share content without marking it as disinformation. On the other hand, the actors adopting policies to combat disinformation are presented as being part of a multi-level governance network, which undertakes both classical regulation and other forms of control such as supervising self-regulation by industry-level branches and co-regulation. By looking at the policies adopted by different actors at different levels (supra-national—EU, national- governments, sub-national—local authorities and private companies), Saurwein and Spencer-Smith[5] conclude that their interaction can be studied through the multi-level governance approach pioneered in EU studies.[6]

In order to exemplify regulation at the national level, Saurwein and Spencer-Smith[7] present the cases of France, Germany and the UK. The first introduced a law allowing judges to immediately order the removal of online content if proved to be disinformation, while the UK adopted a policy paper establishing guidelines for potential future regulation. This included establishing a new regulator of online content and imposing a "duty of care" for the platforms, through a code of practice. Similarly to the EU Commission, UK authorities did not envision a law which could allow the removal of online disinformation, but also a market-oriented approach based on increasing the costs of disinformation through labeling and investing in media literacy and decreasing the costs of good information through subsidizing fact-checking and quality journalism. Alternatively, Germany adopted a tough approach requesting platforms to remove unlawful content such as hate speech in 24 hours after it being reported or seven days in less clear cases, or face considerable fines. Finally, the authors discuss the policies adopted by platforms and market actors, including labeling disinformation and partnering with fact-checkers[8].

Mardsen, Meyer and Brown[9] discuss the way policies against disinformation are adopted and enforced and identify three models: self-regulation, co-regulation and statutory regulation by governments. The authors argue that co-regulation involving a wide and general statute enforced by the state, a regulator setting detailed rules and codes of practices adopted by every private handling online platforms is the best from both an economic and a human rights perspective. Durach, Bargaoanu and Nastasiu[10] discuss the way combating disinformation was handled by the European Commission. Similarly to Mardsen, Meyer and Brown[11], they identify four models of regulation in the area: self-regulation by platforms, co-regulation between supra-national and national authorities on the one hand and private actors on the other, direct regulation and audience centered (or demand-side solutions) that focus the audience such as increasing media literacy and supporting fact-checking. As in the previously mentioned literature, the approach taken by the EU, through the publication of its code of practice is understood as a form of co-regulation, while the laws adopted by France, Hungary, Singapore and Malaysia as a form of direct regulation. Further, the last category of regulation is exemplified by different fact-checking organizations such as *EUvsDisinfo*, but also *Correctiv* (Germany) or *Demagog* (in Czech Republic) and by different media literacy programs which were promoted through "Media Literacy Week". Just like Mardsen, Meyer and Brown[12], Durach, Bargaoanu and Nastasiu[13] recommend co-regulation as a form of avoiding both placing excessive trust in those who have most to lose by regulating themselves (in the case of self-regulation) and the authoritarian pitfalls of direct regulation.

Hedvig Örden[14] argues that EU policies in the area of disinformation have relatively incoherent goals: on the one hand, they look for information coherence and a unified narrative, while on the other, they support content pluralism, but only of "quality content". In her view, this is caused by a competition different epistemic communities taking part in the elaboration of policies: the security/defense establishment defines the referent object of security (the value which must be defended) as informational coherence, while the media/journalistic/fact-checker community sees information pluralism as the relevant referent object. She concludes by arguing that this "hybrid value-constellation" will be carried into the implementation phase, leading to conflicts between these communities.

In an attempt to describe the possible futures of policies combating disinformation, Pherson, Ranta and Cannon[15] elaborate a series of scenarios based on two drivers: the initiators of the policy to combat disinformation and the type of policy to be adopted. The first is divided between government-led proposals and private sector-led proposals, while the second between content-based solutions and user-focused solutions. According to the authors, the former involve verifying actual content, while the latter look at verifying who can post on a particular platform. Based on these two drivers, the authors create a two-by-two matrix and present four possible scenarios: "Pinocchio warnings"—government-mandated warnings on suspicious content, "the Alt-net"—and alternative internet created by the government which one can access only after extensive verification, "Rigid gateways" in which Internet providers establish a protocol for verifying content and a standards board, and the "T-cloud", a space which is handled by internet providers where only certified users can post information and which is accessible for a fee.

Regarding the way policies against disinformation are meant to operate, Saurwein and Spencer-Smith argue that the European Commission conceptualizes the spread of disinformation as a form of market failure, caused by its relatively "low" costs as opposed to the "high" costs of producing high-quality journalism. Thus, the overall policy direction has been to increase the costs of disinformation through automatic detection and labeling and to decrease the cost of high-quality journalism through subsidizing investigative journalism networks[16]

When addressing the effect of platform-level interventions, Ng, Tang and Lee[17] look at two types of restrictions introduced by the Chinese social media platform Sina Weibo. They show that if a post is flagged as fake news, then it is spread in a more centralized fashion, only by influential users who are willing to incur the costs, while others avoid being associated with it. Conversely, if the platforms restricts the forwarding of posts from a specific account or the publication of any information by that account, then fake news posts are spread in a more dispersed way, by a greater number of people. However, before this intervention is taken, the study showed that those with strong ties to the account will not share potentially deceptive posts in order to not attract attention to that account, but those with weak ties will.

The issue of the freedom of speech is also approached by Mardsen, Meyer and Brown[18] who argue that neither governments nor platforms should be responsible for evaluating content, but that co-regulation between all the relevant actors, especially democratically legitimate state regulators should approve measures to combat disinformation and monitor their effectiveness. This is the only way, according to the authors, through which policies directed against disinformation can be compatible with the freedom of speech enshrined in democratic constitutions. Mardsen, Meyer and Brown[19] quote the CJEU judgments in several cases, which disallowed preventative filters which network providers would have had to introduce, on the basis that they might have also blocked legitimate content. A special criticism is reserved to implementing AI-based solutions against disinformation, given its large number of false positives identified. However, in this view, value judgments should not be made by AI only, but by a body of dedicated human verifiers.

In the *Utah Law Review*, Jason Pielemeier[20] analyzes both platform-level, national level and European-level regulation policies against disinformation. He argues that in the case of spreading disinformation it is very difficult show that the elements of a crime (evil action, intention and damages) are manifest. According to him, in order to combat disinformation, platforms have adopted a number of strategies, which do not involve banning disinformation, but decreasing its visibility and making it less profitable. The first has been done through a strategy of limiting reach, while the second through a strategy of disallowing people who share fake news to buy adverts. Finally, at a wider scale, platforms have taken measures to address inauthentic behavior, contextualized fake news shared so that people seeing fake news should also get real news and increased transparency through reporting the number of removals and archiving fake news posts.

Pielemeier[21] also looks to the EU *Code of Practice on Disinformation*, which is seen as part of a "coherent, coordinated and sustained effort to address disinformation" and as a way to determine platforms to self-regulate before any regulation would be imposed on them. The cases of France, Singapore and Singapore are contrasted to the EU approach, as they both involve a strong ban on disinformation and a duty on platforms to remove any content that is reported and verified as misleading.

Concluding this section, one can argue that the literature on policies against disinformation presents the policy outcome as a result of the interplay between actors at different levels and between epistemic communities at any given level. This, according to the authors, leads, many times, to incoherence and vacillation and to over-ambitious aims that might be difficult to achieve. Further, researchers have noted the "light touch" that the Commission adopts, looking to engage the addressees of policies in a form of co-regulation and the market-like approach that European institutions adopt. Finally, from a human-rights perspective, authors have noted the incompatibility between eliminating information from websites and social media platforms and the right to free speech.

3 EU-level policies

The EU Commission first showed interest in the problem of disinformation after the 2016 US elections and Brexit referendum. Given the particularly deleterious effects that disinformation had during these two electoral processes, the Commission decided to approach the problem by commissioning a report on the best ways to combat disinformation. Issued in 2018, the Report of the Independent High level Group on fake news and online disinformation entitled "A multi-dimensional approach to disinformation" looked at the harms that disinformation caused to a democratic society and at the measures previously undertaken by platforms, which, at that particular point in time, were not consistent. Then, the Report proposed five directions of action and policy goals, which, if implemented would considerably decrease the effect of online disinformation.

Transparency, according to the authors of the report, involves a series of measures aimed at increasing the visibility of the source of information and contextualizing information that is known to be false in order to expose those seeing it to high quality information in addition to the fake one. Thus, platforms should clearly mark paid content, even if it is shared by influencers rather than actual advertisements, should create a transparency index for each journalistic source and label disingenuous sources accordingly, and should increase the transparency of the work of fact-checking organizations by implementing a code of practice to which all should adhere. Further, programs aimed at increasing the media literacy should be implemented in schools in order to increase young students' abilities to spot fake news, but also through life-long education programs, which should allow adults to develop these skills. Further, the European Commission should, according to the report, help develop online tools for assessing the veracity of online sources and empower journalists by investing in tools to improve source verification and media innovation. Finally, the report recommends creating a diverse and sustainable media ecosystem by financial support to high quality journalism and by ensuring an absence of interference from governments in editorial policies and content of news media. Concluding, the report foresees the creation of a code of practice which should be adopted through a multi-stakeholder approach involving journalists, online platforms, media organization and fact-checkers.[22]

Relying on this report, the Commission issued several documents, which will be analyzed below. This chapter will focus on each specific document and its implementation reports, aiming to trace the process of how these policies were adopted and put into practice by the relevant authorities. However, given the fact that the Commission issued numerous documents with sometimes overlapping goals, this chapter will present each proposed measure and its follow-up according in a designated table.

The overarching document which sets out the Commission's policies in the area is the "Tackling online disinformation: a European approach"[23] *Communication*, which presents a number of measures which the Commission planned to adopt in the future. The document established four principles on which the future policies will be based: transparency regarding the origin of the information, diversity of information in the information ecosystem so that citizens can make informed decision based on what they see, fostering the credibility of information by showing which sources are trustworthy and fashioning inclusive solutions which will eliminate disinformation in the future by increasing media literacy and raising awareness.[24]

In order to better synthesize the content of the *Communication* and its follow-up report, issued in December 2018[25], the measures proposed will be presented in table form, with the state of implementation mentioned in the follow up report in the second column.

Another crucial document outlining the EU policies on the topics is the *Action plan against disinformation*, which was issued in December 2018[27] and whose implementation was assessed in April 2019[28]. After assessing the previous actions on the topic, as well as the threat that disinformation poses to a democratic society, the Action plan divides policies in four major pillars: improving the capabilities to detect, analyze and expose disinformation, strengthening response, mobilizing the private sector and raising awareness and improving societal resilience. Similarly to the Communication, the implementation of the action plan will be presented as a table.

In order to support the self-regulation of private actors involved in the fight against disinformation, the Commission issued, in 2018, based on the result of the multi-stakeholder forum, the *Code of Practice on Disinformation*[29]. Rather than opting for a direct regulatory approach, the Commission gave private providers indications on what to do and expected them to undertake the relevant measures by themselves. After an assessment by the Commission of the implementation in 2019, the Commission published, in 2021, a guiding document to elaborate on further action and then, in 2022, an improved version of the guide

The *Code of Practice*[30] is divided into five pillars of actions, each including a number of measures to be adopted. The first, improving the scrutiny of ad placements refers to stopping the monetization of fake news by not allowing sites which misrepresent themselves to place ads on the platforms, the second concerns making political and issue-based advertising more transparent by clearly distinguishing it from other content and by disclosing its funders, the third looks at eliminating automated behaviour of fake accounts (bots), the fourth at empowering consumers through making quality content more visible and the last aims to empower the research community through making data on platform use available to independent researchers. Initial evaluation of the implementation of the Code of Practice was mentioned in the 2018 and 2019 reports on the implementation of the Communication and the Action plan against disinformation. However, a comprehensive assessment of the implementation of the *Code of Practice* took place in 2020[31], especially under the pressure of the COVID-19 pandemic and of a new Communication on disinformation in this context.

In its *Assessment of the Code of Practice on Disinformation—Achievements and areas for further improvement*,[32] the Commission was relatively critical of the implementation of the *Code of Practice*. The Commission first observed the progress achieved in all of the pillars, for example rejecting advertisements of their misrepresentation policy (hundreds of thousands of actions by Google and tens of thousands of actions by Twitter) and the verification and labeling of several hundred thousand political ads and the certification of hundreds of official electoral accounts, or, in the case, of Microsoft and Twitter, the complete ban on political advertising. Facebook was praised for eliminating 2.19 billion fake accounts and for cracking down on coordinated inauthentic behavior, while Google and Microsoft for their handling of disinformation during the COVID pandemic, by increasing the visibility of official news sources. On the other hand, Twitter was praised for its action under pillar five of releasing a dataset of 30 million tweets.[33]

Table 23.1 Comparative assessment of the implementation of the "Tackling online Disinformation" Communication

Tackling online disinformation: a European Approach	*Report from the Commission on the implementation of the Communication "Tackling online disinformation: a European Approach"*
Convening a multi-stakeholder forum and creating an EU-wide Code of Practice against disinformation.	The multi-stakeholder forum was convened in May 2018 and the Code of Practice[26] was published in September 2018. It was signed initially by Facebook, Google, Twitter and Mozilla, EDIMA—European Trade association representing online platforms, as well as trade associations of industry and advertisers—EACA (European Association of Communication Agencies), IAB (Interactive Advertising Bureau) Europe, WFA (World Federation of Advertisers) and UBA (United Business Association).
Creating an EU-wide network of fact-checkers Creating a European online platform on disinformation to support the fact-checkers and which allows them to flag inappropriate content.	The Commission initiated a series of workshops with the fact-checking community and supported other initiatives of bodies such as the International Fact-checking Network (IFCN) and of the European Broadcasting Union (EBU). In the technical sphere, "the Commission proposed, under the Connecting Europe Facility work programme 2019, the creation of a new digital service infrastructure for the establishment of a European Platform on Disinformation".
Improving the identification of trustworthy suppliers of information based on electronic identification and authentication of trusted pseudonyms	The Commission discussed with the Horizon 2020 project Co-inform which aims to create a tool to help decision-makers, fact-checkers and citizens can identify safe providers of information.
Investing in research of technology such as artificial intelligence, blockchain and cognitive algorithms that can separate disinformation from high quality information.	The Commission is supporting research projects aimed at identifying fake automated accounts (bots) that spread misinformation, tools to verify audio-visual content and to understand the misperception of information. Also, new funding calls have been proposed, such as "Eunomia (open source solution to identify sources of information), SocialTruth (distributed ecosystem that allows easy access to various verification services), Provenance (intermediary free solution for digital content verification) and WeVerify (content verification challenges through a participatory verification approach)".
Supporting Member states to avoid threats to the integrity of the 2019 European Elections, through high-level conferences and a compendium of best practices.	The Commission convened two conferences on the topic of combating electronic election interference and democracy in Europe.
Supporting media literacy through encouraging independent fact-checkers, including media literacy topics in the SaferInternet4EU42 Campaign, organizing a European Week of Media Literacy, elaborating a report on media literacy in Europe and attempting to include media literacy in the criteria for evaluating school programs through the OECD.	The Commission leveraged the Media Literacy Expert Group for convening an EU-level Media Literacy Week, supported Safer Internet Centers and introduced in the future Audiovisual and Media Services Directive a legal obligation for member states to increase media literacy. Also, negotiations with the OECD are ongoing to introduce media literacy tests in the PISA tests.

Tackling online disinformation: a European Approach	Report from the Commission on the implementation of the Communication "Tackling online disinformation: a European Approach"
Supporting high quality journalism through funding media freedom initiatives and training for journalists and new technologies for news rooms.	The Commission launched a call for projects for the production of high-quality news content through data-driven media and will dedicate a budget on the topic in the 2021–2017 Multi-Annual Framework. It also co-funded the activities of the European Centre for Press and Media Freedom.

Table 23.2 —Comparative assessment of the implementation of the Action plan Against Disinformation

Action plan against disinformation	Report on the implementation of the Action Plan Against Disinformation
Strengthening the Strategic Communication Task Forces and Union delegations, through additional staff and new tools, useful to detect, analyze and expose disinformation. Reviewing the mandates of the Strategic Communication Task Forces for Western Balkans and South	Financial and Human Resources of Strategic Communication Task Forces and European External Action Service were increased.
Establishing a Rapid Alert System for addressing disinformation campaigns Improving communication on Union values and policies Strengthening strategic communication in the Union's neighborhood.	The Rapid Alert System was implemented, and it helped detect several cases of coordinated disinformation, although not a coordinated campaign against European elections. Cases of disinformation were identified in the case of the Notre Dame Cathedral fire and the Austrian political crisis. The Rapid Alert system helped increase cooperation between EU institutions and national authorities.
Monitor the implementation of the Code of practice by private providers of social platforms. In case of non-compliance, the Commission might move to rapid enforcement through regulation.	A closer monitoring of the implementation of the Code of Practice by Online platforms was conducted through the European Regulators Group for Audiovisual Media Services (ERGA). Facebook, Google and Twitter improved the scrutiny of ad placements, increased transparency of political ads and removed YouTube channels and fake accounts especially those engaged in inauthentic behavior.
Organizing targeted campaigns for the public and trainings for media and public opinion shapers to raise awareness of the negative effects of disinformation. Support the creation of teams of multi-disciplinary fact-checkers and researchers to expose disinformation in digital media and social networks Organize the Media Literacy Week and supporting cooperation between media literacy practitioners Monitor the implementation, by Member States of the Elections Package, in preparation for the 2019 European Elections	Seminars, conferences and briefings were held to train journalists and members of the public on the threat of disinformation. Increased communication of EU policies through social media and communication campaigns was held, especially one dedicated to voting in the European elections of 2019. Myth-busting and awareness-raising materials were produced and distributed through the internal Commission Network against disinformation. The European Media Literacy week was organized and an obligation on states to increase media literacy was introduced in the media services directive.

Action plan against disinformation	Report on the implementation of the Action Plan Against Disinformation
	International Fact-Checking Network created a European branch of independent fact checkers in fourteen member States with the support of the EU Commission. The EU Commission financially supports research into technologies for content verification. The Commission launched the Social Observatory for Social Media analysis. The Commission supported measures to increase the integrity of the 2019 European elections through the creation of a network of election authorities and helping them identify gaps and threats, supporting cyber security, preventing the misuse of personal data through clarifying the implementation of the GDPR in an electoral context and boosting the transparency of political advertising.

However, the Commission criticized the platforms for undertaking only a shallow form of verification of ad placements, only focusing on advertisements hosted by themselves and not by other websites which users share. No scrutiny was undertaken of content placed on the websites/ YouTube feeds of influencers or on third party websites who do not misrepresent themselves but are purveyors of well-concealed misinformation. Concerning the second pillar, the Commission criticized the inconsistent verification of issue-based advertising and the existence of loopholes that allow unmarked advertisements to be distributed due to a lack of a uniform system of authorization. Regarding the third pillar, the *Assessment* argues that due to reporting being done only at the aggregate level and only for the actions of foreign actors, it is not possible to evaluate the impact at the EU level. One of the strongest criticisms has been levelled by the Commission against the implementation of the fourth pillar measures, especially at the lack of monitoring of the impact of the tools deployed to increase the visibility of high quality news. Similarly, in the case of the fifth pillar, criticism is levelled against the insufficient cooperation with fact-checkers, which is only done through bilateral agreements and especially the episodic and arbitrary sharing of data with independent researchers.[34]

At the structural level, the *Code* is criticized for a lack of clear definitions, which inhibits coordinated actions in ambiguous areas, the lack of focus on some specific areas such as micro-targeting of political advertising, fairness of access to political advertising and the failure to take into account all possible inauthentic behaviour, such as faked purchase followers, impersonation or the use of influencer.[35]

After witnessing the effects of disinformation during the early part of the COVID-19 pandemic (appropriately called the "infodemic"), the European Commission and the HRVP issued another *Communication* in mid-2020, entitled "Tackling COVID-19 disinformation -getting the facts right".[36]. The document recognizes the need to improve strategic communication aimed at combating disinformation narratives, both inside and outside the EU, to enhance the Rapid Alert system, to increase cooperation between the EU institutions and the Member States, especially in the case of exchanging best practices on issues such as micro-targeting, to better promote information about the assistance the EU offers to third countries, as well as to enter a partnership with the WHO aimed at debunking harmful narratives about the pandemic.

Further, in this *Communication*, platforms are envisioned as having a significant role in the struggle against disinformation, especially through the promotion of authoritative content on the virus, approved by international health agencies, informing users on what policies they have applied to select appropriate information, combating inauthentic behavior, providing data about policies to limit advertisement, supporting fact checkers through open and extensive partnership and through giving proper access to data. According to the document, the Commission will also ensure the emergency measures are not used to stifle freedom of speech and support the resilience of independent media both through the comprehensive recovery package and through funding independent journalism and civil society initiatives in partner countries. Another line of action envisioned by the Commission is raising citizen awareness through increasing media literacy, especially by using its Erasmus+ and European Solidarity Corps project to fund projects dedicated to tackle disinformation, or using the Audiovisual and Media plan to raise awareness among young people.[37]

In May 2021, based on the above-mentioned assessment and on the EU's experience during the pandemic, the Commission published a *Guidance for the Strengthening of the Code of Practice on Disinformation*, which lies at the basis of the 2022 updated code.[38] The document proposes the expansion of the Code's scope to include different types of disinformation, to include misinformation (false but unintentional content) and influence operations by foreign actors, the inclusion private messaging services and other online actors such as advertisement providers and e-commerce platforms. According to the document, the European Digital Media Observatory will play a central role in analyzing disinformation, and the Rapid Alert System will be strengthened. Further, policies to stop advertisements that contain disinformation or that are placed on websites containing disinformation, thus reducing cash flows from disinformation-containing websites. Further, the Guidance foresees strengthening the rules on labeling political and issue-based advertising to make sure labelling remains even if the ad is shared through messaging apps and creating easily accessible repositories for political ads.

Rules for combating automated manipulative behaviour should also be strengthened, especially by creating a common understanding of what is unacceptable and a comprehensive list of what are impermissible techniques. Users should be empowered by enhancing media literacy and making sure that online systems are designed to prevent the spread of disinformation, while false information should be flagged, and users warned. The Guidance envisions a strong role for the fact-checking and research community and a strong duty on platforms to share this data in a GDPR-compliant way, including, in very specific cases, access to identifiably personal data. Also, in the view of the Commission, fact-checkers should have automated access to information and be able to fact-check it easily.

In order to monitor the implementation of the Code, some KPI's (Key Performance Indicators) should be included, both at the level of the service and at the level of the information ecosystem. Among the first, the impact of fact- checking and of measures to display reliable information, as well as the instances of manipulative behaviour detected should be included.

In June 2022, the Commission[39] published a highly updated and reinforced Code of Practice, signed by 33 social media companies and trade federations[40]. Unlike the previous code, the 2022 version is considerably more detailed, including 44 commitments along the five pillars of the first code. Each commitment is detailed into a series of sub-commitments, amounting to a comprehensive list of actions for platforms to undertake. Some of the major novelties of the Strengthened code is the strengthening of the policies against the demonetisation of disinformation, to include a stronger scrutiny of websites buying advertising on the platform, including through the use of fact-checkers (committing to the reporting of the number of third-party audits of buyers of disinformation) and the verification of intermediaries buying advertising in the name of other websites, as well as verifying the content of advertisements placed so that these do not include disinformation inside them.

Regarding the labelling of political advertising, the problem, according to previous policy documents by the Commission, was, among others, definitional, as there could be no agreement on the definition of political and issue-based advertising. Through the enhanced code, parties commit to work together to find such a definition, to put in place mechanisms to clearly distinguish political advertising and paid-for content, even if this is then further shared by users through messaging apps. Sponsors of political ads must be clearly identifiable, and the general outlines of the contracts made with them have to be public information. Political and issue-based ads places must be archived in a repository which should be made public.[41]

Platforms signing on to the code also commit to address a wide array of inauthentic behavior such as malicious deep fakes, hack-and-leak operations, fake accounts and bot-driven amplification, and the use of influencers to, to remove AI-generated content and to use automated verification system that are not prohibited, as well as to exchange information about malicious practices which they have gathered.[42]

Further, platforms commit to support media literacy and critical thinking, including campaigns highlighting the modus operandi of malicious actors and to design systems in such a way as to discourage the spread of disinformation, as well as to prioritize quality content and to transparently publish information about criteria used to prioritize and de-prioritize information, as well as create tools to verify the authenticity of digital content. Further, platforms undertake to offer consistent access to fact-checkers to information, implement systems to allow fact-checkers to flag disinformation, create indicators of trustworthiness, issue warnings from authoritative sources on pieces labelled as disinformation, but also include a transparent appeal mechanism to challenge abusive flagging. Aggregated, non personal data should be easily accessible to interested researchers and even personal data should be made available to vetted persons in relevant cases.[43]

Regarding fact-checking, signatories commit to offering financial contributions to independent fact-checking organizations, and to undertake serious and comprehensive cooperation with fact-checkers, as well as report on the way the decisions of fact-checkers were integrated into their activities. Finally, a much stronger enforcement of the code will be included, especially a Transparency Centre website and a task force to monitor the application of the code.[44]

Information operations/disinformation receive relatively little mention in the Security Union Strategy,[45] where they are addressed in the context of the vulnerabilities exploited by foreign actors, undertaking hybrid attacks by undertaking cyber attacks, attacking critical infrastructure, conducting disinformation campaigns and radicalizing political narratives. The latest *Progress Report on the Implementation of the Strategy*[46] was written in the wake of the Russian invasion of Ukraine an lists a set of measures which have been undertaken, such as the banning of Russian propaganda channels, the expansion of the *Code of Practice* and private actor action against disinformation, as well action by the European Digital Media Observatory to coordinate fact-checkers and to conclude that actors spreading COVID-19 disinformation began to spread pro-Russian narratives. The *Report* refers to the *Strategic Compass* as the key document outlining future policies against disinformation.[47]

The *Strategic Compass*[48], released in March 2022, includes mentions of the actions which the EU will take against disinformation, such as improving the capacity to understand and analyze the threat so that costs can be imposed on perpetrators and supporting free and independent media. Then, the document foresees creating a toolbox which will strengthen the Union's strategic communication capacities, especially of CSDP missions abroad and increasing the use of the Rapid Alert System, as well as creating a data space where information on all relevant incidents will be collected.[49]

4 Discussion—the impact of policies against disinformation

Overall, EU-level policies against disinformation reject a strong, regulation-based approach and prefer to adopt a "nudging" approach, through which they recommend actions and undertake reforms within their own institutions to ensure the coherence of messaging and the fast delivery of appropriate responses to debunk disinformation narratives. Strategic communication is envisioned as a solution to counter information operations conducted by foreign actors and to better communicate the content of EU policies abroad.

Further, the Commission adopted a coordinating role regarding the efforts of several actors such as fact-checkers, independent journalists and NGOs which have programs dedicated to increasing media literacy and improving critical thinking.

Another line of action envisioned by the Commission is funding innovation and media literacy programs, through its research and cooperation programs such as Horizon 2020 and Erasmus+. Thus, funding lines will be introduced, and projects supported that help with the automated detection of disinformation, with combating automated behaviour on social media and with identifying deep fakes produced by artificial intelligence.

The main policy instrument adopted by the Commission is the Code of Practice, which relies on creating a standard of "appropriate behaviour" for private providers of online social media platforms. By giving this standard official sanction, the Commission undertakes the role of a moral authority, which offers prestige-based rewards and sanctions against those who respect it or refuse to do so. This is a type of co-regulation which forestalls direct, mandatory impositions by supra-national authorities, yet aims to lead private actors in the desired direction.

The Code of Practice, similar to other policies, does not look to eliminate content which can be labelled as disinformation, but to make it less desirable, both from a financial and from an informational point of view. The financial aspect is handled through the scrutiny of ad placements, while the informational direction is handled by flagging sources fact-checked as sharing disinformation, making high quality information more visible, eliminating clearly misleading automated behavior and allowing researchers to analyze trends.

As, in many other areas in which it intervenes, the EU Commission seems to envision the "information space" as a market affected by market failures. The failure is caused by the relatively low price of disinformation and the high price of good quality information. The role of regulation is, the view of the European Commission, to address these failures by subsidizing and thus reducing the cost of high quality information and increasing the costs of low quality one. To achieve this, the producers of good quality information (journalists) and its gatekeepers (fact-checkers) are empowered, including through the quasi-mandatory financial contribution of those regulated (platforms), while the purveyors of disinformation are punished (through flagging and the forbidding of monetization).

Alternatively, to the extent that the HRVP and EEAS are involved, a more "state-like" approach is envisioned, with institutions foreseen as taking the lead in handling the "information space", controlling the narratives and debunking disinformation. A special role is given to EU missions abroad, carried out through the CSDP, which act in a "hostile" information space in the countries where they operate. After the Russian invasion of Ukraine, the Council adopted Regulation 2022/350 which expressly forbade the Russian official agencies, Russia Today and Sputnik[50].

Table 23.3 Relationship between low-achieving 15 year olds and online behaviour in the EU and neighbouring countries

	Percentage of low achieving 15-year old in reading in 2018[51]	Seen doubtful information on the internet[52]	Checked truthfulness of information	Checked source of information	Discussed information they saw online	Either checked or discussed or information they saw online	Have not checked the information they saw online because they knew the source was not reliable	Did not check information they saw online because of lack of skills	Did not check the information they saw online because of other reasons
Spearman's correlation between column 1 and current column[53]		-0.62[54] N=35	-0.47 N=35	-0.54 N=35	-0.28 N=35	-0.45 N=35	-0.44 N=35	-0.38 N=35	-0.30 N=35

Source of data: Eurostat

The European Union began conducting studies on individual media behavior only in 2021, so, at the point there is no longitudinal data at country level which could be used to measure the impact of policies to combat disinformation. However, the 2021 assessment showed that citizens are, to some extent, aware of the need to assess the veracity of and double-check information that they encounter online. However, this practice is not uniform across the countries of the European Union.

Given than no information on media-related behaviour has been collected before 2021, a test of the effects of digital literacy policies on media behaviour has to be conducted by using proxy indicators. In this case, the percentage of low- achieving 15 year olds in reading in 2018, measured by the OECD PISA test was selected as a proxy for successful critical reading education in national schools.

The following table shows the relationship between information-related behaviour in the EU and neighbouring countries in 2021 and the percentage of low achieving 15 year olds in reading in 2018 (the 2015 value was used for Spain, due to the lack of data for 2018) (N=35). A correlation analysis was performed between the variables available in the Eurostat data.

The strong correlations obtained between reading achievement and responsible online behaviour shows that countries investing in critical thinking also expect their citizens to evaluate the information they see and cross-check it from multiple sources. This is a strong evidence of the need for EU-wide digital literacy programs.

5 Conclusion

The different approaches of the two European institutions become visible in their handling of the problem of disinformation. While the Commission envisions the "information space" as a market, the HRVP sees it as an area where threats emerge. While the former adopts a solution of "levelling the playing field" between "good" and "bad" actors, the latter aims to become an actor in its own right, entrusted with a responsibility to present the EU's achievements and successes. This has become evident after the Russian invasion of Ukraine, with the very strong measures adopted against Russia Today and Sputnik.

This conclusion can be compared to the way the two institutions regulate in other fields, such as commercial transactions and domestic security policy. This can generate a profile of the policy approaches of the two institutions, which can also be compared to that of the European Parliament or other national or supra-national actors.

Notes

1 European Commission *A multi-dimensional approach to disinformation: Report of the independent high level group on fake news and online disinformation.*
2 Bennett and. Checkel, eds. *Process tracing.*
3 Stoian. "Policy Integration Across Multiple Dimensions: the European Response to Hybrid Warfare."
4 Saurwein, and Spencer-Smith. "Combating disinformation on social media: Multilevel governance and distributed accountability in Europe."
5 Ibid
6 Stephenson "Twenty years of multi-level governance: 'Where does it come from? What is it? Where is it going?'".
7 Saurwein, and Spencer-Smith. "Combating disinformation on social media: Multilevel governance and distributed accountability in Europe."
8 Ibid, p. 836.
9 Marsden, Meyer, Brown. "Platform values and democratic elections: How can the law regulate digital disinformation?

10 Durach, Bargaoanu, and Nastasiu. "Tackling disinformation: EU regulation of the digital space."

11 Marsden, Meyer and Brown. "Platform values and democratic elections".

12 Pielemeier, "Disentangling disinformation".

13 Durach, Bargoanau and Nastasiu, "Tackling disinformation".

14 Ördén. "Deferring substance: EU policy and the information threat."; Ördén. *Securing Judgement: Rethinking security and online information threats.*

15 Pherson, Mort Ranta, Cannon. "Strategies for combating the scourge of digital disinformation.".

16 Ibid, p 835.

17 Ng, Tang, Dongwon Lee. "The effect of platform intervention policies on fake news dissemination and survival: An empirical examination."

18 Marsden, Meyer, and Brown. "Platform values and democratic elections: How can the law regulate digital disinformation?".

19 Ibid.

20 Pielemeier. "Disentangling disinformation: what makes regulating disinformation so difficult?."

21 Ibid, 933–937.

22 European Commission *A multi-dimensional approach to disinformation: Report of the independent high level group on fake news and online disinformation.*

23 European Commission "Communication from the Commission to the European Parliament, the Council, the European Economic and Social Committee and the Committee of the Regions: Tackling Online Disinformation: A European Approach.". *COM* (2018) 236 final.

24 Ibid.

25 European Commission "Report from the Commission the European Parliament, the European Council, the European Economic and Social Committee and the Committee of Regions on the implementation of the Communication "Tackling online disinformation: a European Approach" COM(2018) 794 final.

26 The Code and Its implementation will be discussed below.

27 European Commission and HRVP (2018), "Joint Communication to the European Parliament, the European Council, the European Economic and Social Committee and the Committee of Regions: Action plan against disinformation" 5.12.2018 JOIN(2018) 36 final, 2018.

28 European Commission and HRVP (2019), "Joint Communication to the European Parliament, the European Council, the European Economic and Social Committee and the Committee of Regions Brussels: Report on the implementation of the Action Plan Against Disinformation" 14.6.2019, JOIN (2019) 12 final, 2019

29 European Commission, "Code of Practice on Disinformation", https://digital-strategy.ec.europa.eu/en/library/2018-code-practice-disinformation, 2018 Accessed 25.07.2022

30 Ibid.

31 European Commission "Commission Staff Working Document: Assessment of the Code of Practice on Disinformation—Achievements and areas for further improvement" 10.9.2020 SWD(2020) 180 final

32 Ibid.

33 Ibid.

34 Ibid.

35 Ibid.

36 European Commission and HRVP, "Joint Communication to the European Parliament, the European Council, the European Economic and Social Committee and the Committee of Regions: Tackling COVID-19 disinformation—Getting the facts right" 10.6.2020, JOIN(2020) 8 final, 2020.

37 Ibid.

38 European Commission " Communication from the Commission to the European Parliament, the Council, the European Economic and Social Committee and the Committee of the Regions: European Commission Guidance on Strengthening the Code of Practice on Disinformation" 26.5.2021 COM(2021) 262 final.

39 European Commission "The Strengthened Code of Practice on Disinformation 2022", https://digital-strategy.ec.europa.eu/en/library/2022-strengthened-code-practice-disinformation, 2022 Accessed 25.07.2022

40 European Commission Signatories of the 2022 Strengthened Code of Practice on Disinformation, https://digital-strategy.ec.europa.eu/en/library/signatories-2022-strengthened-code-practice-disinformation, Accessed 25.07.2022

41 European Commission "The Strengthened Code of Practice on Disinformation 2022".
42 Ibid.
43 Ibid.
44 Ibid.
45 European Commission, "Communication from the Commission on the EU Security Union Strategy", Brussels, 24.7.2020, COM(2020) 605 final.
46 European Commission, "Communication from the Commission to the European Parliament and the Council, on the Fourth Progress Report on the implementation of the EU Security Union Strategy", Brussels, 25.5.2022, COM(2022) 252 final.
47 Ibid.
48 European Council "A Strategic Compass for Security and Defense", https://www.eeas.europa.eu/eeas/strategic-compass-security-and-defence-0_en, Accessed 18.08.2022.
49 Ibid.
50 Official Journal of the European Union (2022). COUNCIL REGULATION (EU) 2022/350 of 1 March 2022 amending Regulation (EU) No 833/2014 concerning restrictive measures in view of Russia's actions destabilising the situation in Ukraine. Available at: https://eur-lex.europa.eu/legal-content/EN/TXT/PDF/?uri=OJ:L:2022:065:FULL&from=EN [accessed on 1 July 2022].
51 High numbers represent low achievement at the national level.
52 High numbers represent responsible online behavior.
53 All correlations are statistically significant.
54 Correlations close to 1 or -1 mean a strong association between the two variables.

Bibliography

Bennett, Andrew, and Jeffrey T. Checkel, eds. *Process tracing* (Cambridge University Press, 2015)

Durach, Flavia, Alina Bargaoanu, and Catalina Nastasiu. "Tackling disinformation: EU regulation of the digital space." *Romanian J. Eur. Aff.* 20 (2020): 5–20.

European Commission "*Commission Staff Working Document: Assessment of the Code of Practice on Disinformation—Achievements and areas for further improvement*" 10.9.2020.

SWD(2020) 180 final European Commission "*Communication from the Commission to the European Parliament, the Council, the European Economic and Social Committee and the Committee of the Regions: Tackling Online Disinformation: A European Approach.*". COM (2018) 236 final, 2018

European Commission "*Communication from the Commission to the European Parliament, the Council, the European Economic and Social Committee and the Committee of the Regions: European Commission Guidance on Strengthening the Code of Practice on Disinformation*" 26.5.2021 COM(2021) 262 final

European Commission "*Report from the Commission the European Parliament, the European Council, the European Economic and Social Committee and the Committee of Regions on the implementation of the* Communication "Tackling online disinformation: a European Approach" COM(2018) 794 final, 2018

European Commission "*The Strengthened Code of Practice on Disinformation 2022*", https://digital-strategy.ec.europa.eu/en/library/2022-strengthened-code-practice-disinformation, 2022 Accessed 25. 07.2022,

European Commission and HRVP, "*Joint Communication to the European Parliament, the European Council, the European Economic and Social Committee and the Committee of Regions: Tackling COVID-19 disinformation—Getting the facts right*" 10.6.2020, JOIN(2020) 8 final, 2020

European Commission and HRVP, "*Joint Communication to the European Parliament, the European Council, the European Economic and Social Committee and the Committee of Regions: Action plan against disinformation*" 5.12.2018 JOIN(2018) 36 final, 2018

European Commission and HRVP, "*Joint Communication to the European Parliament, the European Council, the European Economic and Social Committee and the Committee of Regions Brussels: Report on the implementation of the Action Plan Against Disinformation*" 14.6.2019, JOIN(2019) 12 final, 2019

European Commission Signatories of the 2022 Strengthened Code of Practice on Disinformation, https://digital-strategy.ec.europa.eu/en/library/signatories-2022-strengthened-code-practice-disinformation, Accessed 25.07.2022.

European Commission, "*Code of Practice on Disinformation*", https://digital-strategy.ec.europa.eu/en/library/2018-code-practice-disinformation, Accessed 25.07.2022.

European Commission, "*Communication from the Commission on the EU Security Union Strategy*", Brussels, 24.7.2020, COM(2020) 605 final

European Commission, "*Communication from the Commission to the European Parliament and the Council, on the Fourth Progress Report on the implementation of the EU Security Union Strategy*", Brussels, 25.5.2022, COM (2022) 252 final

European Commission, A multi-dimensional approach to disinformation: Report of the independent high level group on fake news and online disinformation. Publications Office of the European Union, 2018.

European Council "*A Strategic Compass for Security and Defense*", https://www.eeas.europa.eu/eeas/strategic-compass-security-and-defence-0_en, Accessed 18. 08.2022.

Marsden, Chris, Trisha Meyer, and Ian Brown. "Platform values and democratic elections: How can the law regulate digital disinformation?." *Computer Law & Security Review* 36 (2020): 105373.

Ng, Ka Chung, Jie Tang, and Dongwon Lee. "The effect of platform intervention policies on fake news dissemination and survival: An empirical examination." *Journal of Management Information Systems* 38.4 (2021): 898–930.

Official Journal of the European Union (2022). *COUNCIL REGULATION (EU) 2022/350 of 1 March 2022 amending Regulation (EU) No 833/2014 concerning restrictive measures in view of Russia's actions destabilising the situation in Ukraine*. Available at: https://eur-lex.europa.eu/legal-content/EN/TXT/PDF/?uri=OJ:L:2022:065:FULL&from=EN [accessed on 1 July 2022].

Ördén, Hedvig. "Deferring substance: EU policy and the information threat." *Intelligence and National Security* 34.3 (2019): 421–437.

Ördén, Hedvig. *Securing Judgement: Rethinking security and online information threats*. Diss. Department of Political Science, Stockholm University, 2020.

Pherson, Randolph H., Penelope Mort Ranta, and Casey Cannon. "Strategies for combating the scourge of digital disinformation." *International Journal of Intelligence and CounterIntelligence 34.2 (2021): 316–341.*

Pielemeier, Jason. "Disentangling disinformation: what makes regulating disinformation so difficult?." *Utah L. Rev.* (2020): 917–940.

Saurwein, Florian, and Charlotte Spencer-Smith. "Combating disinformation on social media: Multilevel governance and distributed accountability in Europe." *Digital Journalism* 8, no 6 (2020): 820–841.

Stephenson, Paul. "Twenty years of multi-level governance: 'Where does it come from? What is it? Where is it going?'." *Journal of European Public Policy* 20.6 (2013): 817–837.

Stoian, Valentin. "Policy Integration Across Multiple Dimensions: the European Response to Hybrid Warfare." Studia Politica. *Romanian Political Science Review* 19, no 3–4 (2019): 97–126.

24

THE STRATEGIC SECURITY ENVIRONMENT AND NATO'S PERSPECTIVES IN DEVELOPING EFFECTIVE STRATEGIC COMMUNICATIONS (2014–2022)

Vira Ratsiborynska

"...the battlefield isn't necessarily a field anymore. It's in the minds of the people."

(Admiral Mike Mullen, 156 Landon Lecture, 2010-)[1]

Introduction

Since the creation of the Alliance in 1949, NATO has been subjected to permanent transformations and political, military, and institutional adaptation. The past two decades have resulted in a new set of security challenges because of the geopolitical shifts taking place in Europe and other emerging security risks stemming from disruptive technologies and hybrid threats. The global strategic environment has dramatically changed and has been transformed because of the many instabilities in NATO's immediate vicinity that have also shaped and accelerated NATO's institutional adaptation.

Russia's war of attrition in Ukraine that has profoundly shocked the West since February 2022 and other fundamental security risks and dynamics coming from the South have influenced NATO's internal institutional dynamics. The emergence of new domains of warfare including space and cyber as well as the problem of emerging and disruptive technologies have prompted NATO to address these threats and to go through adaptations that are reflected not only in its institutional reform processes but also in its strategic communication vision. First, this chapter will address the questions of the strategic security environment and how it impacts NATO's vision and its communication efforts. Secondly, the research will focus on the information environment since 2014 before it concludes with a critical reflection on the strategic communication of NATO.

DOI: 10.4324/9781003190363-29

The strategic security environment shaping NATO's vision

NATO's most recent Strategic Concept underlines many security challenges and risks that affect NATO, government institutions, Member States, and their citizens. As stated in the new 2022 Strategic Concept, "the Euro-Atlantic area is not at peace […]. Strategic competition, pervasive instability and recurrent shocks define our broader security environment".[2] NATO's Strategic Concept devotes much of its attention to the hybrid threats and the authoritarian actors who "interfere in democratic processes and institutions and target the security of our citizens through hybrid tactics, both directly and through proxies. They conduct malicious activities in cyberspace and space, promote disinformation campaigns, instrumentalise migration, manipulate energy supplies and employ economic coercion".[3] The NATO document explicitly designates the Russian Federation as "the most significant and direct threat to Allies' security and to peace and stability in the Euro-Atlantic area" while the People's Republic of China (PRC) is seen as a challenger to Western interests, security, and values.[4]

As stated in the new NATO Strategic Concept, Western democracies are constantly challenged by authoritarian actors while hybrid threats have become a permanent feature of today's security environment and a part of the current NATO security landscape.

Hybrid threats, the spread of disinformation and propaganda, cyber-attacks, etc. are pushing NATO to act with unity and cohesion and engage with different stakeholders on security risk reduction and risk assessment. Acknowledged by many EU and NATO official documents as the EU's Strategic Compass and NATO's Strategic Concept from 2022, the spread of propaganda, disinformation and hybrid threats erodes trust in government institutions, creates and exploits systemic vulnerabilities and societal polarization and affects decision-making processes. In such a volatile security environment, it is important to assess how this security environment shapes NATO's vision, what effects the current information environment has on the citizens and how NATO should strengthen its strategic communication efforts in the future.

Historic background: Hybrid threats and security environment from 2014 till now

After the illegal annexation of Crimea in 2014, which was described by NATO Supreme Allied Commander Philip Breedlove as "the most amazing blitzkrieg we have ever seen in the history of information warfare",[5] NATO acknowledged the need to develop a set of policies to deal with hybrid threats and be able "to prepare for; deter; and if necessary, defend against hybrid warfare".[6] Russian military actions in Ukraine from 2014 onwards became a wake-up call for NATO and a turning point with regard to the new development of its strategic communication activities.[7]

At the Wales NATO Summit in September 2014, the Allies agreed to "ensure that NATO is able to effectively address the specific challenges posed by hybrid warfare threats, including enhancing strategic communications".[8] The objectives of NATO's StratCom were set to "contribute to implementing NATO operations, missions, and activities by incorporating StratCom planning into all operational and policy planning; build public awareness and understanding of NATO".[9]

In January 2014, the NATO Strategic Communications Centre of Excellence in Riga became operational. The main activities of the Centre are, amongst others, supporting the development of a NATO Military Committee strategic communications policy and doctrine, researching early indications of hybrid warfare scenarios, and analyzing the information environment in the Eastern neighborhood.[10] Since 2014 a lot of emphasis has been put by NATO into countering hybrid threats, understanding the information and security environment and collaborating with different stakeholders on security risk assessment.

Since the Warsaw Summit in 2016, NATO has been making progress on its implementation of plans in countering hybrid threats that include such pillars as strategic communications to fight disinformation and propaganda; enhancement of defense and deterrence; cyber defense and strengthening resilience.[11] According to the reflection report "NATO 2030: United for a New Era", "building resilience across Allied populations is the primary responsibility of Allies themselves" and NATO plays a supportive role and "could offer a surge capacity to individual countries whose capabilities may be overwhelmed by e.g. a terrorist attack involving non-conventional means including chemical, biological, or radiological substances".[12] Within collective defense, crisis management and cooperative security (NATO's three core tasks) resilience was identified as a necessary condition "for a robust defensive posture" and as an enabler "for an appropriate engagement of multiple challenges" before any crisis occurs (crisis management).[13]

In 2016 the Member States agreed to monitor security risks related to hybrid threats and to "identify indicators of hybrid threats, incorporate these into early warning and existing risk assessment mechanisms, and share them as appropriate".[14] Special attention has been devoted to the improvement of situational and information awareness and to the enhancement of cooperation on disinformation and propaganda with different stakeholders such as the EU, non-governmental organizations, academia, experts, private sector, and civil society etc. Since 2016 more joint work of NATO and the EU on countering hybrid threats and disinformation and an effective engagement with the NATO Centre of Excellence on Strategic Communications in Riga, the European Center of Excellence on Countering Hybrid warfare in Helsinki and information sharing with partners were put in place. The spread of disinformation by diverse non-state actors and Russia has mobilized NATO and other stakeholders like the EU to act more effectively and to rethink their approaches on how to counter propaganda with facts and information.[15] In 2016 the Alliance underlined the need to "continue to rebut Russian propaganda: not by engaging in it tit-for-tat, but by deconstructing propaganda, debunking Moscow's false historical narrative, by exposing the reality of Russia's actions, and by restating the international rules it is breaking, to tell a compelling story about who we are, what we do, and why we do it. And we must stand united in our actions because actions will always speak louder than words".[16]

Other stakeholders like the European External Action Service Task Forces (East, Western Balkans, South) have become operational since 2015 and have focused their activities on the monitoring of disinformation, debunking disinformation, the enhancement of citizens' awareness and media literacy campaigns.[17] The EU is engaged in diverse work on how to build up societal resilience against disinformation through the EU's strategic communication campaigns "InvestEU" (a Europe that delivers), "EUandME" (a Europe that empowers) and "EU Protects" (a Europe that protects).[18]

The 2018 EU-NATO Joint Declaration stated that the EU and NATO had "increased [.] ability to respond to hybrid threats" and common institutional work has been conducted on the reinforcement of preparedness for crisis and resilience, disinformation and cyber security.[19] Parallel and coordinated exercises (PACE) between the two organizations with the participation of NATO and EU Member States have been taking place every two years since 2016.[20] The exercises include kinetic and non-kinetic elements as disinformation themes and messages and identify lessons on how to strengthen resilience against disinformation and incorporate a whole-of-society approach.[21]

> Civil-military education, training for hybrid warfare [...] with a focus on [rehearsing hybrid style attacks and how to match them, including the full integration of cyber and information warfare], joint conferences, joint working groups, and a maintaining of a balanced force for multiple responses

are essential to successfully deal with the current security environment that is characterized by a wide range of hybrid threats.[22]

Information environment since 2014and its main challenges

As stated by different NATO and EU official sources, the current information environment is complex, diverse, and ambiguous as an array of non-state actors and state actors are acting outside of international norms and are generating an alternative information reality that undermines the effectiveness of governance. Non-state actors and some state actors are seeking an information advantage and want to exert influence across the information space by using deception, disinformation, denial of access to information and by creating ambiguity with competing narratives and messages. Moreover, the complex and contested nature of the current information space is dynamic as it comprises interactive media and immersive extended reality technologies which create "newly ungoverned and under-governed spaces".[23] This information environment that is "characterized by 24/7 news cycle, the rise of social networking sites, and interconnectedness of audiences in and beyond NATO nation territory, directly affects how NATO actions are perceived by key audiences".[24] Robust and well-functioning strategic communication requires a constant information environment assessment and a deep understanding of actors who operate in this environment "in order to contribute to indications and early warnings of hybrid activity and to improve the Alliance's own communications planning and execution".[25]

The Prussian General Carl von Clausewitz once stated in his work "On war" that strategy is described as a tool one uses to achieve the goals of war[26]. Reflecting upon strategy in information space, it is therefore essential to understand how different actors use communication tools to achieve their objectives and how they adapt to changing conditions in an environment dominated by uncertainty. The ability to shape information, generate influence and effectively operate in this complex and contested information space has become one of the priorities of today's statecraft. Furthermore, a better understanding of the information toolkit and of technological trends and their impact on the cognitive processes of a target audience can contribute to a more effective strategic communication.

Analysis of Russia's activities in the information environment indicates that Moscow prioritizes the principles of interconnectivity between military and non-military methods while paying attention to "traditional environments as land, sea, air, space and cyberspace, but also to new ones such as social, digital, energy and others".[27] Domestically, the Russian Federation is closing off its information space, shutting up its independent media, declaring the state as the only legitimate source in the media and strengthening its abilities to "adequately respond" to a potential technologically advanced state-level adversary. Used to operate in the contested information spaces within the grey zones, Russia is trying to achieve "simultaneous effects to the entire depth of enemy territory in all physical media and throughout the information domain".[28] Moreover, Russia is using stated-controlled media companies to promote its narratives domestically and internationally as well as "a wide network of independent-looking websites, political organizations, politicians, businessmen, and companies that have been influenced by journalists, writers, and researchers who are defending Russian views, public relations companies and paid internet commentators".[29]

In the South, ISIL has been conducting successful communication messaging and "its communication tactics are creative and make use of modern technology, particularly when it comes to audiences outside of its territory".[30] ISIL is using diverse online forums and other social media that are difficult to control for governments and other non-formal channels to influence its target audiences. Many sources suggest that ISIL is very "adept in propagating a coherent message that attracts new members, focuses their current force, and shapes the cognitive domain in their favor".[31] ISIL is very strong in shaping the cognitive domain by spreading untruthful information and creating cognitive dissonance in the Islamic community.[32]

China meanwhile employs cognitive warfare and uses a wide array of technological enablers to pursue its information objectives. It asserts control over domestic information and "relies on an extensive influence apparatus that spans a range of print and broadcast media, with varying degrees of attributability, to advance both its domestic monopoly on power and its claims to global leadership".[33]

As can be seen, the information environment as such is very complex and conducive to influence operations. It may be contested by different actors and influencers and can be impacted by different technologies. NATO's StratCom policy needs to be able to anticipate the changes of the information environment and to address all potential deficiencies.

NATO's Strategic Communication, its actors, and NATO's military policy on Strategic Communications

As reflected in many NATO's documents, StratCom is defined as "the purposeful use of communication by an organization to fulfil its mission".[34] StratCom aims to inform, to communicate and to share all relevant messaging to the different target audiences in order to positively shape the perception of those audiences, and to ensure the vision and the strategic objective of the institution are well understood. The 2009 NATO StratCom definition was outlined as "a coordinated and appropriate use of NATO Communication activities and capabilities in support of Alliance policies, operations, and activities in order to advance NATO's aims".[35] According to the strategic document NATO Military policy on Strategic Communications from 2017, StratCom, in the context of the NATO military "is the integration of communication capabilities and information staff function with other military activities, in order to understand and shape the information environment, in support of NATO aims and objectives".[36] Another strategic document, the "Allied Command Operations" Directive 95–2 or "Saceur guidance", refers to StratCom as a mindset coordinating function and a process requiring coherent communication planning and staff.[37]

"StratCom requires the correct use of the information with synchronization among relevant institutions by ensuring unity of efforts towards the result. It acts on the basis of taking into account the socio-cultural structure, history and traditions of the key audience as well as technological factors in the terms of use and transmission of information".[38]

NATO StratCom's objective is to ensure that different audiences get "truthful, accurate and timely communication that will allow them to understand and assess the Alliance's actions and intentions".[39] StratCom activities as well as capabilities include strategic communications, public diplomacy, public affairs, military public affairs, information operations and psychological operations.[40] The North Atlantic Council (NAC) and the Secretary General provide overall direction of public diplomacy and communications.[41] The Deputies Committee provides an overall guidance on strategic communication, while the Committee on Public Diplomacy fulfils the advisory role to the NAC on communication, media and public engagement issues. It formulates recommendations to the NAC how to encourage public understanding of the goals of the Alliance.[42] The Public Diplomacy Division is responsible for public diplomacy programs, and communications, while working with the International Military Staff, the Public Affairs and StratCom Advisor to the Chairman of the Military Committee and with Allied Command Operations and Allied Command Transformation.[43]

The key document of NATO on strategic communication is MC 0628, NATO Military Policy on Strategic Communications, and dates from 26 July 2017. The document moves Strategic Communication from an advisory role to one holding the commander's delegated authority to ensure coherence of NATO actions and words.[44] The document reiterates the importance of strategic communication for the whole functioning of NATO. It clearly defines the authority for communications efforts that should be well coordinated and synchronized.[45] NATO Military Policy also puts emphasis on organizational and structural changes that should be more efficient and provide unity of effort in terms of the communication capabilities. MC 0628 puts attention to the alignment of words and actions, credibility of strategic communication, and empowering of communication at all levels. The policy outlines that StratCom should be put at the heart of all levels of strategic planning and that this requires appropriate communication staff at all levels with assigned responsibility for StratCom implementation.

Overall, MC 0628 is a turning point in the institutional adaptation process that puts strategic communication at the core of NATO's policy and strategy.[46]

As can be seen, NATO has made fundamental progress on strategic communication issues since 2014 but nevertheless shortcomings still do exist and need to be addressed.

Critical reflections on NATO's Strategic Communication

As indicated by numerous security and strategic communication experts, NATO needs to increase its efforts to better understand information manoeuvres of adversaries, to be able to predict and to deter their activities in the information environment.[47] A rapid assessment of the information environment and the development of a corresponding capability on information environment assessment will contribute to the overall effectiveness of NATO's strategic communication. Also, understanding of technological drivers and cognitive warfare and how adversaries use it on the battlefield is highly needed.

NATO's strategic communications efforts need to reflect current and future security risks and challenges and must anticipate them. Raising public awareness of the current security challenges that affect the organization can contribute to a better promotion of the organization amongst the general public. An alignment of threat perception, a unified approach in strategic communication, and common risk assessment in terms of contemporary security threats amongst all NATO Member States can enable better strategic communication coordination of effort.

It is also necessary for NATO to create an environment in which all Member States act together and in accordance with NATO's common values and principles. It is obvious that the multitude of actors involved at StratCom as well as the potential difficulty to reach consensus amongst the Member States on a communication strategy may represent a challenge for NATO and should be considered by NATO in the future.

A more proactive approach in dealing with national communication matters is essential as well as more simplicity in messaging and an alignment of communication with the policy objectives.[48] The development of complementarity and cross-cutting cooperation between the EU and NATO institutions on strategic communication matters and how to address external threats could enhance cooperation and address gaps in understanding the nature of the information environment and hybrid threats. A tailored communication and outreach campaigns to partner nations on hybrid threats and security risks can help to form a better awareness of hybrid threats. An "ethical influencing-working in that grey zone of trying to influence without sliding into manipulation or distortion" is highly important.[49]

Furthermore, NATO and its Member States need to invest in capability development, and need to train and fund their communicators, urgently invest in human resources and create a synthetic training environment to test the audiences.[50] It is essential to "grow the community", engage practitioners and operators to help succeed in communication matters.[51] NATO needs to use an audience-centric approach that has to be implemented in the planning process to support the cognitive effect.[52] "NATO/ACO therefore needs to ensure that, firstly, it has a narrative that resonates with its audiences, and, secondly, its operations and actions are consistent with that narrative".[53]

The most essential elements are to align communication with policy objectives, act in unity to effectively respond to China's and Russia's concepts of information warfare, and to be able to anticipate external challenges. The creation of an unified approach in communication matters amongst all NATO Member States and partner nations within the organization will lead to a success in strategic communication and enhance its effectiveness.

Conclusions

As can be seen from the preceding analysis, NATO's strategic communication activity is driven by policy, strategy and narratives. Right now NATO is going through tremendous changes in its strategy and policy and is trying to adapt to the changing and complex security environment and volatile information environment. There can be no doubt that Russia's military actions in Ukraine, the information contest over narratives coming from strategic level adversaries like China and Russia, and other challenges in the South will remain a new external driver for StratCom's development that may push NATO to implement structural changes and to a better alignment of strategic communication efforts across headquarters.

Under these new circumstances a constant analysis of the information environment is highly necessary. Moreover, a reinforcement of communication practices inside of the organization, better alignment of different lines of effort, and further development of strategic communication capability could increase the effectiveness of NATO's communication efforts in the future.

Notes

1 Admiral Mike Mullen, 156 Landon Lecture, 2010-
2 North Atlantic Treaty Organization. *NATO 2022 Strategic Concept.* Madrid, 2022.
3 Ibid, p. 3.
4 Ibid, p. 5.
5 Missiroli, Antonio, Jan Joel Andersson, Florence Gaub, Nicu Popescu, John-Joseph Wilkins et al. *Strategic Communications East and South.* Paris: EU Institute for Security Studies, 2016.
6 North Atlantic Treaty Organization. "Keynote speech by NATO Secretary General Jens Stoltenberg at the opening of the NATO Transformation Seminar." NATO. 25 March 2015. Accessed July 6, 2022. https://www.nato.int/cps/en/natohq/opinions_118435.htm?selectedLocale=en

7 Karen de Young. "Russia's moves in Ukraine are wake-up call, NATO's Rasmussen says in speech", *The Washington Post*, https://www.washingtonpost.com/world/national-security/russias-moves-in-ukraine-are-wake-up-call-natos-rasmussen-says-in-speech/2-14/03/19/80560d7c-af88-11e3-9627-c65021d6d572_story.html

8 European Parliamentary Research Service. *NATO strategic communications in an evolving battle of narratives*. Brussels: EPRS, 2016.

9 Ibid.

10 Missiroli, Antonio, Jan Joel Andersson, Florence Gaub, Nicu Popescu, John-Joseph Wilkins et al. *Strategic communications East and South*. Paris: EU Institute for Security Studies, 2016.

11 North Atlantic Treaty Organization. "Warsaw Summit." NATO. Accessed July 6, 2022. https://www.nato.int/cps/en/natohq/official_texts_133169.htm

12 North Atlantic Treaty Organization. "NATO 2030: United for a new era, Analysis and recommendations of the reflection group appointed by the NATO Secretary General." NATO. Accessed July 11, 2022. https://www.nato.int/nato_static_fl2014/assets/pdf/2020/12/pdf/201201-Reflection-Group-Final-Report-Uni.pdf

13 North Atlantic Treaty Organization. "Building resilience across the Alliance." HQ SACT. Accessed July 11, 2022.

14 European Commission. "Joint Framework on countering hybrid threats." EUR LEX. Accesses July 11, 2022. https://eur-lex.europa.eu/legal-content/EN/TXT/?uri=CELEX%3A52016JC0018

15 Maronkova, Barbara. "NATO amidst hybrid warfare threats: Effective Strategic Communications as a tool against disinformation and propaganda." In *Disinformation and Fake news,* edited by Shashi Jayakumar, Benjamin Ang and Nur Diyanah Anwar, 117–129. Singapore: Palgrave Macmillan, 2021.

16 North Atlantic Treaty Organization. "Meeting the Strategic Communications Challenge." Public Diplomacy Forum. Accessed July 12, 2022. https://www.nato.int/cps/en/natohq/opinions_117556.htm?selectedLocale=en

17 Ibid.

18 Ibid.

19 Council of the European Union. "EU-NATO Joint Declaration." Consilium Europe. Accessed July 11, 2022. https://www.consilium.europa.eu/en/press/press-releases/2018/07/10/eu-nato-joint-declaration/

20 Ibid.

21 EU and NATO. "Fifth progress report on the implementation of the common set of proposals endorsed by EU and NATO Councils on 6 December 2016 and 5 December 2017." NATO library. Accessed July 11, 2022. https://www.nato.int/nato_static_fl2014/assets/pdf/2020/6/pdf/200615-progress-report-nr5-EU-NATO-eng.pdf

22 Written interview with Prof. Dr. Robert Johnson, Director of the Oxford Changing Character of War Centre, University of Oxford.

23 Friends of Europe. "Europe in 2030: strengthening public-private cooperation in hybrid crises." Friends of Europe, Brussels. Accessed July 13, 2022. https://www.friendsofeurope.org/insights/europe-in-2030-strengthening-public-private-cooperation-in-hybrid-crises/

24 North Atlantic Treaty Organization. "NATO strategic communications policy." NATO Public intelligence. Accessed July 13, 2022. https://info.publicintelligence.net/NATO-STRATCOM-Policy.pdf

25 Maronkova, Barbara. "NATO amidst hybrid warfare threats: Effective Strategic Communications as a tool against disinformation and propaganda." In *Disinformation and Fake news,* edited by Shashi Jayakumar, Benjamin Ang and Nur Diyanah Anwar, 117–129. Singapore: Palgrave Macmillan, 2021.

26 Von Clausewitz, Carl. *On war*. North Charleston, 1984.

27 Alexander Smolovy. "Generator of breakthrough ideas and proposals." *Orugie Rossii,* January 2020. https://www.arms-expo.ru/news/novye-razrabotki/aleksandr-smolovoy-generator-proryvnykh-idey-i-predlozheniy/?mc_cid=d7f9ed37f6&mc_eid=fed21c605f

28 Gerasimov, Valery. "The role of the General staff in the organization of the country's defense in accordance with the new status on the General Staff." *Journal of the Academy of Military Science*, no. 1 (2014).

29 Emrah Uzun. "NATO's Strategic Communication activities during 2014 Ukraine crisis." Uluslararasi Krizve Siyaset Arastirmalari Dergisi 1:154–184.

30 Missiroli, Antonio, Jan Joel Andersson, Florence Gaub, Nicu Popescu, John-Joseph Wilkins et al. *Strategic communications East and South*. Paris: EU Institute for Security Studies, 2016.

31 Small Wars journal. "Defeating ISIL in the information environment." Small Wars Journal Online. Accessed June 13, 2022. https://smallwarsjournal.com/jrnl/art/defeating-isil-in-the-information-environment-0
32 Ibid.
33 Stanford. "New white paper on China's full-spectrum information operations." Stanford Internet Observatory. Accessed July 14, 2022. https://cyber.fsi.stanford.edu/io/news/new-whitepaper-telling-chinas-story#:~:text=China%E2%80%99s%20overt%20messaging%20efforts%20span%20both%20broadcast%20and,media%20outlets%20to%20reassert%20control%20over%20domestic%20information
34 Hallahan, Kirk et al. "Defining Strategic Communication." *International Journal of Strategic Communication,* no.1 (2007):3–4.
35 NATO StratCom CoE. "About Strategic Communications." NATO StratCom CoE. Accessed July 20, 2022. https://stratcomcoe.org/about_us/about-strategic-communications/1
36 North Atlantic Treaty Organization. *NATO Military Policy on Strategic Communications.* NATO, 2017.
37 Interview with NATO StratCom expert, Brussels, May 2022.
38 Emrah Uzun. "NATO's Strategic Communication activities during 2014 Ukraine crisis." *Uluslararasi Krizve Siyaset Arastirmalari Dergisi* 1:154–184
39 Public Intelligence. "Military Concept for NATO Strategic Communications." Public Intelligence. Accessed July 20, 2022. https://publicintelligence.net/nato-stratcom-concept/
40 NATO StratCom CoE. "Improving NATO Strategic Communications terminology. " NATO StratCom CoE. Accessed July 20, 2022. https://stratcomcoe.org/cuploads/pfiles/nato_stratcom_terminology_review_10062019.pdf
41 North Atlantic Treaty Organization. "Communications and public diplomacy." NATO. Accessed July 20, 2022. https://www.nato.int/cps/en/natohq/topics_69275.htm
42 Ibid.
43 Ibid.
44 North Atlantic Treaty Organization. *NATO Military Policy on Strategic Communications.* NATO, 2017.
45 Ibid.
46 Laity, Mark. "The birth and coming of age of NATO StratCom: a personal history." *Defense Strategic Communications,* no.10 (2021): 22–70.
47 Interviews with NATO strategic communication experts, May 2022.
48 Interviews with NATO strategic communication experts, May 2022.
49 Laity, Mark. "The birth and coming of age of NATO StratCom: a personal history." *Defense Strategic Communications,* no.10 (2021): 22–70.
50 Interviews with NATO strategic communication experts, May 2022.
51 Interviews with NATO strategic communication experts, May 2022.
52 Interviews with NATO strategic communication experts, May 2022.
53 Laity, Mark. "The birth and coming of age of NATO StratCom: a personal history." *Defense Strategic Communications,* no.10 (2021): 22–70.

Bibliography

Atkinson, C. 2018. *Hybrid warfare and societal resilience: Implications for Democratic Governance.* https://connections-qj.org/ru/system/files/3906_atkinson_hw_societal_resilience.pdf.

Chivvis, C. S. 2017. *Understanding Russian "Hybrid warfare".* https://www.rand.org/content/dam/rand/pubs/testimonies/CT400/CT468/RAND_CT468.pdf.

Council of the European Union. "*EU-NATO Joint Declaration.*" Consilium Europe. https://www.consilium.europa.eu/en/press/press-releases/2018/07/10/eu-nato-joint-declaration/.

de Young, Karen "Russia's moves in Ukraine are wake-up call, NATO's Rasmussen says in speech", *The Washington Post,* https://www.washingtonpost.com/world/national-security/russias-moves-in-ukraine-are-wake-up-call-natos-rasmussen-says-in-speech/2-14/03/19/80560d7c-af88-11e3-9627-c65021d6d572_story.html.

Emrah, U. "*NATO's Strategic Communication activities during 2014 Ukraine crisis.*" Uluslararasi Krizve Siyaset Arastirmalari Dergisi 1:154–184.

European Commission. "*Joint Framework on countering hybrid threats.*" EUR LEX. https://eur-lex.europa.eu/legal-content/EN/TXT/?uri=CELEX%3A52016JC0018.

EU and NATO. "*Fifth progress report on the implementation of the common set of proposals endorsed by EU and NATO Councils on 6 December 2016 and 5 December 2017.*" NATO library. https://www.nato.int/nato_static_fl2014/assets/pdf/2020/6/pdf/200615-progress-report-nr5-EU-NATO-eng.pdf.

European Parliamentary Research Service. *NATO strategic communications in an evolving battle of narratives.* Brussels: EPRS, 2016.

Gerasimov, V. "The role of the General staff in the organization of the country's defense in accordance with the new status on the General Staff." *Journal of the Academy of Military Science*, no. 1 (2014).

Hallahan, K. et al. "Defining Strategic Communication." *International Journal of Strategic Communication*, no.1 (2007):3–4.

Laity, M. "The birth and coming of age of NATO StratCom: a personal history." *Defense Strategic Communications*, no.10 (2021): 22–70.

Lanoszka, A. 2016. Russian hybrid warfare and extended deterrence in Eastern Europe. *International Affairs*, 92(1), 175–195.

Maronkova, B. 2018. NATO in the New Hybrid Warfare Environment. *UA: Ukraine Analytica.* Issue 1(11).

Maronkova, B. 2018. NATO Amidst Hybrid Warfare—Case Study of Effective Strategic Communications. Presentation at the workshop Disinformation, Online Falsehoods and Fake News, Singapore.

Missiroli, A, et al. *Strategic communications East and South.* Paris: EU Institute for Security Studies, 2016.

NATO. 2022*Strategic Concept.* Madrid, 2022. https://www.nato.int/strategic-concept/index.html.

NATO. *NATO Military Policy on Strategic Communications.* NATO, 2017.

NATO. "*Meeting the Strategic Communications Challenge.*" Public Diplomacy Forum. https://www.nato.int/cps/en/natohq/opinions_117556.htm?selectedLocale=en.

NATO. "*NATO 2030: United for a new era, Analysis and recommendations of the reflection group appointed by the NATO Secretary General.*" NATO. https://www.nato.int/nato_static_fl2014/assets/pdf/2020/12/pdf/201201-Reflection-Group-Final-Report-Uni.pdf.

NATO. "*NATO strategic communications policy.*" NATO Public Intelligence. https://info.publicintelligence.net/NATO-STRATCOM-Policy.pdf.

NATO. "Keynote speech by NATO Secretary General Jens Stoltenberg at the opening of the NATO Transformation Seminar." NATO. 25 March2015. https://www.nato.int/cps/en/natohq/opinions_118435.htm?selectedLocale=en.

NATO. "*Building resilience across the Alliance.*" HQ SACT. Accessed July 11, 2022.

NATO. "*Communications and public diplomacy.*" NATO. https://www.nato.int/cps/en/natohq/topics_69275.htm.

NATO. 2009. *NATO Strategic Communications Policy.* https://info.publicintelligence.net/NATO-STRATCOM-Policy.pdf.

NATO. 2012. *ACO Strategic Communications Directive.* http://www.aco.nato.int/page300302915.aspx.

NATO Stratcom Centre of Excellence. Definition of Strategic Communications. https://www.stratcomcoe.org/about-strategic-communications.

NATO StratCom CoE. "About Strategic Communications." NATO StratCom CoE. https://stratcomcoe.org/about_us/about-strategic-communications/1.

NATO StratCom CoE. "Improving NATO Strategic Communications terminology. " NATO StratCom CoE. https://stratcomcoe.org/cuploads/pfiles/nato_stratcom_terminology_review_10062019.pdf.

NATO Website. Wales Summit Declaration. www.nato.int/cps/ic/natohq/official_texts_112964.htm.

NATO Website. Warsaw Summit Declaration. www.nato.int/cps/en/natohq/official_texts_133169.htm?selectedLocale=en.

NATO Website. Brussels Summit Declaration. https://www.nato.int/cps/ic/natohq/official_texts_156624.htm.

Public Intelligence. "*Military Concept for NATO Strategic Communications.*" Public Intelligence. https://publicintelligence.net/nato-stratcom-concept/.

Smolovy, Alexander. "Generator of breakthrough ideas and proposals." *Oruqie Rossii*, January2020. https://www.arms-expo.ru/news/novye-razrabotki/aleksandr-smolovoy-generator-proryvnykh-idey-i-predlozheniy/?mc_cid=d7f9ed37f6&mc_eid=fed21c605f.

Small Wars Journal. "Defeating ISIL in the information environment." *Small Wars Journal Online.* https://smallwarsjournal.com/jrnl/art/defeating-isil-in-the-information-environment-0.

Stanford. "New white paper on China's full-spectrum information operations." *Stanford Internet Observatory.* https://cyber.fsi.stanford.edu/io/news/new-whitepaper-telling-chinas-story#:~:text=China%E2%80%99s%20overt%20messaging%20efforts%20span%20both%20broadcast%20and,media%20outlets%20to%20reassert%20control%20over%20domestic%20information.

25

THE BALTIC EXPERIENCE IN COUNTERING CONTEMPORARY RUSSIAN DISINFORMATION

Aleksandra Kuczyńska-Zonik

1 Introduction

For several years disinformation, fake news and hate speech have been used against the Baltic states in order to undermine people's confidence in state authority, to weaken credibility and social cohesion within the states. The problem is rooted in the Soviet period which made Lithuania, Latvia and Estonia vulnerable to the dissemination of Kremlin-led disinformation and propaganda. Russia has never fully abandoned the idea of bringing the Baltic states back into its sphere of "privileged interest".[1] Thus, it uses media as the most influential platform to present its values and interests in the region. Its disinformation strategy towards the Baltic states reflects a broad spectrum of techniques: official government statements, state-funded media outlets, proxy websites, bots, false social media personas, cyber-enabled disinformation operations and others.[2] Russia's information influence include narratives, conspiracy theories that combine facts and half-truths, black-and-white thinking, simplification, emotional language and a lack of transparency, both organisational and financial. It offers an alternative point of view claiming they inform "about what others are silent about".[3] Russian propaganda is multidimensional and multi-narrative, encompassing the following themes: Euroscepticism, NATO scepticism, anti-immigration, anti-establishment and memory policy (Lithuania, Latvia and Estonia as "hysteric Russophobes"). Moreover, digital technologies such as professional websites, eye-catching images and video platforms strengthen media attractiveness. Social media, blogs and chats make it possible to interact, create an information discourse and be a part of this digital reality. This is in line with Derina Holtzhausen and Angarsf Zerfass's observation that governments are losing credit of authority multiplied by one-way linear media, which, however, faces the challenge of social media self-created and self-multiplied informational flow.[4] Similarly, Corneliu Bjola and James Pamment warn about the "dark side" of digital diplomacy;[5] Manuel Castells states that national states are losing power in certain fields, but at the same time they form new connections, which lead to new types of influence in global politics and economy.[6] Additionally, Russia expresses criticism and accusation toward the Baltic authorities which is a form of what Edward Lucas called "whataboutism" when Russia tries to deflect international attention away from its authoritarian system to allegedly undemocratic practices within the EU.[7] It negates the

DOI: 10.4324/9781003190363-30

Baltics' legitimacy and undermines the rule of law, which is not, in fact, soft power and public diplomacy approaches, but rather a form of "antidiplomacy"[8] or "sharp power"[9] involving censorship or the use of manipulation to sap the integrity of independent institutions, to take advantage of the asymmetry between free and unfree systems, and to put individual rights and democratic values at risk.

To some extent, Russian propaganda is effective due to the great number of Russian-speaking minorities in the Baltic states. They are accustomed to Russian-speaking media, some of them (particularly the older generations) do not even use the state language. The most influential narratives proposed by Russian propaganda are historic memories which deteriorate relations between the states. The most sensitive and emotional issue concerns the WWII and the Soviet era. Russia denies any hostilities and incorporation of the Baltic states to the Soviet Union. According to Russian authorities, Lithuania, Latvia and Estonia voluntary joined the Soviet Union, based on the bilateral agreements between the mentioned Soviet Union and the governments of the states. As a result, on 9 May thousands of residents in Latvia and Estonia annually celebrate the Victory Day and take part in the sentimental procession of the "Immortal Regiment" to honour the fallen of the Great Patriotic War, carrying photographs of their ancestors.[10] This has a great impact on state authority-minority and majority-minority relations because it may lead to a lower attachment of Russian-speakers to the values and norms of the state where they reside, and, in fact, may threaten social cohesion and integration within the state.[11]

The issue of how a state can counter disinformation activities is a matter of continuous debate. States have got particular possibilities to take special measures and introduce particular policies. National strategies of each country consist of a series of efforts that aim at counteracting threats of manipulation and disinformation, and reinforcing development in the information space. Their counterpropaganda actions include: 1) effective communication and promotion of national values; 2) strengthening the independent media sector and social organisations and 3) increasing social awareness about information manipulation, disinformation and falsification. At the same time states have to cultivate the following normative attributes: 1) truthfulness 2) prudence; 3) accountability; 4) integrity; 5) effectiveness and 5) responsibility.[12] Also, for several years the Baltic states have been establishing comprehensive e-governance ecosystems including public news platform preventing form digital threats. In particular, at the beginning of the pandemic (2020), it was extremely difficult to counter propaganda as a huge amount of fake information has been disseminated by social media. The narratives included conspiracy theories about the origins of the virus ("the virus was invented by biologists and pharmacists in Latvia"[13]), 5G technologies, or stories that the virus was disseminated to control local populations. Numerous coronavirus myths and fake news stories have been circulating on social media. But COVID-19 was not only an obstacle but also an opportunity. It catapulted the search for digital solutions to the very top of policy-makers' agenda. In fact, the Baltic states have made several efforts in digital sphere to alleviate the negative impact of the Russian propaganda striving to create an open, pluralistic environment. But while Lithuania and Latvia introduced bans, censorship and restrictions, only Estonia applied a more liberal approach and avoided censorship.

This article endeavours to offer a critical reflection on the state media environment in a broad regional information complex. The purpose of this paper is to explore the methods of dealing with Russian disinformation in Lithuania, Latvia and Estonia. First, the variety of counterstrategies is examined. Secondly, the author pays attention to the Baltic states to explain the particularities between the states through the perspective of the concept of security and political culture. Here the legal aspects and the regulations of media sector are examined. Then the author turns to the conclusion that while Lithuania, Latvia and Estonia develop a few offensive procedures to fight propaganda in media at both national and international levels, they do not abandon democratic principles and rules such as freedom of speech, pluralism and privacy. This study offers evidence, however, that the independent media environment is perceived as part of a security concern and the states seek to create the most favourable conditions for its development. In fact, the paper finds that the problem must be considered in its relation to state security, political culture, norm and values.

2 Variety of counterstrategies

So far, the Baltic states have applied a variety of policies both national, regional and international levels. At national level these states use a number of comprehensive instruments and tools aimed at counteracting the immediate threats of manipulation and creating favourable conditions for securing the development of information.

First of all, it is strategic communication defined as the purposeful use of communication by various state institution[14] aimed at engaging with intended audiences (including public diplomacy, social communication, information operations and psychological operations) which are carried out by various units and departments. This includes: 1) synchronisation understood by coordination of words and deeds, as well as active consideration of how the state actions and policies are interpreted by the public, and 2) deliberate communication and engagement which cover a wide range of activities: understanding, engaging, informing, influencing, and communicating with people to achieve and maintain social trust. To provide information about policies, plans, and achievements of the institution, and to inform and educate the public about the laws, regulations and all issues that affect the daily lives of the citizens the states may use traditional and social media. The latter has been increasingly important under the digital development.[15] So far the Baltic states, particularly Lithuania and Latvia, have given little attention to it. Disinformation Resilience Index has revealed that some issues are not of good quality in those two countries, including the level of institutional development in the sphere of information security, legal framework, comprehensiveness, existence of state long-term approach to information security as well as the quality of countermeasures by the media community and civil society.[16] The importance of the process is growing, however. Estonia has already developed its strategic communication measures to secure its policy, keep the public informed about the security situation and avoid the panic, neutralise hostile subversive activity, expose fake information and prevent its spread.[17]

On regional level, Lithuania, Latvia and Estonia together with other states from the region agree to build cyber and hybrid threats resilience. The cooperation with EU and NATO in those fields remains their priority. They stress the need to continue developing hitherto multilateral formats and the participation of the United States is still an essential part of their security approach. Moreover, there are other international organizations and agencies such as the European Centre of Excellence for Countering Hybrid Threats in Helsinki (Finland), the NATO Strategic Communications Centre of Excellence (StratCom) in Riga (Latvia), and the Cooperative Cyber Defence Centre of Excellence in Tallinn (Estonia) dealing with countering hybrid threats. Their aim is to raise awareness about the identification of risks and capabilities. While EU is not immune to disinformation, it has intensified efforts to counter disinformation by boosting support for independent media increasing awareness of disinformation activities by propagandists (European Union Institute for Security Studies 2016). Finally there are the Council of Europe, OSCE and other regional structures (Nordic-Baltic Eight, NB8) which support countries to build their capacities by developing legislative proposals organizing regular exercises and providing discussions on countering hybrid threats at ministerial and other levels.

3 Lithuania and Latvia—critical response

In the Russian-speaking media sphere of the Baltic states, there are companies owned by Russian media groups that broadcast Russian content directly in Lithuania, Latvia and Estonia. The second type is the media in Russian, the content of which is organized on the basis of the content of Russia's one. Both of them are usually financed by and depend on Russian sources, which affects their quality, and the way political and social events are presented. In contrast, there are private and public media in the Baltic states broadcasted in Russian language with more liberal approach and anti-Kremlin narratives (Table 1).

While the information threat is recognized as relevant for each of the Baltic states, they have chosen different instruments of taking specific measures within their bounds. Among the Baltic states, Lithuania is the most critical against Russia. Therefore, the Russian-language media is dominated by the narrative of Euroscepticism and negation towards the US and NATO in order to weaken Lithuania's ties with the Western world. The Lithuanian authorities are permanently accused of Russophobia.[18] Additionally, Russia's goal is to radicalize both Russian and Polish minorities. In order to do it Russia builds a coalition of the Polish and Russian-speaking minorities to discredit national elites and weaken social cohesion.[19] In Lithuania, people of Russian descent constitute approximately 6% of the population. As a result, Russian-speaking media are less popular, but Russia continues to improve information influence on the Lithuanian audience. Russian programs and films appear on Lithuanian channels. There are news portals such as Sputnik followed by recipients of Russian and Polish origins. In recent years, the websites have devoted considerable attention to the problems of status of the Polish national minority in the Vilnius region and to (alleged) Polish-Lithuanian conflicts in order to convince the audience that Lithuania discriminates its national minorities. Generally, the Russian-speaking media market in Lithuania is quite small, therefore the effectiveness of Russian propaganda in Lithuania is limited.[20]

In Latvia, where national and ethnic minorities constitute more than 30% of population, Russian-speakers are concentrated in the Latgale region (40% of the population) and the city of Daugavpils (50%). In the capital, Riga, ethnic Russians constitute almost 40% of the population. This makes Russian propaganda more addressable than in Lithuania. In fact, the Russian media in Latvia are well organized. The content is transmitted directly from Russia, or it is broadcasted in Latvia on the basis of Russian one; there are Latvian public media in Russian. Their offer is quite broad and rich including radio programs, particular segments on TV channels and news portals. Latvia is the only Baltic country where three of the five most popular press titles are published in Russian (MK-Latviya, Latvijskie Vesti and Vesti Segodnya. Among Russian TV channels there are well-funded and attractive RTR Planeta, NTV Mir, PBK, REN TV Baltija, and others. It is what Neil Postman called as "infotainment" and information manipulation. According to Postman, the neologism combining the terms "information" and "entertainment" means "information-based media content or programme, with additional entertainment content to increase popularity among the audience.[21]

In Latvia, PBK was the third most watched TV channel with a 7% audience share (in contrast to Lithuania where its daily share was only 2%). The Russian-language media are an effective channel of communication between the Russian-speaking society and political parties and NGOs. Moreover, there are a few information portals such as RuBaltic.ru operating since 2013. Its mission is to reveal "political problems currently expanding in Lithuania,

Latvia and Estonia including an increase of authoritarian tendencies, deterioration of the economic situation and social degradation". It claims that "the Baltic states are a destructive obstacle for the development of post-Soviet states, for deepening Eurasian integration and for normalization of dialogue between Russia and the European Union". This media is highly politicized, too. Russian media in Latvia usually negatively assess Latvia's policy, blaming the authorities for discriminating Russian-speaking people in the country. Latvia is claimed to be a failed state with unsuccessful socio-economic development. The target group is people of pre- and retirement age, with a secondary education and with low socio-economic status. The Russian media are part of Moscow's media policy, which discredit the Baltic states in the international environment and in the opinion of their citizens. They also respond to the information needs of the target group, which blames the Latvian government for its socio-economic status as well as culturally identifies itself with Russia. Such narrative is frequent as Russian-speaking media aims at maintaining ethnic divisions and undermining social cohesion. This limits political, social and cultural stability of the state. On the one hand, due to the popularity of the pro-Kremlin media in Latvia, national minorities live in Russian information sphere separated from Latvian and are poorly integrated within the Latvian community.[22] The group of people using both languages to access information (both in Russian and Latvian) is relatively small. Despite the variety of Russian media operating in Latvian information sphere, their influence on the whole Latvian society is disputable, however. More and more people consume either public media in Latvian or seek alternative sources of information such as the Internet.[23]

There are some laws regulating Lithuanian and Latvian media space where the principles of freedom of speech and media freedom are included. This topic is mentioned in different official documents in Lithuania, such as the National Security Strategy. Media also has tools of self-regulation such as the Commission of Ethics or the Association of Journalists. Their aim is to ensure the development of mass literacy in cooperation with state agencies and institutions. In Latvia several documents concern the financing, monitoring, and management of broadcasting media channels and media companies working inside Latvia, but formally established outside the country, may not be subject to the Latvian regulator. Furthermore, most media channels are dependent on some groups of interests or business within Lithuania's and Latvia's small media markets. This could negatively affect professional standards of mass media.

Both in Lithuania and Latvia there are public media in Russian. Their offer is quite broad and rich including radio programs, particular segments on TV channels and news portals, but not as attractive or well-developed as those financed by Russia. Due to the disinformation and propaganda in the media, both Lithuania and Latvia have restricted the transmission of Russian-language TV channels several times. In 2013, sanctions were first implemented in relation to the PBK TV channel in Lithuania; then, in 2014 to RTR-Planeta and NTV-Mir. The reason was a false message about the role of the Soviet Army in the Baltic States in 1991. In 2015, the RTR-Planeta channel was suspended for three months for promoting hatred between the Ukrainian and Russian people. In 2017, yet another Russian channel in Lithuania, TVCI, was blocked twice, once for a month and later for six months. In 2014, and then in April 2016, Latvia fined Russia-RTR for hate speech and incitement to ethnic conflict. Furthermore, in 2016 Latvian Sputnik lost the right to use the.lv domain due to accusations of propaganda and disinformation. Finally, in 2020 according to the European Union's sanctions on Russian citizens supporting Moscow policy aimed at destabilising Ukraine, Lithuania and Latvia banned Russian TV channels.

4 Estonia's liberal approach

The Estonian media market is more liberal and open than the Lithuanian and Latvian ones. The lack of specific regulations since the 1990s stems from the belief that the media market must be as liberal as possible.[24] Cultural norms which define activity within media environment are of great importance and the overall approach is to avoid over-regulation. On the contrary—they are the results of norms and values as part of the national cultural heritage. Therefore, the media market is still one of the least regulated areas of the economic system in Estonia. Nevertheless, the available legal instruments guarantee transparency of media ownership, editorial autonomy, media participation in the democratic electoral process, and media access for minorities, which is why Estonia is highly ranked concerning media freedom worldwide.[25]

The media market in Estonia is oligopolistic in nature because it is very small and only a few entities operate. Estonian Public Broadcasting (*Eesti Rah-vusringhääling*, ERR), established in 2007, is responsible for three state-owned TV channels (ETV, ETV2, ETV +), radio stations (Vikerraadio, Raadio 2, Klassikaraadio, Raadio 4, Raadio Tallinn) and the streaming platform (Jupiter). The programs are broadcasted in Estonian, but recently the number of hours being translated into Russian has been gradually increasing (including programs with Russian subtitles).

In turn, the most important private media services include the largest media group in the Baltic States—Postimees Grupp (Eesti Meedia until 2019) and Ekspress Group with a turnover of 38 million euro and 24 million euro in 2019 respectively.[26] Postimees Grupp is the owner of the oldest Estonian newspaper Postimees and its news portals in Estonian, Russian and English, several TV channels, radio stations, local magazines and the Baltic News Service (BNS) news agency, which covers all three Baltic states. Postimees Grupp also has leading media platforms outside Estonia: the Latvian news portal TVNet and the LETA news agency, as well as the Lithuanian news portal 15min. On the other hand, Ekspress Group publishes Estonia's largest weeklies—Eesti Ekspress and Maaleht—and the Eesti Päevaleht daily. Its subsidiary—Ekspress Meedia—runs Delfi news portals in Estonia, Latvia and Lithuania. In addition, the Baltijas Mediju Alianse media group is also important (with a turnover of EUR 13.5 million in 2019), whose activities, apart from Estonia, also include Lithuania and Latvia. It specializes in messages addressed to national minorities in Russian. Its headquarters are in Riga, and its branches are in Tallinn and Vilnius.

The most-watched TV channels in Estonia are: ETV, Kanal 2 and TV 3, owned respectively by ERR, Postimees Grupp and TV3 Group (since 2017 owned by the American company Providence Equity Partnerspod, previously—the Swedish company Modern Times Group, MTG).[27] The share of the audience of the abovementioned leading television channels is respectively: 16%, 9% and 9%. The trust in the media is significantly higher than the EU average—around 79% of Estonians trust broadcasters, while the EU average is 66% (according to the Eurobarometer in February 2018).

The market structure and the way media entities operate in Estonia are determined by the size of the market and the specificity of the audience who may use at least one of the two main languages of transmission. Apart from the titular majority, the Russian-speaking minority in Estonia constitutes a significant percentage of the population (28%). It is concentrated in the north-eastern part of the state and in the capital, where ethnic Russians constitute 37% of Tallinn's inhabitants. In turn, in the Ida-Virumaa region, 84% of the population are Russian-speakers, and 47% have Russian citizenship or do not have any citizenship (they are so-called persons with undefined citizenship). The demographic and linguistic environment poses challenges to the modes of operation and development directions of media companies in Estonia. The popularity of the Russian language, a positive attitude to Russian culture and symbols, as well as nostalgia and Soviet resentments create favourable conditions for the Russian-speaking information bubble in the country.

Most of the minority representatives follow the media in their mother tongue while some of them watch programmes in the state language, too. Among the five principles that characterize how Russian-speakers consume Russian media in Estonia Jill Dougherty and Riina Kaljurand[28] mention: 1) Entertainment as primary and news as secondary determinant; 2) Scepticism about any and all news sources; 3) Young people prefer the Internet sources of information; 4) Local news rather than international dominate; 5) Cultural attraction to Russia does not necessarily mean political attraction of the country.

A pro-Kremlin media holding company based in Latvia—Baltijas Mediju Alianse—broadcasts Russian TV channels in Lithuania, Latvia and Estonia, including Pervyy Baltiyskiy Kanal (three versions of the channel for the three countries respectively) and a TV channel based on the media content of the Russian channel NTV-Mir or Ren TV in Estonia. Baltijas Mediju Alianse publishes the Russian-language press MK-Latviya and MK-Estoniya and is the owner of websites in Russian. In 2020, the Sputnik news agency had to resign from operating in Estonia after the EU imposed a sanction on Dmitry Kisseljov, general manager of Rossiya Sevodnya overseeing Sputnik. There are also news portals publishing in Russian but independent form Russian capital, such as Latvian Meduza or Re:Baltica, whose main advantage is objective investigative journalism. The circulation of the Russian press in Estonia is gradually falling. Currently on the market Moskovskij Komsomolec-Estonia and Delovye Vedomostii are present. In August 2020 the Komsomolskaya Pravda v Severnoj Evropie (formerly KP-Baltia) newspaper, published by the Estonian company SKP Media since 2007, was suspended. Previously it was distributed in Estonia, Latvia and Finland with a circulation of 5–12,000 copies. In 2009, Molodoz' Estonii was closed and in 2016—Dz'en' za dnem, and Postimees (in Russian) was limited to online publishing. While the official reasons for the closure of the magazines have been socio-economic reasons: a decline of the readers and an increase of online sources, some representatives of Russian-speakers in Estonia believe that this was another step taken by the national elite against the minority in Estonia.

The Russian-language media struggle with similar problems to the Estonian ones: information websites are poorly structured, short messages dominate, and there is no in-depth analysis, which prove the quality of contemporary journalism is diminished and is in crisis. Some right-wing Estonian politicians (Isamaa) called for limiting the transmission of Russian-language channels, as was the case in Lithuania, Latvia and Ukraine, as an element of the state's information strategy against Russian propaganda. So far Estonia has not taken such a step, however. In contrast, since the Bronze Soldier riots in Tallinn in 2007 related to the relocation of the monument, and especially after Russia's annexation of Crimea in 2014, there has been a lively debate in Estonia about politically motivated journalism. The debate has focused on the Russia's information influence in Estonia as well. For this reason, the Estonian public broadcaster has started broadcasting in Russian for its Russian-speaking residents. In 2015, ETV+ was established. It was successful initially, but it can hardly compete with the content broadcast from Russia, which enjoys great popularity and attractiveness. The Russian-language public media in Estonia concentrates on society and economic issues rather than Russian politics. Currently, ETV + accounts for less than 2% of the daily share of the audience, compared to 11% of the share of the audience of three most popular Russian channels (PBK, RTR Planeta and NTV Mir) in Estonia. It is worth mentioning that similar initiative was made in Latvia in 2015 but in the end it was rejected by the government

Table 25.1 Types of Russian media in the Baltic states

Russian media with Russian content transmitted from Russia	Russian media with content organised on the basis of Russia's one	Private Russian media broadcasted in the Baltic states	Public Russian media broadcasted in the Baltic states
Pro-Kremlin narratives, conservative approach	Pro-Kremlin narratives, conservative approach	Independent narratives, liberal approach	Independent media, liberal approach
For example: PBK TV channel	For example: NTV-Mir and Ren TV channels, RuBaltic.ru and Balt-news websites	For example: Delfi and Postimees websites	For example: ETV+ TV channel, LRT, LSM and ERR news websites

Source: Author.

because of the campaign against promoting the use of the Russian language in Latvia by the National Alliance, one of the governing coalition parties and who had the Minister of Culture at the time.[29]

All state security documents in Estonia are characterized by a wide range of liberalism and the flow of public information being an important element of its legal culture. On the one hand, it helps the media maintain a high degree of independence and resilience from political pressure. On the other hand, this allows pro-Kremlin messages being disseminated more openly and freely than in Lithuania and Latvia. In fact, Estonia has not limited Russian media, confirming that freedom of speech is one of the most important values. But in 2015 it refused the Italian journalist Giulietto Chiesa to come and participate in a meeting of the Impressum club, organized by journalists associated with the Russian daily Komsomolskya Pravda. Chiesa was known for his controversial statements about Russia's policy, he supported Russia's involvement in Georgia and the annexation of Crimea (Italian). Finally, in December 2019, Estonia forced the closure of the local bureau of the news agency Sputnik as an implementation of the EU sanctions imposed in 2014 on Dmitrii Kiselev, who was the head of Russian news agency Rossiya Segodnya.

5 Russian media after Russia's invasion of Ukraine

The situation has changed significantly due to Russia's aggression against Ukraine (2022) when, many pro-Kremlin TV channels in Russian language have been banned from broadcasting in the Baltic states. The Lithuanian Radio and Television Commission (LRTK) has banned the rebroadcasting of six Russian-language TV channels in the country over their incitement of war and propaganda including Belarus 24, NTV Mir, RTR Planeta, Rossiya 24, PBK, and TVCI. Four of them were suspended for five years, and PBK and TVCI for three years. Similarly, Latvia's regulator—National Electronic Mass Media Council (NEMMC)—has informed that there would be a ban on the broadcasts of Rossija RTR for five years, Rossija 24 for four years and TV Centr International for three years in Latvia because these channels represent a threat to Latvia's security. It may be recalled that 41 other Russian channels were taken off the air during the past five years. Finally, in Estonia, the Consumer Protection and Technical Regulatory Authority (TTJA) has initiated proceedings to restrict broadcasting eight channels (RTR-Planeta, RTVI, Rossija 24, REN TV, NTV Mir and PBK) as they have constantly incited hatred and made propaganda. Additionally, two of Estonia's main telecom companies, Telia and Elisa, both announced earlier in the week that they would be terminating agreements with the local franchisee of Kremlin-controlled TV companies, including PBK Estonia, REN TV Estonia and NTV Mir Estonia.

While majority of the political parties have supported a move to ban Russian propaganda channels from the Baltic information sphere, not all politicians voted in favor, bearing in mind that this decision would cause dissatisfaction among Russian-speaking recipients. The pro-Kremlin propaganda is still disseminated on internet portals and in social media, however. Among news portals, it is Baltija.eu (registered in Estonia in 2008, provides information to the Russian diaspora and, in particular, to the Russian-speaking inhabitants of Estonia) and Baltnews (a Russian-language news agency specializing in informing about the socio-political situation in Lithuania, Latvia and Estonia; until 2018, editorial offices were located in the Baltic States, and now—in Moscow). The narratives have concentrated to portray the Baltic states as being poorly managed, Nazi and anti-Russian. As for social media, these are mainly Facebook groups in which their administrators and members actively share content in line with the Kremlin's rhetoric. So far the users of these groups have focused on discrediting the national governments of the Baltic states and improving the image of Russia among the diaspora; currently, they are spreading false information about the war in Ukraine and Ukrainian refugees reaching Lithuania, Latvia and Estonia. While the audience of these news portals and social media groups is not very big, they are still active, and their potential cannot be disregarded.

At the same time, the Baltic states have tried to create an open, mature, independent information environment accessible to Russian-speaking citizens. Moreover, as disinformation is disseminated through social media, development of digital literacy has enabled people to participate in online environments wisely, safely, and ethically. Additionally, in order to build and then maintain trust, governments need to keep the system open, observable, and auditable. Strategies to improve information security in Lithuania, Latvia and Estonia consist in: 1. making their own media offer more attractive, 2. improving the quality of the media by applying new technologies of communication, 3. making media more accessible for national minorities, 4. promoting civic attitudes and cooperation between the state and private media sector as well as non-governmental organizations. Media awareness is growing successfully. The comparison between the Russian-speaking residents' opinion towards president of Russia, Vladimir Putin, and Russia's aggressive policy towards Ukraine in 2014 and 2022 shows that social attitudes are changing. After the annexation of Crimea, 35% of Russian-speaking respondents in Latvia supported Russia while in March 2022 21% were in favour of it. Similarly in 2016, 15% of individuals solidarized with Ukraine, and in 2022—already 22% of them. In 2022 a large group (more than half of Russian-speakers in Latvia) that have not supported either side which may be explained by the information war and the deep identity crisis of the Russians. Both in Latvia and Estonia, different opinions towards Ukraine and Russia between representatives of different ethnic groups are still visible. In turn, the Lithuanian society share the same perspective towards the Russia's aggression—in 2022 91% of the inhabitants were against the military invasion of Ukraine, which is a result of the lower percentage of the Russian-speaking minority and the high degree of integration into Lithuanian society.

6 Conclusions

This study revealed that the Baltic states use several strategies to deal with Russian disinformation from broadcasting bans, censorship and restrictions on media, to measures aimed at creating an open, pluralistic information environment, monitoring human rights violations, developing free and independent media. While Lithuania and Latvia have introduced restrictions toward Russian-speaking media accusing them of disseminating fake news,

Estonia preferred a more liberal approach. Interestingly, after Russia's invasion of Ukraine, all of them banned many Russian media as a gesture of solidarity with Ukraine and also in order to secure their information sphere from deleterious messages from aggressor. In terms of press freedom the Baltic states still rate above the West European countries such as the United Kingdom or France[30] despite the fact that some experts have condemned the limitation of Russian programs as well as the closure of the Russian news agency stated that the sanctions are not a legitimate or appropriate tool for combating it.[31] The issue of effectiveness of pro-Kremlin propaganda in the Baltic states is disputable. On the one hand, Russian disinformation strategy faces natural barriers in the Baltic states, stemming from political (consensus among the parties in terms of pro-European and pro-transatlantic policies), historical (negative experiences with Russian and the Soviet Union regime) and economic (limited presence of Russian capital) factors. On the other hand, it is the demographic structure of the states which is a key issue of its pro-Kremlin media's capability to influence the information sphere in the region.[32] Many old generations still do not know the state language perfectly. So far, open access to information in Russian and high percentage of interlocutors using Russian language in everyday communication (for example in Latgale, Latvia and Ida-Virumaa, Estonia) has made it unnecessary to learn the state language.[33] Thus, ban for Russian media channels, particularly in Latvia and Estonia, may lead to dissatisfaction and frustration among the Russian-speaking community.

Disinformation is still disseminated in social media and the problem is the lack of media literacy, which means it is extremely difficult to react to propaganda in the digital era.[34] Particularly after the annexation of Crimea, Lithuania, Latvia and Estonia made several methods to combat Russian propaganda. Similarly, the Russian–Ukrainian war made the Baltic residents more aware of the destructive potential of Russian propaganda. According to Nerijus Maliukevičius, measures taken in order to improve the information security are based on four principles: awareness-raising, assertiveness, alternative and responsibility.[35] The Baltic states seek to make their own media more attractive and accessible by improving the quality of the content, strengthening civil resilience towards the Kremlin's disinformation and promoting civic sensitiveness. Also, the Baltic states support to create a European Russian-language TV channel as an alternative to TV programs from Russia. However, so far only Estonia has decided to introduce public Russian-language media platform to provide reliable information and shape common national identity based on respect and tolerance, regardless of ethnicity. At the same time, the Baltic states cooperate within trilateral format as well as international organizations. They deepen their cooperation in the areas of combating disinformation, Russian propaganda and cyber threats as the states share the same perception of information threats. Finally, we have been observing that several independent media companies have left Russia in recent years because of threats against their journalists and their work, and have set up operation in Lithuania, Latvia and Estonia which is a promising step for future development of Russian-speaking media in the region.

Finally, the Russian aggression against Ukraine has consolidated the Baltic states' response towards pro-Kremlin disinformation. They implement new approaches and more restrictive procedures. This may be, in fact, an impulse for an increase of scepticism among Russian-speaking communities towards the national authorities. While growing level of distrust due to ineffective methods of delivering information is a global trend[36], Lithuania, Latvia and Estonia are not an exception and social support towards the state institutions has decreased visibly in the recent years.

Notes

1 D. Auers (2015). *Comparative Politics and Government of the Baltic states Estonia, Latvia and Lithuania in the 21st Century*. New York: Palgrave Macmillan.

2 A. Kuczyńska-Zonik (2016). "Russian propaganda—methods of influence in the Baltic States". *Yearbook of the Institute of East-Central Europe*, 14 (2), 43–59.

3 A. Kuczyńska-Zonik (2016). "Antidiplomacy in the Russia's Minority Policy towards the Baltic States". *Baltic Journal of Political Science*, 5, 89–104.

4 D. Holtzhausen & A. Zerfass (2015). "Strategic Communication: Opportunities and Challenges of the Research Area". in: *The Routledge Handbook of Strategic Communication*, D. Holtzhausen, A. Zerfass (eds.), New York.

5 C. Bjola & J. Pamment (2018). *Countering Online Propaganda and Extremism. The Dark Side of Digital Diplomacy*. Routledge.

6 M. Castells (2010). "Globalisation, Networking, Urbanisation: Reflections on the Spatial Dynamics of the Information Age". *Urban Studies*, 2737–2745.

7 E. Lucas (2008). "Whataboutism". *The Economist*. 31 January. https://www.economist.com/europe/2008/01/31/whataboutism.

8 A. Kuczyńska-Zonik (2016). "Antidiplomacy…, 89–104.

9 Ch. Walker (2018). "What Is "Sharp Power'?", *Journal of Democracy*, 29(3), 9–23.

10 E.-C. Onken (2007). "The Baltic States and Moscow's 9 May commemoration: Analysing memory politics in Europe. Europe- Asia Studies, 59(1), 23–46. DOI: 10.1080/ 09668130601072589.

11 I. Birka (2016). "Expressed Attachment to Russia and Social Integration: The Case of Young Russian Speakers in Latvia, 2004–2010". *Journal of Baltic Studies*, 47(2), 219–238. DOI: 10.1080/01629778.2015.1094743.

12 C. Bjola (2018). "The Ethics of Countering Digital Propaganda". *Ethics & International Affairs,* 32(3), 305–315. DOI:10.1017/S0892679418000436.

13 SputnikNews (2020), Кёрёнавирус изёбрели в Латвии? А пёчему бы и нет, 16 March. https://lv.sputniknews.ru/Latvia/20200315/13379140/Koronavirus-izobreli-v-Latvii-A-pochemu-by-i-net.html [28].[07].[2022].

14 K. Hallahan, D. Holtzhausen, B. van Ruler, D. Verčič & K. Sriramesh K. (2007). "Defining Strategic Communication". *International Journal of Strategic Communication,* 1(1), 3–35.

15 Ch.C. Hood & H.Z. Margetts (2007). *The Tools of Government in the Digital Age*, New York: Palgrave Macmillan.

16 DRI (2018). *Disinformation Resilience in Central and Eastern Europe*, The Foreign Policy Council "Ukrainian Prism", The Eurasian States in Transition research center (EAST Center).

17 Republic of Estonia (2017). *Ministry of Defence, National Defence Development Plan for 2017–2026.*

18 V. Denisenko (2020). "The threat of propaganda and the information war for Lithuanian security". in: *Lithuania in the global context: national security and defence policy dilemmas*, G. Česnakas & N. Statkus, Generolo Jono Žemaičio Lietuvos karo akademija, Vilnius; G. Jakštaitė-Confortola (2021). "Russia's "Sharp Power' Manifestations in Lithuania's Mass Media", *Baltic Journal of Law & Politics*, 14(1), 73–102.

19 M. Laurinavičius (2015). "Is Russia Winning the Information War in Lithuania?". CEPA, July. http://cepa.org/index/?id=6fcbe90edddbc8bc99ec9fb4960ef0b9 [10].[10].[2022].

20 M. Morkunas (2022). "Russian propaganda in Baltics: Does it really work?". January 26. https://ssrn.com/abstract=4018976 [10].[10]/[2022].

21 D. Demers (2005). *Dictionary of Mass Communication and Media Research: A Guide for Students, Scholars and Professionals*. Washington State University; L. Langman (2015). "Infotainment". in: *The Wiley Blackwell Encyclopedia of Consumption and Consumer Studies*. D. Thomas Cook, J. M. Ryan (eds.), John Wiley & Sons, Ltd. Published. DOI: 10.1002/9781118989463.wbeccs150.

22 A. Rozukalne (2016). " "All the Necessary Information is Provided by Russia's Channels". Russian-language Radio and TV in Latvia: Audiences and Content". *Baltic Screen Media Review,* 4, 106–124.

23 R. Bambals (2016). "Societal Resilience: The Case of the Russian-speaking Community in Latvia". in: *Societal Security. Inclusion- Exclusion Dilemma. A portrait of the Russian-speaking community in Latvia.* Ž. Ozolina (ed.), Zinātne Publishers, 44–73.

24 A. Jõesaar, S. Rannu & M. Jufereva (2013). "Media for the minorities: Russian language media in Estonia 1990–2012". *Media Transformations,* 118–154.

25 For example, 15th place in the World Press Freedom Index 2021—the highest among the Central and Eastern European countries.

26 D. Donauskaitė, M. Fridrihsone, M. Krancevičiūtė, A. Krūtaine, A. Lastovska, P. Reiljan & A. Tetarenko (2020). *Baltic Media Health Check 2019–2020*. Riga/Tallinn/Vilnius, November. https://www.sseriga.edu/ sites/default/files/2020-11/Baltic_Media_Health_Check_2019_2020.pdf [10].[10].[2022].

27 Kantor Emor (2021). *Teleauditooriumi ülevaade 2021*. https://www.kantaremor.ee/teleauditoorium i-ulevaade/ [10].[10].[2022].

28 J. Dougherty & J. Kaljurand (2015). *Estonia's "Virtual Russian World": The Influence of Russian Media on Estonia's Russian Speakers*. ICDS, October. https://icds.ee/wp-content/uploads/2014/Jill_Dough erty__Riina_Kaljurand_-_Estonia_s__Virtual_Russian_World_.pdf [10].[10].[2022].

29 A. Dimants (2016). "Latvian public service broadcasting (PSB) at a media policy crossroads on the path to public service media (PSM)". *Środkowoeuropejskie Studia Polityczne*, 2, 155–165. DOI:10.14746/ ssp.2016.2.10.

30 World Press Freedom Index (2021). https://rsf.org/en/ranking [21].[09].[2021].

31 Reporters Without Borders (2020). *Baltic countries: Misusing EU sanctions to ban Russian TV channels is not a legitimate tool for promoting reliable information*. 10 July. https://rsf.org/en/news/baltic-countries-m isusing-eu-sanctions-ban-russian-tv-channels-not-legitimate-tool-promoting [10].[10].[2022].

32 G. Kozłowski (2020). "Polityka dezinformacyjna Rosji wobec Estonii". *Sprawy Międzynarodowe*, 72(4), 107–128. DOI:10.35757/SM.2019.72.4.02.

33 T. Vihalemm & V. Kalmus (2009). "Cultural Differentiation of the Russian Minority". *Journal of Baltic Studies*, 40 (1), 95–119. DOI: 10.1080/01629770902722278.

34 R. Hornik (2016). "A strategy to counter propaganda n the digital era". *Yearbook of the Institute of East-Central Europe*, 14 (2), 61–74.

35 N. Maliukevičius (2015). "'Tools of destabilization': Kremlin's media offensive in Lithuania". *Journal on Baltic Security*, 1(1), 117–126.

36 D. Price, T. Bonsaksen, M. Ruffolo, J. Leung, V. Chiu, H. Thygesen, M. Schoultz, A.O. Geirdal (2021). "Trust in Public Authorities Nine Months after the COVID-19 Outbreak: A Cross-National Study". *Social Sciences*, 10.

Bibliography

Auers D. (2015). *Comparative Politics and Government of the Baltic states Estonia, Latvia and Lithuania in the 21st Century*. New York: Palgrave Macmillan.

Bambals R. (2016). "Societal Resilience: The Case of the Russian-speaking Community in Latvia". in: *Societal Security. Inclusion—Exclusion Dilemma. A portrait of the Russian-speaking community in Latvia*. Ž. Ozolina (ed.), Zinātne Publishers, 44–73.

Birka I. (2016). "Expressed Attachment to Russia and Social Integration: The Case of Young Russian Speakers in Latvia, 2004–2010". *Journal of Baltic Studies*, 47 (2), 219–238. doi:10.1080/ 01629778.2015.1094743.

Bjola C. (2018). "The Ethics of Countering Digital Propaganda". *Ethics & International Affairs*, 32 (3), 305–315. doi:10.1017/S0892679418000436..

Bjola C., & Pamment J. (2018). *Countering Online Propaganda and Extremism. The Dark Side of Digital Diplomacy*. Routledge.

Castells M. (2010). "Globalisation, Networking, Urbanisation: Reflections on the Spatial Dynamics of the Information Age". *Urban Studies*, 2737–2745.

Denisenko V. (2020). "The threat of propaganda and the information war for Lithuanian security". in: *Lithuania in the global context: national security and defence policy dilemmas*, G. Česnakas & N. Statkus, Generolo Jono Žemaičio Lietuvos karo akademija, Vilnius.

Demers D. (2005). *Dictionary of Mass Communication and Media Research: A Guide for Students, Scholars and Professionals*. Washington State University.

Dimants A. (2016). "Latvian public service broadcasting (PSB) at a media policy crossroads on the path to public service media (PSM)". *Środkowoeuropejskie Studia Polityczne*, 2, 155–165. doi:10.14746/ssp.2016.2.10..

DRI (2018) *Disinformation Resilience in Central and Eastern Europe*, The Foreign Policy Council "Ukrainian Prism", The Eurasian States in Transition research center (EAST Center).

Donauskaitė D., Fridrihsone M., Krancevičiūtė M., Krūtaine A., Lastovska A., Reiljan P. & Tetarenko A. (2020). *Baltic Media Health Check 2019–2020*. Riga/Tallinn/Vilnius, November. https://www.sseriga. edu/sites/default/files/2020-11/Baltic_Media_Health_Check_2019_2020.pdf.

Dougherty J. & Kaljurand J. (2015). Estonia's "Virtual Russian World": The Influence of Russian Media on Estonia's Russian Speakers. ICDS, October. https://icds.ee/wp-content/uploads/2014/Jill_Dough erty__Riina_Kaljurand_-_Estonia_s__Virtual_Russian_World_.pdf.

European Union Institute for Security Studies (EUISS) (2016). *EU strategic communications With a view to counteracting propaganda*. May. https://www.europarl.europa.eu/RegData/etudes/IDAN/2016/578008/EXPO_IDA(2016)578008_EN.pdf.

Hallahan K., Holtzhausen D., van Ruler B., Verčič D. & Sriramesh K. (2007). "Defining Strategic Communication". *International Journal of Strategic Communication*, 1 (1), 3–35.

Holtzhausen D. & Zerfass A. (2015). "*Strategic Communication: Opportunities and Challenges of the Research Area*". in: *The Routledge Handbook of Strategic Communication*, D. Holtzhausen, A. Zerfass (eds.), New York.

Hood Ch. C., & Margetts H.Z. (2007). *The Tools of Government in the Digital Age*, New York: Palgrave Macmillan.

Hornik R. (2016). "A strategy to counter propaganda n the digital era". *Yearbook of the Institute of East-Central Europe*, 14 (2), 61–74.

Jakštaitė-Confortola G. (2021). "Russia's 'Sharp Power' Manifestations in Lithuania's Mass Media", *Baltic Journal of Law & Politics*, 14 (1), 73–102.

Jõesaar A., Rannu S. & Jufereva M. (2013). "Media for the minorities: Russian language media in Estonia 1990–2012", in: *Media Transformations*, 118–154.

Kantor Emor (2021). *Teleauditooriumi ülevaade 2021*. https://www.kantaremor.ee/teleauditooriumi-ulevaade/.

Kozłowski G. (2020). "Polityka dezinformacyjna Rosji wobec Estonii", *Sprawy Międzynarodowe*, 72 (4), 107–128. doi:10.35757/SM.2019.72.4.02..

Kuczyńska-Zonik A. (2016a). "Russian propaganda—methods of influence in the Baltic States". *Yearbook of the Institute of East-Central Europe*, 14 (2), 43–59.

Kuczyńska-Zonik A. (2016b). "Antidiplomacy in the Russia's Minority Policy towards the Baltic States". *Baltic Journal of Political Science*, 5, 89–104.

Laurinavičius M. (2015). "Is Russia Winning the Information War in Lithuania?". CEPA, July. http://cepa.org/index/?id=6fcbe90edddbc8bc99ec9fb4960ef0b9.

Langman L. (2015). "Infotainment". in: *The Wiley Blackwell Encyclopedia of Consumption and Consumer Studies*. D. Thomas Cook, J. M. Ryan (eds.), John Wiley & Sons, Ltd. Published. doi:10.1002/9781118989463.wbeccs150..

Lucas E. (2008). "Whataboutism", *The Economist*. 31 January. https://www.economist.com/europe/2008/01/31/whataboutism.

Maliukevičius N. (2006). "Geopolitics and Information Warfare: Russia's Approach". *Lithuanian annual strategic review*, 121–146.

Maliukevičius N. (2015). "'Tools of destabilization': Kremlin's media offensive in Lithuania". *Journal on Baltic Security*, 1(1), 117–126.

Morkunas M. (2022). "Russian propaganda in Baltics: Does it really work?". January 26. https://ssrn.com/abstract=4018976.

Onken E.-C. (2007). "The Baltic States and Moscow's 9 May commemoration: Analysing memory politics in Europe". *Europe-Asia Studies*, 59 (1), 23–46. doi:10.1080/09668130601072589.

Price D., Bonsaksen T., Ruffolo M., Leung J., Chiu V., Thygesen H., Schoultz M., & Ostertun Geirdal A. (2021). "Perceived Trust in Public Authorities Nine Months after the COVID-19 Outbreak: A Cross-National Study", *Social Sciences*, 10.

Reporters Without Borders (2020). *Baltic countries: Misusing EU sanctions to ban Russian TV channels is not a legitimate tool for promoting reliable information*. 10 July. https://rsf.org/en/news/baltic-countries-misusing-eu-sanctions-ban-russian-tv-channels-not-legitimate-tool-promoting.

Republic of Estonia (2017). Ministry of Defence, *National Defence Development Plan for 2017–2026*.

Rozukalne A. (2016). "All the Necessary Information is Provided by Russia's Channels". Russian-language Radio and TV in Latvia: Audiences and Content". *Baltic Screen Media Review*, 4, 106–124.

SputnikNews (2020), Кёрёнавирус изёбрели в Латвии? А пёчему бы и нет, 16 March. https://lv.sputniknews.ru/Latvia/20200315/13379140/Koronavirus-izobreli-v-Latvii-A-pochemu-by-i-net.html [28.07.2022].

Vihalemm T., & Kalmus V. (2009). "Cultural Differentiation of the Russian Minority". *Journal of Baltic Studies*, 40 (1), 95–119. doi:10.1080/01629770902722278..

Walker Ch. (2018). "What Is 'Sharp Power'?", *Journal of Democracy*, 29 (3), 9–23.

World Press Freedom Index (2021). https://rsf.org/en/ranking [21.09.2021].

26

THE DISINFORMATION THREAT
LESSONS FROM THE CHINESE EXPERIENCE

Dan Dungaciu and Lucian Dumitrescu

Introduction

Our study focuses on the disinformation used by China in the early stages of the coronavirus pandemic, that is, in the spring of 2020. We aim to demonstrate that China's disinformation has been an integral part of a process of performative construction of China's state and political regime. The narrative of China's "medical exceptionalism" was a totalizing one, in the sense that it had both a systemic and an identity component, but also a public policy one. It is noteworthy that the resilience of China's political system was built in contrast to the political system of liberal democracy, which was portrayed as lacking resilience, through strategic representations and tactical narratives that either stigmatized or securitized he United States Performatively, the state also manifests itself through practices that range "from foreign and security policies to crises of intervention, immigration strategies, the protocols of treaty-making, representative politics of the United Nations, and beyond"[1]. It is already common-place in the field of security studies that foreign policy actions produce and reproduce iden-tity differences[2]. In our view, the strategic representations and tactical narratives conveyed by the *Global Times* in the spring of 2020, which stigmatized the United States, but especially those with a securitization role were part of a process of political contestation that aimed at highlighting the difference between China's political system—presented as resilient—and the one in the United States—presented as non-resilient. Obviously, there may be major differ-ences between the narrative of the resilience of a political system and its current resilience, as has happened in China.

On a systemic level, the narrative of China's "medical exceptionalism" in the spring of 2020 was intended to project the image of a great power that has the resources and the political will[3] to manage a crisis with a global impact. Another systematic goal of China's "medical exceptionalism" narrative has been to give Beijing a surplus of normative power. In the strategic regional and international competition with the United States, China is char-acterized by various vulnerabilities, including the predominance of the extensive model of economic growth, military capabilities and the reduced ability to instil confidence in allies[4]. Especially the latter vulnerability correlates with China's relatively low regulatory power. In general, the normative power of a state lies in its ability to shape a region of interest, in which states have usually conflicting security interests, with the help of (sub) regional

DOI: 10.4324/9781003190363-31

organizations[5]. Thus, the ability of a state to produce "regionness"[6], i.e. to bring the states of the region in the same boat, is an indicator of its normative power. The Shanghai Cooperation Organization, the Asian Infrastructure Investment Bank, and the BRICS[7] Development Bank are regional organizations that prove that China's normative power has developed. The strategic narratives of international cooperation proposed by Beijing, especially during the Xi Jinping[8] administration, are in line with the same logic. With global aid in the spring of 2020, China has tried to define itself as an "indispensable nation", and its humanitarian mission has resembled a "medical Belt and Road Initiative". The narrative of China's "medical exceptionalism" also had an identity component. Through a series of inclusive and exclusive practices—some of the latter directly related to disinformation—Beijing has intended to project the image of a resilient political system mainly internationally, but also domestically. And, simultaneously, a political regime superior to liberal democracy. At the same time, the narrative of China's medical exceptionalism also targeted a public policy component. A number of capabilities were revealed—hospitals, large number of specialists, the ability to mobilize them and the population, material resources, institutional coherence, etc.—through which China could effectively manage a medical crisis. Thus, the narrative of China's "medical exceptionalism" has played a legitimating role both domestically and internationally.

In the first section, the chapter clarifies the concepts and methodology we used. In the second section, it focuses on what we have called the "exclusive practices" of China's "medical exceptionalism" narrative. These are a series of strategic representations and tactical narratives that have either stigmatized or securitized the United States. These exclusive practices are related to disinformation. In the third part, the text discusses the narrative of China's medical exceptionalism in the context of the narratives commonly circulated by Beijing, but also in connection with the strategic thinking of the Xi Jinping administration.

Conceptual and methodological clarifications: political contestation, strategic and tactical narratives

For academic research in the field of security studies, examining the process of disinformation which a state actor is involved implies a number of issues related to research design. In short, these are conceptual and methodological issues. Without a precise conceptualization of disinformation, there is a risk of discussing it descriptively and thus missing its central aspects. Typically, disinformation initiated by a state actor fulfils several of the following conditions: (1) it is part of a larger process of political contestation, (2) it is usually combined with soft power tools, which should be distinguished from disinformation, (3) capitalizes on internal and external strategic (sub) cultures and (4) may change in relation to certain major events in world politics (turning points). As a result, the above elements—or most of them—should be captured especially by research in the field of security studies concerned with the way in which strategic narratives are formed. The research methodology is another important constraint for the academic research of the disinformation in the area of security studies. Clarity is therefore needed in relation to the data required and the way in which this data is collected.

In essence, the study delves into some practices of disinformation that China resorted to in the early phases of the COVID-19 pandemic in order to boost its political legitimacy both at home and abroad. It is already a common place in the realm of security studies that disinformation refers to the spreading of inaccurate information for strategic gains by either state or non-state actors[9]. By contrast, misinformation usually lacks such a "malevolent intent"[10]. The main novelty that our study seeks to bring to the fore is a rather "holistic"[11] conceptualization on China's disinformation practices. In our view, it would be naïve to work on the assumption that China resorted *only* to disinformation practices in order to enhance the legitimacy of its political system at the outset of the COVID-19 pandemic. Whether Chinese authorities employed *mainly* disinformation practices in order to instil more credibility into its political institutions in the early phases of the pandemic, it is something that needs to be looked carefully at. However, this aspect is beyond the scope of this study. As already stated, our study aims at broadening the conceptualization that could be used to scrutinize China's disinformation practices not only with respect to the COVID-19 pandemic but also regarding different other public interventions of the Chinese state. With respect to China, certain experts have already employed to the concept of "sharp power"[12], which combines "hard and soft power instruments of national power"[13]. Apparently, "sharp power" is specific to authoritarian regimes[14], and in the case of the Russian Federation it consists in "active measures, cyberattacks, and disinformation"[15], as "Russia does not want to be loved; it wants to be feared"[16]. The trouble with disinformation is that it could be looked at as a specific "way of action"[17], besides—or along with—diplomacy, economy, soft power, military activities, that any type of political regime could resort to as part of its grand strategy. Especially if that political regime understands grand strategy in the so-called "classical tradition"[18], according to which strategy is about conflict. Arguably, the extent and the specific circumstances under which a political system might resort to disinformation are some of the aspects that need to be carefully scrutinized. These aspects will reveal the level of political development of that political system, as the systematic usage of disinformation tends to be related either to political underdevelopment or democratic backsliding. Our point is that we need a neutral and, at the same time, holistic conceptualization of disinformation if we want to turn it into an object of academic research. Otherwise, disinformation studies may run the risk of securitizing either state or non-state actors. Devoid of academic content, which implies a careful conceptualization and a clear methodology, such studies may end up being used in the political competition among different interest groups[19]. In order to improve the academic research on disinformation, we have employed the concept of "political contestation". Not only that this concept is politically neutral, but it could be applied to the performative actions of any state irrespective of its political system. Political contestation shows that a state may combine different practices in order to get more legitimacy, which is a strategic objective especially in times of rapid social change, as it was the outset of the COVID-19 pandemic. The so-called "medical exceptionalism" that China put on display in the spring of 2020 aimed at reinforcing the legitimacy of its political system, both internally and externally. In our view, China's "medical exceptionalism" was a strategic effort that comprised inclusive and exclusive practices. The former consisted in soft power practices, whereas the latter aimed at the political stigmatisation and securitization of the United States of America. Especially the latter practices, but to a certain extent even the former ones, sought to present the United States as a security threat and, thus, affect its international credibility.

Political warfare refers to "hybrid war without the eventual shooting"[20]. There are a number of issues that need to be addressed in connection with the political contestation First of all, political contestation is a process of undermining the vertical and horizontal legitimacy of a state. That is, on the one hand, the political leaders of the respective state, the public institutions and various public policies, as well as their embed values, are discredited. On the other hand, the national and regional symbols with which the majority of citizens identify are challenged. Secondly, a distinction must be made between electoral political contestation and strategic politicalcontestation. The first is a value of liberal democracy and is frequently practiced by political parties in the interaction of democratic institutions. Strategic political contestation is a coordinated process by which one state uses public resources to defame another state so that the former gains more legitimacy internationally at the expense of the former. The tools of inter-state political contestation are "deception and propaganda, coercive diplomacy and economic leverage, subversion"[21]. Thirdly, the performative construction of the state also involves a process of strategic contestation, i.e. both soft power and disinformation.

Not surprisingly, the narrative of China's "medical exceptionalism" is a "China story"[22] that included both soft power tools and elements of disinformation. The disinformation we are pursuing has involved strategic representations and tactical narratives that have either stigmatized or securitized the United States. There is an important difference between social stigma and a securitization process. Subjected to a systematic process of social stigma, members of a group will never fully integrate into a society. But they will be able to coexist with other social groups. In the case of a securitization process, the social group is not only marginalized, but excluded from society. Therefore, the stigmatization of a state at the international level can have different effects, from increasing the level of ontological insecurity to marginalization within the international community. But securitizing a state can turn it into a security threat to the international community. Hence the risk of isolation or political exclusion. The strategic representations used by China in either stigmatizing or securitizing the United States can be equated with tactical narratives. These may refer to facts but may also contain *fake news*. But even when it comes to facts, tactical narratives are not autonomous. They are subsumed to strategic narratives that give a certain meaning, usually tendentious and emotional, to those facts. That is why we have analyzed Beijing's tactical narratives regarding the United States in the broader context of the strategic narrative of China's "medical exceptionalism".

Strategic narrative is a spectrum of persuasion that combines power and communication. This spectrum of persuasion evolves on a continuum between a rational end and a deep end[23]. The rational end is fed with facts by the tactical narratives, and the deep one with meaning and emotions provided by the strategic narratives. The tactical narrative influences the public's perceptions of a factual issue—which may include *fake news*—while the strategic narrative is more of a foundational nature, shaping "the audience's perception of the very nature of the world."[24] Tactical narratives are usually complement strategic narratives The latter give meaning to the facts or *fake news* delivered through tactical narratives. The total narrative of China's "medical exceptionalism" in the early stages of the coronavirus pandemic combined three types of strategic narratives: systemic, identity, and public policy. *Systemic or constitutive strategic narratives* communicate to the public information about world politics: how it is structured, who are the main actors, the situation of the international system. Systemic strategic narratives are the Cold War, the War on Terror, the rise of China, or Russia's neo-imperialism[25]. China's "medical exceptionalism" has used this systemic dimension. In the spring of 2020, China seemed the only country in the world capable of successfully managing the coronavirus pandemic. *Strategic identity narratives* highlight the central, societal values of an actor and the threats against it. China's "medical exceptionalism" narrative also had a component of political exceptionalism. This was probably the central aspect of China's political contestation process in the early phases of the pandemic. It sought to stress the resilience— and at the same time the superiority—of China's political system over other political systems in the world, but especially over liberal democracy. *The strategic narrative of public policy* can highlight either the normative value of a policy—the protection of the environment for future generations—or the specific institutional capacities held by an actor to implement a specific public policy. The image projected by China at the beginning of the coronavirus pandemic revealed specific resources (hospitals, doctors, material resources, medical expertise, mobilization capacity, coherent policies) for managing a crisis situation.

We have worked from the assumption that disinformation is "an instance of a class of events."[26] That is, a facet of the "China story" proposed by Beijing in the spring of 2020. As a consequence, China's "medical exceptionalism" could be investigated with a case study methodology. As already stated, we address disinformation as part and parcel of China's performative process that was set in motion in the early phases of the pandemic in order to instil more legitimacy into China's political institutions. We have equated China's performative process with a political contestation process for political systems tend to manufacture internal and external alterities in the attempt to define their political identity. In this logic, China's disinformation practices could be explored as a constituent part of a process of political contestation, which has manufactured a totalizing narrative about China's "medical exceptionalism". This totalizing narrative comprised three strategic narratives—systemic, identity and public policy—that were joined by tactical narratives. To explore China's disinformation, we looked at the *Global Times*, one of China's leading foreign policy sites[27] from February to April 2020. The importance of the *Global Times* for such an investigation stems from the fact that it is published by *People's Daily*[28], which takes over the official vision of the Chinese Communist Party[29]. We monitored in the *Global Times* the strategic representations and tactical narratives used by the Chinese political elite. For an in-depth study, the strategic representations contained in the statements of the main political leaders, in government documents, reports of various think tanks, etc. should have been monitored. Because of this limitation, that is, the investigation of a small number of open sources and a small number of cases, our study has a rather exploratory profile.

This methodological section comes with a caveat. Our study has not scrutinized "Chinese media" in order to find out a pattern—if any—of China's disinformation in the early stages of the COVID-19 pandemic. Nor it has explored a potential difference in terms of official rhetoric between "China" and "Chinese media" (although this line of research could be a fruitful one). On the contrary. Due to lack of different resources, we have sought to shorten as much as possible the discursive distance between "China" and "Chinese media". To this end, we have chosen *Global Times* as the most important source of our empirical material, while we also cite the *Xinhua News Agency*. Considering that *Global Times* is published by *People's Daily*, and *People's Daily* is the most influential newspaper published by the Party Central Committee, we have worked from the premise that the discursive distance between "China" and "Chinese media" may have been almost non-existent in the early phases of the pandemic. Doshi argues that the newspapers and magazines published by the Party Central Committee "express official party views, broadcast some Party debates, and also contain detailed authoritative commentaries and official speeches"[30]. Therefore, our article has been premised on the fact that the inclusive and exclusive practices that we have scrutinized come

Table 26.1 China's "medical exceptionalism", a totalizing narrative made up of inclusive and exclusive practices

INCLUSIVE PRACTICES	Systemic narrative	Soft power through tactical narratives
EXCLUSIVE PRACTICES	Identity narrative	Stigmatisation of the United States through tactical narratives
	Public policy narrative	Securitization of the United States through tactical narratives

mainly from "China", as these practices stand for China's "institutionalised discourse"[31]—or at least a fragment of it.

The narrative of China's "medical exceptionalism". Inclusive practices, exclusive practices, disinformation

Into the category of *inclusive practices* falls China's medical assistance to a number of states at a time when the idea of "international solidarity" seemed to be permanently compromised. The *exclusive practices* consist of strategic representations that either stigmatize or securitize the United States, while simultaneously highlighting the resilience of the Xi Jinping administration, which is hence legitimized. Most of the "strategic representations presented by the *Global Times*, a daily newspaper that spreads government propaganda in China, about the United States are rather *soft*. Soft strategic representations tend to stigmatise the United States by revealing different flaws of the American political system. By contrast, hard strategic representations aim at securitizing the United States through the dissemination of either fake news or conspiracy theories. What we have termed soft strategic representations, tend to be predominantly ideological and highlight alleged drawbacks of the American political and administrative system, such as vulnerabilities of the US public health system, the preeminence of the economic vision over public health, human rights violations, errors in the public response to the COVID-19 pandemic, which, in theory, do not fall in line with American exceptionalism. These "negative strategic representations" are not necessarily false. However, they systematically present the Trump administration's failures in managing the coronavirus pandemic. Successes are silenced. But what matters for this report is that the "negative strategic representations" circulated by the *Global Times* about the United States are systematically presented in contrast to the performance of the Xi Jinping administration. The self / other logic is obvious. The self/other logic is usually employed by any political system in an attempt to define its identity. To this end, it manufactures an alterity or other, which consists in either state or non-state actors which depicted as ontologically/politically different. The alterity is defined as being rife with different institutional flaws. In this vein, the drawbacks of the self are either swept under the rug or directly projected onto the other political system. In other words, by stigmatizing and securitizing the American political and economic model, these representations legitimize China's own political and economic model. For this reason, these representations can be considered strategic. Their function is to securitize and legitimize at the same time. As for the exclusive practices that securitized America, the *Global Times* circulated in 2020 a series of representations about the alleged military laboratories of the US military, either in the United States or from military bases in other states. On the verge of *fake news* or even *fake news*, these representations present the United States as a global security threat.

China's medical exceptionalism was manufactured in the spring of 2020 with the help of different *inclusive practices*. Among these, the support that different international experts offered to the "Chinese model" of dealing with the Covid-19 pandemic in its early phases. A world that was beginning to realize its massive dependence on China in relation to relatively common medical equipment, but which proved to be essential in the coronavirus pandemic. What has propaganda got to do with this humanitarian exercise? The connection is pretty obvious. On the one hand, China's humanitarian effort has been highly publicized. And on the other hand, this effort has been linked to the major successes of China, which is portrayed as an "indispensable nation." The involvement of Chinese experts in the fight against coronavirus in Southeast Asia, the Middle East and Africa[32], as well as the medical assistance provided by China to some states in the European Union, the Western Balkans and South America, reveals the global scope of China[33]. As the Chinese public became aware that "China is bigger than the world," Chinese television broadcast images of China's space mission to the moon, a mission launched in December 2018[34]. Beijing's triumphant discourse that promotes the "Chinese model" of managing the coronavirus pandemic is quite clear. That is, a formula of good governance that is based on the massive intervention of the state in society. The logic of this model of good governance is quantitative: it does not matter the human rights; it matters the result obtained. Not surprisingly, China's public health strategy during the coronavirus pandemic was "guaranteed" by international experts. In this regard, the *Global Times* publishes an interview with Richard Horton, editor-in-chief of the British magazine *The Lancet*[35]. Horton stressed that China's measures in Wuhan were correct and timely, which is why the international community should thank Beijing. More important, however, for legitimizing China's medical exceptionalism is the fact that Horton rules out the possibility that the virus was accidentally dropped from the laboratory in Wuhan. "China is not responsible for the pandemic,"[36] he said. It should not be forgotten that the "Chinese model" of managing the coronavirus pandemic has been endorsed by experts and senior officials of the World Health Organization.

China's "medical exceptionalism" has also been manufactured through a series of *exclusive practices* that have included strategic representations that contribute to soft securitization and hard securitization of the United States. The soft ones have stigmatized the United States, while the hard ones suggest that the United States is a security threat. With regard to the soft securitization practices, the *Global Times* is quoting Vice President Pence as noting the March 2020 shortage of tests in the United States. Then the *Global Times* discusses the capacity of Vazyme Biotech in Nanjing. This company would have produced over a million test kits in less than a month and a half[37]. The article also included a statement from the Minister of Industry and Information Technology, discussing China's ability to produce more than 350,000 test kits a day. At the same time, many of these test kits had already been shipped to more than twenty-six countries, including countries with regional powers such as India and Japan. At that time, however, China not only had resources that America apparently did not have. But it owned them in abundance, meaning that he could distribute them to other states, some of which, like Japan, would have expected medical care from Washington rather than Beijing. Other texts in the *Global Times* show the superiority of the public health system in China over that in the United States. The argument is that both testing and treatment are free for patients infected with the new coronavirus[38]. That, in conjunction with Washington's rejection of information provided by Beijing over the new coronavirus, could have triggered a veritable "humanitarian crisis" in the United States, according to the *Global Times*. The daily thus claims China's superiority in all areas of the fight against coronavirus: from testing capacity to the public health system to the administrative measures ordered by the Chinese authorities. *Global Times* reports clearly support China's "medical exceptionalism" in contrast to what happened in the United States at the beginning of the pandemic. Also on the already established model of self / other or in group / out group logic, *Global Times* also presents an alleged case of human rights violations in the United States. This is the case of Amyiaha Cohoon, a high school student from Wisconsin, who had symptoms of coronavirus after a vacation in Florida[39]. Amyiah went to a local hospital, where doctors confirmed that she could be infected with coronavirus, but that at the time of the medical check-up, she did not have the virus. What the *Global Times* suggests is a case of discrimination against a teenager in the United States. According to the Chinese daily, Amyiah Cohoon shared her experience of being infected with the new coronavirus on Twitter. After the posts in question, she was visited by a local police crew who allegedly ordered her to delete the posts on twitter. The reason? That of not generating panic in the community. As a result, the teenager's parents sued the local sheriff. The conclusion of the text is a political one. The Trump administration has proven incompetent in managing the coronavirus pandemic, and this is the explanation for discriminating against those who caught the virus. Second, the United States criticizes China for violating human rights, but does the same on its own territory. Another text published by the *Global Times* discusses what is really behind the Trump administration's challenge to the data provided by Beijing in connection with the coronavirus pandemic[40]. First of all, it would be populism. Specifically, finding a scapegoat like China is meant to mask Washington's clumsiness during the coronavirus crisis. Second, without a scapegoat like China, Chinese journalists believed that President Trump had a good chance of losing the presidential election in 2020. Third, the White House's abrasive discourse on China is seen as a manipulative technique by which Washington seeks to convince the American public of the need to reduce economic and technological dependence on China. Another text presents the Trump administration's so-called "lies" about China[41]. More precisely, it is the "Chinese virus", the "cover-up" of the coronavirus epidemic by the Beijing authorities and the issue of donations made to China by the United States government. The underlying message of the text is the following: the Trump administration is more concerned with saving the financial markets than protecting the citizens from the coronavirus pandemic. As such, the Trump administration is portrayed as a direct security threat to American citizens.

Regarding the "exclusive practices" through which *Global Times* manufactures China's "medical exceptionalism", there are also strategic representations that contribute to a what we have deemed a "hard securitization" of the United States. These strategic representations are *fake news* and suggest that the United States is a security threat. This category includes the accusation made by the spokesperson of the Ministry of Foreign Affairs of China. Zhao Lijian claimed that the new coronavirus was brought to China by the American military who had participated in the World Military Games held in Wuhan in October 2019[42]. Also in the category of "hard securitization" fall the accusations launched by the *Global Times* in connection with the closure, in June 2019, of the biological research laboratory of the American army from Fort Detrick. Chinese journalists claim that the closure of the laboratory was linked to the appearance of a "high number of people who complained of respiratory symptoms, from cough to pneumonia, in two hospitals in a neighboring region in Virginia"[43]. The accusation is "guaranteed" by figures. These are 215 cases of severe pneumonia reported in August by the Centre for Disease Prevention and Control (CPCB). Cases which, according to the US authorities to which the author of the article does not grant credit, were caused by vaping, i.e. the use of electronic cigarettes. Another 805 cases were reported in September in forty-six US states. The article also discusses a petition launched on the White House website, asking the US government to provide information on the reasons that led to the closure of the Fort Detrick military laboratory. As well as research on the new type of coronavirus at that lab. The text is inaccurate about the alleged link between the closure of the Fort Detrick laboratory and the severe pneumonia reported by the CPCB in August and September 2019, respectively. The text launches other allegations. These are alleged medical research laboratories that the United States military has set up in Pakistan, Afghanistan, Kazakhstan, Kyrgyzstan, Uzbekistan and Georgia. The journalist claims that their existence was confirmed by a spokesman for the Ministry of Foreign Affairs of the Russian Federation[44]. The information provided by the article in question appears to be *fake news*. However, the image or strategic representations of the text in question matter more. The image is that of an America that poses a security threat not only to its own citizens or to the citizens of the Russian Federation or China. The article draws on a hard securitization logic and seeks to turn the United States into a global security threat.

Discussion. China's medical exceptionalism in the context of the strategic narratives used by Beijing

Fundamental changes in the international environment tend to change the thinking and strategic action of a state. We already know how the fall of the Berlin Wall, the collapse of the Soviet Union, the 1991 Iraq war, and the 2008 financial crisis changed Beijing's[45] strategic thinking and action. Our study is an exploratory one, but the data presented support the hypothesis that the narrative of China's "medical exceptionalism" was also provoked by a fundamental event—a turning point—such as the coronavirus pandemic. In our view, the fundamental goal of this narrative was to build the resilience of China's political system and to project, internally and internationally, its superiority over liberal democracy.

But that was not the only goal. The narrative of "medical exceptionalism" also aimed at a systemic goal, that of distributing China in the role of a great power. The ability and willingness of a state to project power at the regional and international levels is a powerful behavior[46]. Through global material and medical assistance in the spring of 2020, China has sought to demonstrate its ability and willingness to manage an international medical crisis. We are talking here exclusively about the performative construction of the state. That is, *how* China has self-distributed in the role of the great power. *Why* it self-distributed in this role, respectively the internal and external conditions that allowed it to do this or that it masked with this powerful behavior, is an important but separate discussion. Also in connection with this systemic goal, China wanted to play the role of normative power. In general, normative power effectively combines *logic of consequences* and *logic of appropriateness*[47]. In the sense that it credibly links its strategic interests to universal values. Regional diplomacy, building regional institutions, and multilateral diplomacy are just some of the ways in which China has built an important regulatory extension. Another means is the strategic narratives built especially for the external public. In our view, China's medical exceptionalism is converging with these strategic narratives that are meant to portray China as a normative power. Systematic concerns about soft power emerged relatively late in Beijing, although the narrative of "peaceful coexistence" in the 1960s played an important role in the solidarity of non-aligned countries[48]. Following Hu Jintao's speech at the 17th Chinese Communist Party Congress in 2007, Beijing recognized the importance of soft power. This was the broad context in which the "harmonious society" and the "harmonious world" were discussed, as well as the "peaceful development" of China instead of the "peaceful rise"[49]. The Xi Jinping Administration has released the story "community of common destiny" to the outside world, as well as the story "China, as a harmonious, peaceful nation.[50]" A natural continuation of these narratives dedicated to the external public is the narrative of China's medical exceptionalism. Which reveals China's transformative power, just like the Belt and Road Initiative narrative[51]. All of these narratives legitimize China internationally. But beyond the narratives dedicated mainly to the international audience, "China story" also includes a series of narratives dedicated mainly to the domestic audience. These internal narratives are meant to legitimize the Chinese Communist Party. These include the anti-hegemonic, anti-American narrative, the narrative of the century of national humiliation, as well as the narrative of the essential role played by the Chinese Communist Party in Chinese history[52]. The narratives dedicated to the internal public not only reflect but also systematically reproduce both the popular strategic culture and the strategic culture of the elite, one dominated by nationalist landmarks. The emotions generated by these landmarks can be easily exploited through the strategic representations by which the United States is either stigmatized or securitized. The objective? The legitimation of the Chinese Communist Party in the logic of the narrative of the "besieged fortress". The fact that "China story" has a dual character is not surprising. Narratives for internal and external audiences are in line with China's dual strategic culture[53]. China's "medical exceptionalism" displayed in the spring of 2020 had the same dual character. On the one hand, inclusive practices that have projected China's image as a systemic and normative power internationally and endowed with a resilient political regime in the face of a crisis situation. On the other hand, exclusive practices that included disinformation. That is, strategic representations, tactical narratives, and *fake news* that have stigmatized and securitized the United States in an attempt to emphasize the superiority of China's political regime over liberal democracy.

We argue that the narrative of China's "medical exceptionalism" should be connected not only with the strategic narrative package used by Beijing both domestically and internationally. But especially with the essence of the *China Dream* strategic concept. It was launched in 2012, shortly after Xi Jinping took over as President of the People's Republic of China[54]. In essence, the *China Dream* narrative is based on the "national humiliation" narrative[55]. According to the latter, in the more than one hundred years between the First Opium War of 1840 and the Communist Revolution of 1948, China went through a period of national humiliation caused mainly by Western powers. With this premise in mind, the *China Dream* narrative advocates transforming the 21st century into a China-dominated one. The condition for achieving this goal is national mobilization, a comprehensive process of social and economic modernization, and China's regain of international power. As a result, the *China Dream* narrative links domestic development with China's international expansion. Karmazin and Hynek believe that two elements draw attention to the *China Dream* narrative. First of all, it is an attempt to change the perception of Chinese citizens in relation to China's international perception. Specifically, China should be assessed by its own standards, not by Western standards. This ideological endeavor also includes the reconceptualization of good governance from the perspective of Beijing, which is meant to convince China's important place in the world of many Chinese citizens who are skeptical about this. In fact, in a work already known by researchers in the Western world—*The China Wave. Rise of a civilizational state*—Zhang Weiwei believes that one of China's major vulnerabilities is the internal public that Beijing's (triumphalist) discourse did not convince. For Weiwei, "many Chinese at home are not yet able to understand and appreciate the significance of their country's rise"[56]. The second pillar of Xi Jinping's *China Dream* strategy is to build gigantic infrastructure projects. *Belt and Road Initiative* is the most eloquent expression of *China Dream*. The narrative of China's "medical exceptionalism" in the spring of 2020 is just another facet of *China Dream*

Conclusions

Our study has sought to demonstrate that China has manufactured a narrative of its own "medical exceptionalism" in the early phases of the COVID-19 pandemic. To this end, the Chinese authorities resorted in the spring of 2020 to both inclusive and exclusive practices. While inclusive practices consisted mainly in stressing the humanitarian aid that China disseminated at a global level, exclusive practices aimed at either stigmatizing or securitizing the United States of America. These practices were employed not only to stress the resilience of China's political system in a context of rapid social change, but also to highlight its alleged superiority in comparison to democratic systems, that had a hard time dealing with the COVID-19 pandemic in its early phases. The stigmatizing practices that we have documented in our study implied predominantly ideological representations of the United States, such as different drawbacks of the American political and administrative system. The securitizing practices, however, consisted in strategic fake news that aimed at presenting the United States as a security threat. The trouble with such practices, especially in a context of rapid social change when both vertical and horizontal trust tend to wane rapidly, is that they may subvert the credibility of the political system that is under attack. Under such circumstances, and combined with the abovementioned exclusive practices, even the inclusive practices of China's "medical exceptionalism", may bring their contribution to the disruption of vertical and horizontal trust of the targeted political system. We stress that the difference among the abovementioned exclusive and inclusive practices is only ideal-typical. In practice, however, these practices were combined and, therefore, the so-called inclusive practices could have amplified the effect of the exclusive practices. These aspects need to be carefully and contextually considered, because certain practical effects may be overlooked out of concern to come up with neutral conceptualizations of disinformation. Such conceptualizations are needed for the academic research on disinformation. Absent such conceptualizations, studies on disinformation may end up securitizing certain state or non-state actors. By employing the concept of political contestation, we have demonstrated that "China's story" in the early phase of the COVID-19 pandemic combined soft power with disinformation. However, political contestation may be part and parcel of a strategic exercise. Under such circumstances, even soft power practices are far from being benign and therefore political contestation needs to be identified, exposed and countered.

The "holistic" model that we have employed in order to explore China's disinformation in the early stages of the COVID-19 pandemic, although empirically supported by our study, may not be *the* pattern of disinformation the China has employed in the early phases of the pandemic. We are aware that, except for general and particular patterns, disinformation implies "opportunistic, fragmented, and often contradictory"[57] practices. Moreover, as already in the methodological section, we have mostly scrutinized *Global Times* but not other sources of the Chinese media. Therefore, our study has different inherent limitations that we have been aware of from the very beginning. Nevertheless, the pattern that we have identified could be used as a meta-theory for "holistic" academic approaches on disinformation. Besides examining fake-news, bots and trolls, academic approaches on disinformation could also look at the inclusive and exclusive practices of disinformation employed by a state to boost its legitimacy, whether the latter could be used more than the former in certain contexts, or examining such practices' potential link with certain strategic (sub)cultures of a certain state.

Notes

1 Campbell, *Writing Security*, 25–26.
2 Campbell, *National Deconstruction*; Hansen, *Security as practice*.
3 Onea, The Grand Strategies.
4 Denoon, China's Grand Strategy, 247.
5 Tocci, Who is a normative.
6 Riggirozzi, "Region, regionness, and regionalism".
7 Cai, China's Foreign Policy.
8 Lams, "Examining Strategic Narratives"
9 Bennett and Livingston, "The disinformation order".

10 Corbu et al., "Does fake news".
11 "10 Recommendations"
12 Scobell et al., *China's Grand Strategy*, 43.
13 Ibidem.
14 Ibidem.
15 Krekó, "The Drivers of Disinformation".
16 Ibidem.
17 Colin S. Gray, *The Future of Strategy*.
18 Balzacq et al., *Comparative Grand Strategy*.
19 Rychnovská and Martin Kohút, "The Battle for Truth".
20 Galeotti, Russian Political War, 2.
21 Ibidem. 103.
22 Lams, "Examining Strategic Narratives"
23 Szostek, "Defence and Promotion", 5.
24 Hinck, Cooley, and Kluver, *Global Media*, 4.
25 Miskimmon, O' Loughlin, and Roselle, *Strategic Narratives*.
26 George and Bennett, *Case Studies*, 17.
27 Cai, China's Foreign Policy.
28 Ibidem.
29 Doshi, *The Long Game*.
30 Doshi, *The Long Game*, 42–43.
31 Dunn and Neumann, Undertaking Discourse, 117.
32 https://www.globaltimes.cn/content/1183034.shtml, https://www.globaltimes.cn/content/1183033.shtml, https://www.globaltimes.cn/content/1183026.shtml.
33 https://brazilian.report/latin-america/2020/04/12/china-latin-america-medical-aid-fight-coronavirus/.
34 http://www.xinhuanet.com/english/2020-04/01/c_138937851.htm.
35 https://www.globaltimes.cn/content/1187265.shtml.
36 https://www.globaltimes.cn/content/1187265.shtml.
37 https://www.globaltimes.cn/content/1181827.shtml.
38 https://www.globaltimes.cn/content/1186328.shtml.
39 https://www.globaltimes.cn/content/1187259.shtml.
40 https://www.globaltimes.cn/content/1183776.shtml.
41 https://www.globaltimes.cn/content/1183464.shtml.
42 https://www.military.com/daily-news/2020/03/12/chinese-official-says-us-army-may-have-brought-epidemic-wuhan.html.
43 https://www.globaltimes.cn/content/1187243.shtml.
44 https://www.globaltimes.cn/content/1187243.shtml.
45 Cai, *China's Foreign Policy*; Doshi, The Long Game; Shambaugh, China's Leaders.
46 Onea, The Grand Strategies.
47 Marc and Olsen, "The Institutional Dynamics"
48 Cai, China's Foreign Policy.
49 Shambaugh, *China's Leaders*, 248.
50 Lams, "Examining Strategic Narratives".
51 Miskimmon, O' Loughlin, and Zeng, *One Belt*.
52 Lams, "Examining Strategic Narratives".
53 Rosyidin, "The Dao of"; Schell and Delury, *Wealth and Power*.
54 Karmazin și Hynek, "Russian, US and Chinese".
55 Corr, *Great Powers*, 42.
56 Weiwei, *The China Wave*, 81.
57 Galeotti, "The Gerasimov Doctrine".

Bibliography

Balzacq, Thierry, Peter Dombrowski and Simon Reich (ed.), *Comparative Grand Strategy: A Framework and Cases*. London: Oxford University Press, 2019.

Bennett, Lance W. and Steven Livingston. "The disinformation order: Disruptive communication and the decline of democratic institutions." *European Journal of Communication*, no. 33 (2018): 122–139.

Cai, Kevin G. *China's Foreign Policy Since 1949: Continuity and Change*. New York: Routledge, 2022.

Campbell, David. *Writing Security: United States, foreign policy and the politics of identity*. University of Minnesota Press, 1998.

Campbell, David. *National Deconstruction: Violence, Identity, and Justice in Bosnia*. University of Minnesota Press, 1998.

Corbu, Nicoleta, Alina, Bârgăoanu, Raluca Buturoiu and Oana Ștefăniță. "Does fake news lead to more engaging effects on social media? Evidence from Romania." *Communications*, no. 45 (2020): 694–717.

Corr, Anders. Great Powers, *Grand Strategies: The New Game in the South China Sea*. Annapolis: Naval Institute Press, 2018.

Denoon, David B. H. (ed). *China's Grand Strategy: A Roadmap to Global Power?*. New York: New York University Press, 2021.

Doshi, Rush. *The Long Game: China's Grand Strategy to Displace American Order*. New York: Oxford University Press, 2021.

Dunn, Kevin C., and Iver B.Neumann, *Undertaking Discourse Analysis for Social Research*, Ann Arbor: University of Michigan, 2016.

Galeotti, Mark. "The Gerasimov Doctrine", *Berlin Policy Journal*, 2020, available at: https://berlinpolicy journal.com/the-gerasimov-doctrine/.

Galeotti, Mark. *Russian Political War: Moving Beyond the Hybrid*. New York: Routledge, 2019.

George, Alexander L. and Andrew Bennett. *Case Studies and Theory Development in the Social Sciences*. Cambridge: MIT Press, 2005.

Gray, Colin S. *The Future of Strategy*. Cambridge: Polity Press, 2015.

Hansen, Lene. *Security as practice: Discourse analysis and the Bosnian War*. New York: Routledge, 2006.

Hinck, Robert S., Skye C.Cooley and Randolph Kluver. *Global Media and Strategic Narratives of Contested Democracy: Chinese, Russian, and Arabic Media Narratives of the US Presidential Elections*. New York: Routledge, 2020.

Karmazin, Aleš and Nik Hynek. "Russian, US and Chinese Revisionism: Bridging Domestic and Great Power Politics." *Europe-Asia Studies*, no. 72 (2020): 955–975.

Krekó, Péter. "The Drivers of Disinformation in Central and Eastern Europe and Their Utilization during the Pandemic", available at: https://www.globsec.org/publications/policy-brief-drivers-of-disinforma tion-in-central-and-eastern-europe-and-their-utilization-during-the-pandemic/.

Lams, Lutgard. "Examining Strategic Narratives in Chinese Official Discourse under Xi Jinping." *Journal of Chinese Political Science*, no. 23 (2018): 387–411.

March, James G. and Johan P. Olsen. "The institutional dynamics of international political orders." *International Organization*, no. 52 (1998): 943–969.

Miskimmon, Alister, Ben O' Loughlin and Laura Roselle. *Strategic Narratives: Communication Power and the New World Order*. New York: Routledge, 2013.

Miskimmon, Alister, Ben O'Loughlin and Jinghan Zeng. *One Belt, One Road, One Story? Towards an EU-China Strategic Narrative*. Cham: Palgrave Macmillan, 2021.

Onea, Teodor A. *The Grand Strategies of Great Powers*. New York: Routledge, 2021.

Riggirozzi, Pía. "Region, regionness and regionalism in Latin America: Towards a new synthesis." *New Political Economy*, no. 17 (2012): 421–443.

Royisidin, Mohamad. "The Dao of foreign policy: Understanding China's dual strategy in the South China Sea." *Contemporary Security Policy*, no. 40 (2019): 214–238.

Rychnovská, Dagmar and Martin Kohút. "The Battle for Truth: Mapping the Network of Information War Experts in the Czech Republic." *New Perspectives*, No. 3 (2018): 57–87.

Schell, Orville and John Delury. *Wealth and Power: China's Long March to the Twenty-First Century*. Abacus, 2013.

Shambaugh, David. *China's Leaders: From Mao to Now*. Cambridge: Polity Press, 2021.

Szostek, Joanna. "Defence and Promotion of Desired State Identity in Russia's Strategic Narrative." *Geopolitics*, no. 22 (2017): 571–593.

Tocci, Nathalie (ed.). *Who is a Normative Foreign Policy Actor? The European Union and its Global Partners*. Brussels: Centre for European Policy Studies, 2008.

Weiwei, Zhang. *The China Wave: Rise of a civilizational state*. Hackensack: World Century Publishing Corporation, 2012.

"10 Recommendations by the Taskforce on Disinformation and the War in Ukraine", Available at: https://edmo.eu/2022/06/29/10-recommendations-by-the-taskforce-on-disinformation-and-the-war-in-ukra ine/.

27

THE PERILS OF DISINFORMATION IN LATIN AMERICA

Florina Cristiana (Cris) Matei[1]

The end of the Cold War coupled with the advent of internet and social media have brought about a surge of disinformation throughout the Latin America. On the one hand, populist politicians have used the internet and social media to get elected and/or to govern. On the other hand, non-democratic regimes in the region have capitalized on these tools to stay in power. In parallel, non-democratic regimes both within and outside of the region have resorted to disinformation to influence countries throughout Latin America, and ultimately impact their security and democratic governance. Despite attempts to combat disinformation—both at the country and region levels—the region's security and democracy remain vulnerable to disinformation threats. This chapter explores how disinformation has challenged Latin America's democracy and security since the end of the Cold War. It starts with a theoretical background on disinformation, followed by an assessment of Latin America's experience with disinformation.

DISINFORMATION TRAITS, TOOLS, AND ENABLERS

Misinformation, disinformation, and propaganda are distinct phenomena, yet they are often used interchangeably, as all three activities terms involve untrue or deceptive messages proliferated under "the guise of informative content, whether in the form of elite communication, online messages, advertising, or published articles", as Persily and Tucker inform[2].

Disinformation is understood as false, incomplete or deceiving information that is created and disseminated with an intent to confuse and/or mislead targeted individuals or groups—ranging from social categories to entire populations/countries[3]. Disinformation is created by a wide array of public, private, and social actors (domestic and/or international) and covers a wide spectrum of issues surrounding the public life—including political, historical, health, social, and environmental matters—for such objectives as economic, social, and/or political gain. Disinformation causes and spreads apathy, fear, cynicism, distrust, and paranoia; creates uncertainty and chaos; and prompts individuals and groups to carry out dangerous actions[4]. In this connection, Tumber and Waisbord argue that disinformation "is often intentionally affective and thrives on the generation of feelings of superiority, fear, and anger"[5]. Disinformation in other words has a specific intent.

DOI: 10.4324/9781003190363-32

Disinformation uses an array of interconnected devices, including:

- false stories (or inaccurate news), which are defined as disinformation camouflaged as a media coverage—which actually imitates existing professional news sources (yet using the incorrect captioning of videos and pictures)—in an attempt to garner people's trust;
- fake news, which involves false information that mimics the content of the media, manifested as viral posts from fabricated accounts designed to emulate accurate news, spread for political or economic gain[6] but which lately has been used by politicians as a weapon to attack and delegitimize the media (to include raising skepticism toward real news and real journalists) and weaken trust in institutions in a democracy[7];
- fake reviews, which are used by online commercial platforms to sway the purchasing of specific products and services[8];
- propaganda, which is material developed, either overtly or covertly, by governments, political institutions, or economic organizations to persuade people to adopt or reject a specific political, ideological, or partisan view or position—either by either promoting or challenging such stance[9];
- rumours, which are defined as claims whose power develops from the very process of dissemination[10];
- conspiracy theories, which are beliefs that a clandestine group of powerful people are in control of a society (or societies), and which usually challenge verified scientific discoveries (like, for example, the flat-Earth conspiracy), which can ultimately harm individual citizens, communities, and nations[11].

In general, disinformation aims to be credible and thus it is usually built on some degree of reality—attempting to mirror authentic news channels via such tactics as the fabrication[12] of news and/or manipulation[13] of photos or videos[14]. Several techniques abet the spreading of disinformation, including the following: phishing, which involves theft and misuse of personal and/or confidential information via data breaches;[15] filter bubbles, which are invisible algorithms that can simultaneously magnify and isolate ideas to spread misinformation; echo chambers that are environments that eliminate viewpoints and occur against a background of emotional rather than rational relationship with information to help maximize ideological polarization[16]. Disinformation effects include demonization, radicalization, caricaturisation, sensationalism, polarization of both society and political elites in favour of extreme positions, viralisation, and perception hacking[17].

Numerous actors—domestic and/or international—enable the spread of misinformation and disinformation: foreign states, media (old and new) and corporations; political groups (parties and leaders, including populists, and governments); and ordinary citizens. *Foreign states* use disinformation toward other countries for various purposes, including the following: to promote their interests; to discredit adversary or hostile nations; and, to generate—or accentuate—social polarization and even cause chaos. Non-democratic regimes in particular use sharp power campaigns to target democracies around the world aiming "to attract, distract, and manipulate audiences in democratic countries through their communications outlets, cultural centres, and global learning institutes", as experts from the Global Americans explain[18]. Sharp power disinformation campaigns are usually a joint venture of non-democracies and their media organizations[19].

Some *Media* elements—domestic and international, old and new—generate, use, and spread disinformation often in connection with other political[20] or economic[21] actors. First, journalists help spread fake news even when covering actual news[22]. Second, the online environment provides a fertile ground for nefarious disinformation actors and mechanisms[23]. To be sure, the online environment, and especially social media, does permit microtargeting, i.e. the of monitoring users' identities, interests, hobbies, preferences, and biases and it thus challenges users' ability to identify and discard disinformation consistently to their beliefs[24]. Disinformation actors and enablers capitalize on social media to go viral, and benefit from the social media algorithm-based ability to polarize the public[25].

Political Groups can also originate and disseminate dis- and misinformation. For example, elected and/or appointed politicians and leaders of political parties use social media platforms to distribute their messages and energize their base, and so does the opposition. Nevertheless, populists[26], have a great appetite for and propensity toward disinformation. For them, disinformation is a legitimate strategy and tool in their constant fight against their opponents.[27] Populism has a love-hate relationship with the media. On the one hand, populists regularly attack the media (especially traditional media, which they equate to an enemy) when it challenges their legitimacy or opposes their views. However, populists also capitalize on the media as a sine qua non in reinforcing the division/polarization between the people (which they represent) and the elites, and in communicating with their supporters. Populists maintain close relations with both public and private media organizations that align with their views, and use social media (for Trump, it was Twitter, and for Modi in India it was often WhatsApp)–to bypass critical media. Populists view social media as an ideal vehicle of messaging and communicating to their audiences, because of its independence of traditional elite media that is usually averse to populism. Tumber and Waisbord capture these populist traits:

> These platforms cover and discus hot-button issues, such as immigration, white supremacy, nativism, and climate change denialism, which are at the heart of populist identities and are closely identified with right- wing and far- right groups. Conspiracy theories, absolute lies, fake news, "alternative' facts, and similar informational aberrations are a common presence on these outlets... Populism uncritically accepts the "truths' of allies while it disparages the enemies as liars. "Our' truths are the antipode of "their' lies. Truth encompasses "popular' beliefs, no matter how or whether they are proven... Loyalty to the leader and their policies is often sufficient to determine the truthfulness of any claim[28].

Thus, media is like oxygen to populists.

Governments also generate and spread disinformation. Non-democracies use disinformation to stay in power and control their citizens. The advent of the internet and social media gave rise to what scholars call "digital authoritarianism", understood as "the use of digital technologies to enhance or enable authoritarian governance...[via] such practices as pervasive Internet surveillance and the exercise of tight control over online information flows within a country's borders'[29]. Disinformation coupled with containing the access to, and the flow of, information at the domestic level enable intelligence agencies and government leaders to incessantly monitor their people in order to reduce or eliminate any dissent against the current government[30].

Ordinary citizens who are recipients of disinformation, can, in turn—deliberately or unknowingly—enable propagation of such nefarious information. During crises and uncertainties people tend to be sharing rumours and doubtful information through social media and mobile messaging applications. Likewise, politically engaged people or groups formed around ideologies or identities tend to be more willing to share information that confirms their beliefs or identities than those who are not politically motivated or part of a group; while, at the same time, intentionally ignore content that does not support their views. As Valenzuela et al. argue, "Misperceptions will moderate the relationship between political participation and the spread of misinformation, such that the relationship will be stronger for misinformed users and weaker for informed users"[31].

In sum, the current environment of mis- and disinformation is known as the post-truth engenders social polarization and political sectarism[32]. Regrettably, the COVID-19 pandemic (2020-present) is the perfect example of these intersecting and interacting disinformation devices, tools, and actors at work[33]. Disinformation in all its forms and shapes—including political uses of sensitive data where political opinions discredit science and common sense—has propagated so quickly and widely that it has led to the creation of new COVID-19-specific disinformation words—like infodemic and disinfodemic[34].

LATIN AMERICA'S EXPERIENCE WITH DISINFORMATION

Disinformation has also affected Latin America. Multiple actors—foreign and domestic, state and non-state—produce disinformation in Latin America using multiple ways and means. To begin with, *foreign nations* have been using disinformation campaigns toward Latin American countries to advance their interests in the region (including discrediting the United States and countering the U.S. influence in the region) and generate—or accentuate—social polarization in democratic countries. In this connection, several non-democratic regimes—most notably China and Russia, and to a lesser extent, Cuba, and Venezuela—have used sharp power campaigns to target Latin American and Caribbean democracies; especially, Argentina, Brazil, Chile, Colombia, and Peru, in an attempt to align the leaders of these countries—and leaders in the region general—with their rhetoric and worldview. Chinese and Russian media companies (e.g. Sputnik Mundo, Russia Today (RT), and Xinhua), along with social media companies (e.g. Twitter and Facebook)—create propaganda and disinformation messages that abet these countries' sharp power efforts in the region. These countries' tactics include manipulating and/or omitting data and information; mixing of truths with falsehoods, and conducting propaganda campaigns that are supportive of China and Russia's interests in the region (e.g. regular flow of news that promote their vaccine production and distribution, and denigrate Western countries' vaccination efforts in the region) or propaganda with a political bias (most notably in Argentina, Chile, and Peru), while also attempting to destabilize democracies. And to counter each other's influence in the region. According to experts at Global Americans,

> Chinese and Russian misinformation, disinformation, and propaganda are disproportionately concentrated on thematic targets that lie at the intersection of democratic fault lines, inflaming local political rifts, promoting like-minded and often non-democratic local forces, and portraying China and Russia as benevolent partners and alternatives to the United States throughout the region[35].

Venezuela and Cuba, on the other hand, use both China and Russia-generated disinformation and their own disinformation and misinformation campaign to undermine democracy throughout the region. For example, Venezuela has spread Russia's disinformation about the ineffectiveness of Western vaccines; while praising China's effective handling of the pandemic. Venezuelan politicians have conducted aggressive misinformation and disinformation campaigns against the Colombian government, as a result of Colombia's acceptance of large numbers of Venezuelan refugees and Maduro's political opponents[36]. Cuba's disinformation efforts mostly target U.S. sanctions and seek promoting Cuban innovations in medicine and education[37].

The Latin American *media* (both traditional and new)—which struggles with becoming a profession—has been both an originator and disseminator of disinformation[38]. Indeed, the media has been equally developing and spreading false information and rumours, conspiracy theories, and fake statements (especially by political figures) aimed at denigrating, polarizing, and creating panic and fear throughout the region, or in a particular country[39]. False claims related to coronavirus prevention (for instance, false statements that lemon juice cures COVID-19) emerged in Brazil on social media networks that recommended drinking a daily dose of water with lemon and vitamin C as vaccine cure, and immediately spread to Argentina, then Venezuela, Colombia, Mexico etc[40]. Similar false statements emerged in Mexico, from a pro-Maduro news website (which promoted a combination of ginger, black pepper and lemon as cure for COVID-19)[41]. Conspiracy theories that promoted anti-vax movements[42] and denigrated health workers and institutions[43] also boomed during the pandemic. Inaccurate reporting of the number of tests and cases of COVID completes the COVID-19 disinformation reality[44]. Worrisomely, COVID-19-related disinformation has prompted a rise in violence and threats to violence against healthcare professionals across Latin America (especially in Argentina, Colombia and Mexico), who have been attacked, intimidated (including receiving death threats), and even evicted from their residence[45].

Social media, coupled with the Latin American populations' reliance on social media for news, has been extremely successful in polarizing societies in the region[46]. The spread of social-media-enabled disinformation in Latin America skyrockets during elections, thanks to the lack of content control. The elections in Mexico (2018), Brazil (2018), El Salvador (2019), and Argentina (2019) clearly illustrate this trend. The spread of disinformation through private chat groups on the WhatsApp platform during elections involved a mix of human accounts, bots, and cyborg accounts[47]. According to *New York Times,* "as the system suggests more provocative videos to keep users watching, it can direct them toward extreme content they might otherwise never find"[48]. In Argentina, during the 2019 elections, WhatsApp was equally a major campaign platform and the cause of disinformation and mis-information, targeting both Macri and the other candidates[49]. Social media has thus been cardinal in campaigning in general, and a useful tool for those seeking to undermine the legitimacy of elections in particular[50].

Populists and Populist Governments: From Tele to Techno Presidents

One group of people in particular—populists—has taken advantage of these media lacunae. Pre-social-media populist presidents[51] used the traditional media to establish a direct con-nection to the people which would involve them in the government decision-making and problem-solving processes. The heavy reliance on the media by these leaders got them "the tele-presidents' nickname[52]. Fast forward to the age of internet and social media, the tele-presidents have not only adjusted to but have become quite comfortable with the digital realm which brought about an expansion of their communication and outreach strategies and practices; thus morphing into technopopulists or populists 2.0[53].

In addition, social media's power to reinforce the us-versus-them mentality emboldens Latin American populists to use such platforms as Facebook or Twitter to gain visibility, buttress presidential candidates, comment on a wide spectrum of issues swiftly, provide vitriolic attacks at political opponents, and communicate their views in a one-way-format, with no possibility for debate[54]. In several presidential elections in Latin America in 2018—most notably in El Salvador, Brazil, Costa Rica, and Mexico—the populism-social media nexus fuelled a polarization of the electorate regarding topics such as sexual orientation and social values, and paved the way to election for extremist political/religious politicians in these countries[55]. Bukele's campaign—who "railed against the corruption of the traditional parties and was known for communicating directly with constituents through social media"—spread misinformation widely on social media to increase his chances to getting elected[56]. Bolsonaro, who followed the playbook of the right-wing populist leaders who had recently been elected in other countries (including Moodi in India and Erdogan in Turkey), successfully used social media networks and tools to mobilize a large number of Brazilians against an "enemy" who had to be defeated (mainly the "left'), thus normalizing prejudices[57]. In this connection, Richards and Medeiros explain:

> As a campaign strategy, Bolsonaro bet on the "anti-system" rhetoric, attacking the supposed political "establishment" and invoking different types of... [disinformation]... before and during the campaign. Exploring the fears and prejudices of the average voters, a pervasive social media operation involving misleading, manipulated, and fabricated content was set in motion, which leveraged to its advantage a context of rejection of traditional parties and discredit towards democratic institutions[58].

Despite racist, homophobic, and misogynist rhetoric, Bolsonaro won the 2018 elections easily, thanks to his digital campaign predicated on disinformation and hate speech. Similarly, in Costa Rica, the right-wing candidate Fabricio Alvarado cantered his media campaign against gay marriage, which polarized the election and eliminated other relevant issues from the debate[59]. Likewise, in Mexico, President López Obrador's disinformation against critics of his plans for Mexico's future development has deepened the political polarization in the country[60].

Latin American populists have spread disinformation, especially in its conspiracy theory form, not just to delegitimize opponents but also to distract from their policy fiascos, fend criticism, and even create cohesion. Hugo Chávez in Venezuela and Rafael Correa in Ecuador, for example, used TV and radio channels, as well as newspapers for the mandatory propaganda of their images and messages; they also used Twitter and Facebook to address their citizens directly[61]. In Mexico, Peñabots—bots used by the Mexican government of Enrique Peña Nieto and the PRI—disseminated fake information as a way to divert attention from government failures and taunt criticism[62]. Obrador uses the media to constantly blame the opposition for any major problems in Brazil; the echo chamber effects keep his base energized[63].

In addition, governments in the region have been attacking critics of their handling of the pandemic. Brazil's Bolsonaro, Venezuela's Maduro, Mexico's Obrador, and Nicaragua's Ortega have distinguished themselves as the few presidents in the world who constantly denied the threat posed by the COVID-19 virus. To avoid being blamed for economic crises and harsh lockdowns these presidents have continuously been discounting inconvenient evidence and underrating the pandemic[64]. As Richards and Medeiros note:

> They systematically downplayed the warnings regarding the spread and lethality of the virus, because these recommendations were issued by the populists' traditional

"enemies"—the intellectual/scientific elites and international organizations. Instead, they invoked their own "experts" and pseudo-science, making claims such as that their populations were genetically protected, or that prayer or religious devotionals could prevent an infection[65].

The Bolsonaro Presidency created an office nicknamed the "Office of Hatred", which coordinates the sharing of disinformation, including offensive messages against Bolsonaro's political opponents, including members of his own government[66]. Ultimately, as Tumber and Waisbord note, 21st century populism has enabled both "the politization of the media' and "the mediatization of politics'[67].

In sum, the relationship between populism and disinformation in Latin America is a threat to democratic stability and progress. Indeed, polarization and divide, which populists capitalize on in the region, permit banning certain groups from exercising their freedoms and rights or taking away rights from certain groups altogether (already seen in Brazil, El Salvador, and Mexico)[68]. Even when/if populist movements emerge on a background of real grievances, these grievances are often magnified or distorted in ways that make it easier for politicians to manipulate the masses'—which "is in opposition to the notion of a democracy founded on informed debate and rational voters", as Tumber and Waisbord argue[69]. In this context, disinformation disfigures democracy.

Latin America's Attempts to Combat Disinformation: Ebbs and Flows

Several efforts have been carried out in the region by government institutions and civil society organizations—domestic and international alike—to combat disinformation in Latin America. Some of these endeavours aimed at educating or working with the public to detect and fight disinformation[70]. Other efforts involved bilateral or multilateral (regional) interagency cooperation and collaboration in combating and investigating disinformation[71]. Several governments have enacted or proposed legislations—which are dubbed fake news laws—and policies that would criminalize disinformation[72]. Additional anti-disinformation efforts included endeavours by social media platforms to eliminate false/fabricated information[73].

Furthermore, countries in the region—singly or jointly—created fact checking institutions to combat the spread of disinformation[74]. These Cerberuses of truth have been effective in debunking some conspiracy theories and fake news on social media—including identifying accidental mistakes in spreading rumours or false news during the COVID-19 pandemic[75].

Notwithstanding these endeavours, combating disinformation has only achieved limited success in the region. On the one hand, the enacting of fake news laws has resulted in politicization of security institutions under the pretext of fighting disinformation and misinformation. These agencies have been used by governments du jour to persecute journalists (and even citizens) who criticize the government. For instance, Brazil's LGPD and Argentina's NODIO restrict freedom of speech and individual privacy[76].

On the other hand, the effectiveness of fact checkers' work has also been challenged by insufficient resources and time. Fact checkers' limited personnel and funding have obstructed their capacity to research everything that appears to be disinformation on conspiracy theory in the media and social media. For example, limited resources force fact checking institutions to only focus on national government and candidates, which prevents these agencies from carrying out fact checking of local candidates and policies. In addition, scepticism toward the objectivity of fact checking companies obstructs progress—especially if people have motives to suspect political affinity[77]. Moreover, fact-checking institutions become targets of disinformation themselves, especially by populists and populist governments who want to prevent them from debunking fabricated information generated from within these politicians' circles.

Time also works against fact checkers. Indeed, fact checkers post their evaluation online or share it via traditional media channels; but because it takes time to analyse a social media post, when the assessment is completed, the assessment may not make a significant impact (since a high number of users have already seen the post and will not go on different media channels to check). In addition, fact-checking is more effective on non-encrypted social-media platforms (Facebook and Twitter), and less effective on encrypted social media platforms (WhatsApp), so visible only to the users[78].

CONCLUSION

The preceding discussion reveals the celerity of the spread of disinformation in Latin America. At the domestic level, populists use disinformation to get elected and then to govern; while non-democratic leaders use disinformation along with other repressive and oppressive tools to remain in power. At the international level, non-democracies from within and outside of the region use disinformation to garner support for their policies from countries in the region, while at the same time attempting to dilute democratic progress and stability in the region (by generating social damage and discord, and tainting and delegitimizing public policies and electoral processes)[79].

Latin American countries are starting to experience a Truth Decay—as people tend to take opinions for facts and lost confidence in traditional sources of truth. Truth Decay causes "distrust towards the media and damages the democratic quality of society by encouraging civic apathy, destabilization, chaos, a reduction of pluralism, and a strengthening of polarized communities in which fake news and conspiracy theories are freely propagated", as suggested by Lopez-Garcia et al.[80]. Ultimately, then, disinformation is both a cause and symptom of mistrust in Latin America's democratic institutions and governance.

Notes

1 Disclaimer: The views presented are those of the author and do not necessarily represent the views of the Department of Defense, the Department of the Navy or the Naval Postgraduate School. The author would like to thank and give credit to her colleague, Mr. Christopher Ketponglard, for his research-related assistance.

2 Persily, N., and Tucker, J. A. eds., *Social Media and Democracy Social Media and Democracy The State of the Field, Prospects for Reform,* Cambridge University Press, 2020, pp. 10–11. The literature uses several other terms for inaccurate information. First, is misinformation, which involves unintentional spread of untruthful, partial, or incomplete information and lacks a clearly defined goal. Misinformed citizens are generally unaware that they consume and share false information; yet, although inadvertently disseminated, misinformation can be as misleading as disinformation. Media professionals, for instance, can and do get a story wrong, even if they do not seek to produce a particular effect. Next, is xisinformation, which is that type of false information for which intent cannot be confirmed or attributed (and hence making it difficult to call it disinformation). There is also malinformation, which is intentionally harmful dissemination of sensitive accurate information. Tumber, H., and Waisbord. S., *The Routledge Companion to Media Disinformation and Populism (Routledge Media and Cultural Studies Companions),* Routledge, 2021, pp. 58, 93–95, 135–138; Eveson, R., "The pandemic of disinformation in Latin America", Latin America Bureau, 28 May 2020, https://lab.org.uk/the-pandemic-of-disinformation-in-latin-america/; and Jack, C., "Lexicon of Lies: Terms for Problematic Information", *Data & Society Research Institute,* datasociety.net, pp. 1–20.

3 Disinformation is intentional and can either be verified to be fake, or is unverifiable. The source of disinformation is also dishonest, partial, or deceptive (like, for example, a state actor passes itself off as a regular citizen). Cohen, R.S. et al., "Combating Foreign Disinformation on Social Media. Study Overview and Conclusions", RAND Corporation Report, 2021, pp. 1–119; "Measuring the Impact of Misinformation, Disinformation, and Propaganda in Latin America", *Global Americans*, 2021, pp. 1–306, https://theglobalamericans.org/wp-content/uploads/2021/10/2021.10.28-Global-Americans-Disinformation-Report.pdf.

4 Eveson, R., "The pandemic of disinformation in Latin America", Latin America Bureau, 28 May 2020, https://lab.org.uk/the-pandemic-of-disinformation-in-latin-america/; Lopez-Garcia, G. et al., *Politics of Disinformation,* Wiley, 2021, pp-23–25; and, Valenzuela, S. et al., "The Paradox of Participation Versus Misinformation: Social Media, Political Engagement, and the Spread of Misinformation", *Digital Journalism*, 7:6, 2019, 802–823, DOI: 10.1080/21670811.2019.1623701.

5 Tumber, H., and Waisbord. S., *The Routledge Companion to Media Disinformation and Populism (Routledge Media and Cultural Studies Companions),* Routledge, 2021, pp. 135–138.

6 Gutiérrez-Coba, L. M., Coba-Gutiérrez, P., & Gómez-Díaz, J. A., "The intention behind the fake news about Covid-19: comparative analysis of six Ibero-American countries", *Revista Latina de Comunicación Social*, 2020. 78, 237–264. https://www.doi.org/10.4185/RLCS-2020-1476. Disinformation is fake news insofar as it is deliberately deceiving. As Lopez-Garcia et al. reveal, while "disinformation refers to the strategic dynamics concerning the communicative environment…the catch-all formula "fake news" identifies isolated episodes of fabrications and deception". Lopez-Garcia, G. et al., *Politics of Disinformation, Wiley*, 2021, p. 35. Also, see pp. 23–25. Fake news takes advantage of users" inexperience. Three categories of fake news exist: exposed fabrications, which include "the yellow press and its unverified articles, which, through clickbait…and sensationalist articles, aim to increase its traffic and consequently generate profit"; hoaxes, which are "deliberate fabrication or falsification in the mainstream or social media" and include "rumors, fake graphics or tables, false attribution of authorship, dramatic images, etc." aimed at causing uncertainty by generating messages that bring despair about human behavior, like for instance, the alleged the SARS-COV-2 creation by North American and Chinese doctors in the lab, via genetic manipulation; news satire (or parody), which are "humorous news websites based on irony, often in a mainstream format, such as 'The onion" website". News satire should, however, not be conflated with an imposter website, which deliberately aims to deceive or confuse as it emulates traditional media source. Not all people share fake news to deceive others. Some individuals disseminate fake news for non-deceiving purposes (some people, for example, spread fake news to humor or warn other people, without a deliberate intent to spread fabricated information, or even with the hope that other individuals will confirm or debunk these types of information). The bubble tricks users into equating the likes of the group inside the bubble to the likes of the majority of people. Santos-D'Amorim, K.M., and Fernandes de Oliveira, M.K., "Misinformation, disinformation, and malinformation: Clarifying the definitions and examples in disinfodemic times", *Encontros Bibli: revista eletrônica de biblioteconomia e ciência da informação*, vol. 26, e76900, 2021, January-April, DOI: https://doi.org/10.5007/1518-2924.2021.e76900, pp.1, 2, 3, 4, 5, 6, 7, 8, 9, 10, 11, 12, 13, 14, 15, 16, 17, 18, 19, 20, 21, 22, 23; Sierra Caballero, S. and Sola-Morales, S. "Media Coups and Disinformation in the Digital Era. Irregular War in Latin America", *Comun. Soc*, vol.17, 2020, https://doi.org/10.32870/cys.v2020.7604; Tumber, H., and Waisbord. S., *The Routledge Companion to Media Disinformation and Populism (Routledge Media and Cultural Studies Companions),* Routledge, 2021, pp. 93–98, 217, 234, 626–643; "Measuring the Impact of Misinformation, Disinformation, and Propaganda in Latin America", *Global Americans*, 2021, pp. 1–306, https://theglobalamericans.org/wp-content/uploads/2021/10/2021.10.28-Global-Americans-Disinformation-Report.pdf; and Cohen, R.S. et al., "Combating Foreign Disinformation on Social Media. Study Overview and Conclusions", RAND Corporation Report, 2021, pp. 1–119.

7 Fake news spread on social media via bots (automated computer codes designed to mimic human users) and cyborgs (hybrid human-automated online accounts)—which are built to place hashtags on "Trending" lists, or to silence or bully users, or by posting comments and frequent replies, making their behavior appear human. Trolls—which post viral and bullying messages—are cyborgs, as their accounts are managed by a real users with false identities. Altered audio-visual content creates deepfakes, which are synthetic videos that seem real. pp. "Measuring the Impact of Misinformation, Disinformation, and Propaganda in Latin America", *Global Americans*, 2021, pp. 1–306, https://theglobalamericans.org/wp-content/uploads/2021/10/2021.10.28-Global-Americans-Disinformation-Report.pdf; Cohen, R.S. et al., "Combating Foreign Disinformation on Social Media. Study Overview and Conclusions", RAND Corporation Report, 2021, pp. 1–119; and Tumber, H., and Waisbord. S., *The*

Routledge Companion to Media Disinformation and Populism (Routledge Media and Cultural Studies Companions), Routledge, 2021, pp. 93–98, 626–643.

8 Fake reviews also include fraudulent articles (even if peer-reviewed).

9 In Mexico, for example, Twitter is divided into pro- and anti- Andrés Manuel López Obrador propaganda. "Measuring the Impact of Misinformation, Disinformation, and Propaganda in Latin America", *Global Americans*, 2021, pp. 1–306, https://theglobalamericans.org/wp-content/uploads/2021/10/2021.10.28-Global-Americans-Disinformation-Report.pdf.

10 Persily, N., and Tucker, J. A. eds., *Social Media and Democracy Social Media and Democracy The State of the Field, Prospects for Reform,* Cambridge University Press, 2020, pp. 10–11.

11 Like, for example, the anti-vaccine movement. Santos-D'Amorim, K.M., and Fernandes de Oliveira, M.K., "Misinformation, disinformation, and malinformation: Clarifying the definitions and examples in disinfodemic times", *Encontros Bibli: revista eletrônica de biblioteconomia e ciência da informação*, vol. 26, e76900, 2021, January-April, DOI: https://doi.org/10.5007/1518-2924.2021.e76900, pp.1, 2, 3, 4, 5, 6, 7, 8, 9, 10, 11, 12, 13, 14, 15, 16, 17, 18, 19, 20, 21, 22, 23; Sierra Caballero, S. and Sola-Morales, S. "Media Coups and Disinformation in the Digital Era. Irregular War in Latin America", *Comun. Soc,* vol.17, 2020, https://doi.org/10.32870/cys.v2020.7604; Persily, N., and Tucker, J. A. eds., *Social Media and Democracy Social Media and Democracy The State of the Field, Prospects for Reform,* Cambridge University Press, 2020, pp. 10–11; and Merlan, A., *Republic of Lies. American Conspiracy Theorists and Their Surprising Rise to Power,* Macmillan, 2019.

12 That is "attempting to mimic legitimate news sites to give the audience the impression that the false information is true", according to Global American experts. "Measuring the Impact of Misinformation, Disinformation, and Propaganda in Latin America", *Global Americans*, 2021, pp. 1–306, https://theglobalamericans.org/wp-content/uploads/2021/10/2021.10.28-Global-Americans-Disinformation-Report.pdf.

13 Which takes place when an account disseminates an old image or video from as if it were current; or shares an image or video as if it were from one venue when in fact the shared image or video were taken or recorded in a different venue. "Measuring the Impact of Misinformation, Disinformation, and Propaganda in Latin America", *Global Americans*, 2021, pp. 1–306, https://theglobalamericans.org/wp-content/uploads/2021/10/2021.10.28-Global-Americans-Disinformation-Report.pdf.

14 Cohen, R.S. et al., "Combating Foreign Disinformation on Social Media. Study Overview and Conclusions", RAND Corporation Report, 2021, pp. 1–119. Some people disseminate false information because they truly believe that it is truthful information, which they hope will educate their peers. Valenzuela, S. et al., "The Paradox of Participation Versus Misinformation: Social Media, Political Engagement, and the Spread of Misinformation", *Digital Journalism*, 7:6, 2019, 802–823, DOI: 10.1080/21670811.2019.1623701.

15 Examples include identity theft, attempts to discredit an individual or an institution, profile cloning, denying e-mail access, and economic loss. Santos-D'Amorim, K.M., and Fernandes de Oliveira, M.K., "Misinformation, disinformation, and malinformation: Clarifying the definitions and examples in disinfodemic times", *Encontros Bibli: revista eletrônica de biblioteconomia e ciência da informação*, vol. 26, e76900, 2021, January-April, DOI: https://doi.org/10.5007/1518-2924.2021.e76900, pp.1, 2, 3, 4, 5, 6, 7, 8, 9, 10, 11, 12, 13, 14, 15, 16, 17, 18, 19, 20, 21, 22, 23.

16 Santos-D'Amorim, K.M., and Fernandes de Oliveira, M.K., "Misinformation, disinformation, and malinformation: Clarifying the definitions and examples in disinfodemic times", *Encontros Bibli: revista eletrônica de biblioteconomia e ciência da informação*, vol. 26, e76900, 2021, January-April, DOI: https://doi.org/10.5007/1518-2924.2021.e76900, pp.1, 2, 3, 4, 5, 6, 7, 8, 9, 10, 11, 12, 13, 14, 15, 16, 17, 18, 19, 20, 21, 22, 23.

17 Sierra and Sola-Morales define demonization as a strategy that aims at isolating the opponents by "preventing them from defending themselves, declaring them morally inferior and denying their basic rights". They suggest that radicalization is a strategy whereby "elements of the opponents" speech are taken out of context in order to exaggerate and criminalize their points of view; that caricaturisation constitutes "grotesque representations distorting the image of presidents... and other members of the government, and their qualification as "dictators", "dangerous", "madmen" or "criminals" is justified. They further identify sensationalism as a strategy that turns "anecdotes into, for example, hashtags that trivialize information", polarization in favor of extreme positions, as a strategy that uses "a warlike and conflictive language that accentuates fear and panic among the population and the international public opinion", and viralisation as a strategy that seeks "to increase notoriety or encourage massive debate on certain topics or positions". Sierra Caballero, S. and Sola-Morales, S. "Media Coups and

Disinformation in the Digital Era. Irregular War in Latin America", *Comun. Soc*, vol.17, 2020, https://doi.org/10.32870/cys.v2020.7604. Perception hacking is a strategy whereby "the prospect of an influence operation helps cast doubt on the authenticity of public debate", according to Dwoskin. Dwoskin, E., "Russia is still the biggest player in disinformation, Facebook says", *The Washington Post*, 21 May 2021.

18 Sharp power promotes controversies, which in turn leads to political divisions and swayed audiences. On the other hand, sharp power disinformation campaigns also seek to counterbalance the U.S. influence and policies in a particular region. "Measuring the Impact of Misinformation, Disinformation, and Propaganda in Latin America", *Global Americans*, 2021, pp. 1–306, https://theglobalamericans.org/wp-content/uploads/2021/10/2021.10.28-Global-Americans-Disinformation-Report.pdf.

19 Foreign adversaries (and/or partner nations) also use domestic actors in their disinformation endeavors. Lopez-Garcia, G. et al., *Politics of Disinformation*. Wiley, 2021, pp. 41–42. "Measuring the Impact of Misinformation, Disinformation, and Propaganda in Latin America", *Global Americans*, 2021, pp. 1–306, https://theglobalamericans.org/wp-content/uploads/2021/10/2021.10.28-Global-Americans-Disinformation-Report.pdf; and Cohen, R.S. et al., "Combating Foreign Disinformation on Social Media. Study Overview and Conclusions", RAND Corporation Report, 2021, pp. 1–119.

20 Which will be discussed later in this chapter.

21 Businesses use disinformation to humiliate their competitions or provide their support to political candidates and politicians. Monteiro da Silva Junior, E. and Lima Dutra, M., "A Roadmap for Composing Automatic Literature Reviews: A Text Mining Approach", in Bisset Álvarez, E., ed., *Data and Information in Online Environments,* Second EAI International Conference, DIONE 2021, Virtual Event, March 10–12, 2021, Proceedings, pp. 229–239.

22 In this connection, O'Connor and Weatherall argue that "When journalists share what they take to be most interesting—or of greatest interest to their readers—they can bias what the public sees in ways that ultimately mislead, even if they report only on real events". O'Connor, C. and Weatherall, J.O., *The Misinformation Age: How False Beliefs Spread*, Yale University Press, 2019, pp. 155–179.

23 Tumber, H., and Waisbord. S., *The Routledge Companion to Media Disinformation and Populism (Routledge Media and Cultural Studies Companions),* Routledge, 2021, p. 98.

24 Gutiérrez-Coba, L. M., Coba-Gutiérrez, P., & Gómez-Díaz, J. A., "The intention behind the fake news about Covid-19: comparative analysis of six Ibero-American countries", *Revista Latina de Comunicación Social*, 2020. 78, 237–264. https://www.doi.org/10.4185/RLCS-2020-1476.

25 Bots and cyborgs ultimately aim at creating high numbers of posts and a high number of users, including politicians, influencers, and pundits, to give a false impression of a big movement. They create a synthetic network of websites that contain fabricated news, so that when individuals search online for a piece of fake news they have already seen, these users will find it on a different website, which in turn emulates dissemination of real news. People tend to more easily believe something if they had heard it before (even if untrue). Mobile messaging applications—most notably, WhatsApp, Snapchat, Facebook Messenger, and Vine—allow private informal conversations between small, highly-connected networks, who can share/spread disinformation because these apps lack safeguards against disinformation. In addition, during important political events or crises, bots, cyborgs, and trolls tend to inundate online discussions with unrelated information—another way to spread disinformation. Gutiérrez-Coba, L. M., Coba-Gutiérrez, P., & Gómez-Díaz, J. A., "The intention behind the fake news about Covid-19: comparative analysis of six Ibero-American countries", *Revista Latina de Comunicación Social*, 2020. 78, 237–264. https://www.doi.org/10.4185/RLCS-2020-1476; Tumber, H., and Waisbord. S., *The Routledge Companion to Media Disinformation and Populism (Routledge Media and Cultural Studies Companions),* Routledge, 2021, pp. 93–95, 200–203; "Measuring the Impact of Misinformation, Disinformation, and Propaganda in Latin America", *Global Americans*, 2021, pp. 1–306, https://theglobalamericans.org/wp-content/uploads/2021/10/2021.10.28-Global-Americans-Disinformation-Report.pdf; and, Lopez-Garcia, G. et al., *Politics of Disinformation, Wiley*, 2021, p. 1.

26 There is no universally accepted definition of populism. I use Tumber and Waisbord's definition: "a political movement that both reflects the crisis of liberal democracy and challenges core democratic premises, including freedom of the press, freedom of speech, government accountability, and tolerance of difference". Tumber, H., and Waisbord. S., *The Routledge Companion to Media Disinformation and Populism (Routledge Media and Cultural Studies Companions),* Routledge, 2021, p. 66, and 626–643.

27 Indeed, populists see enemies (media included)—who want to destroy leaders and eliminate movements—everywhere. In addition, populism generally gravitates around leaders who are considered

invincible; so, building "a leadership cult easily devolves into narratives that liberally blend facts, faux facts, proven lies, and absolute fantasies". Tumber, H., and Waisbord. S., *The Routledge Companion to Media Disinformation and Populism (Routledge Media and Cultural Studies Companions),* Routledge, 2021, p. 67.

28 Tumber, H., and Waisbord. S., *The Routledge Companion to Media Disinformation and Populism (Routledge Media and Cultural Studies Companions),* Routledge, 2021, p. 68.

29 China and Russia are the world's leading digital authoritarian nations. https://cyberdefensereview.arm y.mil/Portals/6/Documents/2021_winter_cdr/CDR_Winter_2021.pdf

30 Peters, K.M. and Matei, F.C. "Democratically Technologized intelligence". Paper presented at the International Studies Association Convention, 2021.

31 Valenzuela, S. et al., "The Paradox of Participation Versus Misinformation: Social Media, Political Engagement, and the Spread of Misinformation", *Digital Journalism*, 7:6, 2019, 802–823, DOI: 10.1080/21670811.2019.1623701.

32 Disinformation tends to affect more prominently highly polarized societies, as propaganda (along with other disinformation efforts) spreads information that strengthens one's viewpoints and beliefs, and in turn helps spread incomplete, prejudiced, or untruthful information. Tumber, H., and Waisbord. S., *The Routledge Companion to Media Disinformation and Populism (Routledge Media and Cultural Studies Companions),* Routledge, 2021, pp. 93–97; and "Measuring the Impact of Misinformation, Disinformation, and Propaganda in Latin America", *Global Americans*, 2021, pp. 1–306, https://theglobalam ericans.org/wp-content/uploads/2021/10/2021.10.28-Global-Americans-Disinformation-Report.pdf.

33 The pandemic fuelled an astonishing combination "of facts, fears, rumors, and speculations [that] provided fertile ground for the dissemination of misleading information and cybercrime on the Internet and became an issue to deal with along with the pandemic itself", as Eveson puts it. Eveson, R., "The pandemic of disinformation in Latin America", Latin America Bureau, 28 May 2020, https://lab. org.uk/the-pandemic-of-disinformation-in-latin-america/.

34 Infodemic is defined as a period of "overabundance of the same information, that had to be tackled alongside the pandemic itself", while "disinfodemic is defined as "the falsehoods fuelling the pandemic and its impacts" because "of the huge "viral load" of potentially deadly disinformation that is described ... as a poison, and humanity's other "enemy" in this crisis". Santos-D'Amorim, K.M., and Fernandes de Oliveira, M.K., "Misinformation, disinformation, and malinformation: Clarifying the definitions and examples in disinfodemic times", *Encontros Bibli: revista eletrônica de biblioteconomia e ciência da informação*, vol. 26, e76900, 2021, January-April, DOI: https://doi.org/10.5007/1518-2924. 2021.e76900, pp.1, 2, 3, 4, 5, 6, 7, 8, 9, 10, 11, 12, 13, 14, 15, 16, 17, 18, 19, 20, 21, 22, 23.

35 Russia's disinformation in the region is usually blatant, aimed primarily disrupting democratic progress in these countries via inciting or abetting social unrest and fuelling mistrust in Latin American governments by their citizens (Chile, Peru, Argentina) on the one hand, and undermining the U.S. role as a global and regional hegemon (especially in countries that are long-time U.S. allies and partners, such as Colombia and Chile). Russia has, allegedly, also been meddling with elections in the region (most notably in Mexico). In contrast, China's disinformation campaigns throughout the region are more stealth, aimed at boosting China's image as a philanthropic partner for the region (in particular, Argentina, Chile, Colombia, and Peru—which are important for its geostrategic agenda, as it needs these countries' natural resources), but also at undermining the public trust in democratic institutions (including the media), and challenging the U.S. role as a global and regional hegemon. Chinese propaganda machine uses local media, academia, and government leaders (many of whom visited China). Public media outlets of some countries that have close relationships with China (most notably Argentina and Peru) are very careful not to provide unfavourable coverage toward China and even use Chinese embassies as sources of information. "Measuring the Impact of Misinformation, Disinformation, and Propaganda in Latin America", *Global Americans*, 2021, pp. 1–306, https:// theglobalamericans.org/wp-content/uploads/2021/10/2021.10.28-Global-Americans-Disinformation-Report.pdf.

36 In addition, Venezuela's disinformation endeavors have sought to cause in Colombia. "Measuring the Impact of Misinformation, Disinformation, and Propaganda in Latin America", *Global Americans*, 2021, pp. 1–306, https://theglobalamericans.org/wp-content/uploads/2021/10/2021.10.28-Global-America ns-Disinformation-Report.pdf.

37 "Measuring the Impact of Misinformation, Disinformation, and Propaganda in Latin America", *Global Americans*, 2021, pp. 1–306, https://theglobalamericans.org/wp-content/uploads/2021/10/2021.10. 28-Global-Americans-Disinformation-Report.pdf.

38 The media in the region serves either the governments du jour or various oligarchs. Neither group is interested in truthful or balanced/fair coverage, but rather propaganda or profit (which in turn fuels censorship). Journalists and reporters are either incompetent or threatened to be fired from their jobs if they do not comply with the needs and wants of the media owners. Sierra Caballero, S. and Sola-Morales, S. "Media Coups and Disinformation in the Digital Era. Irregular War in Latin America", *Comun. Soc*, vol.17, 2020, https://doi.org/10.32870/cys.v2020.7604.

39 Mexico's local media, for instance, tends to provide ample time and space to individuals with propensity toward negativity and polarization in spreading disinformation. In Venezuela, the media is subservient to the Maduro administration, and thus is one of the main reasons why the government can spread disinformation. "Measuring the Impact of Misinformation, Disinformation, and Propaganda in Latin America", *Global Americans*, 2021, pp. 1–306, https://theglobalamericans.org/wp-content/uploads/2021/10/2021.10.28-Global-Americans-Disinformation-Report.pdf. In Ecuador, during the pandemic, the media falsely reported that the WHO stated that the Ecuadorian government was incompetent vis-à-vis the COVID-19 pandemic In fact, WHO just noted that Ecuador would be a priority country for COVID-19-related resources. In January 2020, when Australia was affected by fire, media in several Latin American countries—most notably, Colombia, Brazil, and Mexico—disseminated disinformation related to the cause of fire. For example, the media proliferated fake news, conspiracy theories, and rumors that humans versus climate change generated the fire. Eveson, R., "The pandemic of disinformation in Latin America", Latin America Bureau, 28 May 2020, https://lab.org.uk/the-pandemic-of-disinformation-in-latin-america/.

40 Advice to drink bleach as a way to prevent and cure the coronavirus appeared on Mexican social media first, then in Venezuela, and Argentina. Eveson, R., "The pandemic of disinformation in Latin America", Latin America Bureau, 28 May 2020, https://lab.org.uk/the-pandemic-of-disinformation-in-latin-america/.

41 Eveson, R., "The pandemic of disinformation in Latin America", Latin America Bureau, 28 May 2020, https://lab.org.uk/the-pandemic-of-disinformation-in-latin-america/.

42 In Venezuela COVID-19 disinformation included false claims that the Bacillus Calmette–Guérin (BCG) vaccine against tuberculosis is effective against coronavirus. Eveson, R., "The pandemic of disinformation in Latin America", Latin America Bureau, 28 May 2020, https://lab.org.uk/the-pandemic-of-disinformation-in-latin-america/.

43 "…one theory alleges that doctors are extracting a fluid from patients" knees…[another theory alleged that] a Colombian politician claimed that a 'covid cartel' of doctors were unnecessarily admitting COVID-19 patients to intensive care to receive higher payments…Another theory … [alleged] doctors of receiving a cash payment of £13 000 (€14 420; $16 975) for every dead patient with a covid-19 diagnosis". Luke Taylor, "Covid-19 misinformation sparks threats and violence against doctors in Latin America", BMJ 2020; 370 doi: https://doi.org/10.1136/bmj.m3088 (2020).

44 For instance, the Maduro government in Venezuela reported a higher number of tests being administered, as compared to members of the opposition or U.N. observers who reported lower numbers. Eveson, R., "The pandemic of disinformation in Latin America", Latin America Bureau, 28 May 2020, https://lab.org.uk/the-pandemic-of-disinformation-in-latin-america/.

45 Some countries have taken swift actions against abuse of health professionals. Argentina fines such abuses. Luke Taylor, "COVID-19 misinformation sparks threats and violence against doctors in Latin America", BMJ 2020; 370 doi: https://doi.org/10.1136/bmj.m3088 (2020).

46 Latin American citizens are the world's largest consumers of news spread through social media networks and chat applications, like for example, LINE, or WhatsApp groups. Political candidates regularly convey campaign messages via these WhatsApp groups. WhatsApp in particular—which allows large groups of family members, friends, neighbors to share news, exchange information, coordinate courses of actions and events, and discuss issues—has become more prominent in this regard. As these direct messaging platforms have end-to-end encryption, it is very difficult for fact checking watchdogs to monitor them. For example, in Colombia's very polarized society, social media has provided fertile ground for verbal offensiveness and disinformation campaigns between the right and the left supporters and candidates. The accusation and trial of Dilma Rousseff is another example. Opponents used disinformation via social media to fuel anti-Dilma sentiment. Mitchelstein E., Matassi, M., and Boczkowsk, P.J., "Minimal Effects, Maximum Panic: Social Media and Democracy in Latin America", *Social Media + Society,* October-December 2020: 1–11; Bandeira, L. et al. "Disinformation in Democracies: Strengthening Digital Resilience in Latin America", Atlantic Council. March 2019. https://www.atlanticcouncil.org/in-depth-research-reports/report/disinformation-democracies-strength

ening-digital-resilience-latin-america/; Sierra Caballero, S. and Sola-Morales, S. "Media Coups and Disinformation in the Digital Era. Irregular War in Latin America", *Comun. Soc*, vol.17, 2020, https://doi.org/10.32870/cys.v2020.7604; Cohen, R.S. et al., "Combating Foreign Disinformation on Social Media. Study Overview and Conclusions", RAND Corporation Report, 2021, pp. 1–119; Lupu, N. et al., "Social Media Disruption: Messaging Mistrust in Latin America", *Journal of Democracy*, Volume 31, Number 3, July 2020, pp. 160–171, DOI: https://doi.org/10.1353/jod.2020.0038; and, Gutiérrez-Coba, L. M., Coba-Gutiérrez, P., & Gómez-Díaz, J. A., "The intention behind the fake news about Covid-19: comparative analysis of six Ibero-American countries", *Revista Latina de Comunicación Social*, 2020. 78, 237–264. https://www.doi.org/10.4185/RLCS-2020-1476.

47 In Mexico, examples of disinformation included false claims that Nicolás Maduro funded, and Vladimir Putin backed López Obrador's campaign. In Brazil, the oligarch that funded Bolsonaro's campaign allegedly used illegal lists of phone numbers to create WhatsApp groups and share fake news and conspiracy theories about his chief rival, Fernando Haddad of the Workers' Party (for example, false claims that the PT candidate was distributing baby bottles with penis-shaped tops to the population in an effort to combat homophobia. Lupu, N. et al., "Social Media Disruption: Messaging Mistrust in Latin America", *Journal of Democracy*, Volume 31, Number 3, July 2020, pp. 160–171, DOI: https://doi.org/10.1353/jod.2020.0038; Rodríguez-Virgili, J. et al., "Digital Disinformation and Preventive Actions: Perceptions of Users from Argentina, Chile, and Spain", *Media and Communication*, 2021, Volume 9, Issue 1, Pages 323–337, DOI: 10.17645/mac.v9i1.3521.

48 Fisher, M., & Taub, A. (2019, August 11). "How YouTube radicalized Brazil". *The New York Times*.

49 Examples include false claims that immigrants benefited from social-welfare programs, and deepfakes showing candidates making embarrassing statements or appearing inebriated. Lupu, N. et al., "Social Media Disruption: Messaging Mistrust in Latin America", *Journal of Democracy*, Volume 31, Number 3, July 2020, pp. 160–171, DOI: https://doi.org/10.1353/jod.2020.0038; Cabrera-Méndez, M. et al., "Misleading Discourse on Instagram: A Multimodal Study of Latin American Presidential Candidates in the Face of COVID-19", *Anàlisi* 64, 2021 27–47, ISSN 2340–5236.

50 People tend to believe disinformation more quickly when they mistrust their government and when political polarization is high (Argentina, Brazil, and Mexico versus El Salvador). Lupu, N. et al., "Social Media Disruption: Messaging Mistrust in Latin America", *Journal of Democracy*, Volume 31, Number 3, July 2020, pp. 160–171, DOI: https://doi.org/10.1353/jod.2020.0038.

51 Latin America has had a long experience with populism from the early times of Juan Domingo Perón in Argentina, and Getúlio Vargas in Brazil, to the more recent times of Alberto Fujimori in Peru, Fernando Collor de Melo in Brazil, Hugo Chávez in Venezuela, Evo Morales in Bolivia, Jair Bolsonaro in Brazil and Obrador in Mexico.

52 Their governance modus operandi is a form of soft, entertainment-like authoritarianism—via direct communication with the public. Hugo Chávez's television show Aló, Presidente created in 1999 is an example. Tumber, H., and Waisbord. S., *The Routledge Companion to Media Disinformation and Populism (Routledge Media and Cultural Studies Companions)*, Routledge, 2021, pp. 626–643.

53 Facebook, Twitter, and WhatsApp have been cardinal in developing an "epistemic democracy" in Latin American countries "where journalistic values and populist/messianic discourses compete for the attention of digital communities". The selfie taken by Nayib Bukele—the 37-year-old right-wing independent mayor of San Salvador who won the presidential elections in 2018—to connect with his people in El Salvador during his address before the United Nations General Assembly, is a case in point. Tumber, H., and Waisbord. S., *The Routledge Companion to Media Disinformation and Populism (Routledge Media and Cultural Studies Companions)*, Routledge, 2021, pp. 626–643.

54 Tumber, H., and Waisbord. S., *The Routledge Companion to Media Disinformation and Populism (Routledge Media and Cultural Studies Companions)*, Routledge, 2021, pp. 626–643; and Mitchelstein E., Matassi, M., and Boczkowski, P.J., "Minimal Effects, Maximum Panic: Social Media and Democracy in Latin America", *Social Media + Society*, October-December 2020: 1–11.

55 Political candidates used WhatsApp groups to campaign and/or spread disinformation regularly. Lupu, N. et al., "Social Media Disruption: Messaging Mistrust in Latin America", *Journal of Democracy*, Volume 31, Number 3, July 2020, pp. 160–171, DOI: https://doi.org/10.1353/jod.2020.0038. According to Tumber and Waisbord, "Key in this process is the construction of a discursive premise that separates us from them, which is framed around the distinction between traditional values and new threats". Tumber, H., and Waisbord. S., *The Routledge Companion to Media Disinformation and Populism (Routledge Media and Cultural Studies Companions)*, Routledge, 2021, pp. 626–643.

56 For example, false claims that German airline Lufthansa would build a top-notch international airport if Bukele won. Lupu, N. et al., "Social Media Disruption: Messaging Mistrust in Latin America", *Journal of Democracy*, Volume 31, Number 3, July 2020, pp. 160–171, DOI: https://doi.org/10.1353/jod.2020.0038.

57 Lack of trust in the government (due to corruption scandals such as the Car Wash in 2018), coupled with savviness in using social media enabled the spread of disinformation before and during elections. Bandeira, L. et al. "Disinformation in Democracies: Strengthening Digital Resilience in Latin America", Atlantic Council. March 2019. https://www.atlanticcouncil.org/in-depth-research-reports/report/disinformation-democracies-strengthening-digital-resilience-latin-america/.

58 Richards, J. and Medeiros, J., "Using misinformation as a political weapon: COVID-19 and Bolsonaro in Brazil", 17 April 2020, https://misinforeview.hks.harvard.edu/article/using-misinformation-as-a-political-weapon-covid-19-and-bolsonaro-in-brazil/; Bandeira, L. et al. "Disinformation in Democracies: Strengthening Digital Resilience in Latin America", Atlantic Council. March 2019. https://www.atlanticcouncil.org/in-depth-research-reports/report/disinformation-democracies-strengthening-digital-resilience-latin-america/.

59 Tumber, H., and Waisbord. S., *The Routledge Companion to Media Disinformation and Populism (Routledge Media and Cultural Studies Companions),* Routledge, 2021, pp. 626–643.

60 By reiterating the idea that there are only two major groups in society—the people and the elites. "Measuring the Impact of Misinformation, Disinformation, and Propaganda in Latin America", *Global Americans*, 2021, pp. 1–306, https://theglobalamericans.org/wp-content/uploads/2021/10/2021.10.28-Global-Americans-Disinformation-Report.pdf.

61 Tumber, H., and Waisbord. S., *The Routledge Companion to Media Disinformation and Populism (Routledge Media and Cultural Studies Companions),* Routledge, 2021, p. 88; "Measuring the Impact of Misinformation, Disinformation, and Propaganda in Latin America", *Global Americans*, pp. 1–306, 2021, https://theglobalamericans.org/wp-content/uploads/2021/10/2021.10.28-Global-Americans-Disinformation-Report.pdf.

62 Tumber, H., and Waisbord. S., *The Routledge Companion to Media Disinformation and Populism (Routledge Media and Cultural Studies Companions),* Routledge, 2021, pp. 526–544.

63 These populists usually target academia, science, and the media. "Measuring the Impact of Misinformation, Disinformation, and Propaganda in Latin America", *Global Americans*, 2021, pp. 1–306, https://theglobalamericans.org/wp-content/uploads/2021/10/2021.10.28-Global-Americans-Disinformation-Report.pdf. Populists do not care about truth and facts as much as they care about keeping their supporters engaged; so, they use rumors and conspiracy theories to keep their base constantly involved. Richards, J. and Medeiros, J., "Using misinformation as a political weapon: COVID-19 and Bolsonaro in Brazil", 17 April 2020, https://misinforeview.hks.harvard.edu/article/using-misinformation-as-a-political-weapon-covid-19-and-bolsonaro-in-brazil/.

64 Bolsonaro in Brazil and Maduro in Venezuela, for instance, used the COVID-19 pandemic to promote unbased conspiracy theories and fake news, thus promoting social polarization in the country vis-à-vis the pandemic. Venezuela's disinformation also sought denigrating U.S. endeavors to help with vaccines (but lauding those of Russia and China). Bolsonaro equated COVID-19 to a little flu and torpedoed social distancing guidelines, by mingling with his base, while Obrador in Mexico organized rallies in the first weeks of the pandemic, where he hugged and kissed his fans. See Brigitte Weiffen, "Latin America and COVID-19 Political Rights and Presidential Leadership to the Test", *Democratic Theory*, 2020, https://doi.org/10.3167/dt.2020.070208; and "Measuring the Impact of Misinformation, Disinformation, and Propaganda in Latin America", *Global Americans*, 2021, pp. 1–306, https://theglobalamericans.org/wp-content/uploads/2021/10/2021.10.28-Global-Americans-Disinformation-Report.pdf.

65 Richards, J. and Medeiros, J., "Using misinformation as a political weapon: COVID-19 and Bolsonaro in Brazil", 17 April 2020, https://misinforeview.hks.harvard.edu/article/using-misinformation-as-a-political-weapon-covid-19-and-bolsonaro-in-brazil/; Eveson, R., "The pandemic of disinformation in Latin America", Latin America Bureau, 28 May 2020, https://lab.org.uk/the-pandemic-of-disinformation-in-latin-america/.

66 Mendes Tomaz, R., and Mendes Torres T. J., "The Brazilian Presidential Election of 2018 and the relationship between technology and democracy in Latin America", *Journal of Information, Communication and Ethics in Society,* Vol. 18 No. 4, 2020. pp. 497–509, DOI 10.1108/JICES-12-2019-0134; and Richards, J. and Medeiros, J., "Using misinformation as a political weapon: COVID-19 and Bolsonaro

in Brazil", 17 April 2020, https://misinforeview.hks.harvard.edu/article/using-misinformation-as-a-political-weapon-covid-19-and-bolsonaro-in-brazil/.

67 Tumber, H., and Waisbord. S., *The Routledge Companion to Media Disinformation and Populism (Routledge Media and Cultural Studies Companions),* Routledge, 2021, pp. 626–643.

68 Tumber, H., and Waisbord. S., *The Routledge Companion to Media Disinformation and Populism (Routledge Media and Cultural Studies Companions),* Routledge, 2021, pp. 626–643; Peters, K.M., "21st Century Crime: How Malicious Artificial Intelligence Will Impact Homeland Security", Naval Postgraduate School M.A. Thesis, 2019; Walker, R.E., "Combating Strategic Weapons Of Influence On Social Media", Naval Postgraduate School M.A. Thesis, 2019; Mason, K., "Defending American Democracy in the Post-Truth Age: A Roadmap to a Whole-of-Society Approach", Naval Postgraduate School M. A. Thesis, 2020; and Chan, E., "Fighting Bears and Trolls: An Analysis of Social Media Companies and U.S. Government Efforts to Combat Russian Influence Campaigns during the 2020 U.S. Elections", Naval Postgraduate School M.A. Thesis, 2021.

69 Tumber, H., and Waisbord. S., *The Routledge Companion to Media Disinformation and Populism (Routledge Media and Cultural Studies Companions),* Routledge, 2021, p. 159. "The information bubbles promote political polarization [13] to the detriment of diversity—and the diversity is par excellence one of the pillars of democracy." Mendes Tomaz, R., and Mendes Torres T. J., "The Brazilian Presidential Election of 2018 and the relationship between technology and democracy in Latin America", *Journal of Information, Communication and Ethics in Society,* Vol. 18 No. 4, 2020. pp. 497–509, DOI 10.1108/JICES-12-2019-0134.

70 For instance, Global Americans' seminar that brought together journalists, scholars, leaders, civil society representatives, and influencers across Latin America provided insights into how to fight propaganda, fake news, rumors etc. COVID-19: The Brazilian Ministry of Health provided a WhatsApp contact number to the citizens to combat health related disinformation. "Measuring the Impact of Misinformation, Disinformation, and Propaganda in Latin America", *Global Americans,* 2021, pp. 1–306, https://theglobalamericans.org/wp-content/uploads/2021/10/2021.10.28-Global-Americans-Disinformation-Report.pdf; Ponciano da Silva, M. and Godoy Viera, A.F., "The Dilemma of Fake News Criminalization on Social Media", in Bisset Álvarez, E., ed., *Data and Information in Online Environments,* Second EAI International Conference, DIONE 2021, Virtual Event, March 10–12, 2021, Proceedings, pp. 181–196.

71 In 2018, for example, during the presidential elections in Brazil, the Federal Police and the U.S. Federal Bureau of Investigation (FBI) joined forces to investigate cybercrimes and combat fake news. Ponciano da Silva, M. and Godoy Viera, A.F., "The Dilemma of Fake News Criminalization on Social Media", in Bisset Álvarez, E., ed., *Data and Information in Online Environments,* Second EAI International Conference, DIONE 2021, Virtual Event, March 10–12, 2021, Proceedings, pp. 181–196. These collaborative efforts have educated the public on how specific instances of disinformation emerged, "who was affected, who amplified stories, how they spread, and what came out of that circulation". Bandeira, L. et al. "Disinformation in Democracies: Strengthening Digital Resilience in Latin America", Atlantic Council. March 2019. https://www.atlanticcouncil.org/in-depth-research-reports/report/disinformation-democracies-strengthening-digital-resilience-latin-america/.

72 Examples include the 2018 General Law on the Protection of Personal Data (LGPD) in Brazil; and the 2020 Observatory of disinformation and symbolic violence on digital media and platforms (NODIO) in Argentina; the 2020, law against digital crimes "committed through information and communication technologies in Nicaragua". Similarly, in Venezuela, the 2005 revision of the Penal Code stipulates up to five years in prison for anybody who spreads disinformation. Lubianco, J, "11 laws and bills against disinformation in Latin America carrying fines, prison, and censorship", (2020, 16 December). LatAm Journalism Review - The University of Texas at Austin Knight Center for Journalism in the Americas. https://latamjournalismreview.org/articles/laws-and-bills-against-disinformation-in-latin-america/; Tumber, H., and Waisbord. S., *The Routledge Companion to Media Disinformation and Populism (Routledge Media and Cultural Studies Companions),* Routledge, 2021, pp. 790–791; Lopez-Garcia, G. et al., *Politics of Disinformation, Wiley,* 2021, pp. 27–30; and Ponciano da Silva, M. and Godoy Viera, A.F., "The Dilemma of Fake News Criminalization on Social Media", in Bisset Álvarez, E., ed., *Data and Information in Online Environments,* Second EAI International Conference, DIONE 2021, Virtual Event, March 10–12, 2021, Proceedings, pp. 181–196.

73 During the COVID-19 pandemic, platforms—most notably Facebook, Instagram, YouTube, and Twitter—started to delete information that contradicted health guidelines, even if that type of information was shared by politicians (for example, videos showing President Bolsonaro in Brazil, disobeying

rule of social distancing, dismissing the epidemic, and supporting the use of chloroquine; or tweets from Maduro that supported the idea that COVID-19 started in a laboratory). Eveson, R., "The pandemic of disinformation in Latin America", Latin America Bureau, 28 May 2020, https://lab.org.uk/the-pandem ic-of-disinformation-in-latin-america/; and Richards, J. and Medeiros, J., "Using misinformation as a political weapon: COVID-19 and Bolsonaro in Brazil", 17 April 2020, https://misinforeview.hks.harva rd.edu/article/using-misinformation-as-a-political-weapon-covid-19-and-bolsonaro-in-brazil/.

74 In Mexico, more than eighty media outlets established Verificado 2018, funded by civil society and some social media platforms. Verificado fact checked and debunked hundreds of disinformation attempts on social media during the 2018 elections. Similar fact checkers were created in Colombia (Silla Vacia), Brazil (Projeto Comprova) and Argentina (Chequeado and Reverso). Across the region, 22 NGOs from 15 countries created LatamChequea, a fact checking organization that focused on debunking disinformation related to the pandemic. Eveson, R., "The pandemic of disinformation in Latin America", Latin America Bureau, 28 May 2020, https://lab.org.uk/the-pandemic-of-disinforma tion-in-latin-america/; and Lupu, N. et al., "Social Media Disruption: Messaging Mistrust in Latin America", *Journal of Democracy*, Volume 31, Number 3, July 2020, pp. 160–171, DOI: https://doi.org/ 10.1353/jod.2020.0038.

75 These agencies debunked conspiracy theories that linked 5G and the pandemic outbreak, or false COVID-19 cure claims. Eveson, R., "The pandemic of disinformation in Latin America", Latin America Bureau, 28 May 2020, https://lab.org.uk/the-pandemic-of-disinformation-in-latin-america/; Lupu, N. et al., "Social Media Disruption: Messaging Mistrust in Latin America", *Journal of Democracy*, Volume 31, Number 3, July 2020, pp. 160–171, DOI: https://doi.org/10.1353/jod.2020.0038; and Lopez-Garcia, G. et al., *Politics of Disinformation, Wiley*, 2021, p. 35.

76 Argentina's NODIO was used to arrest journalists who reported on government's handling of the COVID-19 pandemic. Ponciano da Silva, M. and Godoy Viera, A.F., "The Dilemma of Fake News Criminalization on Social Media", in Bisset Álvarez, E., ed., *Data and Information in Online Environ- ments,* Second EAI International Conference, DIONE 2021, Virtual Event, March 10–12, 2021, Proceedings, pp. 181–196; and Lubianco, J, "11 laws and bills against disinformation in Latin America carrying fines, prison, and censorship", (2020, 16 December). LatAm Journalism Review - The Uni- versity of Texas at Austin Knight Center for Journalism in the Americas. https://latamjournalism review.org/articles/laws-and-bills-against-disinformation-in-latin-america/.

77 Argentina's Chequeado, which was chaired by the spouse of a mayor connected to the incumbent president, is a case in point. Lupu, N. et al., "Social Media Disruption: Messaging Mistrust in Latin America", *Journal of Democracy*, Volume 31, Number 3, July 2020, pp. 160–171, DOI: https://doi.org/ 10.1353/jod.2020.0038.

78 Lupu, N. et al., "Social Media Disruption: Messaging Mistrust in Latin America", *Journal of Democracy*, Volume 31, Number 3, July 2020, pp. 160–171, DOI: https://doi.org/10.1353/jod.2020.0038.

79 Bandeira, L. et al. "Disinformation in Democracies: Strengthening Digital Resilience in Latin Amer- ica", Atlantic Council. March 2019. https://www.atlanticcouncil.org/in-depth-research-reports/rep ort/disinformation-democracies-strengthening-digital-resilience-latin-america/.

80 Lopez-Garcia, G. et al., *Politics of Disinformation, Wiley*, 2021, p.7.

Bibliography

Bandeira, L. et al. "*Disinformation in Democracies: Strengthening Digital Resilience in Latin America*", Atlantic Council. March 2019. https://www.atlanticcouncil.org/in-depth-research-reports/report/disinforma tion-democracies-strengthening-digital-resilience-latin-america/..

Cabrera-Méndez, M. et al., "Misleading Discourse on Instagram: A Multimodal Study of Latin American Presidential Candidates in the Face of COVID-19", *Anàlisi* 64, 202127–47, ISSN 2340–5236.

Chan, E., "*Fighting Bears and Trolls: An Analysis of Social Media Companies and U.S. Government Efforts to Combat Russian Influence Campaigns during the 2020 U.S. Elections*", Naval Postgraduate School M.A. Thesis, 2021.

Cohen, R.S. et al., "*Combating Foreign Disinformation on Social Media. Study Overview and Conclusions*", RAND Corporation Report, 2021, *pp.* 1–119. https://cyberdefensereview.army.mil/Portals/6/Docum ents/2021_winter_cdr/CDR_Winter_2021.pdf.

Dwoskin, E., "Russia is still the biggest player in disinformation, Facebook says", *The Washington Post*, 21 May2021.

Eveson, R., "The pandemic of disinformation in Latin America", *Latin America Bureau*, 28 May2020, https://lab.org.uk/the-pandemic-of-disinformation-in-latin-america/.

Fisher, M., & Taub, A., "How YouTube radicalized Brazil", *The New York Times*, 11 August2019.

Gutiérrez-Coba, L. M., Coba-Gutiérrez, P., & Gómez-Díaz, J. A., "The intention behind the fake news about Covid-19: comparative analysis of six Ibero-American countries", *Revista Latina de Comunicación Social*, 2020. 78, 237–264. https://www.doi.org/10.4185/RLCS-2020-1476.

Jack, C., "Lexicon of Lies: Terms for Problematic Information", Data & Society Research Institute, datasociety.net, pp. 1–20.

Lopez-Garcia, G. et al., *Politics of Disinformation*. Wiley, 2021.

Lupu, N. et al., "Social Media Disruption: Messaging Mistrust in Latin America", *Journal of Democracy*, Volume 31, Number 3, July 2020, pp. 160–171, DOI: https://doi.org/10.1353/jod.2020.0038.

Lubianco, J, "11 laws and bills against disinformation in Latin America carrying fines, prison, and censorship", (2020, 16 December). *LatAm Journalism Review*. The University of Texas at Austin Knight Center for Journalism in the Americas. https://latamjournalismreview.org/articles/laws-and-bills-against-disinformation-in-latin-america/.

Mason, K., "*Defending American Democracy in the Post-Truth Age: A Roadmap to a Whole-of-Society Approach*", Naval Postgraduate School M. A. Thesis, 2020.

"*Measuring the Impact of Misinformation, Disinformation, and Propaganda in Latin America*", Global Americans, 2021, pp. 1–306, https://theglobalamericans.org/wp-content/uploads/2021/10/2021.10.28-Global-Americans-Disinformation-Report.pdf.

Mendes Tomaz, R., and Mendes Torres T. J., "The Brazilian Presidential Election of 2018 and the relationship between technology and democracy in Latin America", *Journal of Information, Communication and Ethics in Society*, Vol. 18 No. 4, 2020. pp. 497–509, doi:10.1108/JICES-12-2019-0134..

Merlan, A., *Republic of Lies. American Conspiracy Theorists and Their Surprising Rise to Power*, Macmillan, 2019.

Mitchelstein E., Matassi, M., and Boczkowsk, P.J., "Minimal Effects, Maximum Panic: Social Media and Democracy in Latin America", *Social Media + Society*, October-December 2020: 1–11.

Monteiro da Silva Junior, E. and Lima Dutra, M., "A Roadmap for Composing Automatic Literature Reviews: A Text Mining Approach", in Bisset Álvarez, E., ed., *Data and Information in Online Environments, Second EAI International Conference, DIONE 2021*, Virtual Event, March 10–12, 2021, Proceedings, pp. 229–239.

O'Connor, C. and Weatherall, J.O., *The Misinformation Age: How False Beliefs Spread*, Yale University Press, 2019.

Persily, N., and Tucker, J. A. eds., *Social Media and Democracy Social Media and Democracy*.

The State of the Field, Prospects for Reform, Cambridge University Press, 2020.

Peters, K.M., "*21st Century Crime: How Malicious Artificial Intelligence Will Impact Homeland Security*", Naval Postgraduate School M.A. Thesis, 2019.

Peters, K.M. and Matei, F.C. "Democratically Technologized intelligence". Paper presented at the International Studies Association Convention, 2021.

Ponciano da Silva, M. and Godoy Viera, A.F., "The Dilemma of Fake News Criminalization on Social Media", in Bisset Álvarez, E., ed., *Data and Information in Online Environments*, Second EAI International Conference, DIONE 2021, Virtual Event, March 10–12, 2021, Proceedings, pp. 181–196.

Richards, J. and Medeiros, J., "Using misinformation as a political weapon: COVID-19 and Bolsonaro in Brazil", 17 April 2020, https://misinforeview.hks.harvard.edu/article/using-misinformation-as-a-political-weapon-covid-19-and-bolsonaro-in-brazil/.

Rodríguez-Virgili, J. et al., "Digital Disinformation and Preventive Actions: Perceptions of Users from Argentina, Chile, and Spain", *Media and Communication*, 2021, Volume 9, Issue 1, Pages 323–337, doi:10.17645/mac.v9i1.3521..

Santos-D'Amorim, K.M., and Fernandes de Oliveira, M.K., "Misinformation, disinformation, and malinformation: Clarifying the definitions and examples in disinfodemic times", *Encontros Bibli: revista eletrônica de biblioteconomia e ciência da informação*, vol. 26, e76900, 2021, January-April, DOI: https://doi.org/10.5007/1518-2924.2021.e76900, pp.1–23.

Sierra Caballero, S. and Sola-Morales, S. "Media Coups and Disinformation in the Digital Era. Irregular War in Latin America", *Comun. Soc*, vol.17, 2020, https://doi.org/10.32870/cys.v2020.7604.

Taylor, L., "Covid-19 misinformation sparks threats and violence against doctors in Latin America", BMJ 2020; 370 doi: https://doi.org/10.1136/bmj.m3088 (2020).

Tumber, H., and Waisbord. S., *The Routledge Companion to Media Disinformation and Populism (Routledge Media and Cultural Studies Companions)*. Routledge, 2021.

Valenzuela, S. et al., "The Paradox of Participation Versus Misinformation: Social Media, Political Engagement, and the Spread of Misinformation", *Digital Journalism*, 7:6, 2019, 802–823, doi:10.1080/21670811.2019.1623701.

Walker, R.E., "*Combating Strategic Weapons Of Influence On Social Media*", Naval Postgraduate School M.A. Thesis, 2019.

Weiffen, B., "Latin America and COVID-19 Political Rights and Presidential Leadership to the Test", *Democratic Theory*, 2020, https://doi.org/10.3167/dt.2020.070208.

PART VI

A Toolkit for Practitioners

28

THE USE OF DISCOURSE ANALYSIS IN PROPAGANDA DETECTION AND UNDERSTANDING

Julian Richards

Introduction

Recent political events have underlined persistent concerns about the growth of extremist thinking within mainstream political discourse. In France, the April 2022 presidential elections saw *Front National* leader Marine le Pen achieve her party's biggest ever electoral performance with over 40 percent of the votes in the second round, to run the mainstream incumbent, Emmanuel Macron, uncomfortably close for the presidency. Acknowledging the performance of what he described as the "extreme right", Macron admitted that "an answer must be found to the anger and disagreements" that led so many to vote for his rival.[1]

Earlier in the same month, Viktor Orban's *Fidesz* party scored a comfortable victory in the parliamentary elections in Hungary, securing his fourth consecutive win and ensuring the continued dominance of nationalist politics in Hungary for over a decade. With a history of robustly anti-immigrant and authoritarian approaches towards the media and academia,[2] Orban has frequently been accused of being an extremist within Europe's political mainstream.

One of the issues uniting these and many other Western protagonists of the Far-Right in politics is an alignment with the Kremlin and its anti-liberal ideology. Le Pen has often struck a chord with Putin as a NATO sceptic, and once declared that France and Russia share "common civilizational and strategic interests".[3] Orban's electoral victory, meanwhile, was openly welcomed by Putin, recognising the former's rejection of Ukraine's president Zelensky as an "opponent".[4]

Heather Ashby paints a more robust picture, noting not only that Far-Right political extremism is posing a threat much more widely than just in the Western world; but also that the developments constitute a "grave threat to democratic societies".[5] We should also remember that extremist threats should not only be recognised at the level of state politics and governance, but also at sub-state level. Examples of the latter range from the grave threats posed by individual terrorist attackers such as Anders Breivik in Norway, or his disciple, Brenton Tarrant in New Zealand; to broader extremist movements such as Al Qaeda

DOI: 10.4324/9781003190363-33

or the Islamic State, and the myriad of motivated foot-soldiers these movements are able to mobilise around the world.

In these factors, we can start to see the emergence of some of the strands of political extremism and their processes of narratisation and propagandising. A potential nexus is emerging between the fundamentally anti-liberal and essentially anti-Western narrative of authoritarian states such as Russia and China—or, indeed, of Islamist movements such as the Islamic State – and the political and economic fears of ordinary voters who may be attracted to extremist messages by skilfully designed propaganda and information. We can also see in Macron's words a recognition by some mainstream politicians that something is happening in terms of narratives that chime with voters, which may not yet be properly understood.

For these reasons, there is a growing policy imperative, as much as an analytical one, to try to understand the forces at play in these developments and how they are unfolding across polities. In this chapter, a discourse analysis approach is considered to demonstrate how the process of "beliefs from information" can be characterised, understood, and used to form a basis for policy alternatives in the counter-extremism environment. Two examples of extremist narratives are analysed: one, the much-examined treatise by Islamist ideologue, Sayyib Qutb, *Milestones*; and the second, a contemporary communication by the British Far-Right organisation, Britain First. In the analysis, theoretical positions on how propaganda is constructed, and on how discourse analysis identifies themes and modalities in communication, are considered to provide some reflections on policy implications for counter-extremism strategy.

Understanding propaganda

Contemporary references to propaganda cast it as an inherently distasteful, and fundamentally anti-liberal process of information dissemination. At the time of writing, the conflict in Ukraine is causing many commentators in and around the West to lambast the way in which propaganda is used to cloud judgements and enforce authoritarianism. Propaganda is therefore understood in the West as something that others do for nefarious purposes, demonstrating a continuity with the Cold War and the approaches of the Soviet Union. Bakir et al[6] develop the idea by noting how the business of "public relations" (PR) emerged in the West as a wholly legitimate and worthy pursuit, when both PR and propaganda could be said to occupy different positions on the same spectrum of "incentivised" or "coercive" communication.

At root, we are reminded that the original definition of propaganda was simply the *propagation* of ideas. It is generally understood that the Catholic church coined the term in the seventeenth century with the mission to propagate Catholicism in the face of the Reformation. This leads us to consider that, in the words of Welch,[7] propaganda is "ethically neutral": it can be good or bad depending on the circumstances.

The analysis of propaganda as we understand it today began towards the end of the nineteenth century, and was coupled with an interest in how to mobilise and influence populations in increasingly industrialising states. Le Bon's[8] influential writing on "the crowd" took the "popular mind" as its subject and suggested—in rather pejorative terms—that the population was largely ignorant and racist, and easily channelled by skilful propaganda. By the beginning of the twentieth century, interest in the subject was accelerating. In the 1920s, Harold Lasswell[9] "wrote the book on propaganda" in the words of Patrick and Thrall,[10] and effectively launched the study of communications as a university subject. Bernay's "Propaganda"[11] was also published in the early 1920s, and suggested that, since the masses could not possibly understand everything that was happening politically, propaganda was essentially a state duty to help people decide about the big issues. This analysis formed the basis of PR as a modern science.

Debate continues on whether propaganda is a good or bad thing, and how much agency "the masses" have in being influenced by it. Aldous Huxley[12] suggested it depended very much on the situation, while Doob[13] and Postman[14] find it hard to see value in a process that is essentially manipulative.

Jacques Ellul's influential work on propaganda published in the 1960s[15] suggested that propaganda was constituted by a set of false statements that effectively dupe the public, while others disagree, suggesting that "propaganda is far more successful if it sticks to the truth".[16] This is a key factor to which we will return, in terms of the potential disconnect between *belief* and *truth*.

The Western sense that propaganda is what oppressive, totalitarian regimes do, was reflected in the various analyses of the communicative strategies of totalitarian regimes in Europe from the 1930s onwards, where the importance of social control was central to the survival of the regime.[17] In the Cold War that followed, the central importance of information and misinformation in the propaganda war between superpowers, inevitably prompted a great deal of discussion and analysis both before and after the fall of the Berlin Wall. Much of this writing is incorporated within the wider academic field of Foreign Policy Analysis, with Robert Jervis's[18] "Perception and Misperception", and Valerie Hudson's[19] "Foreign Policy Analysis" being particularly useful. Gramsci[20] and Chomsky[21] meanwhile, offered a critical "hegemonic theory" of the West, in which, it is argued, the supposed control of the media by the state was a mechanism to ensure that Western capitalist countries did not succumb to revolution.

It is indeed the case that propaganda has also been an important element of military and intelligence studies, in the shape of its relationship to "covert action" and "psychological operations". Kibbe[22] provides a useful framework of "white, grey and black" propaganda measures in military contexts, relating respectively to official, overt actions; propaganda channelled through proxy or covert channels (such as broadcasts from Radio Free Europe during the Cold War, for example); and deliberately manipulative and fallacious information spread completely covertly to achieve particular ends.

In the contemporary counter-terrorism and counter-extremism research environments, much of the analysis is embodied within discussions about radicalisation and extremism, with Khosrokhavar,[23] Richards,[24] Archetti[25] and Betz[26] offering particularly useful analyses of these concepts and how we might define them. Julia Ebner[27] offers an excellent discussion of the supposed symbiosis between Islamist and Right wing extremism and propaganda, which suggests that the basic principles of propaganda are essentially the same, regardless of the ideological narrative. This presents the second key point in this analysis, namely that communicative mechanisms for propagating extremism are generic and can operate in a range of different ideological environments.

Propaganda and extremism

Recent analyses of extremist discourse have tended to focus, perhaps naturally, on the more organised movements to whom communications are a central part of the overall strategy. Particularly as regards analysis of broad-based Islamist movements of various hues such as Al Qaeda, Islamic State and the Taliban, the word "propaganda" is taken as a normative description of how extremist ideas and sentiments are built and promulgated. Conroy and

Al-Dayel's analysis of ISIS's 2015 "No Respite" video, for example, situates a particular communication within a wider propaganda strategy by the Islamic State organisation. The analysis identifies that the mixture of established details with distorted statistics about history and contemporary society within the video in question, display its "propagandistic and manipulative" nature.[28]

Lakomy makes reference to Islamic State's "propaganda machine"[29] in an analysis of its influential online magazines, *Dabiq* and *Rumiyah*. Alleged "manipulation techniques" in these communications includes a simplistic historical narrative that lays blame for violent acts on the provocations of *kufr*; and uses selective references to the Quran and the *hadiths* to reinforce a notion that violent jihad is a religious obligation.[30]

Similarly, Mehran picks up on the theme of deception and manipulation in an analysis of Taliban videos, noting how "multimodal" communications that combine text with other audio-visual mechanisms can be particularly persuasive, and, accordingly, particularly manipulative.[31] The horror of beheading videos—perhaps the archetypal example of terrorist communication in action—not only provides highly influential "visual facts" shaping public discourse, but also arguably depoliticises debate about the necessity for violent measures taken in response.[32] These are important observations in terms of the development of discourse analysis, as much of the established writing on the methodology has been focused, hitherto, on textual analysis alone.

Many of the analyses pick up on the need for comparative and contextual analysis beyond specific cases. Chiluwa's analysis of internet-based discourse by two Islamist movements in Nigeria, *Boko Haram* and *Ansaru*, notes how the weaving of "intertextual relation" between the communications of regional groups and those of the global jihadist movement, allows for the reinforcement of broader concepts and goals.[33] In an example concerning Far-Right extremism, Enstad notes how "discursive climates" can be reinforcing and contributory to the development of extremist narrative, in his example in relation to the way in which the Russian Far-Right is able to champion Anders Breivik as a hero, under the increasingly ultra-nationalist discursive climate of "Putinism".[34]

Critical Discourse Analysis as a tool

Both of the above examples pick up on the significance of social contextualisation in discourse analysis: an increasing tendency in the field. In an analysis of the case of discussion in the British parliament around the time of the case being made for the Iraq War in 2003, Van Dijk emphasised the importance of the "communicative situation",[35] and, indeed, of the degree of the contextual understanding held by the audience of the communication.[36] In this case, an advanced understanding was shown on all sides of the deeper layers of meaning entailed in the business of "doing politics". In other cases of communication, however, a critical understanding by the receiving party may be less developed or differently motivated.

The methodology of Critical Discourse Analysis (CDA) emerged in the 1970s as an essentially "neo-Marxist/heuristic" approach which tied discourse inextricably to power relations in society, and established that "facts can never be isolated from the domain of values".[37] In this observation, we see a critical link between the concept in propaganda studies that beliefs are more important than facts; and the notion that all discourse is unavoidably linked to a set of political and societal values rather than objective certainties: in a sense, it is all about context.

Early philosophical work on "political myth" recognised the importance of language. While Sorel and Barthes disagreed about the potential agency of political myth, they both agreed that language and discourse were central to its propagation. Sorel proclaimed that you did not have to be a great philosopher to see that language "deceives us constantly as to the true nature of the relationship between things".[38] Barthes, similarly, spoke of the destructive power of "depoliticized speech" which "abolishes the complexity of human acts".[39]

John Austin introduced the idea of Speech Act Theory (SAT), which suggests that language has important agential properties, sometimes by directly initiating actions in the physical world (such as a declaration of war for example).[40] CDA flowed from such ideas, including Fairclough's analysis of the relationship between language and relative positions of power.[41] In security studies, the Copenhagen School of scholars, led by Buzan, Waever and de Wilde, included SAT as a component in their critical thesis of "securitization", which suggested that certain factors could be converted into existential security concerns in the public consciousness by political leaders, through their narratives and utterances.[42]

More recently, much critical analysis has been undertaken of the high-level political discourse following the 9/11 attacks, using SAT and CDA. Jackson, in his extensive critique of the discourse surrounding the "Global War On Terror", notes a similarity with the dichotomous principles of the Cold War.[43] Then, the oppression and lack of democracy in communist states were counter-posed against the West's freedom and respect for human rights, in defining who was right and who was wrong. Similarly, "terrorists" in the contemporary era are characterised as espousing inverse mirror-image values to those of the West, whereby they are barbaric, inhuman and uncivilised.

In times of conflict, a sociology of violence suggests that dehumanization of the out-group is an important device in allowing for violent and murderous acts to be carried out by otherwise ordinary people.[44] Using the rhetoric of "war", furthermore, moves the conflict into a different realm from that of ordinary civil life, where different rules apply, and a more existential conception of the threat and appropriate response can be applied. In all these cases, the words used to describe the threat and identify the enemy are critically important.

As Gastil observed, politics and discourse are "inextricably intertwined".[45] Thus, political leaders can articulate power through their utterances, and can also develop and propagate myth. In his work on language and myth, Edelman noted a complex and symbiotic relationship between the public's notion of what their leaders were there to do, and how leaders and their spokespeople could play on those ideas and propagate mythical constructs which "typically fail to analyse problems adequately and rarely solve them".[46] In this way, lofty ideas such as the fact that government is there to protect the people and ensure their well-being can subtly hide the effect of any actual policy. Similarly, explanations can be proffered for ills befalling the nation, such as the idea that they are because of foreign interference, or immigration, or any number of other factors.

CDA and extremism

The business of applying CDA to any discourse is an approach that considers the "interpretative repertoire" of the promulgator. Structural techniques and mechanisms are applied critically to a contextual understanding of the situation in question.[47] Approaches can vary from fairly quantitative and statistical analyses of content, to looser, qualitative approaches that combine "some descriptive and critical interpretative CDA".[48] Many of the studies described above take such a qualitative approach. Of course, one of the challenges here is that interpretations of the discourse are dangerously open to subjective evaluation[49], but such risks are taken as read in this discussion.

I have written elsewhere that propaganda is constructed around three key factors: **anxiety, influence,** and establishing **narratives of hope and redemption**[50]. It could consequently be argued that the first and most important element of a successful propaganda campaign is identifying and building upon nascent fears and anxieties that are already present, to some degree, in the population. Such propaganda feeds on people's emotions, and stokes angry and emotional responses. In political analysis, this is sometimes called "dog whistle politics"[51]; that is, a politics that touches on darker, deep-seated and peripheral elements in people's consciousness, such as tendencies towards racism or sectarianism, for example.

Doubts about trust and credibility can also be built upon. Anti-Western actors, such as Islamist extremists or indeed Russia, will attempt to build upon notions that the West is essentially hypocritical and just as oppressive as anyone else, despite its claims to a democratic and rules-based order.

On a more domestic level, there is much evidence that trust in the political establishment has declined in the West in recent years[52]. Again, this can form the basis for propaganda for those with an anti-Western agenda, or for those opposed to democracy altogether. Indeed populist ideologues on all sides of the political spectrum will frequently suggest that the political elites have failed them, and they can no longer be trusted: these are, indeed, the issues with which the two examples of France and Hungary opened this chapter.

In all these cases, the underlying fears and anxieties around which the propaganda plays may be unfounded, or a process of "deception by distortion and omission".[53] The important thing is what people believe rather than what is true. (As the arch fascist propagandiser, Bennito Mussolini once said, the "myth of the nation" is a "faith, it is passion. It is not necessary that it should be a reality"[54].)

It should also be noted that, if propaganda can successfully undermine the trust and credibility in its targets, it develops a self-sustainability. Once trust is undermined, then protestations that certain information is untrue or manipulated will fall on deaf ears, and there will be a sense of "they would say that, wouldn't they?" In Far-Right extremism, for example, narratives often focus on the notion that established political elites have allowed uncontrolled immigration into Western countries to fester over several years, but have actively suppressed the true figures and the supposedly dangerous effects of such immigration. Much research has shown that the figures and statistics used in stoking such fears are often considerably exaggerated, but remain remarkably resilient in the face of official attempts to suggest otherwise.[55]

It may often be the case that a fairly simple, dichotomous communication will form the basis of effective propaganda: good versus evil; a question of us versus them. In this way, fears can take on a target object towards which anger and frustration can be directed. This will help to deflect preoccupation with one's own problems and to channel anxiety in another direction. Ideally, there will be a message of hope and redemption. Baker et al[56] described this as a technique of "incentivisation", whereby a persuasive promise is offered for compliance. In many ways, former President Trump's key campaign slogan up to and beyond his 2016 election victory, "Make America Great Again", was a fine example of the art-form. The implication is that Trump supporters could work together to move from a position of anxiety (America no longer being the great power it once was) to a place of redemption in which the former glory was restored. In a very simple message, hope for the future is offered, even if the message does not contain much detail about how it will happen.

In the context of modern media and communications, the information environment in which the message is delivered is particularly important. Many contemporary propagandisers also understand the effectiveness of rapid, short, broad and continual messaging, in capitalising on a cognitive weakness that determine that the more often we hear or read something, the more likely it is to enter our consciousness. (Psychologists call this the "availability" or "frequency" heuristic[57].) While discourse analysis tends to focus more on the content of the message rather than the medium, there is an important point here about the nature of propaganda in the social media environment. Phenomena such as "echo chambers", whereby confirmatory narratives circulate with a particular target community[58], may well lead to the design of more immediate and simplistic messages that can be more readily promulgated in social media.

Case studies

Case Study 1: Sayyid Qutb's "Milestones"

The first case study concerns a short book written by the Egyptian activist, Sayyib Qutb, entitled *Ma'alim fi'l tareeq* (generally translated into English as *Milestones*). This book was published in 1964 in Egypt, and immediately attracted the ire of the authorities, who saw its author as a dangerous Islamist revolutionary with unwelcome connections to the Muslim Brotherhood. Qutb was already in prison at the time, and was executed for sedition not long after the publication of the book, in 1966.

Milestones was not translated into English until 1981, but interest in it resurged after the 9/11 attacks in the US. In the post-9/11 era, the book gained notoriety as one of the key books allegedly consulted for ideological guidance by militants connected with the Al Qaeda movement. Armstrong described Qutb as "the real founder of Islamic fundamentalism in the Sunni world"[59] in her 2002 book about the history of Islam; while Swenson described *Milestones* as "the ideal place to start learning about radical Islam"[60].

The key theme in *Milestones* concerns *jahilliyah*, a concept loosely translated as "ignorance", and specifically the pre-Islamic state of society to which contemporary countries in the Middle East had supposedly fallen back through the influence of the non-Muslim world. General Nasser of Egypt saw this as a dangerous provocation, and it cost Qutb his life.

Milestones has been subjected to CDA on a number of occasions, including most recently by Shehab, who added an extra dimension of analysing "recontextualisation" of the tract through differing translations into English.[61] Taking CDA's emphasis on the importance of context in which a communication is made to an understanding of power relations between writer and reader[62] (and indeed translator), Shehab delivers a structured, syntactical analysis. This suggests that a second translation of the book might offer a deeper understanding of Qutb's original intentions and ideas for those familiar with Islamic concepts and language.[63]

Similarly, Calvert considered Qutb's concept of "Islamist oriented moral regeneration and civilizational confrontation" in terms of Sorelian political myth, noting that this allows for a comparison with other secular, nationalist movements who "likewise live in the realm of myth".[64] Here again we can see the notion of structural similarities at the roots of all extremist discourses, when we deconstruct them systematically.

By applying the proposed analytical framework of **anxiety, influence,** and establishing **narratives of hope and redemption,** *Milestones* demonstrates a number of lexical devices which accord with such a structural analysis; and which offer an over-arching framework of understanding. In this analysis, the text used is an English translation first published in 2002 by the Islamic Book Service in New Delhi, attributed to Abdus Sami, and specifically the opening scene-setting section at the beginning of the book (pages 7–9).

It is also worth mentioning at this stage an important element of some CDA, which relates to identification and analysis of specific "lexical devices" used to develop implicature for the reader. There are many of these used to frame various CDA approaches in different contexts. Richards identifies six that can be used in short CDA exercises, namely: positive and negative descriptors; identifying self and other (us and them); euphemisms, similes and metaphors; certainty and uncertainty of language; and assertions of "givens" and "facts".[65] These can be particularly important in the **influence** part of our three-factor framework, since they can help to embed and deepen acceptance of a particular narrative by using their cognitive effects of persuasion.

The much-quoted opening paragraph of *Milestones* offers an immediate framing around the first element: societal **anxiety**:

> Mankind today is on the brink of a precipice, not because of the danger of complete annihilation which is hanging over its head—this being just a symptom and not the real disease—but because humanity is devoid of those vital values which are necessary not only for its healthy development but also for its real progress. Even the Western world realizes that Western civilization is unable to present any healthy values for the guidance of mankind. It shows that it does not possess anything which will satisfy its own conscience and justify its existence.[66]

These opening words set the primary context for Qutb's thoughts. A historical contextualisation reveals a reference to the Cold War period in which the book was written, and particularly the immediate aftermath of the Cuban Missile Crisis in which many around the world had feared a catastrophic confrontation. And yet, Qutb identifies the real cause of anxiety as the flawed "values" on which Western society is built. The device of "standing on a precipice" as the first words of the book could be said to be a clear framing of a state of anxiety for the intended audience (namely the *ummah*).

Many examples throughout history accord with the initial framing of anxiety in gaining the attention of a target audience. Although a very different context, of course, President Trump's inauguration speech in January 2017 used the same concept:

> For too long, a small group in our nation's capital has reaped the rewards of government while the people have borne the cost. Washington flourished—but the people did not share in its wealth.[67]

In the speech, President Trump conceptualised the key objective of his arrival into office as tackling the loss of power by the American people in favour of political elites, thus setting the scene of anxiety for those voters who had been supposedly disenfranchised.

The second-frame of analysis, **influence**, concerns the development of a compelling narrative which both develops and unpacks the cause of the anxiety, and describes it in ways that can be picked up easily by those wishing to further promulgate the discourse. As described, influential narratives are more effective if they skilfully weave a story between plausible claims and subtly manipulative suggestions, with the main aim being that the target audience identifies with the narrative and accepts it. Use of specific lexical devices such as the confident assertion of "givens" without further explanation, can help certain ideas to be positively received by the intended audience.

In the opening section of *Milestones*, Qutb described in a degree of detail how both the West and the East, between which the Muslim world was caught, could be seen to have failed and to offer scant hope for the people. Regarding the model offered by the Soviet Union, Qutb noted:

But now Marxism is defeated on the plane of thought, and if it is stated that not a single nation in the world is truly Marxist, it will not be an exaggeration. On the whole, this theory conflicts with man's nature and its needs. This ideology prospers only in a degenerate society or in a society which has become cowed as a result of some form of prolonged dictatorship.[68]

Russia, furthermore, had moved from a position of exporting food under the Tsars to having to import it from abroad through sales of gold reserves, due to a "failure of the system of collective farming". Western democracy, meanwhile, had become "infertile" to such an extent that it was having borrow ideas from the socialist bloc. Qutb noted that the scientific advancement of the West which had begun in the Renaissance period had "reached its zenith" and did not "possess a reviving spirit". In more over-arching turns, he suggested that:

All nationalistic and chauvinistic ideologies which appeared in modern times, and all the movements and theories derived from them, have also lost their vitality. In short, all man-made individual or collective theories have proved to be failures.[69]

In this way, a relatively simple narrative is developed in which both Western and Communist ideologies have been seen to have tried in their various ways to offer progress for humanity, but both had either demonstrably failed or atrophied (asserted as "have proved to be failures").

This sets the scene for the final frame, the **narrative of hope and redemption**. Qutb identifies this as the moment in history where Islam steps forward as a third alternative to the failed ideologies of West and East:

At this crucial and bewildering juncture, the turn of Islam and the Muslim community has arrived—the turn of Islam, which does not prohibit material inventions. Indeed, it counts it as an obligation on man from the very beginning of time, when God deputed him as His representative on earth, and regards it under certain conditions as worship of God and one of the purposes of man's creation… Thus the turn of the Muslim community has come to fulfil the task for mankind which God has enjoined upon it.[70]

This last paragraph of the opening section is punctuated with four extracted *surahs* from the Quran, in which, in rather broad terms, the role of man as the earthly and worshipful representative of God is described. As identified earlier in the discussion of ISIS's online magazines, *Dabiq* and *Rumiyah*, a selective use of Quranic references can reinforce in the recipient a notion of solemn religious obligation to follow a particular path of resistance,[71] even if the interpretation drawn from such references can be somewhat subjective.

As Calvert describes, while Qutb did not offer much in the way of specific detail about how an Islamic resistance or jihad should unfold in practical terms, there seems little doubt that many violent Islamist movements across the Muslim world have subsequently used Qutbian ideas to justify what Haddad termed the "Islamic imperative".[72] The historical contextualisation reveals a broader emergence of Political Islam in the Muslim world, at a time when colonial history and the political pressure to align with either West or East were causing frustration and disillusionment in many parts of the world. It is also the case that Qutb's ideas were still very influential amongst some followers of the Muslim Brotherhood many years later, when the organisation precipitated the overthrow of the pro-Western military regime of Hosni Mubarak in Egypt.[73]

Case study 2: Britain First

A somewhat different case is that of the populist, Far-Right organisation, Britain First (BF). At the time of writing, this organisation is primarily an internet-based pressure group, although in 2021 it successfully re-registered as a political party with the UK's Election Commission. If historical precedents are indicative, it is likely that little success will be achieved in either local council or parliamentary elections for the foreseeable future. (The British National Party (BNP), from whom BF later split, remains one of the only Far-Right parties to ever win local council seats in Britain, and none have ever won a parliamentary seat.)

This is somewhat paradoxical, however, as BF shows signs of having a substantial following on social media sites such as Facebook, and thus may attract a wider ideological sympathy than elections suggest.[74] It is also the case that BF shows violent tendencies which cause some concern to policy-makers. This is not least as BF's leader, Paul Golding, carries a number of criminal convictions, including one under the Terrorism Act for refusing to surrender documents and media to UK border officials following a visit to Russia.[75] It is also alleged that Thomas Mair, a Far-Right terrorist who murdered the British MP, Jo Cox, in 2016, shouted "Britain First!" at the time of the attack,[76] thus showing an ideological link between such movements and individual instances of serious security threat.

In this analysis, a statement on the internet entitled "Britain First—Ideology"[77] is examined, using the same three-factor framework of **anxiety, influence**, and **narratives of hope and redemption** applied to the first case study above. Using a loosely structured qualitative CDA approach, this analysis shows that the framework works well as a tool on this ideologically very different strand of extremist thinking. Furthermore, there are notable similarities in rubric and sentiment between the BF and the *Milestones* examples, thus providing further support for the notion that extremist narratives and processes are structurally similar at root.

The statement under analysis provides an explanatory description of BF's ideology and objectives as a movement, presented under the following five categories:

- Right-wing or Left-wing?
- Globalism or Nationalism?
- Family values and morality;
- Race, immigration and demographics;
- Christianity as the foundation.

Each of these sections punctuate text with a selection of images, which are themselves indicative of the narrative, and emphasise the importance of multi-modal forms of persuasive communication. The images tend to fall into a few categories, namely: pictures of Far-Right political leaders who share much of BF's ideological stances (including Donald Trump, Viktor Orban, and Marine le Pen); pictures of the British royal family; pictures of supposed moral degradation, such as a group of primary schoolchildren being read a story by a transvestite; and images of Christian ritual, such as a group of children in a Christian school assembly, and an image of a couple being married in a church. In this analysis, we will focus primarily on the textual, semantic elements of the discourse, but it is clearly the case—perhaps especially in internet-based communications—that a mixture of words and audio-visual mechanisms can add to the persuasive effect.

Interestingly, one of the most immediate similarities between the BF narrative under analysis and *Milestones* is an underlying theme that could be described as focusing on societal values and morals. In *Milestones*, the core theme is that both Western and Communist society do not offer the necessary "life-giving values" to humankind, and that Islam is the only system that can do so. Parallels are drawn between the state of *jahilliyah* that existed before the Prophet Muhammad, in which "moral degeneration" was supposedly rife.[78] Similarly, the BF statement claims to champion "good family values" and to reject "the promotion of hedonism, selfishness and sexual promiscuity". The key difference between these two moralistic narratives, of course, is that the BF statement identifies Christianity as the ideological basis for building a more morally sound society, rather than Islam. Similarly, Qutb explicitly rejects nationalism as a man-made system, while BF proclaims its core ideology to be one of nationalism.

Moral degeneration is one of the key themes in the BF statement establishing a frame of **anxiety**. A related and perhaps more prevalent theme is one that connects with a conspiracy theory sometimes called the "Great Replacement Theory". As Goetz identifies,[79] this is a theory promulgated by "identitarian" movements and other strands of Far-Right ideology, particularly in Europe, which suggests that Christian Europeans are gradually becoming a minority through organised immigration. The BF statement brings together these themes of anxiety a number of times, such as:

> It is no coincidence that the decline in the numbers of people claiming to be Christian—from over 80% of the British population in the 1950s down to just over 53% today—corresponds very closely with a decline in traditional British family values, morals and ethics over the same period.
>
> Also, at a time when demographics suggest that the immigrant-origin population in this country is doubling every twelve to fifteen years, while the population of indigenous Britons is below replacement level, it is not difficult to predict that in a very short period of time—maybe less than two generations—that Christians will become a minority in Britain on the current demographic trajectory
>
> *(sic).*

In previous research into the activities of another Far-Right pressure group in Britain, the English Defence League (EDL), one supporter at an EDL rally claimed that their reason for following the movement was that they did not want their "daughters to grow up having to wear the *burqa*".[80] This small snapshot of sentiment shows how anxiety over cultural change through immigration can translate into a determination to actively participate in an extremist movement. This demonstrates how **influence** can be mobilised on the back of a sense of anxiety.

Significantly, the BF statement suggests that an unspecified community called "globalists" are seeking to ensure that "British people will become a minority within these islands": that is, the relative decline of British (Christian) people is being deliberately engineered through specific policies. The lexical device of asserting a "given" is deployed, as demonstrated in the following extract:

> This seventy-year campaign to dispossess the British people within their homeland amounts to a campaign of genocide, if we are accept the definition of "genocide" as laid down by the United Nations.

It is a deliberate and conscious campaign to dispossess the indigenous peoples of these islands, to make them a numerical minority and relegate them to minority status.

Of course, there is no evidence that any such organised campaign exists in government and has been in place since the 1950s; nor indeed that differential birth-rates and immigration will change the community landscape in the UK to the degree and at the pace asserted by BF. It is also highly fanciful to equate any such process with "genocide", but using such as emotive term is a powerful lexical device. As described, it is irrelevant whether such assertions are truthful: the most important thing is that the theory resonates with the receiving audience and is accepted by it.

Finally, having established a basis of anxiety and influence around the core themes of demographic change and societal degeneration, the BF statements offers a set of **narrative of hope and redemption** by presenting the BF movement as the vanguard of change and resistance:

Only with a return to traditional, healthy moral and family values, underpinned by a strong Christian faith, can we hope to emerge from the present degeneracy and decay with anything that resembles a nation with a future... .

Our policies on immigration will lead to a halt to the demographic dispossession of the British people in these islands and will forever maintain their status as the demographic majority... .

Britain First seeks a return to the natural, healthy nationalist ways of the world and rejects the unworkable, pie-in-the-sky schemes of globalism and the liberal left.

The detail of such policies is not made clear in this statement, but those are perhaps subjects for other publications. At this point, a basic message is promulgated that following this movement is the only way to address the grave situation of societal anxiety that has been depicted. The message is simple, and one that can be easily picked up by those minded to do so.

Policy implications and conclusions

As identified in President Macron's words at the beginning of this chapter, the political implications of rising extremism are partly that significant numbers of voters are starting to identify with extremist narratives to deal with political and economic anxieties they have been facing. While such extremist groups and sentiments are easy to dismiss as unpalatable and unworthy of serious consideration, there is a risk that mainstream political parties will fail to adapt themselves to the anxieties being felt in sections of the electorate, and this could ultimately lead to political surprise. More importantly, dismissing some of these anxieties— such as those over immigration, for example—as products of ignorance or misapprehension, will offer scant comfort when it becomes clear that beliefs are sometimes more important that objective realities in the information people choose to consume.

Propaganda studies show that political beliefs are constructed from carefully-built narratives and discourses, which weave together plausible suggestions, half-truths and manipulative assertions. Language is a very powerful tool in building understanding and belief, and, indeed, in deceiving and manipulating. The job of counter-extremism policy is to understand the cognitive and emotional processes that certain individuals go through in being drawn towards violent ideologies and violent action. This is a process shaped very much by information and ideas. In applying CDA techniques to extremist discourses, counter-extremism officials can attempt to characterise how extremist communicators are developing narratives to which supporters may be drawn. Most importantly of all, this understanding will allow the development of counter-narratives that aim to neutralise the cognitive power of the extremists.

This chapter has presented a basic, and readily-deployed model for the CDA of extremist narratives, which offers a high-level, qualitative analysis of discourse. Further research can usefully evaluate how well such models work in analytical and policy environments, and how best they could form the basis of training for counter-extremism policy makers.

Notes

1 Kirby, "French election".
2 Ashby, "Far-Right Extremism".
3 Aarup, "Le Pen".
4 BBC News, "Hungary election".
5 See note 2 above.
6 Bakir et al, "Organized Persuasive Communication", 312.
7 Welch, *Opening Pandora's Box,* 17.
8 Le Bon, *The Crowd.*
9 Laswell, *Propaganda Technique.*
10 Patrick and Thrall, "Beyond Hegemony", 99.
11 Bernays, *Crystallizing Public Opinion.*
12 Cited in Postman, "Propaganda", 132.
13 Doob, *Propaganda,* 3.
14 Ibid.
15 Ellul, *Propaganda.*
16 Nelson, "Propaganda, Behaviorism and Conscience", 48.
17 Welch, *Propaganda: Power and Persuasion.*
18 Jervis, *Perception and Misperception.*
19 Hudson, *Foreign Policy Analysis.*
20 Gramsci, *Selections.*
21 Chomsky, *Necessary Illusions.*
22 Kibbe, "Covert Action".
23 Khosrokhavar, *Radicalization.*
24 Richards, *Extremism, Radicalization and Security.*
25 Archetti, "(Mis)Communication Wars".
26 Betz, "The Virtual Dimension", 510–40.
27 Ibid.
28 Conroy and Al-Dayel, "Identity Construction", 13.
29 Lakomy, "Recruitment and Incitement", 577.
30 Ibid., 573.
31 Mehran et al., "Deep Analysis", 3.
32 Friis, "Beyond Anything", 746.
33 Chiluwa, "The Discourse", 334.
34 Enstad, "Glory", 780.
35 Van Dijk, *Discourse and Context,* 219.
36 Ibid., 225.
37 Chiluwa, "The Discourse", 322.
38 Sorel, *Réflexions,* 251.
39 Barthes, *Mythologies,* 143.
40 Austin, *How to do Things.*
41 Fairclough, *Language and Power.*
42 Buzan et al., *Security.*

43 Jackson, *Writing*.
44 Malešević, *Sociology*, 142–3.
45 Gastil, "Undemocratic Discourse", 469.
46 Edelman, "Language, Myths and Rhetoric", 131.
47 Wetherell and Potter, "Rhetoric", 172.
48 Chiluwa, "The Discourse", 327.
49 Wetherell and Potter, "Rhetoric", 183.
50 Richards, "Propaganda".
51 López, *Dog Whistle Politics*.
52 Guibernau, *Migration*.
53 Conroy and Al Dayel, "Identity Construction", 13.
54 Cited in Finer, *Mussolini's Italy*, 218.
55 See for example: Migration Observatory. https://migrationobservatory.ox.ac.uk/resources/briefings/m igrants-in-the-uk-an-overview/
56 Baker et al., "A useful methodological synergy?", 321.
57 Tversky and Kahneman, "Availability".
58 Richards, "Fake News", 105–6.
59 Armstrong, *Islam*, 169.
60 Cited in Shehab, "Islamic Discourse", 177.
61 Ibid., 178.
62 Fairclough, 2008: 67, cited in Shehab p.177.
63 Shehab, "Islamic Discourse", 201.
64 Calvert, "Sayyid Qutb", 510.
65 Richards, "Disseminating the Intelligence", 111.
66 Qutb, *Milestones*, 7.
67 https://www.aljazeera.com/news/2017/1/20/transcript-donald-trump-inauguration-speech-in-full
68 Qutb, *Milestones*, 7.
69 Ibid., 8.
70 Ibid., 9.
71 Lakomy, "Recruitment and Incitement", 573.
72 Cited in Calvert, "Sayyid Qutb", 510.
73 Bayet, 2013: 593.
74 Copsey, "The curious case".
75 Dearden, "Britain First".
76 Boyle and Akkoc, "Labour MP".
77 https://www.britainfirst.org/ideology
78 Qutb, *Milestones*, 30.
79 Goetz, "The Great Replacement".
80 Richards, "Reactive community mobilization", 183.

Bibliography

Aarup, Sarah Anne. "Le Pen vows to keep Russia close to prevent an alliance with China". *Politico*, April 18, 2022. https://www.politico.eu/article/le-pen-vows-to-keep-russia-close-to-prevent-an-alliance-with-china/.

Archetti, Cristina. "(Mis)Communication Wars: Terrorism, Counter-Terrorism and the Media." In *Propaganda, Power and Persuasion: From World War 1 to Wikileaks, edited by David Welch*. London: IB Tauris

Armstrong, Karen. *Islam: A Short History*. New York: Random House, 2002.

Ashby, Heather. "Far-Right Extremism is a Global Problem: And it is time to treat it like one". *Foreign Policy*, January 15, 2021. https://foreignpolicy.com/2021/01/15/far-right-extremism-global-problem-worldwide-solutions/?msclkid=801cec48c6ee11ecbafe6a58b4a135eb.

Austin, John. *How to do Things with Words*. Wotton-under-Edge: Clarendon Press, 1962.

Baker, Paul, Costas Gabrielatos, Majid Khosravinik, Michal Krzyzanowski, Tony McEnery, and Ruth Wodak. "A useful methodological synergy? Combining critical discourse analysis and corpus linguistics to examine discourses of refugees and asylum seekers in the UK press." *Discourse and Society* 19, no. 3 (2008): 273–306.

Bakir, Vian, Eric Herring, David Miller, and Piers Robinson. "Organized Persuasive Communication: A new conceptual framework for research on public relations, propaganda and promotional culture." *Critical Sociology* 45, no. 3 (2019): 311–328.

Barthes, Roland. *Mythologies*. Translated by A. Lavers. New York: Hill and Wang, 1972.

BBC News. "Hungary election: Viktor Orban's victory hailed by Putin". April 4, 2022. https://www.bbc.co.uk/news/world-europe-60981648.

Bernays, Edward. *Crystallizing Public Opinion*. New York: Boni and Liverlight, 1923.

Betz, David. "The Virtual Dimension of Contemporary Insurgency and Counter-Insurgency". *Small Wars and Insurgencies* 19, No. 4 (2008): 510–540.

Boyle, Danny, and Raziye Akkoc. "Labour MP Jo Cox dies after being shot and stabbed as husband urges people to 'fight against the hate' that killed her." *The Telegraph*, June 17, 2016. https://www.telegraph.co.uk/news/2016/06/17/labour-mp-jo-cox-shot-in-leeds-witnesses-report/.

Buzan, Barry, Ole Waever, and Jaap De Wilde. *Security: A New Framework for Analysis*. Coulder CA: Lynne Rienner, 1998.

Calvert, John. "Sayyid Qutb and the Power of Political Myth: Insights from Sorel." *Historical Reflections* 30, no. 3 (2004): 509–528.

Chiluwa, Innocent. "The Discourse of Terror Threats: Assessing Online Written Threats by Nigerian Terrorist Groups." *Studies in Conflict and Terrorism* 40, no. 4 (2017): 318–338.

Chomsky, Noam. *Necessary Illusions: Thought Control in Democratic States*. Boston: South End Press, 1989.

Conroy, Meghan, and Nadia Al-Dayel. "Identity Construction through Discourse: A Case Study of ISIS's No Respite Video." *Studies in Conflict and Terrorism*, doi:10.1080/1057610X.2020.1738683 (2020): 1–26.

Doob, Leonard W. *Propaganda: Its Psychology and Technique*. New York: Henry Holt and Co, 1935.

Copsey, Nigel. "The curious case of Britain First: wildly popular on Facebook, but a flop in elections." *Democratic Audit UK*, July 17, 2017. https://www.democraticaudit.com/2017/07/17/the-curious-case-of-britain-first-wildly-popular-on-facebook-but-a-flop-in-elections/.

Dearden, Lizzie. "Britain First leader Paul Golding convicted of terror offence." *The Independent*, May 20, 2020. https://www.independent.co.uk/news/uk/crime/britain-first-conviction-news-paul-golding-terrorism-act-a9524351.html.

Edelman, Murray. "Language, Myths and Rhetoric." *Society* 35, no. 2 (1998); 131–139.

Ellul, Jacques. *Propaganda: The Formation of Men's Attitudes*. Translated by K. Keller, and Jean Lerner. New York: Vintage Books, 1973.

Enstad, Johannes Due. "'Glory to Breivik'! The Russian Far-Right and the 2011 Norway Attacks." *Terrorism and Political Violence* 29, no. 5 (2017): 773–792.

Fairclough, Norman. *Language and Power*. London: Longman, 1989.

Finer, Herman. *Mussolini's Italy*. London: Victor Gollancz, 1935.

Friis, Simone Molin. "'Beyond Anything We Have Ever Seen': Beheading Videos and the Visibility of Violence in the War against ISIS," *International Affairs* 91, no. 4 (2015): 725–746.

Gastil, John. "Undemocratic Discourse: A Review of Theory and Research on Political Discourse". *Discourse and Society* 3, no.4 (1992): 469–500.

Goetz, Judith. "'The Great Replacement'—Reproduction and population policies of the far right, taking the Identitarians as an example." *Journal of Diversity and Gender Studies* 8, no. 1 (2021): 60–74.

Gramsci, Antonio. *Selections from the prison notebooks*. New York: International Publishers, 1971.

Guibernau, Montserrat. *Migration and the Rise of the Radical Right: Social Malaise and the Failure of Mainstream Politics*. London: Policy Network, 2010.

Hudson, Valerie. *Foreign Policy Analysis: Classic and Contemporary Theory*. Lanham MD: Rowman and Littlefield, 2014.

Ieţcu-Fairclough, Irena. "Critical Discourse Analysis and Translational Studies: Translation, Recontextualization, Ideology." *Diacronia* 5, no. 2 (2008): 68–71.

Jackson, Richard. *Writing the War on Terrorism: Language, Politics and Counter-Terrorism*. Manchester: Manchester University Press, 2005.

Jervis, Robert. *Perception and Misperception in World Politics*. Princeton NJ: Princeton University Press, 1976.

Khosrokhavar, Farhad. *Radicalization: Why Some People Choose the Path of Violence*. New York: The New Press, 2015.

Kibbe, Jennifer D. "Covert Action". *Oxford Research Encyclopaedia of International Studies*, 2017. https://oxfordre.com/view/10.1093/acrefore/9780190846626.001.0001/acrefore-9780190846626-e-135.

Kirby, Paul. "French election result: Macron defeats Le Pen and vows to unite divided France". BBC News, April 25, 2022. https://www.bbc.co.uk/news/world-europe-61209058.

Lakomy, Miron. "Recruitment and Incitement to Violence in the Islamic State's Online Propaganda: Comparative Analysis of *Dabiq* and *Rumiyah*." *Studies in Conflict and Terrorism* 44, no. 7 (2021): 565–580.

Laswell, Harold D. *Propaganda Technique in the World War*. New York: Knopf, 1927.

Le Bon, Gustave. *The Crowd: A Study of the Popular Mind*. London: T Fisher and Unwin, 1896

López I.H. *Dog Whistle Politics: How Code Racist Appeals have Reinvented Racism and Wrecked the Middle Class*. Oxford: Oxford University Press, 2015.

Malešević, Sinisa. *The Sociology of War and Violence*. Cambridge: Cambridge University Press, 2010.

Mehran, Weeda, Umniah Al Bayati, Matthew Mottet, and Anthony F. Lemieux. "Deep Analysis of Taliban Videos: Differential Use of Multimodal, Visual and Sonic Forms across Strategic Themes." *Studies in Conflict and Terrorism*, doi:10.1080/1057610X.2020.1866739,2021: 1–23.

Nelson, Richard. "Propaganda, Behaviorism and Conscience". *Journal of Thought* 15, No. 1 (1980): 45–51.

Patrick, Brian A. and Thrall, Trevor A. "Beyond Hegemony: Classical Propaganda Theory and Presidential Communication Strategy After the Invasion of Iraq". *Mass Communications and Society* 10, No. 1 (2007): 95–118.

Postman, Neil. "Propaganda". *ETC: A Review of General Semantics* 36, No. 2 (1979): 128–133.

Qutb, Sayyid. *Milestones (Ma'alim fi'l tareeq)*. Translated by Abdus Sami. New Delhi: Islamic Book Service, 2002.

Richards, Julian. "Reactive community mobilization in Europe: The case of the English Defence League." *Behavioral Sciences of Terrorism and Political Aggression* 5, no. 3 (2013): 177–193.

Richards, Julian. *Extremism, Radicalization and Security: An Identity Theory Approach*. London: Palgrave Macmillan, 2017.

Richards, Julian. "Disseminating the Intelligence: A Briefing Exercise Using Critical Discourse Analysis." In *The Art of Intelligence: More Simulations, Exercises and Games*, edited by Rubén Arcos and William J. Lahneman, 100–119. Lanham: Rowman and Littlefield, 2019.

Richards, Julian. "Propaganda i ekstremistisk sammenhæng". Nationalt Center for Forebyggelse af Ekstremisme, *Forskningsartikel* no.2 (2019). https://stopekstremisme.dk/publikationer/propaganda-i-en-ekstremistisk-sammenhaeng.

Richards, Julian. "Fake news, disinformation and the democratic state: a case study of the UK government's narrative." *Icono 14* 19, no. 1 (2020): 95–122.

Shehab, Ekrama. "Islamic Discourse, Ideology and Translation: Sayyid Qutb's Milestones as a Model." *Journal of Translation Languages* 19, no. 2 (2020): 174–203.

Sorel, Georges. *Réflexions sur la violence*. Translated by T.E. Hulme and J. Roth. New York: Collier Books, 1961.

Tversky, Amos, and Daniel Kahneman. "Availability: A heuristic for judging frequency and probability." *Cognitive Psychology* 5, no. 2 (1973): 207–232.

van Dijk, Teun A. *Discourse and Context: A Sociocognitive Approach*. Cambridge: Cambridge University Press, 2008.

Welch, David. "*Opening Pandora's Box: Propaganda, Power and Persuasion*." In *Propaganda, Power and Persuasion: From World War 1 to Wikileaks*, edited by David Welch, 3–18. London: IB Tauris, 2015.

Wetherell, Margaret, and Jonathan Potter. "Rhetoric and Ideology." Chap. 12 in *Analysing Everyday Explanation: A Casebook of Methods*, edited by Charles Antaki, 168–183. London: Sage, 1988.

29
ANTICIPATORY APPROACHES TO DISINFORMATION, WARNING AND SUPPORTING TECHNOLOGIES

Rubén Arcos and Cristina M. Arribas

Countering disinformation is not an easy endeavour for different reasons: (1) the speed of dissemination of manipulative contents through different channels, including through private messaging apps and offline interactions that influence the reception process of contents by users; (2) disinformation by authoritarian states, non-state actors like criminal and terrorist organizations, and proxy political actors of the infosphere that exploit existing specific vulnerabilities identified in targeted groups and societies that make them more prone to accept mis- and disinformation pieces and hostile narratives; (3) fragmentation of the information environment—a dynamic environment in which new social media platforms can join the digital ecosystem of channels—which makes difficult to early identify malicious pieces by security and fact-checking practitioners; (4) the absence of a verifiable continuum between the identity of users across the offline and online spectrum: digital identity is something fuzzy in which a correspondence digital–real is difficult to establish, and a real identity can be hidden or manipulated; (5) existing political polarization mediates and biases the access to news stories and opinions by individuals in a digital ecosystem characterized by the shift of the traditional gatekeeping role from the news organization to the consumer of content. This non-exhaustive list of factors suggests some specific challenges of practitioners and authorities when addressing disinformation and foreign information manipulations.

Anticipation to emerging issues and proactive communications has always been the hallmark of strategic issue management by communication practitioners in industry. This chapter argues that, similarly, for combatting disinformation as a specific national security threat and challenge requires practitioners to adopt an anticipatory approach consisting in providing situational awareness and intelligence to facilitate the decisions and actions of authorities and practitioners, including through informing preventive communication strategies. It argues that the analyses and assessments of activities in the information environment by foreign states and non-state actors, the narratives employed, and their place in grand strategies, constitute a prerequisite for the formulation and implementation of defensive measures and preventive communications as part of the response. The chapter

DOI: 10.4324/9781003190363-34

explores different perspectives on how disinformation and foreign information manipulations impact the intelligence function in democracies and the contribution of intelligence in addressing the threat. It argues that, vulnerabilities and threat assessments, foresight and early warning on hostile information and influencing activities, coupled with holistic approaches (including a set of government and nongovernment actors such as fact-checking organizations) in information sharing and in the preventive communication response, impact the capability of being anticipatory and the effectiveness of the response. The chapter also explores the role of technology in supporting analysis and early warning with this regard, as well as in supporting fact-checking activities.

Anticipatory management and issue management

Issue management and crisis management are important management and communication practices, providing organisations with processes and techniques to "identify risks and issues early; take planned action to influence the course of those issues; respond effectively if issues develop into crises; and protect organisational reputation during and after a crisis"[1]. For a problem to be qualified as an "issue", its significance for and impacts on the organization are essential[2]. An issue is "any development—usually in the public arena—which, if it continues, could have a SIGNIFICANT impact on the operation or future interests of the organization"[3].

Matera and Artigue have pointed out on the similarity, from a processual perspective, of issue management campaigns with other public relations campaigns—the PR process encompasses research, planning, implementation, and evaluation (RPIE)—and on the importance of both, early identification of the issue as it is emerging, and how the outcome can be influenced by "the proper amount of carefully planned and well-timed communication strategies"[4, 5].

Crable and Vibbert discussed on the relationship of three forces—government, citizens, and business—in relation to the categories of authority and influence over public policy:

> What is clear is that "citizens" and "business" have no co-equal authority. What they do have is influence, and that influence affects dramatically the authority of government to do its work. [...] The task of citizens and business is not to "make" the policies, but to influence them [...] Organizations, having no "authority" over public policy, are limited to exerting "influence" over policies enacted by those who do have the authority. This influence, as defined here, is why "issue management" has importance: Issue management can permit an organization, with no actual authority, to influence public policy.[6]

The issue management process encompasses different steps, though these steps vary depending on the existing models described in the public relations scholarship[7]. William Renfro, for instance, identified four main stages:

1 Identifying potential issues by scanning the horizon and beyond of the corporation's current and planned operating environment and peripheral environment.
2 Researching the background, future, and potential impacts of these issues.
3 Evaluating issues competing for anticipatory operations and actions programs
4 Developing strategies for these anticipatory operations.[8]

While in the Chase/Jones model there is a fifth evaluation step:

1   Issue identification,
2   Issue Analysis,
3   Issue change strategy options,
4   Issue Action Programming, and
5   Evaluation of results"[9].

Together with the issue process model the issue life cycle is regarded as the other principal framework for managing issues[10].

Crable and Vibbert introduced the idea of "issue status" throughout a "life cycle": (1) potential, (2) imminent (or emergent, as referred as well in the PR scholarship), (3) current, (4) critical, and (5) dormant.[11] Eli Sopow's workbook "The Critical Issues Audit" similarly discussed on anticipating issues in advance by considering the different issue stages across a predictable pattern of progression and how "taking pre-emptive action can neutralize your problem before it mushrooms out of proportion and out of your control"[12]. In Sopow's issue progression model, issues have three different stages (emerging, active, and ebbing) and managers have opportunities to gather information from different sources as the issue's intensity increases (See Table 29.1).

Timothy Coombs (2002) introduced the idea of issue contagions enabled by the Internet and how their rapid spread from person to person, like a virus, may alter the probability of impact of an issue, jumping from the Internet to mainstream traditional media, and hence shifting issue priorities and requiring decisions and actions. Coombs also suggested an issue threat analysis perspective, recommending to assess impact ("the amount of damage an issue can inflict on an organisation") and likelihood, where "likelihood can be conceptualised as a function of issue legitimacy and issue manager power"[13].

Table 29.1. Stages in issue progression

Stage	Intensity of signals/signposts and over time (visibility)	Information sources
Emerging	Low	Technical journals, reports, regional news outlets meetings' minutes, public opinion surveys, content analysis
Active 1	Medium	Specialized publications as in the previous stage and local news outlets
Active 2	High to Peak	Mainstream national news media, stakeholders that affect or affected by the issue, public opinion polling and content analysis
Ebbing	High to low	Newspaper editorials and other journalistic opinion pieces like op-eds and letters to the editor

Source: Authors. Adapted from Sopow's issue progression curve (1994): 65–68.

Anticipatory intelligence and warning on disinformation and hostile influencing

The anticipatory approach of issue management for dealing with public issues provides a useful perspective for dealing with disinformation activities by hostile actors. We are not however arguing for a direct transference but for adapting anticipatory management tools and underlying "philosophy" of issue management and proactive and preventive communication techniques.

Disinformation and different forms of information manipulations are necessarily about something (an event, development, institution, organization, or a person). It may aim to alter the conditions and context of reception of rightful information by groups and individuals, create confusion, mobilize basic emotions like fear or anger about something, justify and legitimise decision and actions, and set misunderstanding about decisions and developments. A very clear example is the activities in the information domain conducted by the Russian Federation's state-funded media like RT and Sputnik in relation to the aggressions against Ukraine and other unacknowledged interference activities targeting foreign countries.

A potential approach when addressing coordinated disinformation activities is assessing the probability of a foreign entity's exploitation of an existing vulnerability in a targeted democratic society and by what means. This entails anticipating the risk of an attack in the information environment by assessing the hostile entities' capabilities, intentions, and activities, and the target vulnerability, since risk is "a function of threat (adversaries' capabilities, intentions, and opportunities) and vulnerability (the inherent susceptibility to attack)"[14]. From this perspective, the analysis of vulnerabilities may examine existing political, social, and historical cleavages, opportunities like an economic crisis and a pandemic, or interrogate for questions such as what keeps us together, or separate us as society, what divides us, or what motivates us[15]. Foreign entities and hostile actors with no authority but with the capability of influencing may activate latent issues in target societies, of which they might already have gathered information, and disseminate and amplify narratives towards those issues. A potential tool for assessing vulnerabilities is elaborating a matrix anticipating domain (STEEP, PMESII, or Hybrid Threat targeted domains) specific issues likely to escalate throughout the life cycle when activated by foreign entities through information operations (See Table 29.2).

At the same time, an analysis of the capabilities of adversaries for hostile influencing will gather information and map their media ecosystem of owned/affiliated/covertly sponsored outlets, and social media platforms or assets likely to be used, as well as identify patterns of behaviour and past practices, and anticipate likely ad-hoc alliances with local actors in particular situations.

Forecasting disinformation for warning intelligence

The analysis of capabilities and intentions of adversaries, coupled with a target vulnerability assessment can be part as well of an indications and warning system. In the work "Forecasting Terrorism" (2004), Sundri Khalsa developed a forecasting methodology that provides a useful point of departure.

This approach may develop scenarios as well by examining driving forces, and then develop a system of indicators for each scenario. By combining foresight techniques with a forecasting system of indicators, authorities and strategic communicators can develop both countermeasures and preventive communications like pre-bunking activities against disinformation.

Table 29.2. Vulnerability assessment matrix to potential issues emerging in different domains with examples.

Question: How vulnerable are we (country X) to foreign/hostile information manipulations regarding these specific issues?

Domain (STEEP PMESII, or (13) Hybrid Threat domains)	Issue	Issue description	Vulnerability assessed score (0 to 10)	Potential early signals of activity in the information environment
1.Environmental	1.1.Climate change	Negationist discourses threaten societal understanding and behaviour in relation to anthropogenic climate change	[Insert assessed score for prioritisation]	Low reach accounts in fringe social media replaying narratives from digital outlets associated to a foreign entity that misinform on scientific consensus about climate change or questions.
2.Political	2.1. NATO	Information manipulations on the nature of NATO threatens internal political consensus in the response to a foreign state aggression against a friend or allied country		A former military officer covertly funded by a foreign state publishing a paper in a predatory journal alleging that NATO is an offensive alliance with an imperialistic grand strategy

Source: authors

Cynthia Grabo explained the difference between indicators and indications in the following way:

> an indicator is a known or theoretical, step which the enemy should or may take in preparation of hostilities. It is something which we anticipate may occur, and which we therefore usually incorporate into a list of things to be watched for, known as "indicators list" [...] (Information that any of these steps is actually being implemented constitutes an indication).[16]

This difference between expectation (indicator) and actuality (indications) is important to keep in mind when developing a warning system against disinformation (See Table 29.3). At the same time, the actuality of disinformation activity may be an indicator of something likely to happen in the foreseeable future, like for example future military actions.

On the other hand, there are special events that present opportunities for malign influencing like international summits, caucuses and elections, councils of ministers, sport and cultural events, and others. Since there is advance publicity on when these events will be held,

Table 29.3. Example of list of indicators associated with a potential disinformation scenario

Scenario: On the road to an important legislative debate on reforming a set of norms for harmonization at the European level, an authoritarian competing state interferes in the process through a coordinated campaign in several European countries.

Capability indicators

1. New social media accounts (fake personas) with an unusual high number of followers in mainstream social media

2. Setting of state-funded media broadcasting in a language different than the originating state and shares on private media companies in countries X, Y and Z

3. New fringe social media platforms

4. Flagship industry and academic forums for international debates

5. Publishers with a focus on social sciences publishing in foreign languages

n.

Intention indicators

1. Public statements by the adversary state's political leaders on the issue

2. Increase of public diplomacy activities of the competitor state through state-funded media and social channels

3. Meetings with industry leaders and opposition leaders of target countries

4. Un-attributed cyberattacks to institutional websites

5. Leak of private messages of political leaders positioned on the issue

n.

Vulnerability indicators

1. Low level of public understanding on the topic theme under discussion

2. Lack of sufficient levels of awareness on hostile influencing

3. Lack of trust in democratic institutions

4. Lack of a system of pluralistic independent media

5. Consumption patterns of news stories in social media

n.

Source: Authors based on Khalsa 2004

hostile entities may size the opportunity for re-activating latent issues, framing, or conditioning the public conversation around those events. It is a fact that advance public statements by foreign political leaders on those events have the potential to influence the news media agenda. At the same time, a previous set of activities in the information domain can provide context for interpretation of the news by individuals and stakeholders and shape perceptions, attitudes, and opinions.

An intelligence function with competence in disinformation and hostile interference is useful for disrupting and neutralizing these attempts. Foreign threat actors can employ their intelligence organizations in covert influencing activities and the involvement of intelligence in countering such unacknowledged activities is key.

The war in Ukraine has shown examples of how Western intelligence made use of preventive communication both by releasing sanitized intelligence with the public and by publishing regular intelligence updates on the situation in different regions suffering the attacks of the Russian Federation. Public communication by security and intelligence services against disinformation and the attempts of foreign entities to manage perceptions of civilians around important developments is useful in that it helps individuals to understand the situation and distinguish known facts from all sorts of speculations, provide context for interpretation of new information and help professionals in the media and fact-checkers in contrasting information for verification.

Technologies supporting open-source intelligence and against disinformation

Countering disinformation through analytic tasks like detection and analysis of disinformation content can be facilitated through Artificial Intelligence methods and tools. A report of NATO StratCom COE explained how AI can be used for assisting activities like detecting bots in social media platforms, screening content, tracking hostile narratives, or identifying AI-generated audiovisual and textual content[17]. The Government Communication Headquarters (GCHQ), which is one of the leading agencies of the UK intelligence machinery, has acknowledged how Artificial Intelligence (AI) will be fundamental to its mission and "ability to deal with the ever-increasing volume and complexity of data, and to develop the capabilities needed to defend against AI-enabled threats by malicious actors."[18] More specifically, with regards to foreign disinformation, the GCHQ has pointed that "AI enabled tools could be deployed for machine-assisted fact checking through validation against trusted sources and to detect deepfake media content," enhance capabilities in the detection and blocking of botnets, and the identification of "troll farms" and other sources of mis- and disinformation.[19] Natural language processing in support of automated fact-checking has been one of the key focus of innovations against disinformation.

Mis- and disinformation are also aggravating factors of what Gregory Treverton called "threats without threateners", that can come as an aggregate effect of individual actions[20], like, a for example, a pandemic. In this context, an information epidemic or infodemic, including the spread of deliberately false or misleading content[21] can adversely impact the understanding and behaviour of individuals and AI-based innovations can facilitate the processing and analysis of information flows in the digital communication environment, and "identify infodemic signals".[22]

AI-based technologies supporting intelligence, fact-checking and argumentation-checking

Interestingly, research and development on AI systems and models is a phenomenon contemporary with the rise of the mis- and disinformation in the last years. Disinformation is not a new phenomenon but now it can be enabled by AI techniques as well, and hence it can potentially increase the capabilities and ways in which threat actors produce and disseminate malicious content and advance their goals and interests. AI-based systems have been implemented both with malicious ends to support the spread of disinformation, but also to assist in tasks related to detection and debunking.

The United States' National Security Commission on Artificial Intelligence refers to three paths where throughout which AI and related disciplines and technologies such as Machine Learning (ML) or Natural Language Learning (NPL) can be employed by hostile actors in information operations: message, audience, and medium[23]:

- Message. ML, together with NLP, can generate original text-based content, predict the most likely word sequences for automatic response, understand the natural language inputs, and construct realistic natural language contents[24]. Also, IA can be used to create audio-visual forgeries (deepfakes) by employing generative adversarial networks and reinforcement learning (RL)[25].

- Audience. ML systems can be used to segment target audiences and identify common patterns for profiling and customization of key messages according to the beliefs and ideologies of individuals[26].
- Medium. According to NSCAI, "AI can be embedded within platforms, such as through ranking algorithms"[27]. Additionally, private messaging platforms present a high risk of being employed for the dissemination of malicious information due to their viralization capacity and the difficulty of detection with these channels. Complex conversational AI, that makes use of various technologies—ML, NLP, Natural language understanding (NLU), Deep learning, and predictive analytics,[28] can exacerbate the problem. Regarding bots—automated accounts that mimic human behaviour—that can be used to spread inflammatory content with the aim of promoting a specific view or stance, ML may soon enable both a greater amplification potential and avoid their detection. Some companies have already started offering services of synthetic content and amplification bots that can follow, retweet, and like tweets[29].

On the positive side, AI has also a great potential for supporting the countering of hostile information operations, tackling mis- and disinformation, and "preserve peace and security by helping guarantee the integrity of democratic elections [...] provide real time forecasts of mass violence in volatile contexts"[30]. Risks related to mis- and disinformation may be exacerbated by several trends:

> the blurring lines between foreign and domestic influence operations, the outsourcing of these operations to private companies that provide influence as a service, and the conflict over distinguishing harmful disinformation and protected speech.[31]

This AI-based support for countering disinformation can be placed in the following categories:

Digital environment monitoring tools

There exist many commercial tools and platforms on the market that allow real-time monitoring of the digital environment, supporting the identification of latent issues, the detection of disinformation campaigns, as well as the monitoring of how communication plans are being implemented in areas like counteracting the effects of mis- and disinformation.

Although these tools began to appear in the wake of the explosion of social networks and smartphones in the late 2000s, currently, these platforms have improved substantially thanks to the use of ML and NLP, which allows greater refinement of the results, analysis and data visualization precision[32].

A selection of information sources and keywords allows an *ad hoc* configuration of these tools focused on to the issue/topic selected. The source configuration allows to include social media platform accounts, not only mainstream ones such as Twitter, Facebook, Instagram, or YouTube, but also newer or more regional ones—depending on the platform—such as VK, Telegram, Webchat or TikTok. Likewise, the crawler that they incorporate allows documents to be obtained from websites such as digital newspapers and other news media outlets, web portals, forums, and blogs. The multilingual support allows working simultaneously in different languages depending on the configuration of terms made and the topic of interest.

ML and its different functions allow text extraction and analysis, data analysis and visualization. Also, ML-based support allows the identification of the logic behind past user's search strategies, which can impact the relevance and accuracy of posterior searches. This avoids the so-called blacklists—lists of terms that were manually inserted to exclude unwanted results—and the number of false positives is substantially reduced.

These tools in turn allow to identify within the configuration and user interface those terms and hashtags that are most repeated among the accounts or sources analysed; this way it is possible to predict within an issue those factors or elements that may be more controversial and susceptible to be used by malicious actors. Similarly, these tools help in the identification of topics that can exacerbate polarization and that precisely for this reason are likely to be exploited in the future by bad actors.

Another functionality of great interest provided by ML is the ability to conduct social network analysis (SNA) within a platform, build knowledge graphs, identify influencers, communities, and associations between groups that sometimes show opposing or very different interests and ideologies.

On the other hand, by working from the API provided by the social networking platforms, it is possible to extract information related to demographic data, geolocation, professional profile as well as the interest of the users, hence allowing practitioners to identify the targeting strategy employed by bad actors when conducting information operations.

Sentiment analysis, on the other hand, makes it possible to identify not only whether the sentiment of a post is negative or positive, but whether it agrees with a broader idea (i.e. extremist stances towards a range of political issues, including foreign policy issues) through the creation of labels by the user. In the same way that the user can refine the results through machine learning, sentiment analysis can be fine-tuned thanks to machine learning by reproducing the user's behaviour.

The most advanced tools also include image, audio, and video analysis functions allowing the recognition of objects or people, and the conversion of audio content to a text file through text mining.

Fact-checking tools

AI has also allowed the development of automatic fact-checking tools or platforms, through the implementation of algorithmic models based on deep learning, machine learning, natural language processing (NLP), and big data[33]. These tools differ in their approach, and it is possible to establish different categorizations among them[34].

According to Choras et al. (2020)[35], ML has been applied for the detection of fake news, by either focusing on text analysis, reputational analysis, network analysis, and image manipulation recognition. Following Lucas Graves, it is possible the describe the following activities within fact-checking that can be supported by AI: (1) claim identification, consisting of the extraction of claims from a text through a combination of natural language processing and machine learning; (2) claim verification, which is still a developing field for AI, and requires analysing inferences, context, and supporting evidence for these claims[36]. There are two related tasks that automation supports: matching claims against a library of previously verified statements, and against a reliable source[37]. Following this taxonomy, we find platforms that integrate several modular solutions based on algorithms. Among the most famous is ClaimBuster, a platform that detects factual claims[38]. The tool uses an API to launch queries to search engines (Google) and databases (Wolfram Alpha). It also allows the user to identify if a content has already been verified[39]. Similar platforms allowing the identification and comparison of contents are CheckThat (CLEF Initiative) and Full Fact Live.[40]

Detection of bots and trolls

The use of bots for legitimate purposes by social media platforms has been common among companies and government agencies to help users provide them with information or services[41]. Research has shown that social bots represent an important percentage of the total social accounts. A 2016 study found that between a 5 and 9% of Twitter accounts corresponded to social bots, which produced the 24% of the total tweets[42]. Current findings from a 2022 study conducted by CHEQ, an Israeli cybersecurity company specializing in "advanced bot mitigation technology"[43] showed that from the 5.21 million website visits analysed that came from Twitter, a 11.71% were from bot accounts[44].

On the other hand, the use of bots to amplify mis- and disinformation is well known. Detecting unusual activity from new accounts, or existing accounts, that start posting about a topic that has been identified as potentially viral can be a good indicator that a malicious information campaign is ongoing. Numerous research have explored the behavior of these automated accounts[45]. It is worth mentioning the number of investigations focused on the analysis of specific activities, such as the Russian interference in the 2016 United States presidential elections, anti-vaccine misinformation before the COVID-19 pandemic, or COVID-19 related conspiracy theories. The relationship between bots and trolls has also been examined. A 2018 study showed how in the vaccine debates, a coordinated action was observed between Russian bots that spread messages with anti-vaccine content and trolls that tried to sow discord and erode the social consensus on vaccination[46].

The scholarship focused on bot detection has produced a number of studies[47]. After examining the existing literature, Van Der Walt and Eloff pointed out various approaches for detecting bots. Among others, these approaches consist of examining fake content linked to the account; profile image and other identity attributes; specific circumstances such as the timeframe between account creation and first post; frequency of activity; stance and sentiment towards particular topic issues; relationship with similar accounts or target accounts.

The analysis of trolls is more challenging than bots' identification. Stylometry assisted by ML—the study of the distinctive aspects of someone's writing style—is one the approaches that has been applied to this purpose[48]. One example of this kind of approach is observed in the analysis conducted by Marcellino et al., (2020)[49] on the tweets disseminated during the 2016 US presidential campaign.

Argument-checking

There are, however, other important challenges that could further assist these efforts through a more proactive and anticipatory approaches by facilitating early identification and inoculation through the dissemination of counter messages and narratives[50]. With this regard, tools assisting the process of mapping claims, underlaying argumentation and supporting factual statements disseminated online can facilitate the fast crafting of counter messages against misleading content and conspiracy theories.

The Artificial Intelligence for Decision Making Initiative in Australia has called attention to argument mining for its potential analytical applications in security and defence[51].As explained by The Hunt Laboratory for Intelligence Research of the University of Melbourne, argument mining consists of automatically extracting the reasoning structure from texts, being "a rapidly-developing area of artificial intelligence, in the sub-field of natural language processing." (See https://huntlab.science.unimelb.edu.au/home/research/argument/). Together with its potential application to intelligence analysis[52] argument mining can greatly assist in dealing with mis- and disinformation. Visser, Lawrence and Reed argued for supplementing fact-checking with "reason-checking":

> Successfully recognizing fake news depends not only on understanding whether factual statements are true, but also on interpreting and critically assessing the reasoning and arguments provided in support of conclusions. It is, after all, very possible to produce fake news by starting from true factual statements and drawing false conclusions by applying skewed, biased, or otherwise defective reasoning.[53]

Lippi and Torroni (2016) described the web system MARGOT[54] (an acronym that stands for Mining ARGuments frOm Text), design to make argument mining accessible outside the argumentation mining community, and that "performs argument mining by exploiting a combination of advanced machine learning and natural language processing techniques."[55]

Conclusion

Countering disinformation and influence activities in the infosphere requires not only reactive strategies when situations escalate and malicious contents have greater reach and potential to influence beliefs, attitudes, and behaviors of target audiences, but also anticipatory approaches able to early identify emerging issues that can be the object of disinformation. Frameworks and tools from issue management may orient the activity of practitioners when addressing disinformation from an anticipatory perspective.

AI-based technologies already assist fact-checking and intelligence efforts against disinformation and research suggest the role of AI will be even more important in the future with this regard. However subject matter expertise will keep being relevant for risk assessment and warning on disinformation.

There are promising and relevant areas of research like argument mining in which technology can further assist counter influencing activities. Tracking the spread of content across platforms is an important area for further research to understand how disinformation originates, spread and influence target audiences.

Notes

1 Jaques Tony, *Issue and Crisis Management: Exploring Issues, Crises, Risk and Reputation.* (Oxford University Press Australia & New Zealand. Kindle edition, 2014), 323.
2 Ibid., 22.
3 Ibid., 19.
4 Matera, Fran R. and Ray J. Artigue. *Public relations campaigns and techniques: building bridges into the 21st century.* (Needham Heights, MA: Allyn and Bacon, 2000), 165.
5 See also: Xifra, Jordi. *Comunicación proactiva. La gestión de conflictos potenciales en las organizaciones.* (Barcelona: Gedisa, 2009).
6 Crable, Richard E., and Vibbert, Steven L. "Managing Issues and Influencing Public Policy," Public Relations Review 11, no. 2, (1985): 4.
7 Arcos, Rubén. "Public relations strategic intelligence, intelligence analysis, communication and influence," *Public Relations Review* 42 (2016): 264–270.
8 Renfro, William L. *Issue management in strategic planning.* (Westport, CT: Greenwood Publishing Group, 1993), 67.
9 Chase, W. Howard. *Issue management: Origins of the future.* (Stamford, CA: Issue Action Publications, 1984), 36.
10 Jaques, Tony. *Issue and Crisis Management: Exploring Issues, Crises, Risk and Reputation.* OUP Australia & New Zealand. Kindle edition, 2014.

11 Crabble, Richard E., and Vibbert, Steven L. "Managing Issues and Influencing Public Policy," *Public Relations Review* 11, no. 2, (1985): 5–7.

12 Sopow, Eli. *The Critical Issues Audit*, (Leesburg, VA: Issue Action Publications, 1994), 65.

13 Coombs, W. Timothy "Assessing online issue threats: Issue contagions and their effect on issue prioritisation," *Journal of Public Affairs* 2, no. 4, (2002): 220.

14 Goldman, Jan and Susan Maret. Intelligence and information policy for national security: key terms and concepts, (Lanham, MD: Rowman and Littlefield, 2016), 487.

15 Ivan, Cristina, Chiru, Irena, Arcos, Rubén. "A whole of society intelligence approach: critical reassessment of the tools and means used to counter information warfare in the digital age." *Intelligence and National Security Intelligence* 36, no. 4, (2021).

16 Grabo, Cynthia, with Jan Goldman. *Handbook of Warning Intelligence: complete and declassified edition* (Lanham, MD: Rowman & Littlefield, 2015): 10.

17 Juršėnas, Alfonsas, Karlauskas, Kasparas, Ledinauskas, Eimantas, Maskeliūnas, Gediminas, Rondomanskas, Donatas, Ruseckas, Julius. "The Role of AI in the Battle Against Disinformation." Riga: NATO Strategic Communications Centre of Excellence, 2022, https://stratcomcoe.org/pdfjs/?file=/p ublications/download/The-Role-of-AI-DIGITAL.pdf?zoom=page-fit

18 GCHQ. "The Ethics of Artificial Intelligence: Pioneering a New National Security" 2021: 6.

19 Ibid., 35.

20 Treverton, Gregory F. *Approaching threat convergence from an intelligence perspective, in Unconventional Weapons and International Terrorism Challenges and new approaches.* Edited by Magnus Ranstorp Magnus Normark, (London and New York: Routledge. Kindle Edition, 2009): 141–162.

21 Gruzd, Anatoliy and Mai, Philip. "Going viral: How a single tweet spawned a COVID-19 conspiracy theory on Twitter." *Big Data & Society* 7, no 2, (2020).

22 World Health Organization (WHO). "Digital solutions to health risks raised by the COVID-19 infodemic: policy brief." Copenhagen: WHO Regional Office for Europe, (2022): IV https://www.who.int/europe/publications/m/item/digital-solutions-to-health-risks-raised-by-the-covid-19-infodemic.-synthesis-report vi

23 National Security Commission on Artificial Intelligence. Final Report (2021). https://www.nscai.gov/wp-content/uploads/2021/03/Full-Report-Digital-1.pdf

24 Suta, P.; Lan, X., Biting, W., Mongkolnam, P., Chan, J. H. "An Overview of Machine Learning in Chatbots." *International Journal of Mechanical Engineering and Robotics Research* 9, no. 4, (April 2020): 502–511.

25 See: NSCAI 2021: 47.

26 Ibid., 48.

27 Ibid., 48.

28 Sedova, Katerina, McNeill, Christine, Johnson, Aurora, Joshi, Aditi, Wulkan, Ido. "AI and the Future of Disinformation Campaigns Part 2: A Threat Model" Center for Security and Emergency Technology, 2021.

29 Ibid.

30 Yankoski, Michael, Weninger, Tim and Scheirer, Walter. "An AI early warning system to monitor online disinformation, stop violence, and protect elections." Bulletin of the Atomic Scientists, 76, no. 2, (2020): 85.

31 Sedova et al. Ibid., 1

32 Yip, Pedro and Blaclard, Vincent. "Social Listening Is Revolutionizing New Product Development. New techniques better capture consumer intelligence in real time." MIT Sloan Managent Review, 2009

33 See: Huynh, Viet-Phi and Papotti, Paolo. "A Benchmark for Fact Checking Algorithms Built on Knowledge Bases." In Proceedings of the 28th ACM International Conference on Information and Knowledge Management, (2019): 689–698. See also: Miranda, Sebastião, Nogueira, David, Mendes, Afonso, Vlachos, Andreas, Secker, Andrew, Garrett, Rebecca, Mitchel, Jeff, Marinho, Zita. "Automated Fact Checking in the News Room." (2019) García-Marín, David. "Modelos algorítmicos y fact-checking automatizado. Revisión sistemática de la literatura." Documentación de Ciencias de la Información 45, no. 1, (2022): 7–16. Andreas Hanselowski. "A Machine-Learning-Based Pipeline Approach to Automated Fact-Checking." PhD diss., Technical University of Denmark, 2020.

34 Huynh, Viet-Phi and Papotti, Paolo. "A Benchmark for Fact Checking Algorithms Built on Knowledge Bases." Proceedings of the 28th ACM International Conference on Information and Knowledge Management, (2019): 689–698.

35 Choraś, Michał, Demestichas, Konstantinos, Giełczyk, Agata, Herrero, Álvaro, Ksieniewicz, Paweł, Remoundou, Konstantina, Urda, Daniel, Woźniak, Michał. "Advanced Machine Learning techniques for fake news (online disinformation) detection: A systematic mapping study." *Applied Soft Computing* 101 (March 2021).

36 Graves, Lucas. "Factsheet: Understanding the Promise and Limits of Automated Fact-Checking." Accessed November 13, 2022

37 Ibid

38 Miranda, Sebastião, Nogueira, David, Mendes, Afonso, Vlachos, Andreas, Secker, Andrew, Garrett, Rebecca, Mitchel, Jeff, Marinho, Zita. "Automated Fact Checking in the News Room." (2019) García-Marín, David. "Modelos algorítmicos y fact-checking automatizado. Revisión sistemática de la literatura." *Documentación de Ciencias de la Información* 45, no. 1, (2022): 7–16.

39 Ibid

40 Huynh, Viet-Phi and Papotti, Paolo. "A Benchmark for Fact Checking Algorithms Built on Knowledge Bases." In Proceedings of the 28th ACM International Conference on Information and Knowledge Management, (2019): 689–698.

41 Abeer Aldayel, Walid Magdy. "Characterizing the role of bots in polarized stance on social media." *Social Network Analysis and Mining* 12, no. 30 (2022),

42 Morstatter Fred, Wu Liang, Nazer, Tahora H, Carley Kathleen N, Liu Huan. "A new approach to bot detection: Striking the balance between precision and recall." In: IEEE/ACM Conference on Advances in Social Networks Analysis and Mining (ASONAM), (2016): 533–540.

43 https://cheq.ai/technology/

44 Johansen, Alyson Grace. "What's a Twitter bot and how to spot one." Norton Emergency Trends, September 5, 2022.

45 Abeer Aldayel, Walid Magdy. "Characterizing the role of bots in polarized stance on social media." Social Network Analysis and Mining 12, no. 30 (2022), Assenmacher, Dennis, Clever, Lena, Frischlich, Lena, Quandt, Thorsten, Trautmann, Heike and Grimme, Christian. "Demystifying Social Bots: On the Intelligence of Automated Social Media Actors." Social Media + Society 6, no. 3 (2020). Badawy, Adam, Ferrara, Emilio and Lerman, Kristina "Analyzing the Digital Traces of Political Manipulation: The 2016 Russian Interference Twitter Campaign," 2018 IEEE/ACM International Conference on Advances in Social Networks Analysis and Mining (ASONAM), (2018): 258–265.

46 Broniatowski, David A., Jamison, Amelia M., Qi, SiHua, AlKulaib, Lulwah, Chen, Tao, Benton, Adrian, Quinn, Sandra C. and Dredze, Mark. "Weaponized Health Communication: Twitter Bots and Russian Trolls Amplify the Vaccine Debate." *American Journal of Public Health* 108, (October 2018): 1378–1384.

47 Orabi, Mariam, Mouheb, Djedjiga, Al Aghbari, Zaher, Kamel, Ibrahim. "Detection of Bots in Social Media: A Systematic Review." *Information Processing & Management* 57, no. 4, (July 2014). See also: Cresci, Stefano, Pietro, Roberto D., Petrocchi, Marinella, Spognardi, Angelo and Tesconi Maurizio. "Fame for sale: Efficient detection of fake Twitter followers." Decision Support Systems 80, (December 2015): 56–71; and Cresci, Stefano. Pietro, Roberto D., Petrocchi, Marinella, Spognardi, Angelo and Tesconi Maurizio. "Social Fingerprinting: Detection of Spambot Groups Through DNA-Inspired Behavioral Modeling," *IEEE Transactions on Dependable and Secure Computing* 1, no. 4, (July-Aug. 2018): 561–576. See: Rheault, Ludovic, Musulan, Andreea. "Efficient detection of online communities and social bot activity during electoral campaigns." *Journal of Information Technology & Politics* 18, no.3, (2021).

48 Miao, Lin, Last, Mark, Litvak, Marina. "Detecting Troll Tweets in a Bilingual Corpus" *Proceedings of the 12th Conference on Language Resources and Evaluation* (LREC 2020), (2020): 6247–6254.

49 Marcellino, William, Cox, Kate, Galai, Katerina, Slapakova, Linda, Jaycocks, Amber, Harris, Ruth. "Human-machine detection of online-based malign information." RAND Europe, 2020. Accessed November 13, 2022.

50 Guo, Zhijiang; Schlichtkrull, Michael, and Vlachos, Andreas. "A Survey on Automated Fact-Checking," Transactions of the Association for Computational Linguistics, 10 (2022): 178–206.

51 Defence Science and Technology Group. "Call for Pilot Project Proposals Artificial Intelligence for Decision Making 2022 Initiative." 2022.

52 Thorburn, Luke. "Enhancing human analysis of large corpora via argument mining and backlinks—a prototype." (March 2021).

53 Visser, Jacky, Lawrence, John, and Chris Reed. "Reason-Checking Fake News" *Communications of the ACM* 63, no. 11, (November 2020): 39.

54 http://margot.disi.unibo.it See also: https://penelope.vub.be
55 Lippi, Marco, Torroni, Paolo. "MARGOT: A web server for argumentation mining, Expert Systems with Applications." 65 (2016): 292–303.

Bibliography

Abeer Aldayel, WalidMagdy. "Characterizing the role of bots' in polarized stance on social media." *Social Network Analysis and Mining* 12, no. 30 (2022), https://doi.org/10.1007/s13278-022-00858-z.

Arcos, Rubén "Public relations strategic intelligence: Intelligence analysis, communication and influence." *Public Relations Review* 42 (2016): 264–270.

Assenmacher, Dennis, Clever, Lena, Frischlich, Lena, Quandt, Thorsten, Trautmann, Heike and Grimme, Christian. "Demystifying Social Bots: On the Intelligence of Automated Social Media Actors." *Social Media + Society* 6, no. 3 (2020), https://doi.org/10.1177/2056305120939264.

Badawy, Adam, Ferrara, Emilio and Lerman, Kristina "Analyzing the Digital Traces of Political Manipulation: The 2016 Russian Interference Twitter Campaign," *2018 IEEE/ACM International Conference on Advances in Social Networks Analysis and Mining (ASONAM)*, (2018): 258–265, https://doi.org/10.1109/ASONAM.2018.8508646.

Broniatowski, David A., Jamison, Amelia M., Qi, SiHua, AlKulaib, Lulwah, Chen, Tao, Benton, Adrian, Quinn, Sandra C. and Dredze, Mark. "Weaponized Health Communication: Twitter Bots and Russian Trolls Amplify the Vaccine Debate." *American Journal of Public Health* 108, (October2018): 1378–1384, https://doi.org/10.2105/AJPH.2018.304567.

Chase, W. Howard. *Issue management: Origins of the future.* Stamford, CA: Issue Action Publications, 1984.

Cresci, Stefano, Pietro, Roberto D., Petrocchi, Marinella, Spognardi, Angelo and Tesconi Maurizio. "Fame for sale: Efficient detection of fake Twitter followers." *Decision Support Systems* 80, (December2015): 56–71, https://doi.org/10.1016/j.dss.2015.09.003.

Cresci, Stefano. Pietro, Roberto D., Petrocchi, Marinella, Spognardi, Angelo and Tesconi Maurizio. "Social Fingerprinting: Detection of Spambot Groups Through DNA-Inspired Behavioral Modeling," *IEEE Transactions on Dependable and Secure Computing* 1, no. 4, (July-Aug. 2018): 561–576, https://doi.org/10.1109/TDSC.2017.2681672.

Choraś, Michał, Demestichas, Konstantinos, Giełczyk, Agata, Herrero, Álvaro, Ksieniewicz, Paweł, Remoundou, Konstantina, Urda, Daniel, Woźniak, Michał. "Advanced Machine Learning techniques for fake news (online disinformation) detection: A systematic mapping study." *Applied Soft Computing* 101 (March2021), https://www.doi.org/10.1016/j.asoc.2020.107050.

Coombs, W. Timothy "Assessing online issue threats: Issue contagions and their effect on issue prioritisation," *Journal of Public Affairs* 2, no. 4, (2002): 215–229. https://doi.org/10.1002/pa.115.

Crable, Richard E., and Vibbert, Steven L. "Managing Issues and Influencing Public Policy," *Public Relations Review* 11, no. 2, (1985): 3–16. https://doi.org/10.1016/S0363-8111(82)80114–80118.

Defence Science and Technology Group. *"Call for Pilot Project Proposals Artificial Intelligence for Decision Making 2022 Initiative."* Accessed November 12, 2022. https://www.dst.defence.gov.au/sites/default/files/opportunities/documents/AI4DM_Challenges.pdf.

GCHQ. *"The Ethics of Artificial Intelligence: Pioneering a New National Security"* 2021. Accessed November 12, 2022. https://www.gchq.gov.uk/files/GCHQAIPaper.pdf.

García-Marín, David. "Modelos algorítmicos y fact-checking automatizado. Revisión sistemática de la literatura." *Documentación de Ciencias de la Información* 45, no. 1, (2022): 7–16. https://doi.org/10.5209/dcin.77472https://revistas.ucm.es/index.php/DCIN/article/view/77472.

Goldman, Jan and Susan Maret. *Intelligence and information policy for national security: key terms and concepts.* Lanham, MD: Rowman and Littlefield, 2016.

Graves, Lucas. *"Factsheet: Understanding the Promise and Limits of Automated Fact-Checking."* Accessed November 13, 2022 https://reutersinstitute.politics.ox.ac.uk/sites/default/files/2018-02/graves_factsheet_180226%20FINAL.pdf.

Grabo, Cynthia, with Jan Goldman. *Handbook of Warning Intelligence: complete and declassified edition.* Lanham, MD: Rowman & Littlefield, 2015.

Gruzd, Anatoliy and Mai, Philip. "Going viral: How a single tweet spawned a COVID-19 conspiracy theory on Twitter." *Big Data & Society* 7, no 2, (2020). https://doi.org/10.1177/2053951720938405.

Guo, Zhijiang; Schlichtkrull, Michael, and Vlachos, Andreas. "A Survey on Automated Fact-Checking," *Transactions of the Association for Computational Linguistics*, 10 (2022): 178–206. https://doi.org/10.1162/tacl_a_00454.

Andreas Hanselowski. *"A Machine-Learning-Based Pipeline Approach to Automated Fact-Checking."* PhD diss., Technical University of Denmark, 2020. https://doi.org/10.25534/tuprints-00014136.

Huynh, Viet-Phi and Papotti, Paolo. "A Benchmark for Fact Checking Algorithms Built on Knowledge Bases." In *Proceedings of the 28th ACM International Conference on Information and Knowledge Management*, (2019): 689–698. https://doi.org/10.1145/3357384.3358036.

Ivan, Cristina, Chiru, Irena, Arcos, Rubén. "A whole of society intelligence approach: critical reassessment of the tools and means used to counter information warfare in the digital age." *Intelligence and National Security Intelligence* 36, no. 4, (2021).

Jaques, Tony. *Don't Just Stand There: The Do-it Plan for Effective Issue Management.* Melbourne: Issue Outcomes, 2000.

Jaques, Tony. *Issue and Crisis Management: Exploring Issues, Crises, Risk and Reputation.* OUP Australia & New Zealand. Kindle edition, 2014.

Johansen, Alyson Grace. "What's a Twitter bot and how to spot one." Norton Emergency Trends, September 5, 2022, https://us.norton.com/blog/emerging-threats/what-are-twitter-bots-and-how-to-spot-them#.

Juršėnas, Alfonsas, Karlauskas, Kasparas, Ledinauskas, Eimantas, Maskeliūnas, Gediminas, Rondomanskas, Donatas, Ruseckas, Julius. "The Role of AI in the Battle Against Disinformation." Riga: NATO Strategic Communications Centre of Excellence, 2022, https://stratcomcoe.org/pdfjs/?file=/publications/download/The-Role-of-AI-DIGITAL.pdf?zoom=page-fit.

Khalsa, Sundri. *Forecasting terrorism: indicators and proven analytic techniques.* Lanham: Scarecrow Press, 2004.

Lippi, Marco, Torroni, Paolo. "MARGOT: A web server for argumentation mining, Expert Systems with Applications." 65 (2016): 292–303, https://doi.org/10.1016/j.eswa.2016.08.050.

Marcellino, William, Cox, Kate, Galai, Katerina, Slapakova, Linda, Jaycocks, Amber, Harris, Ruth. *"Human-machine detection of online-based malign information."* RAND Europe, 2020. Accesed November 13, 2022 https://www.rand.org/pubs/research_reports/RRA519-1.html.

Matera, Fran R. and Ray J.Artigue. *Public relations campaigns and techniques: building bridges into the 21st century.* Needham Heights, MA: Allyn and Bacon, 2000.

Miao, Lin, Last, Mark, Litvak, Marina. "Detecting Troll Tweets in a Bilingual Corpus" *Proceedings of the 12th Conference on Language Resources and Evaluation (LREC 2020)*, (2020): 6247–6254. https://aclanthology.org/2020.lrec-1.766.pdf.

Miranda, Sebastião, Nogueira, David, Mendes, Afonso, Vlachos, Andreas, Secker, Andrew, Garrett, Rebecca, Mitchel, Jeff, Marinho, Zita. *"Automated Fact Checking in the News Room."* (2019) arXiv. https://doi.org/10.48550/arXiv.1904.02037.

Morstatter, Fred, Wu, Liang, Nazer, Tahora H, Carley, Kathleen N, Liu, Huan. "A new approach to bot detection: Striking the balance between precision and recall." In: *IEEE/ACM Conference on Advances in Social Networks Analysis and Mining (ASONAM)*, (2016): 533–540. https:// ieeexplore-ieee-org.proxy.lib.sfu.ca/document/7752287

National Security Commission on Artificial Intelligence. Final Report (2021). Accessed November 12, 2022, https://www.nscai.gov/wp-content/uploads/2021/03/Full-Report-Digital-1.pdf.

Orabi, Mariam, Mouheb, Djedjiga, Al Aghbari, Zaher, Kamel, Ibrahim. "Detection of Bots in Social Media: A Systematic Review." *Information Processing & Management* 57, no. 4, (July2014), https://doi.org/10.1016/j.ipm.2020.102250.

Renfro, William L. *Issue management in strategic planning.* Westport, CT: Greenwood Publishing Group, 1993.

Rheault, Ludovic, Musulan, Andreea. "Efficient detection of online communities and social bot activity during electoral campaigns." *Journal of Information Technology & Politics* 18, no.3, (2021), https://doi.org/10.1080/19331681.2021.1879705.

Sedova, Katerina, McNeill, Christine, Johnson, Aurora, Joshi, Aditi, Wulkan, Ido. *"AI and the Future of Disinformation Campaigns Part 2: A Threat Model"* Center for Security and Emergency Technology, 2021https://cset.georgetown.edu/wp-content/uploads/CSET-AI-and-the-Future-of-Disinformation-Campaigns-Part-2.pdf.

Suta, P.; Lan, X., Biting, W., Mongkolnam, P., Chan, J. H. "An Overview of Machine Learning in Chatbots." *International Journal of Mechanical Engineering and Robotics Research* 9, no. 4, (April2020): 502–511, https://doi.org/ 10.18178/ijmerr.9.4.502–510. http://www.ijmerr.com/uploadfile/2020/0312/20200312023706525.pdf.

Thorburn, Luke. *"Enhancing human analysis of large corpora via argument mining and backlinks—a prototype."* (March2021). Accessed November 13, 2021. https://cpb-ap-se2.wpmucdn.com/blogs.unimelb.edu.au/dist/8/401/files/2021/03/ArgumentMiningBacklinks.pdf.

Treverton, Gregory F. *Approaching threat convergence from an intelligence perspective, in Unconventional Weapons and International Terrorism Challenges and new approaches.* Edited by Magnus Ranstorp and Magnus Normark, London and New York: Routledge. Kindle Edition, 2009: 141–162.

van der Walt, Estée and Eloff, J.H.P. (2018). Are attributes on Social Media Platforms usable for assisting in the automatic detection of Identity Deception? In: *Proceedings of the Twelfth International Symposium on Human Aspects of Information Security and Assurance (HAISA 2018).* 57–66.

Varol, Onur, Ferrara, Emilio, Davis, Clayton. B., A., Menczer, Filippo, Flammini, Alessandro. "*Online human-bot interactions: Detection, estimation, and characterization.*" Proceedings of the Eleventh International AAAI Conference on Web and Social Media, (2017): 280–289.

Visser, Jacky, Lawrence, John, and Chris Reed. "Reason-Checking Fake News" Communications of the ACM 63, no. 11, (November2020): 38–40. https://cacm.acm.org/magazines/2020/11/248201-reason-checking-fake news/abstract.

World Health Organization (WHO). "*Digital solutions to health risks raised by the COVID-19 infodemic: policy brief.*" Copenhagen: WHO Regional Office for Europe; 2022. https://www.who.int/europe/publications/m/item/digital-solutions-to-health-risks-raised-by-the-covid-19-infodemic.-synthesis-report.

Xifra, Jordi. *Comunicación proactiva. La gestión de conflictos potenciales en las organizaciones.* Barcelona: Gedisa, 2009.

Yankoski, Michael, Weninger, Tim and Scheirer, Walter." An AI early warning system to monitor online disinformation, stop violence, and protect elections." *Bulletin of the Atomic Scientists*, 76, no. 2, (2020): 85–90, https://doi.org/ 10.1080/00963402.2020.1728976.

Yip, Pedro and Blaclard, Vincent. "*Social Listening Is Revolutionizing New Product Development. New techniques better capture consumer intelligence in real time.*" MIT Sloan Managent Review, 2009https://sloanreview.mit.edu/article/social-listening-is-revolutionizing-new-product-development/.

30

AI TECHNOLOGIES TO SUPPORT DETECTION OF INFORMATION MANIPULATION ON SOCIAL NETWORKS AND ONLINE MEDIA: A QUICK OVERVIEW

Frédérique Segond

Introduction

According to Wikipedia, manipulation in general is a method deliberately used with the aim of controlling or influencing a person's thoughts, choices, and actions, via a relationship of power or influence (suggestions, constraints). The idea behind is to distort or orient the perception of the people's reality.

Online media use different ways to manipulate readers like shifting their attention to insignificant news in order to divert them from important issues, using readers' emotions to reduce their critical sense or promoting mediocrity as fashionable in order to flatter readers' egos. On social networks the means for manipulation are similar in part, with the difference of being even more audience targeted and individual centric. The fact that social networks are divided into groups, ideological communities where people exchange among themselves, within their own community, facilitates targeted manipulation.

Another way to manipulate readers is to provide them with distorted information. Information distortion is often referred to as information manipulation. Information manipulation usually consists in disseminating fake or distorted news on purpose. Information manipulation is not a new phenomenon, what is new is the impact it can have in the Internet era. Internet is an echo chamber where distorted information travels faster and impacts the entire world.

They are many studies and articles from social humanities presenting, for instance, the different psychological manipulation methods, rhetoric or technics of propaganda. This paper will not address the latter. This paper presents various technologies coming from artificial intelligence (AI) that can be used to support the detection of information manipulation on the Internet, focusing on online media and social networks.

While information manipulation is as old as information itself, the digital era which permits the instantaneous diffusion of information increases its effects. Indeed, online media and social networks are information distribution channels that are now at the heart of information

DOI: 10.4324/9781003190363-35

with almost half of the population getting at least some information from them. These channels, with the large volume of information at their fingertip and large potential audience, allow an effect of resonance and informational remanence, with the particularity of also blurring the geographical boundaries and temporal constraints of information. In other words, online media and social networks are privileged vectors to manipulate people's opinion to influence them at a large scale in order to serve the cause or interest of different groups. Information manipulation on the Internet is mostly done by spreading fake news and related conspiracy theories, supporting, for instance, recruitment by radicalized groups, bullying, harassment in general or propaganda of any type. Artificial Intelligence can provide technologies that support the detection of information manipulation by automatically identifying and locating them.

While societal issues are complex and often require a human interpretation, the amount of data produced in real time on the Internet, together with technologies coming from artificial intelligence, represent both a challenge and an opportunity for social scientists and computer scientists to work together on social issues. Social networks represent a goldmine for social and computer scientists interested in information manipulation. Social networks and on-line media are daily producing huge amounts of data. Facebook by itself yields around 4 petabytes of data a day. The advent of Big Data goes hand in hand with recent developments in the field of Artificial Intelligence. In computer science, researchers and engineers are working on new ways to collect, store, index, analyze, link, visualize, and interpret data of different kinds, structured and unstructured, using different technics.

AI technologies able to detect fake news are of different nature. Fake news detection can be done by detecting communities, bots, writing style, information spreading in real time, special type of discourse on textual content, veracity, image and video manipulation, manipulation of audio, emotions etc. The list is long and we cannot cover all these aspects in one paper. We will focus on technologies based on artificial intelligence such as graph analysis, natural language, audio, image, and video processing. For each of them we briefly describe their role in fake news detection, explain what they can do today and suggest what the remaining challenges to be addressed are to improve their impact on information manipulation detection. Rather than exhaustivity, this paper seeks to provide lay readers who are not familiar with AI technologies used to detect information manipulation with a panorama of what these technologies are, what they are used for and what are the remaining difficulties.

As mentioned before, from a technological point of view, detection of information manipulation implies numerous AI technologies including text, image, video, speech, and sound processing as well as graph analysis. In what follows we briefly present the role these technologies can play in supporting the detection of manipulation, through fake news and deep fake detection. We have chosen to split these technologies along three lines: whether they act on structure, content, or the impact of information contained in social network and medias.

1. Fake news detection

There are two generic ways of detecting fake news: analyzing their content and analysing dissemination routes. While *mis*information or the involuntary dissemination of false information, and *dis*information, the voluntary dissemination of false information (produced intentionally), need to be distinct from the technologies used to detect fake news, these two are still often confused. This confusion is mainly due to the fact that the automatic detection of the intent behind the publication of a fake remains very difficult.

As mentioned above, (automatically) detecting fake news can be done either by analyzing its dissemination/broadcasting network, its content, or by doing both. We present hereafter the different AI technologies which can support dissemination, impact, and content analysis.

1.1. Dissemination and Impact

One possibility is to concentrate on the sources of information and their path of dissemination in order to answer questions such as: which site spread this information first? What is the "tendency", the "belief" associated with this first broadcaster? How can we qualify it? Is it related to a terrorist group? Or to radicalized communities?

The detection of the first site which disseminated a given piece of information is done by following the "route" in the dissemination graph to search for its origin. This is interesting but not necessary enough to qualify a piece of news. Indeed, knowing which site first disseminated a news item does not necessary mean knowing "who" is behind it, and therefore does not support fake news detection. While some sites are well known as fake news disseminators, there are many unknown sites which disseminate fake news every day. This is why it is equally important, when finding the first broadcaster of a news, to also try to qualify it. A simple way of qualifying the first broadcaster account is to understand what topics are discussed in the news and on the site in general. This can be done by simply counting the number of occurrences of words contained in the news and associating them with their meaning. For instance, if one detects that news published on a site contains words such as "vaccines, lies, big pharma groups, government" chances are high that is a conspiracy site, while commonly used words such as "abortion, illegal foetal rights" will indicate a radical (anti-abortion) site, etc. As a result, each site is associated with a list of concepts, topics that are themselves associated with different tendencies (conspiracist, radical, terrorist etc.). This site classification allows to infer the probability that a given site spreads fake news without even knowing exactly who is behind them.

Link to source identification of a news item constitutes the detection of whether a piece of news has been generated automatically by a bot, or written by a real person. There are different ways of detecting a bot consisting mainly in tracking the behavior of a bot compared to the one of a human. This is what is used, for instance, by web sites when they ask visitors to first answer to questions such as the results of a mathematical calculation or an image comparison. Another way to detect bots is to track the so called "Coordinated Inauthentic Behavior". "Coordinated Inauthentic Behavior" refers to information that appears simultaneously on different social networks without any link and without any reason. This behavior often reflects the presence of a group that "directs" this information and is a strong indicator of the diffusion of false information. Detecting such behavior implies working at three different levels: spatiality in the infosphere, temporality and content comparison. This automatic detection mostly consists in detecting spaces where information appears. This "space" detection is supported by analysis and comparison of its content (text, images, video) on different sources (spaces). Detecting the temporal aspect of diffusion remains difficult. The most efficient approach would be to be able to combine all these aspects.

Popularity of news is also a good indicator of fake news. Indeed, studies show that false news travels faster than true stories. Popularity of a news item can be detected by the analysis of links between internet users, number of shares, retweet, etc., The idea here is to detect the dissemination, to predict the propagation of a fake news in order to be able to measure its potential impact. It is interesting to combined dissemination with the profile of people who reshare the news. Obviously, if an influencer on a given internet community shares some news, on whatever topic, chances are higher that this item will be reshared and have a stronger impact (linked to influencers' detection).

2. Content

2.1. Text

Another revealing piece of information to be used for fake news detection is Internet users' comments on news: do they agree, do they disagree, are they positive, negative, angry, etc. The general idea is that fake news triggers more reactions from both sides: people who believe they are true and people who do not. In this case the idea is to use AI technologies to support textual content analysis. The current trend, called stance detection, consists in determining the position of a person towards a target (a concept, idea, event, etc.). The object of analysis in this case focuses on people's opinion about news headlines but also in related associated comments. For instance, in the following comment "it's useless because climate change doesn't exist. The proof is that it rained and was cold this summer in France", the AI method can detect a disagreement, a posture of refusal (it's useless, climate change doesn't exist).

Detecting feelings and emotions is also important for fake news detection because manipulation techniques considerably play upon elementary human feelings, emotions of readers. This is done, for instance, through facts exaggeration, events invention, or false correlations. The fact is that fake news are often relatively long texts allows an even better manipulation of emotions and feelings of the readers.

Detecting feelings—e.g. opinions, sentiments (positive, negative, neutral), emotions (anger, disgust, fear, happiness, sadness and surprise)—in textual content produced on the Internet consists in determining the "feeling" polarity of a text. This detection can be performed at a document—news on Media—or message—social networks—level, without a specific target. In such a case the goal is to evaluate the general feeling of a person on a given text (positive, negative or neutral). But it is often the case that different feelings are expressed on different topics within a given message or a text. In other words, opinions, sentiments, and emotions have a target and messages can be positive and negative at the same time regarding different, targets. Therefore, it is important to determine to what the opinion, sentiment, or emotion expressed relates exactly. Like in

> Why would we ever get a Covid shot? Pfizer even said they never tested the Covid shot to see if it would even prevent transmission of COVID-19? A healthy eating program is better protection towards the virus in foods like coriander that remove toxins from the body in preference to adding toxic proteins that cause damage just like the virus.

The above comment expresses negative opinion about Covid vaccine and positive opinion about a "healthy eating program". Even though the consideration of emotion intensity can also be of interest for fake news detection, this remains difficult. Emotion can be detected in videos, images, and in speech as we will see below. Combining emotion detection in different modalities seems to be the one that gives the best results.

AI technology can also support the detection of the veracity of a news analysis content itself, and as before, there again are different ways to evaluate news content veracity on different media.

On textual content it can be based on the qualification of symbols or words used. Indeed, punctuation like exclamation or question marks or emoticons are very frequent in fake news. Fake news also makes a great use of personal pronouns as well as of emotional and colloquial vocabulary like in "Do **you** know any sick people? Where are the dead from Covid? Are **we** sure they all died from Covid? For **me**, it's a manipulation of numbers, a **big trick**". Detecting these even at a simple lexical level and counting them can greatly support the evaluation of veracity. This is usually referred to as style detection. Something interesting that could support fake news detection is the detection of irony, sarcasm, or even humor, but this remains very complicated, and their automatic detection is an open challenging research topic..

More sophisticated ways to evaluate the veracity of textual content of a news consist in comparing it with other on-line articles referring to the same news. This mainly involves looking for articles directly associated with the headline and for other articles which refer to the same news item under the same or different title or/and using content from Wikipedia as a comparison (for instance regarding the way a given event is presented compared to its description in Wikipedia). To that end it is necessary to compute textual similarity with other news about the same event. Computation of similarity between news can be based on titles, summaries or on core content. Nowadays this is mainly done at lexical level. Current research investigates how to increase the use of the semantics of content and compare it to the textual content of a news piece. For instance, make a comparison between the topics, events discussed, emotions expressed on certain topics and the expressions used (in connection with the detection of rhetoric and influencers).

At the core of comparing content news there is the capacity of automatically detecting the different topics they address and grouping them accordingly in order to compare them but also going further in analyzing them and detecting the events they refer to. Event detection is therefore an important task to support news comparison. Indeed, once a system has detected similar events mentioned in different news, it can analyze comments expressed about these events, compare opinions and emotion.

Detecting an event is detecting something significant that happens at a specific time and place, usually involving people. Event detection calls for tools able to analyze the syntax/ grammar of a sentence as well as its meaning. The higher the semantic level, the more complicated Semantic analysis focuses on meaning of both individual words and words combined together. While syntax concentrates on grammatical issues, semantics goal is to reach the meaning, the intent of sentences. Semantic analysis is complicated at the level of a sentence and it becomes even more complicated between different sentences. Today, systems give relatively good results on a very "syntactic" level. As events are usually represented by action verbs, like "kill, attack, explode" etc. The tasks for AI consist in finding the event itself (usually the verb called the predicate) and its related attributes, in other words its subject, its object; its complements within a sentence. For instance, for the sentence "a car exploded this morning in the center of Kabul", the system returns that the event is "explode", its subject is "car", and its complements of time and place are respectively "this morning" and "center of Kabul." This information already contains some semantics as it is necessary to know that "center of Kabul" refers to a location and "morning" refers to a moment in time.

However, more semantics is needed as they are different ways to express an event as shown in the following examples which both refer to the event "explosion:"

A.This morning a car explosion in the center of Kabul killed three people and injured many others.

B.A car explodes in the center of Kabul in the morning killing three people and injuring many.

In the above examples, the expected results for these two sentences would be:

a   Event: explosion
b   what: car
c   location: central Kabul
d   date: today (morning)
e   victims: 3 dead, many injured

This shows the need for merging all attributes around the same event. For example, when the same event is described in several sentences (for instance in headlines or in the core of articles or post) as in the following sentences: "This morning a car explosion in the center of Kabul killed three people and injured many others. The Islamic State claimed responsibility for the attack. The son of the minister of education was among the victims".

There are many other routes to explore about textual content analysis of fake news, but they are less mature. For instance, it would be interesting to consider contradictions present in internet users' positions, or the credibility of the person writing about a given topic. The first one is related to discourse analysis and can be carried on social networks both at message level and at discussion thread levels. Evaluating a person's credibility is even more difficult as it can be based on many different aspects, such as general positions on diverse topics on different networks, institution belongings, style, etc. This implies the need to fuse different information sources and also link these to influencer detection. Both topics are interesting current research topics.

It is also important to mention that all this can be done at a multilingual level as a news item can be and is often published in different languages. The comments may themselves also be in different languages and contradictory within or across languages etc. This aspect multiplies the difficulty by the number of languages and call the need for techniques such as machine translation and cross-lingual search tools.

2.2. *Image video (deepfake)*

Fake news detection can also be supported by video and images analysis. This concerns the detection of manipulation mainly in still images as videos are often decomposed into a succession of still images to be analyzed. Fakes news detection in images is called deepfake detection. Images manipulation can consist in different aspect like changing the context, manipulating people's faces ad bodies or both. In most cases deep fakes are mainly about face manipulation. Research on the detection of deepfakes is linked to the research on the generation of false images insofar as the detectors must be able to detect any new type of manipulation.

Today, thanks to deep learning, based on neural networks, it is easy to produce synthetic faces either by starting or not from real faces whose attributes or expression will be modified. There are four main types of face manipulation present in still images and videos: change of expression, manipulation of attributes (e.g. eye color, nose shape), synthesis of an entire face, change of identity.

Deepfake detectors work relatively well on high resolution still images and are able to detect expression changes. More difficult, however, is the detection of attribute changes and it is very difficult to detect identification changes. This work consists of the detection of attributes such as mouth shape, nose size, eye color but also of age or gender and then in comparing them across different images.

2.2.1. Detection of different attributes to detect fake images/videos

DETECTION OF EXPRESSIONS

This involves detecting the replacement of a facial expression (source) by another (target) on a given face. This manipulation can be done on the same face or on a different face. But the manipulation only affects the expression, not the whole face (see identity) nor any of its attributes. For example, on an image of Macron smiling, it is possible to change the smile to a yawn, or put Macron's smile on Biden's face.

Detecting a change in identity: this involves detecting the replacement of a face (target image) by another (source image). For example, replace Putin's face by Macron's in a video.

Detection of synthetic faces: this involves detecting the "synthetic" part only as the faces to detect, by definition, does not exist in reality.

COMPARISON OF IMAGES, VIDEOS (ADDITION OF OBJECTS, PEOPLE, SCENES ETC.)

Detecting deepFakes is becoming increasingly difficult since it depends on the ability to generate quality deepfakes. Indeed, the more image and video manipulation methods are developed, the more complicated the detection becomes. Since the methods employed are learning and neural network based, the development of efficient deepfake detectors is highly dependent on the learning corpus. In other words, to obtain good results one needs to have access to a lot of data as input to the learning algorithms. This makes it difficult to detect new types of manipulation as the data on which to learn, the corpora, does not exist by definition. It is important to evolve the learning corpus in parallel with the evolution of deepfakes generation methods.

As seen before for textual content, detecting events in images can support deepfake detection. The aim here is to detect events in still images or videos. For example, to detect an explosion or a demonstration. As for text it is difficult to detect events in images. Some of the issues to solve are the access to an ontology of events and probably also an ontology for each type of event in order to know what the visual characteristics are and to search for them. For example, an image that contains an explosion must have specific characteristic localized in a particular part of the image. In videos, interesting events to consider are, for instance, the abrupt changes of scene, coupled with the soundtrack. The performance of these detection modules greatly depend on the quality of the images, the videos (and therefore also sound and image), the conditions of the shooting and the sound.

There are many places for improvement in deepfake detection such as:

- Improvement of robustness in general (e.g. to low quality images, compression etc.)
- Improvement of attribute manipulation detection
- Improvement of identity manipulation detection
- Improved detection of deepfakes in videos
- Combination of visual and audio features. For example, consistency between mouth movement and speech
- Considering common sense knowledge: humans have two eyes, no horns etc

- Evasion of DeepFake Detection: to escape from the detection of deepfakes, for example by replacing the fake face by another one which will be more difficult to detect by a system (the fake face is, in some way, hidden).

2.3. Speech

The aim here is to detect manipulation in voice. This consists, for instance, in identifying voice of a speaker and checking if it corresponds to her/his real voice or in detecting event in sound tracks (like explosion, start of a car engine or cries) or even emotions in speakers' voices to determine their emotional state. It involves the detection of a predefined set of key words, prosody or the pitch of the voice. This can be done after transcription (we then come back to the detection of emotions in text), or by detecting directly these features in the sound track.

The first issue of identifying a person's voice, is relatively easy. Each person's voice is associated with its "signature" which consist in the combination of different characteristics which is then compared to the signature of the voice on the soundtrack. However, this means having access to a database with enough samples of people voice.

The objective of automatic sound event detection methods is to recognize what is happening (event) in an audio signal and when it occurs. In practice, the goal is to recognize at which temporal instances different sounds (events) are active in an audio signal.

This is a very difficult task for two main reasons:

Firstly, the nature of the sounds to be detected and the way they appear in natural environments. Sound events are very diverse and unlimited (music, birds singing, explosion, train entering a station etc.) and cannot be described, for example, in an ontology. In the case of transcription, for instance, the recognition task is simpler because the set of phonemes is finite, and a language model can detect them. For sound elements, this set is infinite.

The second difficulty is that the target sound events are often very far from the microphone; they appear in the middle of other sounds which are often louder than them. All this makes data collection and annotation very difficult (we come back to the notion of ontology of sound events which does not exist).

As for emotion detection directly on the sound track, this remains difficult, constituting an important research topic.

Conclusion

In this chapter we have given an overview of AI technologies that can support the detection of information manipulation. We have shown that they can be used at different levels: dissemination and impact, content

Different AI technologies are in play to support detection of information manipulation: graphs, textual, video and still image and speech/voice analysis. All of them address similar issues and can be combined to reach higher performance level and support humans in detection of information manipulation. to get the best of fake news detection.

Bibliography

Cao, J., P. Qi, Q. Sheng, T. Yang, J. Guo, and J. Li. 2020. Review of *Exploring the Role of Visual Content in Fake News Detection*. Edited by K. Shu, S. Wang, D. Lee, and H. Liu. *Disinformation, Misinformation, and Fake News in Social Media. Lecture Notes in Social Networks.*

Chanclu, Anaïs, Laurianne Georgeton, Corinne Fredouille, and Jean-Francois Bonastre. 2020. Review of PTSVOX: Une Base de Données Pour La Comparaison de Voix Dans Le Cadre Judiciaire (PTSVOX: A Speech Database for Forensic Voice Comparison). In *Actes de La 6e Conférence Conjointe Journées d'Études Sur La Parole (JEP, 33e Édition), Traitement Automatique Des Langues Naturelles (TALN, 27e Édition), Rencontre Des Étudiants Chercheurs En Informatique Pour Le Traitement Automatique Des Langues (RÉCITAL, 22e Édition).*, 1: 73–81. ATALA & AFCP.

Delobelle, Jérôme, Amaury Delamaire, Elena Cabrio, Ramón Ruti, and Serena Villata. 2020. Review of Sifting the Arguments in Fake News to Boost a Disinformation Analysis Tool. In *Proceedings of NL4AI 2020–2024th Workshop on Natural Language for Artificial Intelligence*.

Küçük, Dilek, and Fazli Can. 2020. "Stance Detection." *ACM Computing Surveys* 53 (1): 1–37. https://doi.org/10.1145/3369026.

Mesaros, Annamaria, Toni Heittola, Tuomas Virtanen, and Mark Plumbley. 2021. Review of *Sound Event Detection: A Tutorial*. In *IEEE Signal Processing Magazine*, 38:67–83. IEEE. https://doi.org/10.48550/arXiv.2107.05463.

Nakov, Preslav, Alberto Barrón-Cedeño, Giovanni da San Martino, Firoj Alam, Julia Maria Struß, Thomas Mandl, Rubén Míguez, et al. 2022. "Overview of the CLEF–2022 CheckThat! Lab on Fighting the COVID-19 Infodemic and Fake News Detection." *Lecture Notes in Computer Science*, 495–520. https://doi.org/10.1007/978-3-031-13643-6_29.

Nandwani, Pansy, and Rupali Verma. 2021. "A Review on Sentiment Analysis and Emotion Detection from Text." *Social Network Analysis and Mining* 11 (1). https://doi.org/10.1007/s13278-021-00776-6.

Orabi, Mariam, Djedjiga Mouheb, Zaher Al Aghbari, and Ibrahim Kamel. 2020. "Detection of Bots in Social Media: A Systematic Review." *Information Processing & Management* 57 (4): 102250. https://doi.org/10.1016/j.ipm.2020.102250.

Srivastava, Akhilesh Kumar, and Rijwan Khan. 2022. "Fake News Detection System Using Stance Detection and Machine Learning Approaches." *International Journal of Forensic Software Engineering* 1 (4): 378. https://doi.org/10.1504/ijfse.2022.123982.

Schuster, Tal, Roei Schuster, Darsh J. Shah, and Regina Barzilay. 2020. "The Limitations of Stylometry for Detecting Machine-Generated Fake News." *Computational Linguistics*, March, 1–18. https://doi.org/10.1162/coli_a_00380.

Su, Sen, Yakun Wang, Zhongbao Zhang, Cheng Chang, and Muhammad Azam Zia. 2017. "Identifying and Tracking Topic-Level Influencers in the Microblog Streams." *Machine Learning* 107 (3): 551–578. https://doi.org/10.1007/s10994-017-5665-1.

Vorakit Vorakitphan, Elena Cabrio, Serena Villata. (2022). "PROTECT: A Pipeline for Propaganda Detection and Classification". CLiC-it 2021—Italian Conference on Computational Linguistics, Jan 2022, Milan, Italy.

Vosoughi, Soroush, Deb Roy, and Sinan Aral. 2018. "The Spread of True and False News Online." *Science* 359 (6380): 1146–1151. https://doi.org/10.1126/science.aap9559.

Wang, Yaohui, and Antitza Dantcheva. 2020. "A Video Is Worth More than 1000 Lies. Comparing 3DCNN Approaches for Detecting Deepfakes." IEEE Xplore. November 1, 2020. https://doi.org/10.1109/FG47880.2020.00089.

31

WARGAMING DISINFORMATION CAMPAIGNS

Roger Mason

Introduction

Propaganda is ancient. For the past century belligerent nations have more consciously and systematically employed information as a weapon: often *dis*information. Starting with the First World War, combatant nations recognized the corrosive effect of disinformation on the will and morale of an enemy, and countries like the Axis powers of World War Two put it to effective use. For example, Germany established a Ministry of Propaganda, with Minister of Propaganda Joseph Goebbels becoming one of the most powerful men in Nazi Germany.

Disinformation was also a valuable weapon for the Allied powers. British intelligence services developed a radio propaganda operation called "Soldatensender Calais." The radio station in England broadcast news, sports, and popular music from Germany. Speeches from Hitler were rebroadcast. There were public service announcements regarding military travel safety. The station only broadcast at night but became very popular. Gradually, anti-Nazi messaging was introduced as the station's popularity grew[1].

By the start of the 21st century, the West's military and technological superiority was a challenge to nations like Russia, China, and Iran. They recognized their inability to succeed in direct confrontations with conventional forces. In 2013 the Russian Federation Army Chief of Staff Valery Gerasimov offered a solution: far more centrality for information as a weapon[2].

In a 2013 speech titled "The Value of Science Is in the Foresight: New challenges demand rethinking the forms and methods of carrying out combat operations" Gerasimov signalled a change in the Russian approach to war. Gerasimov said, "The very rules of war have changed. The role of nonmilitary means of achieving political and strategic goals has grown, and, in many cases, they have exceeded the power of force of weapons in their effectiveness. The information space opens wide asymmetrical possibilities for reducing the fighting potential of the enemy"[3].

Gerasimov described a flexible force set of disinformation, cyber warfare, and military operations where each option could operate individually and independently or in concert with the others. Gerasimov proposed a multi-step process including political subversion, proxy sanctuary, coercive deterrence, military intervention, and coercive deterrence.[4] The elevation of disinformation and cyber operations to an equivalence with kinetic action allows

DOI: 10.4324/9781003190363-36

for continuous operations. While ground forces resupply and rest, disinformation and cyber operations can continue.

Defending against disinformation campaigns is challenging. Attacks often come from a variety of sources and proxies masking the attacker's identity. Identifying the source of the attack is often difficult. The problem of attribution has resulted in fears of escalation. Countries victimized by disinformation and cyberattacks are oftentimes limited to offering claims of false accusations and installing more firewalls[5].

The problem has grown with the explosion of social media. There has never been a point in history where so many individuals have easy access and exposure to information. How the information is consumed and manipulated directly influences its impact[6]. This has led to confusion and deception by hostile actors. Automated bot armies and trolls have been responsible for manipulating public opinion[7].

The threat to Western nations is growing with the availability to enemies and competitors of improved tactics and technology[8]. Propaganda and the malicious use of information have been repurposed with devastating effects to match political and societal change. This strategy was particularly effective in the 2015 Russian seizure of Crimea. The Russians established a convincing narrative to accompany their unconventional invasion[9].

As new weapons and combat capabilities are developed questions about countering them emerge. From the Macedonian phalanx to new generation warfare, military leaders have sought methods to understand new threats, develop countermeasures, and train their armies. This will be true in combat by disinformation. Relying on direct experience has been and will continue to be costly. Throughout history, leaders observed that experience gained through battle costs lives and destruction and often provided little opportunity for analysis. So, military leaders sought another way. In the Middle Ages, a new approach, called wargaming, was developed.

The Origins of Wargaming

Practical wargaming appeared in Europe during the mid-16th century. Reinhard zu Solms designed a prototype wargame based on cards depicting various types of military units. The cards were color-coded and differentiated by size to indicate their relative strengths. The game was provided to young noblemen as an introduction to military tactics[10].

In 1644 Christopher Weikmann developed a board game with military playing pieces called The King's Game. In 1790 Ludwig Hellwig invented a wargame based on a modified chess board. He expanded the standard 64 squares on a traditional chess board to 1617 squares with four types of terrain features: (1) open space, (2) mountains, (3) forests. His game is noteworthy for the use of fictional and historical playing scenarios. Ludwig Hellwig named his invention the "*Kriegspiel*" or wargame[11].

Hellweg's game caught the attention of Georg Venturini. Venturini transferred Hellweg's design to game boards employing road maps with a grid system. Each grid square represented a mile. Venturini added additional variables like weather and logistics. The improved game was called *Neues Kriegspiel* or New Wargame[12]. In the same year as Hellweg's wargame John Clerk invented the first practical naval wargame system, based on Royal Navy tactics. He developed a miniature-based wargame that allowed players to replay historical battles and evaluate naval tactics[13].

In the 19th century, wargame development peaked when Georg Von Ressiwitz and his son developed a game that included most of the previous wargame innovations. They named it *Kriegspiel*. Their game was a complete system that was scalable to size of unit and played on variable terrain. The game was eventually developed into a gaming system with standardized parts. Copies of the game were distributed to units of the Prussian army for training[14].

The US Navy adopted wargaming in the early part of the 20th century. Wargames became a staple at the Naval War College. The primary focus was on a potential war with Japan. In the Cold War era, wargames were used to explore scenarios that could lead to nuclear war, and to some extent the possibilities for fighting or holding in check nuclear war[15]. Interest in wargames steadily increased through the Gulf War conflicts to the current warfare in eastern Europe. Wargames have a long history of providing training, analysis, and problem-solving. The first step in applying wargaming to disinformation warfare is understanding wargaming's strengths and limitations.

The Strengths and Limitations of Wargaming

Wargaming is a process that offers tremendous advantages but with some limitations. There are three areas where wargaming can help to understand disinformation warfare: analysis, evaluating existing courses of action, and planning[16].

Wargaming and Analysis

The Second World War was the birthplace of operations research. Operations research applied scientific methods to a variety of military and scientific problems. The research goal was to provide a quantitative basis for decision-making. RAND Corporation employed wargaming in the 1950s to understand the problems of nuclear warfare[17].

The RAND scientists gradually discovered their wargames were very good for understanding many of the human variables related to decision-making; with the games, factors that did not fit within a mathematical model could be understood and explored[18]. The limitation was that the RAND wargames were not scientific. They involved human decision-making that was subject to many variables, so outcomes could not be quantified or replicated[19]. If wargames are not scientific (in this sense of "scientific"), what can they offer in the way of analysis?

There are three factors where wargames can offer unique analytic insights. They are the interactions of systems, the decisions related to the interactions, and the environment in which this occurs. Let's take this "this" apart, "unpack it," as our colleagues in the humanities say. Wargames typically contain systems comprised of and in a sense representing the individual conflict actors. As these systems interact, decisions get made by those actors. Wargames are well suited to model the systems and simulate the decision-making. Wargames can also help to understand how the environment surrounding the systems will impact their behavior.

Real-time Gaming

A related form of wargaming analysis is real-time gaming. Real-time gaming involves employing a team of wargamers to play through various scenarios as real-world operations unfold. There are several notable examples of real-time gaming in World War Two. On June 6, 1944 German Seventh Army commander Fredrich Dollmann had scheduled a wargame for his corps and division commanders. The game's purpose was to develop potential courses of action for a possible invasion of Normandy. When the attack began on 6 June, Dollmann ordered the game to continue as a real-time planning and analysis method[20].

A wargame system that can track and analyze real-time campaigns is extremely useful.

The UK Ministry of Defense has been experimenting with real-time or rapid campaign analysis called "RCAT" or Rapid Campaign Assessment Toolset. The development of RCAT is a joint project of Cranfield University and the UK Centre for Defense Enterprise[21]. "RCAT is a UK- developed wargaming method designed to produce manual-tabletop wargames at the operational level"[22]. The purpose of the system is to provide capabilities-based analysis employing manual wargames. The system can be applied to historical campaigns, potential conflicts, or real-world situations such as an emerging disinformation campaign. The next option that wargames offer is evaluating courses of action.

Evaluating a Course of Action

The US military defines a Course of Action (COA) as

> a plan describing the selected strategies and operational actions designed to accomplish the mission according to the commander's intent. Modeling strategies and operational actions then simulating them in a variety of contexts is where wargames shine. The game can be focused on specific interactions, decision points, and environmental changes. The COA can be replayed to evaluate the impact of differing variables.[23].

The UK Centre for Defense Enterprise's Rapid Campaign Assessment Tool Kit (RCAT) research has expanded to evaluating COAs. The designers identified the ability of RCAT to identify and evaluate courses of action during development playtests. "The play sessions suggest that RCAT, by enabling the players to concentrate on discussions about the potential actions they could take during their turn, could be an effective way to examine courses of action"[24].

Developing a Course of Action

Sometimes a plan or course of action does not exist. Typically, there are four steps to developing a course of action. The first step is developing a scenario by identifying the threats and hazards and then prioritizing them. The second step is developing a timeline for these priorities and determining the time needed to respond to them. The third step involves developing decision points for decisions and action. The fourth step encompasses connecting the decision points with the threats and hazards in a continuous plan. The nuclear war planning of the cold war refined this technique[25]. During the development of the plan a wargame can be conducted to evaluate each step. When the plan is completed, it has already been pressure tested.

Developing Your Disinformation Model

In wargaming the model represents the system or situation you wish to recreate. The simulation is how you manipulate the model. All of this begins by developing the model.

Level of the Game

The first step in developing a specific game model is determining at what level the game will occur. The broader the game setting, the more significant the simulation's options for analytical width and depth. For example, a strategic disinformation game might target the entire population of a country. This allows for a wide variety of outcomes and impacts within game play. A tactical game will be limited to the local forces and populations directly engaged. The game level will determine what weapons and actions will be included.

In the real world a tactical engagement can escalate into strategic warfare, but games are scientific at least this far: they simplify for clarity; and are limited in time. For that "first step," we encourage separate development of tactical and strategic, with the potential for inviting participants in a tactical game back for a strategic game with the tactical set-up as a kind of narrative backstory.

Weapons and Operations

The second issue is what types of disinformation weapons will be included in the game. These must be appropriate to the level of the game. It is relatively easy to determine the type and impact of kinetic weapons. Such decisions are more difficult with cyber and information weapons. where effects can be profound, but the causes are often intangible.

One solution is to develop a weapons profile employing a combination of sources. These include both what the hostile actor claims their weapons can do and what those weapons have done in any real world examples. This requires a careful balancing act. For example prior to the Russia-Ukraine war, the Russian military made bold predictions regarding their weapons platforms of all sorts. Unfortunately for the Russians, these systems underachieved when employed in battle.

Delivery Methods

Understanding disinformation campaigns is essential to developing an accurate model for a wargame. A variety of tactics are used to deliver disinformation. One of the most successful tactics is the synchronization of disinformation and cyber operations. Lucas Hauser noted, "Cyber attacks and disinformation are an effective combination for coordinated attacks"[26].

Hauser proposes three phases of a synchronized attack. The first phase often involves long-term preparation: the attacker establishes topical priorities and avenues for exploitation. The second phase consists of cyber-attacks to disrupt the adversaries' electronic and physical infrastructure. The final phase is exploiting disorder by promoting disinformation. Hauser observed this approach is cost-effective and creates confusion and prevents attribution.

Operational Elements

Wargaming a disinformation campaign means you probably will be developing a model employing military concepts. The use of some military terms is useful. It provides a means of describing and organizing a mostly intangible domain. Without some means to understand notions such as "changing national opinions," it is difficult to offer context. The three most basic aspects of military operations are terrain, maneuver, and combat.

Terrain

Info space is the terrain where disinformation warfare is fought. Info space is a notional void filled with narratives, opinions, and ideas. Narratives, opinions, and ideas represent the terrain points the game will be played on. The objective of the wargame will be to dominate the terrain (as defined), and gain control of the space.

Concepts and systems are two parts of info space terrain. Concepts can range from how people think to how they feel about an idea or topic. Systems represent how these concepts and ideas translate into real world activities. Some of the systems that appear and interact in info space might include political and social systems and institutions in the general society, and the over-all culture of the military. Controlling key terrain in info space allows disinformation to change how people think and influence how they will act.

When wargaming info space, it may be helpful to think of a void with pieces of a mosaic spread across the expanse. The game might begin with isolated narrative points your opponent does not control. Each player would develop a strategy to gain control of various narrative points and gradually fill the void by controlling narratives and inserting ideas and opinions.

Maneuver

Maneuver involves using movement to place an enemy in a position of disadvantage. In ground warfare, moving to an unoccupied terrain point is usually easily accomplished. Maneuvering while in contact with enemy forces or to an occupied terrain point is more complicated in information warfare, as in infantry warfare. This requires the strength to eject your adversary or overwhelm their control—in this case of an idea like a specific narrative.

Combat

Conventional combat and disinformation combat in the info space also have much in common. First, there are direct attacks where a concept or narrative is introduced to justify conventional action. (e.g., a false flag operation to provide justification for an invasion.) These attacks represent frontal assaults.

Indirect attacks can be used to deny legitimacy to enemy actions or undermine confidence in the enemy's leadership. Counterattacks can be used to change the narrative or introduce a competing idea. Flanking attacks can occur when the attacker employs proxies or automated capabilities such as bot armies. Reconnaissance occurs when the attacker uses small forces to probe the enemy for weakness or narrative sensitivity.

Attrition and unrestricted warfare are common in disinformation campaigns. Attackers can wear down their opponent by continual attacks on specific narrative points. This corrosive effect can spread to other terrain points, making them vulnerable. Unrestricted warfare means that no narratives or topics are off the table. This emphasizes a hostile actor's willingness to take any action to achieve victory.

Combat by massed effects can occur. This is when the attacker employs all available forces to strike their opponent. The intent may be to overcome one important terrain point or degrade the enemy's capabilities. Massed effects can include direct media messaging, social media attacks, use of non-state actors, criminals, and proxies. This type of attack can be referred to with the recent rhetorical term "gish gallop," where one side attempts to overwhelm their opponent[27]. There are more colorful, and arguably more helpful, terms for the tactic: introducing argument after argument, however relevant or irrelevant, however fallacious, or misleading; the goal here is distraction.

Game Architecture

When designing a wargame, it is essential to establish the goals and objectives for the game. The goal represents the object of the overall effort. The objectives reside, so to speak, within the goal of the game. Combined with the goal and objectives is the context in which they will appear. For example, is this game intended for analysis, planning, or training? Each of the applications provide different outcomes and require unique design elements.

Simply declaring a wargame is about disinformation provides little specific understanding regarding what role disinformation will play in the game. Is this game only about disinformation operations? Is disinformation a force alternative alongside a limited number of other factors, like kinetic operations or cyber warfare? Or is disinformation one of many options that players can employ in the game?

Scenario Development

A wargame scenario is an immersive environment containing the background, game objectives, and resources. It provides the players a general context of space and time and establishes the battle space. "The scenario sets the scene or narrative for players and establishes the world in which they will make decisions."[28]

The US Army Strategic Wargame Handbook states that a scenario guides players as they examine a particular problem or issue[29]. The scenario provides the players context including who they are, what resources they control, their objectives, and the nature of the simulated environment. This allows the players to develop their strategy, solve problems, make decisions, and, within the world of the game, take actions.

The scenario environment is established at the start of the game. A combat environment is dynamic. The player's actions can alter the environment, interactions between opposing forces, or change related to context of the environment (e.g., a night turn that reduces movement and combat capabilities in a tactical wargame). Scenarios may include injects and random events.

Injects are information or changes within the game environment and are introduced with a specific effect in mind. Random events may be introduced in the game to provide players with sudden changes that may benefit or hinder their play. Random events provide a measure of uncertainty to the players. They can alter available resources or introduce an unexpected problem (e.g., The unexpected death of a high-profile influencer).

Fitting the Battle Space

The scenario must fit within the confines of your battle space. First, a wargame needs sufficient time to model the weapons and actions realistically. It requires enough space for maneuver on a strategic or tactical scale. Finally, it requires enough time for events to occur. Providing sufficient time is critical for the players to solve problems, make decisions, and take actions.

Using a Storyboard

When assembling a wargame, designers face two obstacles. What should I include in the game and where should they appear during the game? The artist, Pablo Picasso, began his career as a draftsman. He often prepared dozens of preparatory sketches of his paintings. Picasso would piece together the final design for his larger paintings by arranging his drawings of the various sections[30]. Similarly, employing visualization is very useful in wargame design. Movie directors developed the use of the storyboard to capture actions and events within the continuous narrative of a movie. The storyboard displays action, movement, and dialog as they will appear in the movie. This allows the director to divide the action into sections that can be evaluated and manipulated. It also provides a strategic perspective enabling you to view the movie in its entirety.

Storyboards work well for designing wargames. They allow the designers to evaluate the various parts of the game and see where those parts should occur in the narrative. For example, if the game is based on a historical scenario, you can ensure certain events match how they unfolded historically. For an analytic game, a storyboard allows the designers to visualize where to place "environmental" changes such as random events or injects. This technique is particularly useful for disinformation games. Disinformation operations can be complex to model with concurrent or overlapping actions and effects. The storyboard allows designers to place each one and evaluate their potential impact on the game.

Validation

Can a wargame about disinformation be validated? The answer is no, wargames are not scientific instruments, and the results cannot be duplicated[31]. If a wargame cannot be scientifically validated, should we be concerned about this? The answer is yes. Game designers should be careful to make clear their limitations, and to ameliorate them: it *is* possible to provide forms of validation that improve the quality and confidence in your game.

Three types of validation should be considered. They are construct, content, and face validity. Construct validity of a wargame explores whether the model is an accurately represents

real-world systems or situations. For example, after experiencing the game will the player have an understanding how disinformation campaigns work? This does not mean the game must include every facet and variable, but it must accurately model the topic.

Content validity of a wargame assesses if a game contains the vital aspects of the real-world system or situation. If construct validity is at the macro view, content validity is more of the micro view. Content validity underpins the individual points that support construct validity. Content validity is especially essential because critics of the game will be looking for an important factor or variable the designer missed, often in a conscious attempt to devalue the design.

Face validity is more informal than the other forms of validity. It is the weakest form of validity because it is subjectively based. Still, while face validity may be the weakest validation, it is also the point where the first evaluation of validity occurs. If the game does not appear to be realistic and representative of the topic, potential users will reject it, and the other forms of validity will not be considered.

The first step toward achieving validity is understanding what you are modeling. Some wargamers believe you must be a subject matter expert in every concept involved in the game. That level of expertise would be valuable but is unnecessary. For example, wargamers design games that include air operations. For purposes of the game design, they do not need to be able to pilot an F-16 fighter. They need to understand its performance parameters, weapons capabilities, and how an F-16 is employed.

The foundation for establishing validity is the understanding of what you are modeling. What are the systems, actors, decisions, and environment where this situation occurs? What makes it like other types of situations and what makes it unique? The perceived validity of your game—if a legitimate perception—will depend upon the depth and quality of your research.

Exercises

Modeling and simulation tools can be adopted at levels below the threshold of a fully developed wargame. Some examples include workshops and tabletop exercises.

Workshops

A workshop involves a discussion-based analysis of a topic or problem. "Workshops are typically narrowly focused and are intended to produce a discrete product such as a model or framework"[32]. A workshop can be useful in the initial stages of determining what type of game or exercises is appropriate. These models often serve as the basis for a wargame.

Tabletop Exercise

The US War College defines a tabletop top exercise as "a discussion-based wargame where players sit at tables and interact with one another to address the key issues of the wargame"[33]. Tabletop exercises are valuable in introducing people to new subject matter. The tabletop exercise is a flexible tool and can be useful for exploring unfamiliar topics, which for the next few years will include disinformation warfare.

Tabletop exercises can capture much of the advantages of wargaming but on a more limited scale. This type of exercise can range from a seminar discussion to focused problem-solving. Because of the limitations of scale, the depth of the outcomes does not match a wargame. Tabletop exercises do offer the limited ability to analyze or evaluate problems without the design and development requirements of a wargame.

Summary

In February of 2022 the Russian Federation invaded Ukraine. The conventional operations were supported by new generation tactics including cyber-attacks and a disinformation campaign[34]. The disinformation campaign pushed outward in three directions; internal consumption in the Russian Federation, international media, and the Ukrainian population.

From a disinformation perspective the campaign has been unique in its intensity and scale on the Russian side. NATO and Ukraine have conducted their own aggressive information operation to blunt the impact of Russian disinformation. These facts point to a pair of conclusions. First, disinformation as a strategy is not likely to disappear. Secondly, an important part of countering the threat is the development of tools that allow us to model and analyze its framework.

This is not just a new threat but one in a form both amorphous and abstruse. Wargaming can provide the structure needed to pin down a nebulous system. Wargames can provide the methods to understand the potential of disinformation warfare and how to counter it. In the 1950s cold warriors considered the threat of nuclear war as unthinkable and impossible to comprehend. Disinformation does not threaten our physical harm but the possibility of controlling how we think and act. Just as it did seventy years ago wargaming is available to address threats that are unthinkable and impossible to comprehend.

Notes

1 Delmer, 1962.
2 Gerasimov, 2016, 23–29.
3 Mattsson, 2015, 61–70.

4 Karber, 2015.
5 Lewandowsky, 2013
6 Helmus et al., 2018.
7 Katz, 2021.
8 Bushwick, 2022.
9 Brown, 2022.
10 Wintjes, 2015.
11 Vego, 2012.
12 Sayre, 1908.
13 Clerk, 2008.
14 Caffrey, 2000.
15 Emery, 2021, 11–31.
16 Grant, 2021.
17 Emery, 2021.
18 Bracken & Shubik, 2001.
19 Perla & McGrady, 2007.
20 Vego, 2012, 1–42.
21 Lee, 2021.
22 Lee, 2021.
23 Hannay et al., 2015, 12–16.
24 Lee, 2021.
25 Bracken & Shubik, 2001
26 Hauser, 2022.
27 Paul & Matthews, 2016, 29–46.
28 Appelget, Burks, & Cameron, 2020.
29 Markley, 2015.
30 Lloyd, 2018.
31 Perla & McGrady, 2007.
32 Markley, 2020.
33 Markley, 2020.
34 Diepeveen, Borodyna, & Tindall, 2022.

Bibliography

Appelget, J., Burks, R., & Cameron. F. (2020). *The Craft of Wargaming: A detailed planning guide for defense panners and analysts*. Naval Institute Press.

Bracken, P., Shubik, M., (2001). Wargaming the Information Age: Theory and purpose. *Naval War College Review*, Vol. 54. #2. 1–14. https://digitalcommons.usnwc.edu/cgi/viewcontent.cgi?referer=&httpsredir=1&article=2490&context=nwc-review.

Brown, S., (2022). *In Russia Ukraine War Social Media Stokes Ingenuity and Disinformation*. MIT Sloan School of Management. https://mitsloan.mit.edu/ideas-made-to-matter/russia-ukraine-war-social-media-stokes-ingenuity-disinformation.

Bruvoll, S., Hannay, J., Svendsen, G., Asprusten, M., Kvernelv, V., Fauske, K., Luvlid, R., and Hyndoy, J. (2015). *Simulation Supported Wargaming for the Analysis of Plans*. NATO. MSG-133/RSY-025, 12–16. https://www.sto.nato.int/publications/STO%20Meeting%20Proceedings/STO-MP-MSG-133/MP-MSG-133-12.pdf

Bushwick, S., (2022). Russia's Information War is Being Waged on Social Media Platforms. *Scientific American*. https://www.scientificamerican.com/article/russia-is-having-less-success-at-spreading-social-media-disinformation/.

Caffrey, M. (2000). History of wargames: Toward a history-based doctrine of wargaming. *Air and Space Power Journal*. 15. 33–56. https://www.airuniversity.af.edu/Portals/10/ASPJ/journals/Chronicles/caffrey.pdf.

Clerk, J. (2008). *An essay on naval tactics: Systematic and historic, 1790*. A reprint of the first edition. Naval and Military Press.

Delmer, S., (1962) *The Black Boomerang*. Viking Press.

Diepeveen, S., Borodyna, O., Tindall, T., (2022). *A War on Many Fronts: Disinformation around the Russia-Ukraine war*. Overseas Development Institute. https://odi.org/en/insights/a-war-on-many-fronts-disinformation-around-the-russia-ukraine-war/

Emery, J., (2021). Moral Choices Without Moral Language: 1950s political military wargaming at the RAND Corporation. *Texas National Security Review*. Volume 4, Issue 4, 11–31. ISSN 2576–1153. https://tnsr.org/2021/09/moral-choices-without-moral-language-1950s-political-military-wargaming-at-the-rand-corporation/.

Gerasimov, V., (2016). The Value of Science is in the Forethought: New challenges demand Rethinking forms and methods of carrying out combat operations. *Military Review*. January-February, 23–29. https://www.armyupress.army.mil/portals/7/militaryreview/archives/english/militaryreview_20160228_art008.pdf.

Grant, V., (2021). Wargames: Simulating wartime decisions helps prepare for the real thing. *National Security and Science*. Los Alamos National Laboratory. https://discover.lanl.gov/publications/national-security-science/2021-summer/wargames.

Hauser, L., (2022). *Coordinated Chaos: Synchronized cyberwarfare and disinformation attacks*. The Project on International Peace and Security. The College of William and Mary.

Helmus, T., Bodine-Baron, E., Radin, R., Magnuson, M., Mendelsohm, J., Marcellino, W., Bega, A., &Winkelmin, Z., (2018). *Russian Social Media Influence: Understanding Russian propaganda in eastern Europe*. RAND Corporation, pp. 61–70.

Karber, P., (2015). Russia's New Generation Warfare. NGA. https://www.nga.mil/news/Russias_New_Generation_Warfare.html

Katz, E., (2021). Liar's War" Protecting civilians from disinformation during armed conflict. *International Review of the Red Cross*. 102, (914). 659–682. https://international-review.icrc.org/articles/protecting-civilians-from-disinformation-during-armed-conflict-914.

Lee. A., (2021). *An Exploration of the Rapid Campaign Assessment Toolset*. Defense and Research Development Canada. https://cradpdf.drdc-rddc.gc.ca/PDFS/unc361/p813183_A1b.pdf.

Lewandowsky, S., (2013). Misinformation, Disinformation, and Violent Conflict. *American Psychologist*. 69 (7), 487–501. https://psycnet.apa.org/record/2013-35832-001.

Lloyd, C., (2018). *Picasso and the Art of Drawing*. Fine Art Books.

Markley, J., (2015). *Strategic Wargaming Series Handbook*. US Army War College. 16. https://csl.armywarcollege.edu/DSE/StrategicWargamingDivision/publications/USAWC%20Wargame%20Handbook%201%20July%2015.pdf.

Mattsson, O., (2015). Russian Military Thinking: A new generation of warfare. *Journal on Baltic Security*. Vol. 1. Issue 1, 61–70.

Miser, H., (1981). Operational Research and Systems Analysis. *Science*. Volume 209. (4).

Paul, C., & Matthews, M. (2016). *The Russian Firehose of Falsehood Propaganda Model RAND*, 29–46. https://doi.org/10.7249/PE198.

Perla, P., Markowitz, M., Weuve, C., Duggan, K., & Woodard, L. (2004). *Wargame Creation Skills and Wargame Construction Kit*. CNA. CRM D0007042.A3.1–111. https://www.cna.org/archive/CNA_Files/pdf/d0007042.a3.pdf.

Perla, P., McGrady, E., (2007). *Wargaming and Analysis*. CNA.

Sayre, F. (1908). *Map maneuvers and tactical rides*. Ft. Leavenworth, KS: Army Service School Press.

Vego, M., (2012). German Wargaming. *Naval War College Review*. Vol. 65. (34), 1–42.

Wintjes, J. (2015). Europe's Earliest Kriegspiel? *British Journal of Military History*, 2 (1), 15–33.

CONCLUSION: ANALYSING AND ASSESSING DISINFORMATION AND HOSTILE INFLUENCING

François Fisher

Dear Readers,

As Director of the Permanent Secretariat of the Intelligence College in Europe (ICE), I am highly pleased to be given the opportunity to add some personal remarks to the closing words of this handbook.

The Handbook of Disinformation and National Security is a major milestone, which I strongly support. Disinformation is nowadays one of the biggest challenges to our democratic and liberal society, as could recently be observed during the COVID-19 pandemic or the Russian "special military operation" in Ukraine.

The topic of disinformation has a more personal relevance for me as I contributed to the creation of the *EU Hybrid Fusion* Cell (EU HFC) as part of the *EU Intelligence and Situation Centre (EU INTCEN)*. This cell was established in May 2016 to fight against hybrid threats endangering the EU. For this purpose, the *EU Hybrid Fusion Cell* works closely with various international actors such as NATO or the *European Centre of Excellence for Countering Hybrid Threats (Hybrid CoE)*, based in Helsinki[1].

Another example of close European collaboration against disinformation is the dynamic *Strategic Communication Division* (especially the East *Stratcom TF*) of the *European External Action Service (EEAS)*, which has been working on tackling foreign disinformation, information manipulation, and interference, as well as on strengthening its strategic communications in the Eastern Partnership, the Southern Neighbourhood, and the Western Balkans.

Given these examples on already existing European and transatlantic cooperation, it becomes clear that countering disinformation and building resilient societies and institutions are common goals, which can only be achieved when working hand in hand. Joining our forces as European democratic countries allows us to increase our capacities to detect and understand malicious activities at an early stage and to improve our ability to withstand and recover from attacks. This *Handbook of Disinformation and National Security* reflects this common endeavour and is perfectly in line with the *"raison d'être"* of the ICE: contributing to a common strategic security culture in Europe.

The Intelligence College in Europe is a unique intergovernmental initiative, which was launched on 5th March 2019 in Paris and brings together 23 Member countries[2] as well as 7 Partner countries[3]. The ICE connects all the intelligence communities of European countries with national and European decision-makers, as well as with the academic world. The

DOI: 10.4324/9781003190363-37

College intends to stimulate strategic thinking, thus developing a common intelligence culture. Furthermore, it facilitates exchanges of professional and academic views on a wide range of intelligence-related topics, and disseminates these works in order to develop a non-prescriptive strategic intelligence culture in Europe. In order to achieve this, the College organises every year thematic seminars, training programmes and outreach sessions on different subjects of common concern such as hybrid threats, disinformation or artificial intelligence.

The content of this joint Spanish and Romanian initiative shows, if need be, how decisive these subjects are for our democracies. I feel sure that the readers, be they from the intelligence and security world, communication and media, academic world or otherwise, were pleased to immerse themselves in this complex topic through very clear and well-articulated articles, written on all the major topics by the best contributors, covering all the main malign actors, as well as all the most important vectors of disinformation. This *Handbook* is definitively not just reserved for the relatively "happy few" dedicating their time to this fight, but is contributing to the education and the "acculturation" of the general public. As the *Intelligence College in Europe* has already taken up this view on many other security issues; protecting our democratic values will force us to think in terms of "total" or "global defence" with the resilience and awareness of our democratic societies as a core element.

Notes

1 The *Hybrid CoE* supports EU and NATO staffs working on hybrid threats and by this strengthens the cooperation between EU and NATO in this domain.
2 Austria, Belgium, Croatia, Cyprus, Czech Republic, Denmark, Estonia, Finland, France, Germany, Hungary, Italy, Latvia, Lithuania, Malta, Netherlands, Norway, Portugal, Romania, Slovenia, Spain, Sweden and the United Kingdom.
3 Bulgaria, Greece, Ireland, Luxemburg, Poland, Slovakia and Switzerland.

INDEX

For Product Safety Concerns and Information please contact our EU
representative GPSR@taylorandfrancis.com
Taylor & Francis Verlag GmbH, Kaufingerstraße 24, 80331 München, Germany

www.ingramcontent.com/pod-product-compliance
Lightning Source LLC
Chambersburg PA
CBHW081222220326
41598CB00037B/6862